B

EXS 56:
Experientia Supplementum,
Vol. 56

REGULATORY PEPTIDES

Edited by
J. M. Polak

1989
Birkhäuser Verlag
Basel · Boston · Berlin

Editor:

Prof. Julia M. Polak
Royal Postgraduate Medical School
Department of Histochemistry
Hammersmith Hospital
Du Cane Road
London W12 0HS
England

Library of Congress Cataloging in Publication Data

Regulatory peptides.

   (EXS; 56)
   Includes index.
   1. Peptide hormones. 2. Neuropeptides. I. Polak,
Julia M. II. Series: Experientia. Supplementum;
v. 56.
QP572.P4R45   1989   599′.01927   88-22126
ISBN 3-7643-1976-3

CIP-Kurztitelaufnahme der Deutschen Bibliothek

Regulatory peptides/ed. by J. M. Polak.–Basel; Boston;
Berlin: Birkhäuser, 1989
   (Experientia : Supplementum ; Vol. 56)
   ISBN 3-7643-1976-3 (Basel ...) Pb.
   ISBN 0-8176-1976-3 (Boston) Pb.
NE: Polak, Julia M. [Hrsg.]; Experientia/Supplementum

© 1989 Birkhäuser Verlag
        P.O. Box 133
        4010 Basel
        Switzerland

Printed in Germany
ISBN 3–7643–1976–3
ISBN 0–8176–1976–3

Contributors

Aguzzi, A., Department of Human Pathology, University of Pavia, 27100 Pavia, Italy

Allen, T. G. J., Department of Anatomy and Developmental Biology and Centre for Neuroscience, University College London, Gower Street, London WC1E 6BT, England

Andrews, P. C., Department of Biochemistry, Purdue University, West Lafayette, Indiana 47907, USA

Barnes, P. J., Department of Medicine, Cardiothoracic Institute, Brompton Hospital, London SW3, England

Bartfai, T., Department of Biochemistry, Arrhenius Laboratory, University of Stockholm, 10691 Stockholm, Sweden

Bloom, S. R., Department of Medicine, Royal Postgraduate Medical School, Hammersmith Hospital, Du Cane Road, London W12 0HS, England

Brayton, K. A., Department of Biochemistry, Purdue University, West Lafayette, Indiana 47907, USA

Buffa, R., Diagnostic Histopathology Center, University of Pavia at Varese, 21100 Varese, Italy

Burnstock, G., Department of Anatomy and Developmental Biology and Centre for Neuroscience, University College London, Gower Street, London WC1E 6BT, England

Burrin, J. M., Department of Medicine, Royal Postgraduate Medical School, Hammersmith Hospital, Du Cane Road, London W12 0HS, England

Ceccatelli, S., Department of Histology, Karolinska Institute, P.O. Box 60400, 10401 Stockholm, Sweden

Coghlan, J. P., Howard Florey Institute of Experimental Physiology and Medicine, University of Melbourne Parkville, Victoria, 3052 Australia

Darby, I. A., Howard Florey Institute of Experimental Physiology and Medicine, University of Melbourne Parkville, Victoria, 3052 Australia

Darling, P. E., Howard Florey Institute of Experimental Physiology and Medicine, University of Melbourne Parkville, Victoria, 3052 Australia

Davenport, A. P., Parke-Davis Research Unit, Addenbrookes Hospital Site, Hills Road, Cambridge CB2 2QB, England

Dietl, M. M., Preclinical Research, Sandoz Ltd, 4002 Basel, Switzerland

Dixon, J. E., Department of Biochemistry, Purdue University, West Lafayette, Indiana 47907, USA

Fahrenkrug, J., Department of Clinical Pathology, Bispebjerg Hospital, Copenhagen, Denmark

Goodlad, R. A., Cancer Research Campaign Cell Proliferation Unit, Department of Histopathology, Royal Postgraduate Medical School, Hammersmith Hospital, Du Cane Road, London W12 0HS, England

de Groat, W. C., Department of Pharmacology and Center for Neuroscience, University of Pittsburgh, Pittsburgh, Pennsylvania 15261, USA

Gulbenkian, S., Department of Histochemistry, Royal Postgraduate Medical School, Hammersmith Hospital, Du Cane Road, London W12 ONN, England

Haralambidis, J., Howard Florey Institute of Experimental Physiology and Medicine, University of Melbourne Parkville, Victoria, 3052 Australia

Hassall, C. J. S., Department of Anatomy and Developmental Biology and Centre for Neuroscience, University College London, Gower Street, London WC1E 6BT, England

Hill, R. G., Parke-Davis Research Unit, Addenbrookes Hospital Site, Hills Road, Cambridge CB2 2QB, England

Hökfelt, T., Department of Histology, Karolinska Institute, P.O. Box 60400, 10401 Stockholm, Sweden

Hughes, J., Parke-Davis Research Unit, Addenbrookes Hospital Site, Hills Road, Cambridge CB2 2QB, England

Jørgensen, J., Department of Clinical Pathology, Bispebjerg Hospital, Copenhagen, Denmark

Kuwayama, Y., Department of Ophthalmology, Osaka University Medical School, Osaka, Japan

Laties, A. M., Department of Ophthalmology, University of Pennsylvania School of Medicine, Scheie Eye Institute, Philadelphia, Pennsylvania, USA

Lindh, B., Department of Histology, Karolinska Institute, P.O. Box 60400, 10401 Stockholm, Sweden

Llewellyn-Smith, I. J., Department of Anatomy and Histology and Centre for Neuroscience, School of Medicine, Flinders University, Bedford Park, S.A. 5042, Australia

McGregor, G. P., Department of Medicine, Royal Postgraduate Medical School, Hammersmith Hospital, Du Cane Road, London W12 0HS, England

Meister, B., Department of Histology, Karolinska Institute, P.O. Box 60400, 10401 Stockholm, Sweden

Melander, T., Department of Histology, Karolinska Institute, P.O. Box 60400, 10401 Stockholm, Sweden

Millhorn, D., Department of Histology, Karolinska Institute, P.O. Box 60400, 10401 Stockholm, Sweden

Ottesen, B., Department of Clinical Pathology, Bispebjerg Hospital, Copenhagen, Denmark

Palacios, J. M., Preclinical Research, Sandoz Ltd, 4002 Basel, Switzerland

Palle, C., Department of Clinical Pathology, Bispebjerg Hospital, Copenhagen, Denmark

Penschow, J. D., Howard Florey Institute of Experimental Physiology and Medicine, University of Melbourne Parkville, Victoria, 3052 Australia

Pittam, B. S., Department of Anatomy and Developmental Biology and Centre for Neuroscience, University College London, Gower Street, London WC1E 6BT, England

Polak, J. M., Department of Histochemistry, Royal Postgraduate Medical School, Hammersmith Hospital, Du Cane Road, London W12 0NN, England

Rindi, G., Department of Human Pathology, University of Pavia, 27100 Pavia, Italy

Schalling, M., Department of Histology, Karolinska Institute, P.O. Box 60400, 10401 Stockholm, Sweden

Seroogy, K., Department of Histology, Karolinska Institute, P.O. Box 60400, 10401 Stockholm, Sweden

Silini, E., Department of Human Pathology, University of Pavia, 27100 Pavia, Italy

Solcia, E., Department of Human Pathology, University of Pavia, 27100 Pavia, Italy

Stone, R. A., Department of Ophthalmology, University of Pennsylvania School of Medicine, Scheie Eye Institute, Philidelphia, Pennsylvania, USA

Su, H.C., Department of Histochemistry, Royal Postgraduate Medical School, Hammersmith Hospital, Du Cane Road, London W12 0HS, England

Terenius, L., Department of Pharmacology, Biomedicum, Uppsala University, P.O. Box 573, 75123 Uppsala, Sweden

Tregear, G. W., Howard Florey Institute of Experimental Physiology and Medicine, University of Melbourne Parkville, Victoria, 3052 Australia

Tsuruo Y., Department of Histology, Karolinska Institute, P.O. Box 60400, 10401 Stockholm, Sweden

Usellini, L., Diagnostic Histopathology Center, University of Pavia at Varese, 21100 Varese, Italy

Uttenthal, L. O., Department of Medicine, Royal Postgraduate Medical School, Hammersmith Hospital, Du Cane Road, London W12 0HS, England

Van Noorden, S., Histochemistry Unit, Histopathology Department, Royal Postgraduate Medical School, Hammersmith Hospital, Du Cane Road, London W12 0NN, England

Varndell, I. M., Histochemistry Unit, Histopathology Department, Royal Postgraduate Medical School, Hammersmith Hospital, Du Cane Road, London W12 0NN, England

Villani, L., IRCCS Policlinico S. Matteo, 27100 Pavia, Italy

Wharton, J., Department of Histochemistry, Royal Postgraduate Medical School, Hammersmith Hospital, Du Cane Road, London W12 0NN, England

Wintour, E. M., Howard Florey Institute of Experimental Physiology and Medicine, University of Melbourne Parkville, Victoria, 3052 Australia

Wright, N. A., Cancer Research Campaign Cell Proliferation Unit, Department of Histopathology, Royal Postgraduate Medical School, Hammersmith Hospital, Du Cane Road, London W12 0HS, England

Contents

x

Individual organs

Abbreviations for peptides used in the following reviews

A II	angiotensin II
ACTH	adrenocorticotropic hormone
ANF	atrial natriuretic factor
ANP	atrial natriuretic peptide
AP III	atriopeptin III
AVP	arginine vasopressin
BK	bradykinin
BN	bombesin
CCK	cholecystokinin
CGRP	calcitonin gene-related peptide
CLIP	corticotropin-like intermediary lobe peptide
CPON	C-terminal-flanking peptide of NPY
CRF	corticotropin-releasing factor
CT	calcitonin
DYN	dynorphin
EGF	epidermal growth factor
ELE	eledoisin
END	endorphin
ENK	enkephalin
GAL	galanin
GH	growth hormone
GIP	gastric inhibitory polypeptide
GLI	glucagon-like immunoreactivity
GLP	glucagon-like peptide
GRF	growth-hormone releasing factor
GRP	gastrin-releasing peptide (mammalian bombesin)
GRPP	glicentin-related pancreatic peptide
HCG	human chorionic gonadotropin
INS	insulin
KAS	kassinin
LEU	leucine-enkephalin
LHRH	luteinizing hormone-releasing hormone
LPH	lipotropin
MET	methionine enkephalin
MOT	motilin
MPGF	major proglucagon fragment

MSH	melanocyte-stimulating hormone
NKA	neurokinin A (see also SK, substance K)
NKB	neurokinin B
NPY	neuropeptide Y
NT	neurotensin
OXT	oxytocin
PDN	C-terminal-flanking peptide of calcitonin
PGP 9.5	protein gene product 9.5
PHI	peptide histidine isoleucine amide
PHM	peptide histidine methionine
PHV	prepro-VIP 91-122
PHY	physalaemin
POMC	proopiomelanocortin
PP (or) PPP	pancreatic polypeptide
PYY	peptide tyrosine tyrosine
SECR	secretin
SK	substance K (see also NKA, neurokinin A)
SOM	somatostatin (also SS)
SP	substance P
SS	somatostatin (also SOM)
TRH	thyrotropin-releasing hormone
TSH	thyrotropin
VIP	vasoactive intestinal peptide

Introduction

J. M. Polak and S. R. Bloom

For some time Experientia has published, as a unique feature, interdisciplinary multi-author reviews, giving a comprehensive overview of subjects regarded as 'growing edges' of science.

The enthusiasm shown by the readers was contagious and thus it was felt necessary to compile a special volume dealing with the novel aspects of regulatory peptides. This book covers some of the growing areas in regulatory peptide research and, although it is based on the original volume of Experientia, it is expanded and updated.

The topic of 'regulatory peptides' is relatively young and has grown at an unprecedented pace, from the embryonic conception of 'gut hormones' or 'brain neuropeptides' some 15 years ago to the realisation that these active peptides are found, almost without exception, in every part of the body in all vertebrate[18] and many invertebrate species[23].

Why the term 'regulatory peptides'? It represents a convenient label encompassing both the active peptides present in nerves, which are released as (putative) neurotransmitters, and those in endocrine cells, which act locally or at a distance as circulating hormones, these being the main components of the so-called diffuse neuroendocrine[18] or APUD system[17]. Morphological studies support this physiological viewpoint. Endocrine cells in many respects resemble neural cells, since they contain a 'receptive site', sometimes in the form of microvilli (as in endocrine cells of the gut), and a 'transmitter area', on occasions represented by an elongation of the cytoplasm, as described so elegantly for somatostatin cells by the work of Larsson[12]. Furthermore, similar neurosecretory granules are present in the cytoplasm of both endocrine and neural cells (figs 1 and 2). Knowledge of the extent of the regulatory peptide-containing system is by no means complete. New peptides continue to be discovered at a rapid rate from such diverse origins as the skin of amphibians[5] (e.g. bombesin) or extracts of primitive invertebrates[23] (e.g. *Hydra* head-activating peptide) or from the disclosure of a novel mammalian cDNA structure[19] (e.g. calcitonin gene-related peptide) or new peptides identified by the characteristic amide at the C-terminus[22] (e.g. NPY, PYY, galanin).

The anatomical and physiological confines of regulatory peptides are becoming blurred, encroaching on all areas of the body and all biological

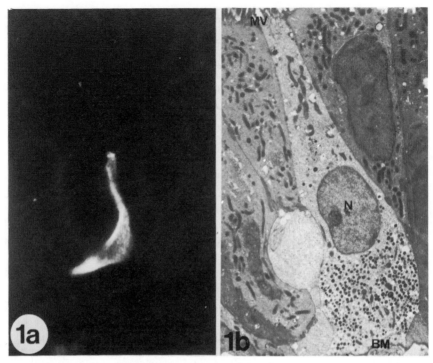

Figure 1. *a* An endocrine cell of the human small intestinal mucosa revealed by immunostaining of enteroglucagon. Indirect immunofluorescence technique ($\times 300$). *b* An electron micrograph showing the typical appearance of an endocrine cell with electron-dense secretory granules polarised toward the basement membrane (BM), below the nucleus (N), and microvilli (MV) reaching into the lumen ($\times 6000$).

disciplines. This is illustrated by the recently discovered 56 KD neurotrophic peptide termed neuroleukin[8]. This peptide, which promotes neuronal growth, was extracted and cloned from mouse salivary gland and later shown also to be a lymphokine product of lectin-stimulated T-cells, hence its name. The amino acid sequence of neuroleukin is partly homologous to highly conserved regions of the external envelope protein of the *h*uman *T l*eukaemia *v*irus III (HTLV III), the retrovirus responsible for AIDS.

Analysis of the genetic code for regulatory peptides has provided interesting information. It is now, for instance, well recognised that a single gene may encode more than one regulatory peptide (e.g. CGRP and calcitonin, VIP and PHM); likewise, more than one gene has recently been shown for the same or a closely similar peptide (α and β CGRP, the neurokinins etc.).

The processing of peptides from larger precursor molecules has attracted considerable interest lately and although the precise enzymic machinery involved in each step is still poorly understood, considerable

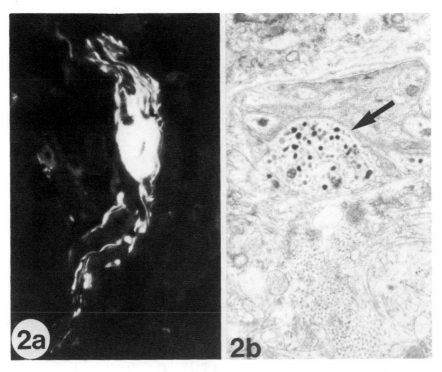

Figure 2. *a* Peptide-containing ganglion cell and fibres of the submucous plexus of porcine colon revealed by an immunostain for peptide histidine isoleucine. Indirect immunofluorescence technique. (× 300). [Reproduced, by permission, from Bishop et al. Peptides *5* (1984) 255–259]. *b* An ultrastructural profile of a p-type or peptidergic nerve (arrow) with characteristic large, electron-dense granules (× 10,000).

information may be gained from the application of advanced procedures for chemical characterisation (chromatography, fast atom bombardment techniques) and the use of region-specific antibodies. Illustrative examples include the tissue-specific expression of the two different forms of CGRP, α and β. Using region-specific antibodies, chromatographic analysis and molecular biological techniques (e.g. Northern gel analysis, *in situ* hybridisation), it was possible to determine that α CGRP is mostly a 'sensory' neuropeptide whereas β CGRP is expressed in neurons present for instance in the gut wall. Since the application of 8-methyl-N-vanillyl 5-nonenamide (capsaicin), a compound found in extracts of red pepper and known to deplete and damage primary sensory afferents (see below), has no effect on the expression of β CGRP in the lower gastrointestinal tract, it is possible to infer the non-sensory nature of β CGRP in this location[14] (figs 3 and 4). Another illustrative example is given by the tissue-specific processing of two separate but chemically related peptides VIP/PHI in different areas; for instance, in the stomach, nasal mucosa and genital tract, PHI, which is derived from the same precursor as VIP,

4

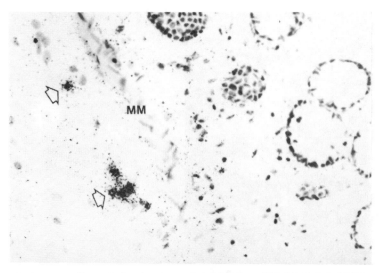

Figure 3. An autoradiograph showing *in situ* hybridisation of ^{32}P-labelled β-CGRP-specific c-RNA probe to a section of rat colon. Intense labelling (arrows) can be seen in submucous ganglia beneath the muscularis mucosae (mm). (× 120).

is not processed to the original 27 amino acid peptide but rather into a larger 42 amino acid peptide termed PHV[25] (fig. 5).

Regulatory peptide localisation has moved from the descriptive, static stage to an inspiring state of dynamic studies. The seminal work of Orci and colleagues has revealed an exciting 'functional morphology' of the β cell of pancreatic islets[15,16]. By the use of modern electron microscopical immunocytochemistry using antibodies recognising particular epitopes of proinsulin and insulin itself and by measuring intracellular pH, Orci and colleagues have been able to monitor and compartmentalise the intracellular events that take place during insulin biosynthesis. This approach, together with the increasing use of *in situ* hybridisation (fig. 6) for the localisation of peptide mRNA species and of *in vitro* autoradiography for the visualisation of peptide binding sites, will permit further disclosure of the morphological events involved in peptide synthesis, release and subsequent receptor activation.

Pharmacological manipulation of the regulatory peptide system has permitted further understanding of the putative roles played by its components. For instance, the use of capsaicin, as mentioned earlier, has provided evidence supporting the sensory nature of nerves containing peptides such as substance P and CGRP[7]. Furthermore, the use of 6-hydroxydopamine and reserpine has demonstrated that the novel peptide NPY coexists with catecholamines in the sympathetic nervous system[1]. The search for specific peptide blockers is avid and although progress has been slow some advances have recently been made[3,6]. Thus, substance P

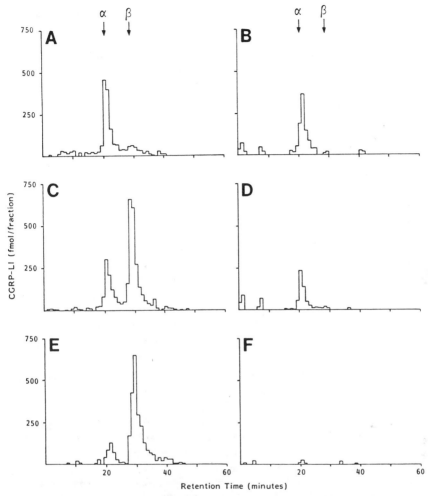

Figure 4. Representative cation exchange-radioimmunoassay profiles of CGRP-like im-munoreactivity in extracts of normal rat dorsal root ganglia. (*A*, *B*); normal rat colon (*C*, *D*) and colon from capsaicin-treated rats (*E*, *F*). Each pair of profiles is from the same set of column fractions assayed either with the total CGRP assay (*A*, *C*, *E*) or with the α-CGRP assay (*B*, *D*, *F*). The positions on the profiles corresponding to the peaks produced by synthetic rat α-CGRP and β-CGRP are indicated. [Reproduced, by permission, from ref. 14]

and bombesin analogues and antagonists and CCK antagonists have been incorporated into the list of peptide blockers originally comprising only naloxone, which blocks the actions of enkephalin.

Growth-promoting properties are being recognised increasingly as part of the armamentarium of actions displayed by regulatory peptides[26]. One illustration of this is the production and secretion of bombesin and bombesin-related peptides by the rapidly growing tumour, small cell carcinoma of pulmonary and non-pulmonary origin[2,9] (fig. 7). Bombesin,

6

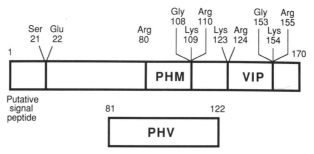

Figure 5. Diagram showing the structure of pre-pro-VIP and the positions of three derivative peptides: peptide histidine methionine (PHM human equivalent of PHI), vasoactive intestinal peptide (VIP) and pre-pro-VIP 91-122 (PHV-42).

orginally extracted from the skin of the amphibian *Bombina bombina*, is found in mammals[2]. The alternative name of gastrin-releasing peptide (GRP) has recently been adopted[20]. Bombesin enhances the growth of tumour cell lines maintained in culture or in nude mice and specific antibodies to bombesin suppress this growth[4]. Binding sites for bombesin have been found on the surface of malignant tumour cells both biochemically[13] and morphologically[11].

The construction of hybrid genes (e.g. SV40 plus the promoter region of a regulatory peptide gene) has been used for some time[10]. This has led

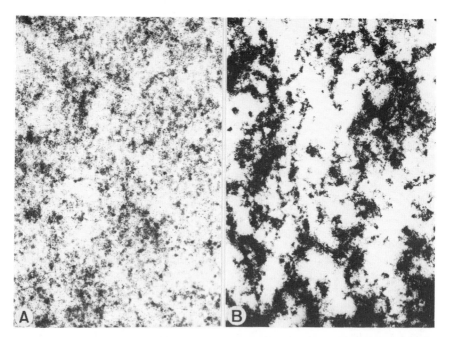

Figure 6. *In situ* hybridisation of prolactin mRNA in pituitaries of *A*) control and *B*) lactating rats using a ^{32}P-labelled cRNA probe for the coding region of the prolactin gene. (× 120).

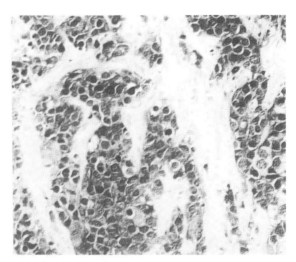

Figure 7. Small cell carcinoma of the lung immunostained with antiserum to the N-terminus of the C-terminal peptide of pro-GRP (gastrin-releasing peptide or human bombesin). Strong immunostaining is present in the tumour cells. Peroxidase anti-peroxidase method with haematoxylin counterstain, photographed using Nomarski optics ($\times 380$).

to the controlled production of endocrine tumours in experimental animals which have reproduced all the biological and pathological features of their human counterparts (fig. 8). In the future, with more advanced knowledge of the structure of RNA species, it will be possible to insert antisense RNA sequences[24] which will block the expression of a mRNA in a given tissue. This could lead only to further areas of research into the possible roles of regulatory peptides but also to new potential therapeutic approaches.

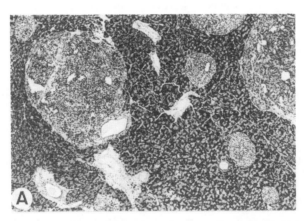

Figure 8. *A* Low power micrograph showing a haematoxylin and eosin-stained section of pancreas from an arginine-vasopressin-simian 40 (AVP-SV40) transgenic mouse. Dysplastic, enlarged islets can be seen, one of which is budding off a duct ($\times 90$).

8

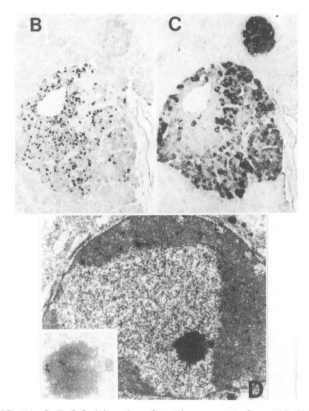

Figure 8. (*Continued*) *B–C* Serial sections from the pancreas of an AVP-SV40 transgenic mouse showing dysplastic (centre) and normal (top right) islets near a duct (× 200). *B* Immunostain for large T-antigen, a marker for SV40. The protein is localised to dysplastic nuclei only. *C* Immunostain for insulin showing denser peptide immunoreactivity in normal cells with no nuclear large T-antigen immunoreactivity. *D* An electron micrograph of a pancreatic tumour from an AVP-SV40 transgenic mouse. Large T-antigen immunoreactivity is localised to the heterochromatin using the indirect immunogold labelling method with 20-nm gold particles. (× 22,400, insert: × 35,200). [Reproduced by permission from Rindi et al., Virchows Archiv A Path. Anat. Histopath. *412* (1986) 255–266]

This book addresses some of the most fundamental questions in this field.

1 Allen, J. M., Rodrigo, J., Yates, J. C., Savage, A. P., Polak, J. M., and Bloom, S. R., Vascular distribution of neuropeptide Y (NPY) and effect on blood pressure. Clin. exp. Hypertens. *A6* (1984) 1879–1882.
2 Anastasi, A., Erspamer, V., and Bucci, M., Isolation and structure of bombesin and alytesin, two analogous active peptides from the skin of the European frogs Bombina and Alytes. Experientia *27* (1971) 166–167.
3 Chang, R. S. L., and Lotti, V. J., Biochemical and pharmacological characterisation of an extremely potent and selective non-peptide cholecystokinin antagonist. Proc. natl. Acad. Sci. *83* (1986) 4923–4926.

4 Cuttitta, F., Carney, D. N., Mulshine, J., Moody, T. W., Fedorko, J., Fischler, A., and Minna, J. D., Bombesin-like peptides can function as autocrine growth factors in human small-cell lung cancer. Nature *316* (1985) 823–826.

5 Erspamer, V., and Melchiorri, P., Active polypeptides: from amphibian skin to gastrointestinal tract and brain of mammals. Trends pharm. Sci. *1* (1980) 391–394.

6 Evans, B. E., Bock, M. G., Rittle, K. E., Di Pardo, R. M., Whitter, W. L., Veber, D. F., Anderson, P. S., and Freidinger, R. M., Design of potent, orally effective, non-peptidal antagonist of the peptide hormone cholecystokinin. Proc. natl. Acad. Sci. *83* (1986) 4918–4922.

7 Gibson, S. J., Polak, J. M., Bloom, S. R., Sabate, I. M., Mulderry, P. M., Ghatei, M. A., McGregor G. P., Morrison, J. F. B., Kelly, J. S., Evans, R. M., and Rosenfeld, M. G., Calcitonin gene-related peptide immunoreactivity in the spinal cord of man and of eight other species. J. Neurosci. *4* (1984) 3101–3111.

8 Gurney, M. E., Heinrich, S. P., Lee, M. R., and Hsiang-Shu, Y., Molecular cloning and expression of neuroleukin, a neurotrophic factor for spinal and sensory neurons. Science *234* (1986) 566–574.

9 Hamid, Q. A., Addis, B. J., Springall, D. R., Ibrahim, N. B. N., Ghatei, M. A., Bloom, S. R., and Polak, J. M., Expression of the C-terminal peptide of human pro-bombesin in 361 lung endocrine tumours, a reliable marker and possible prognostic indicator for small cell carcinoma. Virchows Arch. A.*411* (1987) 185–192.

10 Hanahan, D., Heritable formation of pancreatic β-cell tumours in transgenic mice expressing recombinant insulin/simian virus 40 oncogenes. Nature *315* (1985) 115–122.

11 Lackie, P. M., Cuttitta, F., Minna, J. D., Bloom, S. R., and Polak, J. M., Localisation of receptors using a dimeric ligand and electron immunocytochemistry. Histochemistry *83* (1985) 57–59.

12 Larsson, L. I., Goltermann, N., de Magistris, L., Rehfeld, J., and Schwartz, T. W., Somatostatin cell processes as pathways for paracrine secretion. Science *205* (1979) 1393–1395.

13 Moody, T. W., Bertness, V., and Carney, D. N., Bombesin-like peptides and receptors in human tumour cell lines. Peptides *4* (1983) 683–686.

14 Mulderry, P. K., Ghatei, M. A., Spokes, R. A., Jones, P. M., Pierson, A. M., Hamid, Q. A., Kanse, S., Amara, S. G., Burrin, J. M., Legon, S., Polak, J. M., and Bloom, S. R., Differential expression of α-CGRP and β-CGRP by primary sensory neurons and enteric autonomic neurons. Neuroscience **25** (1988) 195–205.

15 Orci, L., Ravazzola, M., Amherdt, M., Madsen, O., Perrelet, A., Vassalli, J.-D., and Anderson, R. G. W., Conversion of proinsulin to insulin occurs coordinately with acidification of maturing secretory vesicles. J. Cell Biol. *103* (1986) 2273–2281.

16 Orci, L., Ravazzola M., Amherdt, M., Madsen, O., Vassalli, J.-D., and Perrelet A., Direct identification of the prohormone conversion site in insulin-secreting cells. Cell *42* (1985) 671–681.

17 Pearse, A. G. E., The cytochemistry and ultrastructure of polypeptide hormone-producing cells of the APUD series, and the embryologic, physiologic and pathologic implications of the concept. J. Histochem. Cytochem. *17* (1969) 303–313.

18 Polak, J. M., and Bloom, S. R., Immunocytochemistry of the diffuse neuroendocrine system, in: Immunocytochemistry: Modern Methods and Applications, 2nd edn, pp. 328–348. Eds J. M. Polak and S. Van Noorden. John Wright & Sons Ltd., Bristol 1986.

19 Rosenfeld, M. G., Mermod, J. J., Amara, S. G., Swanson, L. W., Sawchenko P. E., Rivier, J., Vale, W. W., and Evans R. M., Production of a novel neuropeptide encoded by the calcitonin gene via tissue-specific RNA processing. Nature *304* (1983) 129–135.

20 Spindel, E., Mammalian bombesin-like peptides. Trends Neurosci. *9* (1986) 130–166.

21 Springall, D. R., Ibrahim, N. B. N., Rode, J., Sharpe, M. S., Bloom, S. R., and Polak, J. M., Endocrine differentiation of extrapulmonary small cell carcinoma demonstrated by immunohistochemistry using antibodies to PGP 9.5, neuron-specific enolase and the C-flanking peptide of human pro-bombesin. J. Path. *150* (1986) 151–162.

22 Tatemoto, K., Isolation of new peptides from brain and intestine, in: Frontiers of Hormone Research, vol. 12, Interdisciplinary Neuroendocrinology, pp. 27–30. Eds M. Ratzenhofer, H. Höfler and G. F. Walter. Karger, Basel 1983.

23 Van Noorden, S., The neuroendocrine system in protostomian and deuterostomian invertebrates and lower vertebrates, in: Evolution and Tumour Pathology of the Neuroendocrine System, pp. 7–38. Eds S. Falkmer, R. Hakanson and F. Sundler. Elsevier Science Publishers BV, Amsterdam 1984.

10

24 Weintraub, H., Izant, J. G., Harland, and R. M., Anti-sense RNA as a molecular tool for genetic analysis. Trends Genet. January (1985) 22–25.
25 Yiangou, Y., Di Marzo, V., Panico, M., Morris, H., and Bloom, S. R., Purification and sequence analysis of a novel 42-amino acid peptide (PHV-42) that is contained in prepro-VIP. Reg. Pept. (Abst.) *15* (1986) 199.
26 Zachary, I., Woll, P. J., and Rozengurt, E., A rôle for neuropeptides in the control of cell proliferation. Devl Biol. *124* (1987) 295–298.

Regulatory peptide immunocytochemistry at light and electron microscopical levels

S. Van Noorden and I. M. Varndell

Summary. Immunocytochemical techniques applied at both light and electron microscopical levels are valuable in the study of regulatory peptide distribution in normal and diseased tissue, whether in the form of sections or whole cell preparations. Successful immunolocalisation depends on 1) adequate preservation of the peptide antigen and the tissue structure in which it resides; 2) a suitably specific and sensitive labelled antibody detecting system. In general, peptides are stable molecules, most of which retain their antigenicity after conventional cross-linking fixation and tissue processing, allowing standard immunocytochemical methods to be used for light and electron microscopy. Regulatory peptides are derived from precursor molecules and several 'families' of structurally similar peptides are now generally recognised. Region-specific antibodies may be needed to overcome problems of cross-reactivity or to identify a bioactive form in the presence of its precursor. Multiple co-localisation of different related and unrelated peptides in the same cell or even storage granule is now recognised and can be identified by immunocytochemistry.

It is almost fifty years since Albert Coons introduced labelled antibodies as tools for the localisation of antigens in situ[14]. As with most fundamental discoveries, the significance of his work was not fully appreciated for some years. New labelling techniques were introduced in the 1960s and early 1970s, adaptations for ultrastructural work followed, and from the late 1970s to date the number and quality of antisera available has mushroomed. Immunocytochemistry, whether by fluorescence-, enzyme-, radio- or metal particle-labelled antibodies, has become an essential technique in any study involving localisation or characterisation of regulatory peptides in nerves or endocrine cells. The standard immunocytochemical methods are now known well enough not to need rehearsing here[15,16,43,52–54,64,72,74,75]. Rather, we shall discuss special problems associated with (but not necessarily confined to) peptide immunolocalisation, ways of overcoming them and how to decide what method is most appropriate.

Advantages of peptides as candidates for immunolocalisation

Stability

To begin on a positive note, peptides are remarkably resistant to autolysis. Thus it is possible, though undesirable, to use tissue from animals that have been dead for several hours and still achieve informative results. This stability may be due to the post-translational processing and packaging of peptides into membrane-bound secretory/storage granules resulting in their protection from proteolytic enzymes. Thus, peptide-containing cells can frequently be recognised when other morphological or immunocytochemical clues have become lost or are barely accessible due to poor fixation, postmortem deterioration, etc. Despite this, the best preservation is undoubtedly obtained by using tissue as fresh as possible and preferably fixed by perfusion shortly after death. Once fixed, peptides remain immunoreactive for years, whether the tissue is kept in fixative or in alcohol, processed to wax or resin blocks, stored frozen, or even as sectioned material, though results from long-stored sections may be sub-optimal. Oxidising agents such as acid fixatives, ultraviolet radiation, high temperatures, etc. should be avoided.

Phylogenetic conservation

Another advantage is that many peptides have retained much of their molecular make-up over the course of evolution, although the biological actions of chemically similar molecules in various species may differ widely[71]. This means that an antibody raised to a peptide extracted from rat tissue may well react with a similar peptide in human tissue, or even in fish or snail. This phylogenetic conservation of immunoreactive amino acid sequences has had important effects on regulatory peptide studies, allowing a judicious amount of extrapolation of experimental results from laboratory mammals to man without the necessity for raising antibodies to native peptides from each species. Nevertheless, a word of caution must be introduced here. The only way to be certain of the structure of a peptide identified by immunocytochemistry in a heterologous species, even if it reacts with an antibody known to be 'specific' for a particular peptide, is to extract it and sequence it. Hence, peptides identified only by immunoreactivity are usually referred to as 'gastrin-like', 'glucagon-like', etc. It is rare to find peptides that are absolutely identical in any two species, and indeed a peptide may appear in more than one molecular form in the same species. We are still only at the beginning of discovering the functions of the various forms of peptides.

Problems

Size

Peptides are relatively small molecules, the smallest consisting of only three amino acids (e.g. thyrotrophin-releasing hormone). Because of their small size, it may be difficult to raise an antibody using the peptide alone as immunogen, and a larger, carrier protein may have to be incorporated to make the molecule immunogenic. Antibodies will then be produced not only against the peptide (or hapten) or parts of it, but also against the carrier protein. This is no obstacle to successful immunostaining, provided that the carrier protein is not present in the tissue to be stained. Examples of popular carrier proteins are albumin, keyhole limpet haemocyanin and thyroglobulin. Antibodies to albumin as a carrier protein may react with albumin in the tissues, but can be removed from the antibody solution by addition of albumin. Thyroglobulin is not usually present in tissues but might become a nuisance if the antibody was required to stain thyroid tissue[79] or identify a metastasis for which thyroid was a possible source. Similarly, limpet haemocyanin is no problem unless molluscan tissue is to be stained. It is therefore important to know what carrier protein, if any, has been used in the immunisation procedure so that appropriate absorption of possible interfering antibodies can be carried out if necessary.

Solubility

Peptides are soluble and this may lead to loss of stainable material during processing of the tissue or during the immunostaining procedure. Some method of chemical 'fixation' is needed to make the peptide insoluble and also to retain the integrity of the tissue structure. Unfortunately, most traditional fixation methods that provide good tissue preservation damage the antigenicity of peptides to some extent so that immunoreactive sites of the molecule may be irreversibly altered. Methods were therefore developed for fixation using substances that are less strongly cross-linking than the conventional aldehyde fixatives or alternatively by snap-freezing to immobilise the peptides in the tissue, denaturing the peptides by drying the frozen tissue then using a fixative in a vapour phase. Inevitably some compromise has to be made between antigen reaction and tissue preservation, and it must be borne in mind that the peptide antigen may be partly soluble even after fixation. Improved antibodies and more sensitive methods now allow revelation of many peptides in conventionally fixed and processed tissue (see Section 'Fixation and processing').

Peptide families and cross-reactivity

Most, if not all regulatory peptides are members of groups of structurally similar molecules that have probably evolved by gene duplication and mutation from an ancestral form[23]. The resulting peptides are likely to share some amino acid sequences and probably some biological actions, although each will produce a full effect only after reaction with its own specific receptor. Along with the peptides, the receptors have surely also diversified, the combination resulting in a finely tuned range of specific binding properties and responses. However, the immunogenic parts of the molecule are not necessarily involved in this adjustment and if they have not been subject to genetic variation they may be shared by several members of the family. Thus an antibody to the N-terminal of glucagon may have some cross-reactivity with the N-terminal parts of vasoactive intestinal polypeptide, gastric inhibitory peptide, secretin, peptide histidine isoleucine and growth hormone releasing hormone (fig. 1), and an antibody to the C-terminal of gastrin will cross-react with the C-terminal of cholecystokinin (fig. 2).

In general, short amino acid sequences are likely to be common to several peptides, whether or not they belong to the same 'family', and polyclonal antisera raised to the whole molecule of any one of these peptides may well contain antibodies that react with the same sequence(s) in another. For example, antibodies to the molluscan cardioexcitatory peptide, FMRF-amide, may also react with the mammalian peptides, pancreatic polypeptide and γ_1MSH which have the same C-terminal-RF amide, although the peptides are quite different and have no known functional similarity.

It may be possible to remove the cross-reacting antibodies, and thus to 'clean up' the antiserum, by absorption with the appropriate amino acid

```
VIP       H S D A V F T D N Y T R L R K Q M A V K K Y L N S I L N -NH₂
Secretin  H S D G T F T S E L S R L R D S A R L Q R L L Q G L V -NH₂
GIP       Y A E G T F I S D Y S I A M D K I R Q Q D F V N W L L A Q².....
Glucagon  H S Q G G F T S A Y S K Y L D S R R A Q D F V Q W L M D T -OH
PHI       H A D G V F T S D F S R L L G Q L S A K K Y L E S L I -NH₂
GHRH      Y A D A Y F T N S Y R K V L G Q L S A R K L L Q D I M S R².....
```

Figure 1. Amino acid sequences of peptides of the VIP family (N-terminal or complete molecule), using the single-letter code for amino acids. VIP, vasoactive intestinal polypeptide; GIP, gastric inhibitory peptide; PHI, (porcine) peptide with histidine and isoleucine; GHRH, growth hormone-releasing hormone.

```
Gastrin   ....W L E E E E E A Y G W M D F -NH₂

CCK       ....H R I S D R D Y M G W M D F -NH₂
```

Figure 2. C-terminal amino acid sequences of gastrin and cholecystokinin (CCK).

sequences. However, this will, to some extent, diminish the binding capacity of the antibody and may result in weak staining. A better ploy would be to immunise initially with an unshared portion of the molecule so that the resulting antibodies would be 'region-specific' and non-cross-reacting. This approach was used to produce antibodies that would distinguish cholecystokinin from gastrin in areas of the intestine where both were present[11] (fig. 2). A more modern approach is to develop monoclonal antibodies to unshared portions of the molecule.

Peptide precursors

In recent years the use of recombinant DNA techniques has enabled mRNA sequences encoding the final transcriptional form of regulatory peptides to be determined. It has become clear that bioactive peptides, probably without exception, are synthesised in precursor (prohormone) form (see Dixon, this volume). Some prohormones contain multiple copies of the bioactive fragment (e.g. pro-enkephalin contains one leucine-enkephalin and six methionine-enkephalin molecules[68]) or may express single copies of several peptides which have biological activity (e.g. proglucagon is composed of the peptide hormones glucagon, glicentin and the glucagon-like peptides GLP-1 and GLP-2 [6,44,73]). These so-called polyproteins present a challenge to the peptide immunocytochemist because antibodies cannot always distinguish between the peptide as a free entity and the same sequence when incorporated in its precursor molecule, unless it is possible to select antibodies that recognise only the free terminals of the peptide. Under ordinary circumstances there is no way of knowing with an antibody to ACTH, for instance, whether ACTH itself is being localised, or pro-opiomelanocortin, the precursor of several peptides including ACTH, endorphin and several types of MSH.

In some cases immunocytochemical tests for a peptide may be negative, despite the cell's potential ability to produce that peptide. It may be possible to demonstrate this by exogenous proteolytic cleavage. For example, only after application of cathepsin B was it possible to demonstrate the active form of pancreatic glucagon in enteroglucagon cells of the gut with an antibody that did not recognise the precursor form, glicentin[58]. This confirmed the identity of the enteroglucagon cell but does not mean that it naturally produces biologically active pancreatic glucagon. This technique has not been widely applied, and it must be recognised that uncontrolled use of proteolytic enzymes may give rise to falsely positive localisation.

Even if it is known that the free peptide is being localised, there is no guarantee that the peptide is in a normal biologically active state. Tumours may produce abnormal molecular forms[33,63] or the peptide may be prevented from its normal actions by the abnormal metabolic state of the tissue in which it was produced (secretin in coeliac patients is produced by but not released from the S cells[51]). Thus the finding of peptide by immunocytochemistry (or even by radioimmunoassay) in a tumour provides a suggestion, but no proof, that the tumour is actively secreting that peptide; obviously the clinical data must be taken into account.

Quantity of stored peptide

The amount of peptide stored in epithelial endocrine cells is easily accessible to antibodies and can be visualised by many immunocyto-chemical methods. When the peptide is present in nerves it provides more of a problem because of the very small quantities in any cross-section of an axon down which the peptide is being transported and the sinuous path of a nerve fibre offers a further impediment to achieving a full section. There are several methods of overcoming this difficulty—the most usual one being to use quite thick sections (12–20 μm), either from pre-fixed cryostat blocks or unfrozen, fixed, free-floating sections from a Vibratome (Oxford Instruments, USA) or whole-mount preparations of suitable tissues such as layers of the gut wall[17] or stretch preparations of the iris diaphragm of the eye[67]. Because of the thickness of the preparations it is usually necessary to include some method of removing lipids from the tissue to allow antibodies to penetrate. This can be done by including detergents in the rinsing buffers or in the antibody solution or by taking the tissue preparations through solvents and back to water before carrying out the staining technique[17].

Identification of peptides in neuronal cell bodies in experimental animals may be made easier by prior application of an axonal transport blocker such as colchicine[29,61]. This results in a build-up of peptide in the cell body, allowing firm localisation by immunocytochemistry.

Choice of method

The decision as to which technique of fixation, processing and immunostaining should be used depends mainly on whether the peptide to be localised is present in nerves or in endocrine cells, and on foreknowledge of how well it resists fixation damage. Fixation by perfusion is preferred but immersion fixation must obviously suffice in many instances. Generally, tissue for immunocytochemical examination should be obtained as fresh as possible and should be processed to an inert state (snap-frozen, paraffin wax or resin block) without undue delay. High temperatures (greater than 55–60°C) should be avoided and all solutions, whenever practicable, should be neutral-buffered although Eldred et al.[24]

indicated some advantages in using two pH (acid-base) fixation. Osmolarity of the fixative and washing buffers should be tailored to 'physiological' levels and thus the osmotic status of the tissues should be known or determined prior to fixation. Indeed, for electron microscopical examination careful consideration of vehicle osmolarity and pH is of paramount importance.

Representative tissue samples should be taken to minimise the risk of errors induced by uneven distribution of antigens. A range of fixatives and processing protocols should be used for each issue. Whenever possible, complementary data should be obtained from biochemical (radioimmunoassay, chromatography), physiological or pharmacological techniques.

Fixation and processing

It is paradoxical that good morphological preservation prevents maximal expression of antigen immunoreactivity, presumably by fixation-induced changes to the antigen as a whole, to specific epitopes, or to adjacent molecules in the tissue. Thus, a compromise must be achieved to enable optimal reaction product deposition with acceptable architectural preservation. Unfortunately, a 'best' fixative cannot be recommended for regulatory peptides.

Light microscopy. Fixation with some kind of formaldehyde (phosphate-buffered 10% formalin, buffered picric acid-formalin, Bouin's fluid) followed by dehydration and wax embedding in the conventional manner is ideal for morphology and luckily most of the known neuroendocrine peptides are immunoreactive after this treatment. Problems may arise with particularly susceptible peptides like vasoactive intestinal polypeptide and substance P which can often not be visualised in paraffin sections, particularly when they are present in nerve fibres. The remedy for this is to use pre-fixed cryostat sections or, even better, pre-fixed, unfrozen sections from a Vibratome.

Greater immunoreactivity of all peptides is probably retained after fixation in parabenzoquinone. This is a mildly cross-linking reagent that can be used in solution (0.4% in phosphate-buffered saline (PBS)) for a short period ($\frac{1}{4}$–2 h depending on size of sample), followed by thorough washing in PBS with 15% sucrose (as cryoprotectant) and freezing for cryostat sections. Parabenzoquinone may also be used as a hot vapour (60°C for 3 h) after the tissue has been freeze-dried[9,49]. The freeze-dried, vapour-fixed tissue is subsequently embedded in wax and sectioned. The tissue becomes brown after treatment with benzoquinone, and this provides a dark, non-fluorescent background against which positive immunofluorescence with fluorescein stands out well. Variation of the fixation conditions may improve the result[12]. It is important to use pure-crystalline parabenzoquinone.

Freeze-dried material, whether fixed in parabenzoquinone or formaldehyde vapour, is excellent for immunocytochemistry of peptides in endocrine cells but is not suitable for localising peptides in nerve fibres because it is difficult to cut wax sections of sufficient thickness to be useful.

Parabenzoquinone solution, though a good preserver of peptide im-munoreactivity, is not a good tissue fixative and because the structure of the tissue is often unaesthetic, immunofluorescence tends to be the method of choice for this material rather than immunoperoxidase or some other light microscopical technique.

Many laboratories have achieved excellent immunofluorescence of neuropeptides in cryostat sections of formaldehyde fixed material[61]. This has the advantages of preserving tissue structure but may also have the disadvantage of induced fluorescence of biogenic amines (e.g. dopamine, adrenaline, serotonin) which might detract from the specific immunofluorescence image. Many other tissue components are also autofluorescent after formalin fixation, and non-immunostained controls are an essential requirement where formaldehyde has been used as a fixative. Immunoenzyme methods may be used to overcome this problem.

Alcohol and acetone are inadequate fixatives for peptides as precipita-tion alone does not overcome the solubility problem, and some form of cross-linking is required. Thus cryostat sections of fresh frozen material are far from ideal for peptide immunocytochemistry as the sections thaw on being picked up on a slide and the peptide diffuses away from its original site.

Protease treatment for formalin-fixed wax sections has been advocated for revealing 'over-fixed' antigens[37]. This is not usually useful for peptide immunocytochemistry—perhaps because the molecules are relatively small, compared to immunoglobulins for example.

Immunostaining at the light microscope level may be carried out on wax or epoxy resin sections (after removal of the embedding medium and rehydration of the sections) or on pre-fixed cryostat sections mounted on slides or free-floating. Stretched membrane preparations, free-floating Vibratome sections, whole cells, free or attached to a substrate and gut nerve plexus preparations may also be used. The only preparations that are almost always unsatisfactory are unfixed cryostat sections (see above).

As much surplus fixative as possible should be rinsed out of the tissue before it is processed further, whether by dehydration or by freezing. Formaldehyde fixation is, to some extent, reversible, so that cross-linked groups may be released by washing for reaction with antibodies. How-ever, prolonged washing can be deleterious in that soluble compounds, including peptides, may be lost from the tissue. A cryoprotectant such as sucrose (15%) should be incorporated in the rinsing fluid when cryostat blocks are to be prepared.

Sections. Both cryostat sections and paraffin sections should be mounted on slides that have been pre-coated with adhesive. Glycerine-albumen may be adequate for wax sections and formol-gelatine or chrome-gelatine for cryostat sections, but an excellent 'all-purpose glue' is poly-L-lysine (mol.wt 150,000–300,000, 1 mg/ml) applied to the slide like a blood smear in a very thin coat. The difference in charge between the tissue and the coated slide ensures that the section stays on throughout the rigorous incubation and washing steps of the immunoreaction[38]. One disadvantage is that poly-L-lysine coated slides are unsuitable for parallel investigations by silver impregnation techniques as they attract a non-specific silver deposit. This caution also applies to the immunogold-silver staining method and other methods of silver enhancement.

Pre-fixed cryostat sections (10–50 μm) should be air-dried at room temperature for 30 min to 3 h before commencement of immunostaining. Longer drying may reduce the immunoreactivity of the peptides.

Paraffin sections should be thoroughly dried in an oven at 37°C overnight and should not be put on a hot-plate. Paraffin sections may be stored for many months without deterioration of peptide immunoreactivity. Even old histologically stained sections may often be de-stained and subjected successfully to immunocytochemistry[32]. The same treatment may be applied to an immunostained section which gave a negative result, if it is desired to attempt demonstration of a different antigen.

Electron microscopy. Formaldehyde and glutaraldehyde, either alone or in mixtures (e.g. paraformaldehyde-lysine-periodic acid, Karnovsky's fixative) have been used routinely in the majority of electron immunocytochemical procedures for the localisation of regulatory peptides. Conventional processing procedures, with the exception of osmication, may be used although some antigens, notably vasoactive intestinal polypeptide (VIP), are heat-labile and alternative methods for resin embedment need to be evaluated[56,57].

A second paradox is that osmium tetroxide acts as an excellent membrane fixative and contrasting agent but also efficiently masks many peptide antigens. There is some evidence to suggest that the action of osmium tetroxide is reversible[8] which could make retrospective electron microscopical immunocytochemistry possible.

Some peptide antigens, notably the pancreatic and pituitary hormones do survive osmication (for example see fig. 5).

General conditions for fixation. The following considerations should be proposed:

1) Tissue should be as fresh as possible—fixation delay should be minimised.

2) The fixative should effectively cross-link proteins but not to the exclusion of subsequent immunoreagent access.

3) Fixation time should be optimised—excessive periods serve no purpose, too short periods may lead to morphological deterioration.

4) Fixation should at least be commenced at ambient temperature, or close to that of the tissue in order to avoid temperature-induced artifacts.

5) Formaldehyde fixation is reversible—tissue should not be left in buffer alone for prolonged periods of time. An attempt should be made to minimise the time the tissue spends in a liquid phase.

6) pH and osmolarity are important parameters and may be critical for some antigens.

7) For ultrastructural studies avoid freezing and process some tissue without osmication.

Methods

Light microscopy

The choice of method is mainly a matter of personal preference, at present, peroxidase is still the most usual label for endocrine cells and fluorescein isothiocyanate for nerves in cryostat sections, but other fluorescent labels, alternative enzymes, immunogold and radiolabels have also been used. There is now a vast literature on immunocytochemical methodology.

Sensitivity. The sensitivity of immunocytochemical marking methods depends on the relation between the amount of antigen detected and the intensity of the reaction. A highly sensitive reaction detects a very small amount of antigen. This is particularly important when one considers the small amounts of peptides found in nerve sections. Nevertheless the relatively insensitive indirect immunofluorescence method is often used for showing nerves, probably due to the high receptivity of the eye for the vivid green fluorescence of fluorescein against a dark background. The discussion below encompasses any immunoreaction and is not exclusive to peptides. Primary (polyclonal) antibodies should be diluted as far as is compatible with a good reaction, in order to reduce the level of background staining due to heterologous antibodies (as well as for economy). Incubating with the primary antibody for 4 h to overnight increases the extent to which it can be diluted and still reach an acceptable equilibrium with the antigen in the tissue. The same reasoning would apply to the second (and third) layer antibodies in the sequence, but these are usually purified to some extent and are unlikely to react with tissue components if non-specific binding sites have been blocked with normal serum or ovalbumin or gelatine prior to beginning the staining process. It is usual to incubate for 30 min to 1 h with the second (and third) immuno-reagents.

The sensitivity of a reaction may be increased by stepping up the amount of label attached to the combining site. This can be achieved by using a three-layer technique such as the peroxidase anti-peroxidase (PAP) technique[65] or one of the avidin-labelled biotin methods[13] or simply by repeating layers so that a build-up of label is achieved[69,70].

The sensitivity may also be increased by enhancing the contrast of the final reaction product with the background. If peroxidase is the label, the addition of imidazole to the enzyme development solution containing diaminobenzidine and hydrogen peroxide can give a darker reaction product. Osmication after the reaction is another way of doing this, and various other methods involving addition of heavy metal salts to the incubating medium can result in a blue-black rather than a brown reaction product[36] There also exist several ways of building on the primary reaction product by silver or gold precipitation[35].

As an alternative, the immunogold-silver staining method may be used, avoiding enzyme labels and using a colloidal gold-labelled second antibody which is made visible in strong contrast by further reaction with silver lactate[35]. The intensity of reaction is often greater with this method than with any peroxidase method (fig. 3), as was shown by the relative ease with which a 'difficult' peptide, vasoactive intestinal polypeptide, was demonstrated in conventionally fixed tumours[31]. Sensitivity has recently been reviewed by Scopsi and Larsson[62].

Electron microscopy

Direct and indirect techniques of differing complexity and 'sensitivity' are available to the electron microscopical immunocytochemist. Full details of these techniques are given elsewhere[54,75]. Here, it is relevant to consider the two major methods used to identify regulatory peptides as the ultrastructural level.

Pre-embedding method. In this method the tissue is reacted with antibody before or after fixation but prior to embedding with ultrathin sectioning. This is the technique of choice for cell surface antigen and receptor immunolocalisation[21,22,40,46,59] and for the detection of antigens prone to solubilisation or denaturation by dehydrating agents and resin components. It is also the most suitable technique for electron microscopical studies of large and highly heterogenous tissues, such as the mammalian brain[41,55].

Briefly, thick slices (20-several hundred μm) of fresh or fixed tissue are cut on a Vibratome or tissue chopper. Alternatively, suitable tissues may be treated for whole mount immunostaining (e.g. isolated crypts of

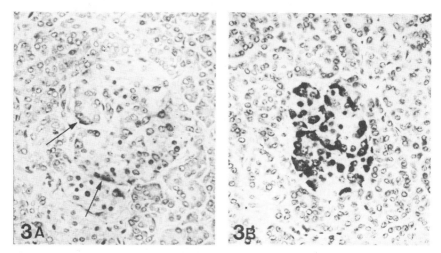

Figure 3. Immunostaining for glucagon in human pancreatic islets (Bouin's-fixed, paraffin-embedded, 4-μ sections). The optimal dilution for the primary antibody, rabbit anti-glucagon, is 1:5000 for the peroxidase anti-peroxidase technique. Here, similar sections were exposed to the antibody at a dilution of 1:80,000 for 16 h at 4°C, then developed *A*) by the peroxidase anti-peroxidase technique and *B*) by immunogold-silver staining. In *A* a few immunostained cells are just visible (arrows) but in *B* many cells are intensely stained, showing the increased sensitivity that can be achieved with this method. × 320.

Lieberkühn, retinal sheets, etc.). The tissues are then immunostained using a modified indirect immunoperoxidase technique, usually in the presence of Triton X-100 or a similar detergent which will aid penetration, post-fixed and contrasted with osmium tetroxide. The tissue may then be dehydrated and embedded in resin. Considerable care must be taken during sectioning as the optimal reaction deposit is found in a narrow band some 2–4 μm below the exposed surface of the tissue, which is independent of total thickness. Penetration may be further aided by the use of F(ab')$_2$ or Fab fragments at any stage in the procedure.

The pre-embedding method allows the identification of peptide-containing nerves (fig. 4) and/or cells or perikarya before ultrastructural examination. On ultrathin sections the immunostained elements may be readily observed. A great deal of tissue structure and spatial information may be obtained, although no precise antigen localisation data will be gained. In combination with autoradiography, the interrelationships of multiple transmitters have been observed simultaneously (see review by Pickel and Beaudet[50]). Conventional double immunostaining procedures for studying co-existence of peptides and other transmitters cannot be carried out using pre-embedding procedures alone. To date, particulate markers have been found to penetrate multicellular layers of tissues very poorly, although some encouraging results have been reported recently.

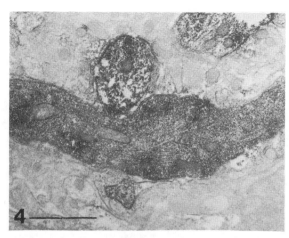

Figure 4. Electron micrograph of a neurophysin-immunoreactive nerve fibre in the mouse supraoptic nucleus. Pre-embedding immunoperoxidase method. Oxidised diaminobenzidine reaction product is diffusely distributed throughout the fibre. Scale bar = 1 μm. Figure provided by courtesy of Drs M. Castel, J. Morris and F. Shaw and reproduced with permission from Polak and Van Noorden[53].

Post-embedding method. Two distinct procedures may be considered:

a) *Semithin-thin procedure.* Alternate semithin (0.5–2.0 μm) and thin (60–100 nm) sections are cut from resin-embedded tissue blocks. The semithin sections are mounted on glass slides and immunostained using any of the procedures outlined previously. The serial thin sections are viewed by transmission electron microscopy. In this way immunohistochemistry and conventional electron microscopy images may be correlated. Although peptide-containing cells at the light microscope level can be correlated with their electron microscopical appearance the major disadvantage is that the actual subcellular site of immunoreactivity cannot be visualised. The semithin-thin procedure is not compatible with neuropeptide localisation as single fibres are not readily correlated.

b) *On-grid immunostaining procedure.* The immunocytochemical reaction is performed directly on ultrathin, grid-mounted tissue sections. Enzyme and colloidal particle-linked markers can be used in on-grid procedures. However, the use of gold particles as the marker of choice for high resolution antigen localisation studies is generally accepted (fig. 5). Several techniques, including gold-labelled antigen detection[42], protein A-gold[26,60], immunoglobulin-gold[20], avidin-biotin-gold[10] and hapten sandwich methods[48] are now available and each has its advocates and critics. Homogeneous gold particle populations can be made with relative ease and elegant serial and multiple immunolabelling procedures have been reported.

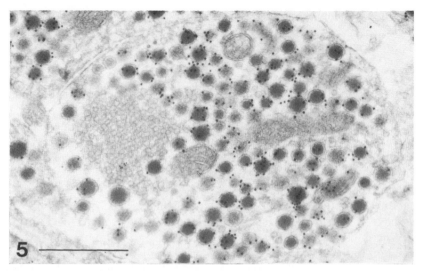

Figure 5. Electron micrograph of vasopressin immunoreactivity in neurosecretory vesicles in a nerve terminal from rat neurohypophysis. Immunogold staining procedure with 20-nm gold particles. Agranular vesicles, unstained, are clearly visible. Scale bar = 500 nm.

Specificity

Comparative immunoabsorption plays a large part in establishing peptide immunocytochemical specificity, but all immunocytochemical findings must be backed up by biochemistry. This was illustrated by the continual finding of pancreatic polypeptide (PP) immunoreactivity in glucagon cells of the intestine, despite the fact that no PP could be extracted from that region. Subsequently the PP-related peptide, peptide tyrosine tyrosine (PYY), was discovered and it transpired that the antibody to PP cross-reacted with PYY, with which PP shares several amino acid sequences. PYY was extractable in large quantities from the intestine and it is now accepted that it co-exists with glucagon in some gut endocrine cells.[3]

Very careful attention must be paid to absorptions and cross-absorptions in any peptide immunocytochemical study. The use of monoclonal antibodies will allow a reactive sequence to be better defined, but will nevertheless not distinguish between primary antigen and a known or unknown related peptide that shares the sequence. Methods for identifying the reactive epitope recognised by antibody molecules are now becoming available, and this will help considerably in our understanding of 'specificity'[27].

Multiple staining

Light microscopy

It is becoming increasingly apparent that the different peptides contained in the diffuse neuroendocrine system are intimately connected and integrated into a widespread control system. Several peptides may be present in the same structural component, be it an endocrine cell or a nerve, and even in the same secretory granule, particularly if they are products of proteolytic cleavage of a larger precursor molecule. Consequently it has become important to develop methods for visualising the different peptides simultaneously with differently coloured reaction products in order to determine their relationship to each other and to the tissue as a whole. Highly specific antibodies are a prerequisite.

Where possible, comparison of serial sections immunostained for different antigens is probably the most accurate method of comparison, but this is limited to materal suitable for thin (1–2 μm) sections which pass through the same structures, and to tissue with prominent 'landmarks' to facilitate comparison. It is frequently necessary to stain for several peptides or other substances in the same preparation. Here particular attention must be paid to avoidance of background staining which will detract from the clarity of the reaction. Lack of cross-reaction between the systems used to reveal two peptides in the same preparation is vital.

The simplest method is double direct immunostaining in which two antibodies to different peptides are labelled with different markers (enzymes, radiolabel, fluorochrome or gold particle) and are applied simultaneously or sequentially to the tissue preparation, at their predetermined optimal dilutions. Thus each of two peptides will be differently visualised. If they are present in the same structure, and assuming that one reaction does not mask the other, a mixture of labels will be present on that structure. If fluorescein and rhodamine are used as different fluorescent labels, viewing the preparation alternately with the appropriate filters will reveal single or double staining. Photography on the same frame with different filters will show double stained areas in orange and the single stained areas in red or green.

If two different enzymes are used, e.g. peroxidase and alkaline phosphatase, they can be separately developed in contrasting colours, e.g. brown and blue. Mixed colours show purplish-grey in contrast.

When indirect methods are used it becomes necessary to ensure that the second layer antibodies do not cross-react with the immunoglobulins from the species providing the first layer antibodies, which might result in a confused picture. Ways of overcoming the problem are to use primary antibodies raised in different species, revealed by non-cross-reacting, species-specific second layer antibodies[45] or to carry out the reaction sequentially instead of simultaneously, blocking the binding properties of the first immunoglobulins by exposure to hot formaldehyde vapour before carrying out the second reaction[78].

The first method of all employed three antibodies, all raised in rabbits and localised sequentially by indirect immunoperoxidase developed in different colours[16] but this is a cumbersome method compared with later ones, requiring elution of the immunoreactants and a complex series of controls to ensure specificity.

There are only a few of the strategies that have been adopted to stain several peptides in a single preparation. Combinations of autoradiography with enzyme labels[39] or gold/silver with enzyme[19,35] may also be cited.

Multiple staining of peptides has revealed the co-existence of substance P and CGRP in a proportion of neurones in the dorsal root ganglion[28] and of neuropeptide Y with met-enkephalin in the adrenal medulla[76]. Many further combinations surely await detection.

Electron microscopy

Similarly, at ultrastructural level, one logical extension to the capacity for localising a single antigen or groups of similar antigens is the development of a reliable procedure which allows discrimination between two or

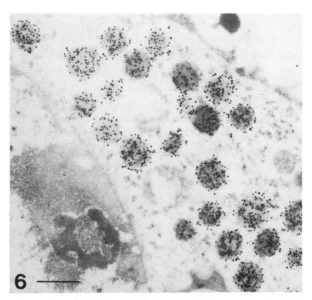

Figure 6. Electron micrograph of human pancreatic D (somatostatin-containing) cell doubly immunostained to reveal immunoreactivity for rat somatostatin cryptic peptide (RSCP, 20-nm gold) and somatostatin-28-(1–12) (10-nm gold) using the two-surface immunoglobulin-gold method. Scale bar = 500 nm.

more distinct, but co-existing or neighbouring, antigens. Various combinations of immunoenzyme, immunoferritin and immunogold procedures have been described for the ultrastructural demonstration of multiple peptide antigens. There are several compelling reasons for knowing whether multiple peptides derived from a single precursor or from different genes are co-produced, co-packaged and co-secreted. In cases of peptide-producing tumours, successful treatment of the patient may require knowledge of each peptide elaborated by the tumour. Neuromodulation may be achieved by co-release of multiple neuropeptides; alternatively, release of separately packaged peptides may act in a similar, but time-dependent, way.

Co-localisation. The localisation of multiple peptide antigens to single organelles has been demonstrated on several occasions[1,2,4,5,7,30,66,76]. Conversely, secretory granules within one cell have been shown to express different components of the pro-somatostatin molecule[77] (fig. 6). Each of the cited publications contains details of the double immunostaining procedures (see also Polak and Varndell[54]).

Some problems in applications of peptide immunocytochemistry to biological systems

Cell cultures

With the increasing use of peptide-expressing cell lines maintained in culture and the ease with which such cells may be manipulated by pharmacological agents, immunocytochemical procedures will soon be in demand. Cultured cells do pose unique problems.

Tissue processing. Adherent cells may be grown on solvent-resistant plastic coverslips which may then be processed into resin following pre-embedding immunostaining or prior to on-grid immunostaining. Suspension cultures may be pelleted and fixed into an agar/gelatin matrix before immunolabelling. There are, however, may other preparative procedures and specialist texts should be consulted.

Whole cells may need to be treated with a permeabilising agent to enable immunoreagent access to the cytoplasmic organelles. Permeabilisation must not be allowed to destroy ultrastructural morphology and thus a compromise must again be attained.

Immunocytochemistry. Cultured cells are suitable candidates for both pre- and post-embedding procedures but frequently poor results have been obtained. Ultrastructural examination of these cells has revealed that they are frequently poorly differentiated with few secretory granules, although high supernatant peptide levels may be measured. There is some evidence to implicate direct release of peptide from the rER/Golgi without secretory granule formation. The processed state of the secreted product is not currently known, i.e. are proteolytic enzymes co-secreted or is processing complete before secretion? These problems remain to be resolved, though it must be stressed that some cell lines in culture mimic the parent cell in most respects.

Regulatory peptide-producing tumours

In an immunocytochemical study of 125 pancreatic endocrine tumours, Heitz et al.[34] recorded that 50 of 95 'active' tumours (with a clinical syndrome) and 15 of 30 initially designated 'non-secreting' tumours were found to be of a mixed cell type. However, despite the presence of various peptide combinations in the majority of the tumours investigated, the clinical syndrome was most often attributed to the inappropriate secretion of only one of the products. It is important to note that in this series the distribution of immunoreactive cells was irregular and thus subject to sampling error. A large majority of the tumours contained immunoreactive peptides but in all tumours there were cells that did not display immunoreactivity to the entire panel of fifteen antibodies applied. As a rule, the hormone responsible for the clinical symptoms could be localised but with variable intensity which was apparently independent of the plasma concentration of the peptide.

There are several factors which could account for this:

1) Poor sampling. The patchy distribution of immunoreactive tumour cells could certainly be a source of sampling error. Thus a 'negative' result obtained from immunostaining or radioimmunoassay of an extract of part of a tumour does not exclude the presence of a particular peptide in the whole tumour. Similarly, the number or intensity of immunoreactive cells of part of a tumour sample cannot be extrapolated to the whole tumour.

2) Decreased storage capacity. Creutzfeldt et al.[18] suggested that the storage capability of tumour cells may be significantly reduced compared to normal peptide-producing cells. Indeed, it is frequently found that peptide-producing cells are poorly granulated, despite the fact that they often possess extensive endoplasmic reticulum, numerous mitochondria and a well-developed Golgi apparatus. In their study of pancreatic endocrine tumours Heitz et al.[34] reported that 25 of the tumours could not be diagnosed by electron microscopy because the majority of cells were devoid of secretory granules. A complicating factor here is the ultrastructural identification of secretory granules containing the non-dominant peptide product. One example of this is pancreatic polypeptide (PP) which is commonly found in vasoactive intestinal peptide (VIP) produc-

ing pancreatic tumours (VIPomas). VIP and PP secretory granules are morphologically similar and cannot be distinguished without resort to electron microscopical immunocytochemistry.

3) Molecular forms. It is now apparent that regulatory peptides, possibly without exception, are cleaved from larger pro-molecules. This post-translational processing is initiated in the rER and continues into the maturing secretory granules. Transcriptional errors and incomplete translational processing could result in molecular forms of a peptide being produced which may be recognised by particular immunoreagents but which may, or may not, have normal biological activity. It is equally possible that an abnormal molecular form of a peptide could be produced which is not, or is only partially, identifiable using the routine range of antibodies. One of the most important aspects of immunocytochemistry in general is the application of appropriate specificity checks and controls.

Conclusion

In conclusion, the choice of method for immunostaining a peptide depends on the nature of the peptide, the localisation, the type of tissue and the questions needing answers. In order to understand how peptides work, we need to know where they are localised, how they are stored and when they are released. The most valuable techniques at present to answer these questions have their basis in immunochemistry and there is now available a dizzying choice of suitable methods. No one method will suffice for all peptides in all situations, but the choice is wide and the techniques are increasingly well documented. There seems to be little doubt that future review articles will combine immunocytochemistry with *in situ* hybridisation histochemistry to prove the link between peptide synthesis and peptide storage and processing.

1 Adachi, T., Hisano, S., and Daikoku, S., Intragranular colocalization of immunoreactive methionine-enkephalin and oxytocin within the nerve terminals in the posterior pituitary. J. Histochem. Cytochem. *33* (1985) 891–899.
2 Ali-Rachedi, A., Ferri, G.-L., Varndell, I. M., Van Noorden, S., Schot, P. C., Ling, N., Bloom, S. R., and Polak, J. M., Immunocytochemical evidence for the presence of gamma$_1$-MSH-like immunoreactivity in pituitary corticotrophs and ACTH-producing tumours. Neuroendocrinology *37* (1983) 427–433.
3 Ali-Rachedi, A., Varndell, I. M., Adrian, T. E., Gapp, D. A., Van Noorden, S., Bloom, S. R., and Polak, J. M., Peptide YY (PYY) immunoreactivity is co-stored with glucagon-related immunoreactants in endocrine cells of the gut and pancreas. Histochemistry *80* (1984) 487–491.
4 Ali-Rachedi, A., Varndell, I. M., Facer, P., Hillyard, C. J., Craig, R. K., MacIntyre, I., and Polak, J. M., Immunocytochemical localization of katacalcin, a calcium-lowering hormone cleaved from the human calcitonin precursor. J. clin. Endocr. Metab. *57* (1983) 680–682.

5 Anderson, J. V., Christofides, N. D., Vinas, P., Wharton, J., Varndell, I. M., Polak, J. M., and Bloom, S. R., Radioimmunoassay of alpha rat atrial natriuretic peptide. Neuropeptides 7 (1986) 159–172.

6 Bell, G. I., Sanchez-Pescadar, R., Laybourn, P. J., and Najarian, R. C., Exon duplication and divergence in the human preproglucagon gene. Nature 304 (1983) 368–371.

7 Bendayan, M., Double immunocytochemical labelling applying the protein A-gold technique. J. Histochem. Cytochem. 30 (1982) 81–85.

8 Bendayan, M., and Zollinger, M., Ultrastructural localization of antigenic sites on osmium-fixed tissues applying the protein A-gold technique. J. Histochem. Cytochem. 31 (1983) 101–109.

9 Bishop, A. E., Polak, J. M., Bloom, S. R., and Pearse, A. G. E., A new universal technique for the immunocytochemical localisation of peptidergic innervation. J. Endocr. 77 (1978) 25P–26P.

10 Bonnard, C., Papermaster, D. S., and Kraehenbuhl, J.-P., The streptavidin-biotin bridge technique: Application in light and electron microscope immunocytochemistry, in: Immunolabelling for Electron Microscopy, pp. 95–111. Eds J. M. Polak and I. M. Varndell. Elsevier Science Publishers, Amsterdam 1984.

11 Buchan, A. M. J., Polak, J. M., Solcia, E., and Pearse, A. G. E., Localisation of intestinal gastrin in a distinct endocrine cell type. Nature 277 (1979) 138–140.

12 Bu'Lock, A., Vaillant, C., and Dockray, G. J., Immunohistochemical localisation of peptidergic nerve cell bodies in the gut following rational improvements to fixation with parabenzoquinone. Reg. Peptides 3 (1982) 67.

13 Coggi, G., Dell'Orto, P., and Viale, G., Avidin-biotin methods, in: Immunocytochemistry, Modern Methods and Applications, 2nd edn, pp. 54–70. Eds J. M. Polak, and S. Van Noorden. John Wright and Sons, Bristol 1986.

14 Coons, A. H., Creech, H. J., and Jones, R. N., Immunological properties of an antibody containing a fluorescent group. Proc. Soc. exp. Biol. Med. 47 (1941) 200–202.

15 Coons, A. H., and Kaplan, M. H., Localization of antigen in tissue sections. J. exp. Med. 91 (1950) 1–13.

16 Coons, A. H., Leduc, E. H., and Connolly, J. M., Studies on antibody production. I. A method for the histochemical demonstration of specific antibody and its application to a study of the hyperimmune rabbit. J. exp. Med. 102 (1955) 49–60.

17 Costa, M., Buffa, R., Furness, J. B., and Solcia, E., Immunohistochemical localization of polypeptide in peripheral autonomic nerves using whole mount preparations. Histochemistry 6 (1980) 157–165.

18 Creutzfeldt, W., Endocrine tumors of the pancreas, in: The Diabetic Pancreas, pp. 551. Eds B. W. Volk and K. F. Wellmann. Bailliere-Tindall, London 1977.

19 De Mey, J., Hacker, G. W., De Waele, M., and Springall, D. R., Gold probes in light microscopy, in: Immunocytochemistry, Modern Methods and Applications, 2nd edn, pp. 71–88. Eds J. M. Polak and S. Van Noorden. John Wright and Sons, Bristol 1986.

20 De Mey, J., Moeremans, M., Geuens, G., Nuydens, R., and De Brabander, M., High resolution light and electron microscopic localization of tubulin with the IGS (Immuno-Gold Staining) method. Cell Biol. int. Rep. 5 (1981) 889–899.

21 De Waele, M., Haematological electron immunocytochemistry, in: Immunolabelling for Electron Microscopy, pp. 267–288. Eds J. M. Polak and I. M. Varndell. Elsevier Science Publishers, Amsterdam 1984.

22 De Waele, M., De Mey, J., Moeremans, M., De Brabander, M., and Van Camp, B., Immunogold staining method for the detection of cell surface antigens with monoclonal antibodies, in: Techniques in Immunocytochemistry, vol. 2, pp. 1–3. Eds G. R. Bullock and P. Petrusz. Academic Press, London 1983.

23 Dockray, G. J., Molecular evolution of gut hormones: application of comparative studies on the regulation of digestion. Gastroenterology 72 (1977) 344–358.

24 Eldred, W. D., Zucker, C., Karten, H. J., and Yazulla, S., Comparison of fixation and penetration enhancement techniques for use in ultrastructural immunocytochemistry. J. Histochem. Cytochem. 31 (1983) 285–292.

25 Gallyas, F., Görcs, T., and Merchenthaler, I., High grade intensification of the end-product of the diaminobenzidine reaction for peroxidase histochemistry. J. Histochem. Cytochem. 30 (1982) 183–184.

26 Geuze, H. J., Slot, J. W., Van der Ley, P. A., and Scheffer, R. C. T., Use of colloidal gold particles in double labelling immunoelectron microscopy of ultrathin frozen tissue sections. J. Cell Biol. *89* (1981) 653–665.

27 Geysen, H. M., Barteling, S. J., and Meloen, R. H., Small peptides induce antibodies with a sequence and structural requirement for binding antigen comparable to antibodies realised against the native protein. Proc. natn. Acad. Sci. USA *82* (1985) 178–182.

28 Gibson, S. J., and Polak, J. M., Neurochemistry of the spinal cord, in: Immunocytochemistry, Modern Methods and Applications, 2nd edn, pp. 360–389. Eds. J. M. Polak and S. Van Noorden. John Wright and Sons, Bristol 1986.

29 Gibson, S. J., and Polak, J. M., Anand, P., Blank, M. A., Morrison, J. F. B., Kelly, J. S., and Bloom, S. R., The distribution and origin of VIP in the spinal cord of six mammalian species. Peptides *5* (1984) 201–207.

30 Gulbenkian, S., Merighi, A., Wharton, J., Varndell, I. M., and Polak, J. M., Ultrastructural evidence for the co-existence of calcitonin gene-related peptide and substance P in secretory vesicles of peripheral nerves in the guinea pig. J. Neurocyt. *15* (1986) 535–542.

31 Hacker, G. W., Springall, D. R., Van Noorden, S., Bishop, A. E., Grimelius, L., and Polak, J. M., The immunogold-silver staining method: a powerful tool in histopathology. Virchows Arch. Path. Anat. *406* (1985) 449–461.

32 Halmi, N. S., Immunostaining of growth hormone and prolactin in paraffin embedded and stored or previously sectioned material. J. Histochem. Cytochem. *26* (1978) 486–495.

33 Hamid, Q., Bishop, A. E., Sikri, K. L., Varndell, I. M., Bloom, S. R., and Polak, J. M., Immunocytochemical characterization of 10 pancreatic tumours, associated with the glucagonoma syndrome, using antibodies to separate regions of the pro-glucagon molecule and other neuroendocrine markers. Histopathology *10* (1986) 119–133.

34 Heitz, P. U., Kasper, M., Polak, J. M., and Klöppel, G., Pancreatic endocrine tumours: Immunocytochemical analysis of 125 tumors. Hum. Path. *13* (1982) 263–271.

35 Holgate. C., Jackson, P., Cowen, P., and Bird, C., Immunogold-silver staining: new method of immunostaining with enhanced sensitivity. J. Histochem. Cytochem. *31* (1983) 938–944.

36 Hsu, S.-M., and Soban, E., Color modification of diaminobenzidine (DAB) precipitation by metallic ions and its application to double immunohistochemistry. J. Histochem. Cytochem. *30* (1982) 1079–1082.

37 Huang, S., Minassian, H., and More, J. D., Application of immunofluorescent staining in paraffin sections improved by trypsin digestion. Lab. Invest. *35* (1976) 383–391.

38 Huang, W. M., Gibson, S. J., Facer, P., Gu, J., and Polak, J. M., Improved section adhesion for immunocytochemistry using high molecular weight polymers of L-lysine as a slide coating. Histochemistry *77* (1983) 275–279.

39 Hunt, S. P., and Mantyh, P. W., Radioimmunocytochemistry with [³H] biotin. Brain Res. *291* (1984) 203–217.

40 Lackie, P. M., Cuttitta, F., Minna, J. D., Bloom, S. R., and Polak, J. M., Localisation of receptors using a dimeric ligand and electron immunocytochemisty. Histochemistry *83* (1985) 57–59.

41 Langley, O. K., Ghandour, M. S., Vincendon, G., and Gombos, G., An ultrastructural immunocytochemical study of nerve-specific protein in rat cerebellum. J. Neurocyt. *9* (1980) 783–798.

42 Larsson, L.-I., Simultaneous ultrastructural demonstration of multiple peptides in endocrine cells by a novel immunocytochemical method. Nature *282* (1979) 743–746.

43 Larsson, L.-I., Peptide immunocytochemistry. Prog. Histochem. Cytochem. *13* (1981) No. 4.

44 Lopez, L. C., Frazier, M. L., Su, C.-J., Kumar, A., and Saunders, G. F., Mammalian pancreatic preproglucagon contains three glucagon-related peptides. Proc. natn. Acad. Sci. USA *80* (1983) 5485–5489.

45 Mason, D. Y., and Sammons, R. E., Alkaline phosphatase and peroxidase for double immunoenzymatic labelling of cellular constituents. J. clin. Path. *31* (1978) 454–462.

46 Matutes, E., and Catovsky, D., The fine structure of normal lymphocyte subpopulations—a study with monoclonal antibodies and the immunogold technique. Clin. exp. Immun. *50* (1982) 416–425.

47 Nakane, P. K., Simultaneous localization of multiple tissue antigens using the peroxidase-labeled antibody method: a study on pituitary glands of the rat. J. Histochem. Cytochem. *16* (1968) 557–560.

32

48 Newman, G., and Jasani, B., Post-embedding immunoenzyme techniques, in: Immunolabelling for Electron Microscopy, pp. 53–70. Eds J. M. Polak and I. M. Varndell. Elsevier Science Publishers, Amsterdam 1984.
49 Pearse, A. G. E., and Polak, J. M., Bifunctional reagents as vapour and liquid phase fixatives for immunohistochemistry. Histochem. J. 7 (1975) 179–186.
50 Pickel, V. M., and Beaudet, A., Combined use of autoradiography and immunocytochemical methods to show synaptic interactions between chemically defined neurons, in: Immunolabelling for Electron Microscopy, pp. 259–266. Eds J. M. Polak and I. M. Varndell. Elsevier Science Publishers, Amsterdam 1984.
51 Polak, J. M., Pearse, A. G. E., Van Noorden, S., Bloom, S. R., and Rossiter, M. A., Secretin cells in coeliac disease. Gut 14 (1973) 870–874.
52 Polak, J. M., and Van Noorden, S., Eds, Immunocytochemistry, Practical Applications in Pathology and Biology. John Wright and Sons, Bristol 1983.
53 Polak, J. M., and Van Noorden, S., Eds, Immunocytochemistry, Modern Methods and Applications, 2nd edn. John Wright and Sons, Bristol 1986.
54 Polak, J. M., and Varndell, I. M., Eds, Immunolabelling for Electron Microscopy. Elsevier Science Publishers, Amsterdam 1984.
55 Priestley, J. V., Pre-embedding ultrastructural immunocytochemistry: Immunoenzyme techniques, in: Immunolabelling for Electron Microscopy, pp. 37–52. Eds J. M. Polak and I. M. Varndell. Elsevier Science Publishers, Amsterdam 1984.
56 Probert, L., De Mey, J., and Polak, J. M., Distinct subpopulations of enteric p-type neurones contain substance P and vasoactive intestinal peptide. Nature 294 (1981) 470–471.
57 Probert, L., De Mey, J., and Polak, J. M., Ultrastructural localization of four different neuropeptides within separate populations of p-type nerves in the guinea pig colon. Gastroenterology 85 (1983) 1094–1104.
58 Ravazzola, M., and Orci, L., Transformation of glicentin-containing L-cells into glucagon-containing cells by enzymatic digestion. Diabetes 29 (1980) 156–158.
59 Robinson, D., Tavares de Castro, J., Polli, N., O'Brien, M., and Catovsky, D., Simultaneous demonstration of membrane antigens and cytochemistry at ultrastructural level: A study with the immunogold method, acid phosphatase and myeloperoxidase. Br. J. Haemat. 56 (1984) 617–631.
60 Roth, J., The preparation of protein A-gold complexes with 3 nm and 15 nm gold particles and their use in labelling multiple antigens on ultrathin sections. Histochem. J. 14 (1982) 791–801.
61 Schultzberg, M., Hökfelt, T., Nilsson, G., Terenius, L., Rehfeld, J. F., Brown, M. Elde, R., Goldstein, M., and Said, S. I., Distribution of peptide- and catecholamine-containing neurones in the gastrointestinal tract of rat and guinea pig: Immunohistochemical studies with antisera to substance P, vasoactive intestinal polypeptides, enkephalins, somatostatin, gastrin/cholecystokinin, neurotesin and dopamine β-hydroxylase. Neuroscience 5 (1980) 689–744.
62 Scopsi, L., and Larsson, L.-I., Increased sensitivity in peroxidase immunocytochemistry. A comparative study of a number of peroxidase visualisation methods employing a model system. Histochemistry 84 (1986) 221–280.
63 Steiner, D. F., Docherty, K., Hofmann, C., Madsden, O., Nahum, A., Labrecque, A., and Carroll, R., Biosynthesis of neuroendocrine peptides in normal and tumour cells, in: Endocrine Tumours, the Pathobiology of Regulatory Peptide-Producing Tumours, pp. 38–56. Eds J. M. Polak and S. R. Bloom. Churchill Livingstone, Edinburgh 1985.
64 Sternberger, L. A., Ed., Immunocytochemistry, 2nd edn. John Wiley & Son, Inc., New York 1979.
65 Sternberger, L. A., Hardy, P. H. Jr, Cuculis, J. J., and Meyer, H. G., The unlabeled antibody-enzyme method of immunohistochemistry. Preparation and properties of soluble antigen-antibody complex (horseradish peroxidase-antihorseradish peroxidase) and its use in identification of spirochetes. J. Histochem. Cytochem. 18 (1970) 315–333.
66 Tapia, F. J., Varndell, I. M., Probert, L., De Mey, J., and Polak, J. M., Double immunogold staining method for the simultaneous ultrastructural localization of regulatory peptides. J. Histochem. Cytochem. 31 (1983) 977–981.
67 Terenghi, G., Polak, J. M., Ghatei, M. A., Mulderry, P. K., Butler, J. M., Unger, W. G., and Bloom, S. R., Distribution and origin of calcitonin gene-related peptide (CGRP)

immunoreactivity in the sensory innervation of the mammalian eye. J. comp. Neurol. *233* (1985) 506–516.

68 Udenfriend, S., and Kilpatrick, D. L., Proenkephalin and the products of its processing: Chemistry and biology, in: The Peptides, vol. 6. Eds S. Udenfriend and J. Meienhofer. Academic Press, London 1984.

69 Vacca, L. L., 'Double bridge' techniques of immunocytochemistry, in: Techniques in Immunocytochemistry, vol. 1, pp. 155–182. Eds G. R. Bullock and P. Petrusz. Academic Press, London 1982.

70 Vacca, L. L., Abrahams, S. J., and Naftchi, N. E., A modified peroxidase-antiperoxidase procedure for improved localization of tissue antigens. J. Histochem. Cytochem. *28* (1980) 297–307.

71 Van Noorden, S., and Falkmer, S., Gut-islet endocrinology—some evolutionary aspects. Invest. Cell Path. *3* (1980) 21–25.

72 Van Noorden, S., and Polak, J. M., Immunocytochemistry of regulatory peptides, in: Techniques in Immunocytochemistry, vol. 3, pp. 115–154. Eds G. R. Bullock and P. Petrusz. Academic Press, London 1985.

73 Varndell, I. M., Bishop, A. E., Sikri, K. L., Uttenthal, L. O., Bloom, S. R., and Polak, J. M., Localization of glucagon-like peptide (GLP) immunoreactants in human gut and pancreas using light and electron microscopic immunocytochemistry. J. Histochem. Cytochem. *33* (1985) 1080–1086.

74 Varndell, I. M., and Polak, J. M., Immunocytochemistry, in: Endocrine Tumours, The Pathobiology of Regulatory Peptide-producing Tumours, pp. 116–143. Eds J. M. Polak and S. R. Bloom. Churchill Livingstone, Edinburgh 1985.

75 Varndell, I. M., and Polak, J. M., Electron microscopical immunocytochemistry, in: Immunocytochemistry, Modern Methods and Applications, 2nd edn, pp. 146–166. Eds J. M. Polak and S. Van Noorden. John Wright and Sons, Bristol 1986.

76 Varndell, I. M., Polak, J. M., Allen, J. M., Terenghi, G., and Bloom, S. R., Neuropeptide tyrosine (NPY) immunoreactivity in norepinephrine containing cells and nerves of the mammalian adrenal gland. Endocrinology *114* (1984) 1460–1462.

77 Varndell, I. M., Sikri, K. L., Hennessy, R. J., Kalina, M., Goodman, R. H., Benoit, R., Diani, A. R., and Polak, J. M., Somatostatin-containing D cells exhibit immunoreactivity for rat somatostatin cryptic peptide in six mammalian species. An electron-microscopical study. Cell Tiss. Res. *246* (1986) 197–204.

78 Wang, B.-L., and Larsson, L.-I., Simultaneous demonstration of multiple antigens by indirect immunofluorescence or immunogold staining. Histochemistry *83* (1985) 47–56.

79 Williams, E. D., Immunocytochemistry in the diagnosis of thyroid diseases, in: Immunocytochemistry, Modern Methods and Applications, 2nd edn, pp. 533–546. Eds J. M. Polak and S. Van Noorden. John Wright and Sons, Bristol 1986.

Aspects of measurement and analysis of regulatory peptides

J. M. Burrin, L. O. Uttenthal, G. P. McGregor and S. R. Bloom

Summary. Although almost all methods of mass measurement of regulatory peptides still depend on the high affinity antibody, the traditional Yalow and Berson radioimmunoassay technique is becoming outdated. Pure monoclonal antibodies allow excess antibody two-site assay techniques with a variety of different labels (preferentially non-radioactive) of great sensitivity and speed. The large amounts of particular monoclonal antibodies available allow several different laboratories to use the same reagents and have increased comparability. Unfortunately many regulatory peptides exist in multiple molecular forms and attention must be paid to antibody region specificity. Improved methods of extraction of regulatory peptides from plasma and tissue allow more accurate quantitation. New techniques for rapid high resolution chromatography make distinction of different molecular forms much easier than hitherto. Better education in techniques and/or attention to inter-assay standards are necessary to improve the comparability of regulatory peptide measurement in the future.

For the last two and a half decades regulatory peptides have been measured by a standard radioimmunoassay technique as first outlined by Yalow and Berson[34]. This technique has served us well. It is highly sensitive and quite adequately specific to allow detection of very small quantities of regulatory peptides, both in their tissue of origin and in various body fluids. Coupled with separation techniques it has been possible to show that many of these peptides exist in multiple forms, often localised to particular tissues. Unfortunately not all problems have been overcome. Firstly it is still not all that easy to obtain sufficient ligand to develop an antibody of sufficient specificity and sensitivity, to couple sufficient radioactive iodine 125 without damaging the ligand (in the conventional system) and define conditions of assay in which the ligand is sufficiently protected from damage. Secondly even when it is possible to overcome all these difficulties in an individual laboratory, the results are frequently at variance with those produced by other laboratories. The technology required to set up the conventional radioimmunoassay is now well documented but extremely tedious and many workers have neither the time nor the resources to adequately characterise and optimise each step. Added to this, the use of antibodies with different region-specificity to measure multiple forms of regulatory peptides that are ill-characterised and subject to unknown degradative influences contributed to frequent technical failure. The advent of monoclonal antibodies, improved assay technology, easier and more reliable separation techniques and a better understanding of the chemical nature of regulatory peptides promises a steady improvement in the future. Meanwhile great attention to detail, experience in the field and strict application of standard procedures, including chromatographic analysis of samples, interchange of standards and antibody between laboratories and simultaneous use of different antisera, for example, will improve matters today. The subject is vast and this article will touch on a few points of current interest which throw light on the nature of the problem.

Assay techniques

The description of radioimmunoassay (RIA) by Berson and Yalow[34] and of saturation analysis techniques by Ekins[9] in the early 1960's had a major impact on the assay of hormones. Although immunoassay methods, using unlabelled reagents, had been in use for many years, the introduction of a radioisotopic label to discriminate between the antibody bound fraction and the free fraction to the antigen increased the sensitivity of previous physico-chemical procedures from approximately 10^{-8} moles to 10^{-12} moles, thus considerably extending the range of substances which could be quantitated directly.

In RIA, antigen (Ag) and antibody (Ab) interact reversibly to form a soluble antigen-antibody complex. Radiolabelled and unlabelled Ag in the standard or sample compete for binding to the limited number of antibody binding sites. The process obeys the law of mass action, so that the greater the quantity of unlabelled Ag present, the less labelled Ag is bound. Free Ag is then separated from bound Ag and the distribution of radioactivity between these two fractions is measured.

In principle, radioimmunoassay can be used for the quantitative determination of any substance available in pure form, and to which an antibody can be raised. The technique has found widespread application in the measurement of peptide and non-peptide hormones and over the last twenty years the subject has been extensively reviewed[6,7,23]. The popularity of RIA has been maintained due to its sensitivity, specificity and widespread application. However, conventional RIA does suffer from a variety of disadvantages. These include the hazards associated with the use of radioactivity both in the preparation of the labels and with regard to waste disposal. Other drawbacks include the limited shelf-life of the label, the requirement for expensive detection equipment, the limited number of radioisotopes that can be used practicably, and the virtual impossibility of developing non-separation or homogeneous assays. This last disadvantage has also prevented automation of RIA in most cases.

Inevitably, alternative approaches to hormone assay have been sought which overcome these disadvantages of RIA. These alternative approaches have used non-labelled techniques, non-isotopic labels and non-competitive or reagent excess systems.

Non-labelled immunoassays use the classical precipitation reaction between Ag and Ab as an indicator reaction. This group of assay techniques includes radial immunodiffusion, nephelometry, turbidimetry and nephelometric inhibition[13]. The lower detection limit for these procedures is about 10 μg/l and is restricted because all reactants must be present in mutually compatible concentrations. These techniques were originally limited to the assay of proteins present at relatively high concentrations, since only in such circumstances were the Ag/Ab complexes of a sufficient size to be detected. New developments in this area followed the introduction of particle counting immunoassay (PACIA) and it has proved possible to assay low molecular weight haptens and to increase the sensitivity of the method by coupling the Ag or the Ab to latex particles of about 0.8 μm in diameter[4]. The presence of the corresponding reaction partner in the assay mixture causes agglutination of the latex particles. Quantitation is performed using a particle counter with a threshold set at 1.2 μm so that only unagglutinated particles are counted. Technicon (Technicon Instruments Corporation, Tarrytown, New York 10591) are using PACIA reagents on the Random Access Analyser RA1000 for the assay of T4. In this method latex particles coated with Ab are caused to agglutinate by a polyvalent T4-ficoll conjugate. Agglutination is inhibited by the presence of Ag in the standards or samples. PACIA methods for T4, T3, AFP, ferritin, TSH and HCG are also available from Acade Diagnostic Systems, Brussels, Belgium under the name of IMPACT. The sensitivity of these methods is claimed to be 10^{-12}–10^{-15} moles. The advantages of PACIA are that no separation step is required, the incubation times are short and the system can be fully automated. In addition to particles, a variety of labels have been proposed as alternatives to radioisotopes. Perhaps the most popular labels for immunoassays have been enzymes and fluorophores and more recently the potential of luminescent labels has begun to be realised.

Enzymes have found extensive application as labels in immunoassays and there are several reviews dealing with enzyme immunoassay[2,20,30]. Some of the earliest uses of enzymes simply substituted them for the radioisotopes in conventional RIA[29]. Later developments resulted in the enzyme-linked immunosorbent assay (ELISA) which combined the virtues of solid phase technology with the merits of an enzyme-labelled immunoreagent. Unlike RIA, where only a few isotopes (^{125}I and ^{3}H) can be used, the choice of enzymes is virtually unlimited, provided that the enzyme is stable, cheap, has a high turnover rate and can be linked to an immunoreagent. Enzymes which have been used include alkaline phosphatase, galactosidase, peroxidase and glucose oxidase. The detection system used to monitor the presence and concentration of the enzyme label may be spectrophotometry, fluorescence or luminescence.

ELISA methods are heterogeneous requiring separation steps since the enzyme activity is unaltered by the Ag-Ab reaction. Homogeneous enzyme assays have also been developed which do not require separation steps because the assay depends on inhibition or activation of the enzyme label by antibody binding. Perhaps the best example of homogeneous enzyme-labelled hapten assays is the enzyme-multiplied immunoassay test (EMIT) commercialised by the Syva Corporation, Palo Alto, California, USA[5]. Conjugation of the enzyme to the hapten does not destroy the enzyme activity. However, binding of hapten-specific antibody to the label results in inhibition of enzyme activity. Free hapten in the standards or samples relieves this inhibition by competing for antibody. Thus in the presence of antibody the enzyme activity is proportional to the concentration of free hapten. A wide variety of drugs have been measured by this technique which is rapid, can be automated and requires only a small sample volume.

The field of application of enzyme immunoassays in their various forms corresponds largely to that of radioimmunoassay. A number of heterogeneous enzyme immunoassays have similar detection limits to RIA but homogeneous assay systems are less sensitive. The sensitivity for peptide hormones using EIA was generally lower than that of RIA until the recent introduction of enzyme amplification techniques. In this method the enzyme label in the immunoassay is used to provide a trigger substance for a secondary system which generates a larger quantity of coloured product[14]. The enzyme-amplified immunoassay differs from the conventional type in that the product from the enzyme label need not, in itself, be measurable but instead can act catalytically on the secondary system which remains essentially silent until activated in this way. Using this type of system, highly sensitive assays for progesterone, HCG and TSH have been described, the latter demonstrating a considerable increase in sensitivity over conventional RIA[24].

Fluorophores, i.e. labels that fluoresce on appropriate excitation, have also been used to develop hormone immunoassays[27]. Initially they were less sensitive than RIAs because of the background fluorescence from many proteins and other constituents in biological samples. Many of the first assays using fluorophores were therefore heterogeneous and required separation steps so that background and interference problems were made manageable. As with enzymes, fluorophores have been used as straight replacements for radiolabels in competitive fluorimmunoassay, immunofluorometric assay and two-site immunofluorometric assay. These types of fluorescent immunoassays (FIA) have been applied to a wide range of hormones, although most assays are about ten times less sensitive than RIA because of the high background fluorescence associated with proteins in the sample and materials such as plastic and glass. Fluorophores do, however, have some advantages over enzymes as non-isotopic labels. Any separation method can be used since the fluorescent label is small. Complete separation of bound and free is not necessary, it is only required to clear the bound or free fraction from the light path, and, as with RIA, immediate end point detection is possible on completion of the immunological reaction. This is in contrast to both EIA and assays employing chemiluminescence where an additional chemical step is required. However, like enzymes, fluorescent labels have also been used to develop separation free assays by at least five different techniques.

In 1977 an enhancement fluorimmunoassay for thyroxine was described[26]. Fluorescein-labelled T4 was found to give an abnormally low fluorescence yield. Upon binding by antibody, the T4 group is apparently held away from the fluorophore and the quenching is relieved, thus resulting in an actual enhancement of the fluorescence. No labelling of the antibody is required and the fluorescence signal decreases with added unlabelled T4. The original method was seriously limited because enhancement also occurred on non-specific binding of the conjugate to serum proteins. A modified conjugate was therefore synthesised, which completely eliminated the unwanted fluorescence modulation and gave excellent performance.

In direct contrast to this are FIAs which depend upon antibody binding leading to a reduction in fluorescence. In these types of direct-quenching FIA the amount of labelled Ag bound and therefore the extent of quenching is inversely related to the amount of unlabelled Ag present. Direct-quenching FIA is applicable only to the assay of haptens since Ab binding of a protein labelled with a fluorophore rarely results in any significant change in fluorescence intensity, probably because of the distance between the antigenic determinant and the site of label attachment.

However, indirect-quenching FIA or fluorescence protection immunoassay has been applied to proteins. In this type of assay, Ag, Ag labelled with fluorescence and Ab are incubated then anti-fluorescein Ab added, which quenches any label in the free fraction but because of steric hindrance by the first Ab is unable to bind the fluorescein groups in the bound fraction, which continue to fluoresce.

Fluorescence excitation transfer immunoassay involves the use of Ag labelled with fluorescein and Ab labelled with rhodamine which acts as a quencher. When these reactants combine the fluorescence intensity is reduced. When unlabelled hapten or Ag is present some of the quencher-labelled Ab will be used up and be unavailable for binding to the fluorescein-labelled Ag. The fluorescence intensity thus increases with increasing concentration of free antigen. This type of assay has been applied to both haptens and proteins, and can be performed using a rate protocol so that assays are extremely rapid[28].

An alternative approach for polyvalent antigens was to put both labels on the Ab. This simplifies the protocol, since a single combined reagent is added to the sample. Binding of the fluorescent and quencher labelled Abs to the antigen brings them into close proximity and effects quenching. With increasing concentration of Ag, quenching is less likely to occur.

One way of increasing the sensitivity of measurement of all these fluorescent immunoassays has been by means of instrumentation. In time-resolved fluorometers, a fast light pulse which excites the probe is used and fluorescence is measured after a certain time has elapsed from the moment of excitation[16]. Fluorescence due to non-specific background usually has a short decay time of less than 10 ns. Thus by using fluorescent probes whose excited state has a long decay time, e.g. chelates or rare earth metals such as europium, and measuring the fluorescence after, for example, 100 ns, this type of interference can be completely removed. LKB, Turku, Finland, have already marketed kits for HCG, TSH, AFP and ferritin based on this type of approach. The LKB method known as DELFIA (dissociation enhanced lanthanide fluorimmunoassay) involves the use of an antibody labelled with an europium (a lanthanide metal) chelate and a first antibody immobilised onto the surface of microtitration wells. The fluorescence of the europium ion is developed and highly intensified by the addition of enhancement solution after the immunoreaction has been completed. Light emission is then measured in the time-resolved fluorometer. Time-resolved fluorometry thus offers a considerable improvement in sensitivity over conventional methods of measurement. The speed, simplicity and precision of end-point detection indicate that FIA will play an increasingly important role, especially as improvement in methodology (such as multi-labelling) and in instrumentation (such as time-resolved fluorescence) result in a marked improvement in the sensitivity that can be achieved.

Luminescent substances such as chemi- or bioluminescent molecules have been used in immunoassays, both directly as labels for Ag or Ab and indirectly for the luminescent quantitation of enzyme and co-factor labelled ligands[15,33]. They have advantages over fluorescent labels since few luminescent substances occur in biological materials and so little background interference occurs. Luminescent immunoassay (LIA) is analogous to RIA and has been used for quantitation of steroid hormones by both homogeneous and heterogeneous techniques. In homogeneous assays binding of the luminescent label to the Ab results in enhancement of the light emission. This is reduced on addition of sample to the reaction mixture. Luminescent-labelled antibodies have also been used for immunoluminometric assays for AFP and TSH, the latter resulting in an increased sensitivity over RIA[32].

Luminescent assays for enzymes such as peroxidase and glucose oxidase are considerably more sensitive than are conventional colorimetric assays, a factor that has been exploited in the quantitation of enzyme conjugates in luminescent enzyme immunoassay.

Similarly EMIT assays may be monitored using the NADH-dependent bacterial luminescence system with an increase in the sensitivity achieved over colorimetric detection. The bacterial luminescence system has also been used to monitor ligand-cofactor cycling reactions in luminescence cofactor immunoassay. Both NAD and ATP conjugates have been employed and homogeneous assays developed. Luminol, isoluminol, luminol derivatives, and lucigenin derivatives have all been used as chemiluminescent labels. They are cheap, readily available and able to bind to both antigens and antibodies. They are very stable and chemiluminescent steroid labels have been used for up to one year with no decrease in efficiency.

Recently the use of acridinium esters as chemiluminescent labels for immunoassay systems has also been reported[32]. These compounds are more luminescent than luminol and can be stimulated to produce their chemiluminescence under much milder conditions. Acridinium ester-labelled antibodies have been used in two-site immunoassays for AFP and TSH with excellent results.

There is no doubt that many LIAs have already produced results comparable with that of RIA, in terms of sensitivity and reliability. It is probable that by combining the usefulness of chemiluminescent labels with the inherently high sensitivity that can be achieved by reagent excess assays using labelled antibodies, it will be possible to greatly improve the performance of peptide hormone assay systems. However, the acceptance of LIA in the clinical laboratory depends on the availability of reliable and cheap luminometers.

It is thus evident that many different labels may, in principle, be used in immunoassay systems. Until recently few of them compared favourably with radioisotopes with regard to their sensitivity and freedom from background interference. In practice, therefore, non-isotopic labels were restricted in their application to the measurement of analytes present at relatively high concentrations and where high assay sensitivity was not a vital requirement. However, with the advent of monoclonal antibody production methods non-competitive immunoassay systems became more readily available and this type of assay system is potentially capable of achieving improvements in sensitivity of many order of magnitude. Thus by combining non-isotopic labels with labelled antibody (IRMA) systems, RIA no longer has the edge in terms of sensitivity. Indeed the full sensitivity potential of non-competitive immunoassay systems can only be fully exploited by using non-radioisotopic antibody labels[10].

IRMA assays were first introduced in the 1960's and were originally claimed to offer increased sensitivity over RIA[18]. The main disadvantage of this approach was the need to isolate and label relatively pure antibodies and the high consumption of antibody. These disadvantages can now be overcome with the use of monoclonal antibodies, which are by definition directed against a single antigenic determinant, and are theoretically an indefinite supply of Ab with constant characteristics. In this type of assay the Ag is reacted with a labelled monoclonal Ab and a second monoclonal Ab attached to a solid support and directed against a different antigenic determinant added later. This system may be described as non-competitive or reagent excess and has the advantage of shortening assay incubation times. Other advantages which arise from the use of labelled Abs include their greater stability and the avoidance of problems sometimes associated with the preparation of labelled Ags.

RIA is no longer the only method for measuring peptides; many of the alternative approaches can now compete with RIA in terms of sensitivity and also overcome the disadvantages of RIA. In addition they may offer advantages in terms of the precision of detection, the development of separation-free assays more suitable for automation and the increased speed of labelled antibody assays. The final choice of method for measurement of a particular substance now depends, to a great extent, upon the prejudice of the consumer and not the limitations of the method.

Antibodies

The major factor limiting assay sensitivity is the binding avidity of the antiserum used. This is true even in excess antibody two site measuring systems (see above) although different kinetics apply. The latter technique is much easier to apply with pure antibody and is therefore most suitable for use with monoclonal technology. In general it is still easier for individual scientists to develop the traditional radioimmunoassay and this therefore tends to be the first measuring technique used with a novel regulatory peptide. Once a mass demand develops, there is more justification for the greater investment of time and effort required to raise and develop monoclonal antibody systems and this is often best undertaken commercially.

Rabbits are the favourite animal in which to raise polyvalent antisera as they are large enough to provide a reasonable supply yet small enough to be economical in large numbers. Raising polyvalent antibodies is more of an art than a science and many different approaches have been successful. Our own laboratory has employed the following technique which, while successful, has no particular theoretical background. Rabbits are injected with water-soluble antigen (or aqueous suspension) emsulsified in Freund's adjuvant. Care is taken that the oil phase is continuous and approximately 0.6 ml Freund's adjuvant is used with 0.4 ml aqueous phase per animal, the primary being followed by a boost injection at three months, four months, five months and six months (using half the amount to boost that was used in the primary). If at the end of six months the animal has not responded immunisation is discontinued. Responding animals are then rested for three months and unless the antibody affinity and avidity is very high, the way the hapten is presented is changed. Four methods of presentation are routinely used (a quarter of the animals by each method). Thus the hapten is administered unconjugated, conjugated via carboxyl groups and primary amino groups using the bifunctional agent, carbodiimide[11], by primary amino groups alone using glutaraldehyde[22], and tyrosyl or histidyl residues using bis-diazonium salts. The advantage of the four different approaches lies in the unknown nature of the antigenic sites on the hapten. By coupling at various points (or not coupling at all), the chance of successfully developing antisera and having a range of region specificity is increased. The carrier used is bovine serum albumin which is cheap and easy to handle. In subsequent boosts switching to more exotic carrier proteins, such as limpet haemocyanin or bovine thyroglobulin may be tried and sometimes results in a dramatic improvement in antibody titre or affinity[21].

Monoclonal antibodies differ from polyvalent antibodies in being highly specific but usually of lower affinity. Their main difference is their invariance, purity and mass production. Thus antibody-producing B lymphocytes are harvested from rodent spleen at a variable time after the animals have been immunised (using the same technology as above) and fused with rodent myeloma lines to achieve immortality[1,31]. Once fused (usually with polyethylene glycol) the problem is screening the numerous hybridomas. However this is done, it will take a considerable amount of time and effort. To achieve a high affinity monoclonal antibody of particular specifity may require very many animals and literally years of work. The reward will be a mass-produced reagent capable of producing standard answers in many different laboratories. It can be assumed that this is the way of the future.

Peptide extraction factors influencing quantitative recovery

Measurement of regulatory peptides in plasma is of importance when they act in a hormonal role. Plasma contains a number of potential interfering and degradative factors. Some assays are sufficiently sensitive so that only a very small amount of plasma (for example 10 μl) need be added and these interfering factors are so diluted that they become irrelevant. With less sensitive assays it may be necessary to extract the peptide, for example by absorbing onto a Sep Pak and washing off most of the plasma constituents before eluting the peptide with acetonitrile which is then removed by evaporation. Such a 'purification step' however introduces its own problems as recovery will be variable and not all molecular forms of the peptide will necessarily behave identically. Fortunately antibody avidity is such that it is usually possible to detect plasma concentrations when these are biologically meaningful.

For the precise quantitative analysis by radioimmunoassay of regulatory peptides in tissues properly validated extraction procedures are essential. The failure to consider all the factors which may prevent maximal recovery of peptide immunoreactivity from tissues possibly accounts for the frequently observed inter-laboratory variation in reported tissue concentrations. We have found that boiling in aqueous acetic acid (0.5 M) (or in specific cases, water) of finely cut tissue obtained immediately following dissection will usually provide maximal recovery of peptides so far investigated but attention has to be paid to the possiblity of alterations in immunoreactivity of individual peptides. Destruction of the peptide-degrading enzymes is achieved by the treatment of tissue in boiling extraction medium, usually contained in a polypropylene tube held in a boiling water bath.

For accurate measurement of tissue peptide levels, it is necessary to consider their post-mortem stability in tissue and the procedures to extract all the immunoreactive peptide for radioimmunoassay. Substantial losses of peptide can occur during extraction procedures as a result of enzymic and non-specific adsorptive processes. Because of the specificity of radioimmunoassay, minimal purification is required. Extraction methods require a means of solubilising the peptide and maintaining it in a solution which itself does not interfere with the radioimmunoassay. Boiling of tissue in aqueous media or the use of organic solvents are the most frequently used methods of inactivating peptide-degrading enzymes[19]. The former is recommended since the latter involves more processing steps, thereby increasing the possibility of adsorptive losses. The biological activity and immunoreactivity of regulatory peptides are fortunately usually heat stable.

It is recognised that in the purification of proteins substantial losses can be incurred as a result of adsorption to structural proteins present in the extract solution. Obviously, for quantitative analysis of peptides, this needs to be avoided. Losses of a peptide by adsorption to cellular material such as solubilised structural proteins appear pH-dependent and this determines the choice of acid or water for extraction. Most peptides are most soluble in acid and may exhibit a low solubility in water extracts but for the few acidic peptides, such as the gastrins and the CCK-peptides, the reverse is true.

Irrespective of pH, peptide loss can occur if the tissue is homogenised (fig. 1), presumably due to adsorption to sites which become exposed following cellular disruption. This is an important observation since most workers involved in peptide analysis of tissue use techniques of cellular disruption. Our laboratory has found that significant losses of peptide can occur even using the non-mechanical method of ultrasonication. It is possible to achieve maximal recovery of peptide by boiling finely diced tissue pieces without homogenisation.

The post-mortem stability of peptides in tissue has attracted particular attention because of the potential value in pathological investigations of measuring human tissue peptide levels. Already in neuropathology, there has been much interest in altered levels of brain peptides as possible neurochemical markers of certain disorders[8,12,25]. Because it is usually

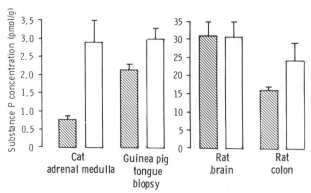

Figure 1. Tissue concentrations of substance P-immunoreactivity (pmol/g as means ± SEM) were estimated from assay of the same tissue extract samples boiled in 0.5 M acetic acid (10 ml/g) both before (open histogram) and after (hatched histogram) homogenisation.

impossible to obtain fresh pathological samples for peptide radio-immunoassay it is important to be aware of the post-mortem stability of the specific peptides of interest.

Data from a number of studies suggest that peptides are reasonably stable in the post-mortem brain[3]. However, in this laboratory investigations of peptide levels in post-mortem mouse brain, stored at 4°C or 24°C over prolonged periods revealed that a particularly labile pool of peptide may be present in addition to a larger much more stable pool[17]. Figure 2 shows that in the post-mortem mouse brain there is a very rapid decrease in substance P at 4°C and this is probably due to degradation of the extra-vesicular peptide leaving the protected vesicular pool which is stable for as long as 48 h. The apparently slower rate of degradation at the

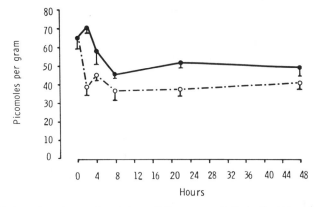

Figure 2. Post-mortem changes in substance P-immunoreactivity in the mouse brain kept at 28°C (continuous line) and at 4°C (broken line) for up to 48 h. Values are given as mean concentrations ± SEM (n = 4) in pmol/g wet weight.

higher temperature suggests that in the post-mortem brain there may be still some post-translational synthesis of peptide from a non-immuno-reactive precursor form and that the effect of this is more significant at the higher temperature. Because of these early post-mortem changes, imme-diate extraction is preferred.

Snap-freezing and frozen storage is possible without any significant effect on recovery of peptide from tissue but there is then the attendant problem of possible thawing which will cause very significant loss[17]. This is a particular problem when dealing with small tissue specimens of biopsy size which, if frozen, will inevitably thaw before extraction.

In addition to quantitative losses, chemical changes, such as oxida-tion or deamidation may occur during extraction which may effect immunoreactivity and thus not react in the assay. The influence of such chemical changes on quantitative peptide measurement needs to be checked for each individual peptide. Reverse-phase high-pressure liquid

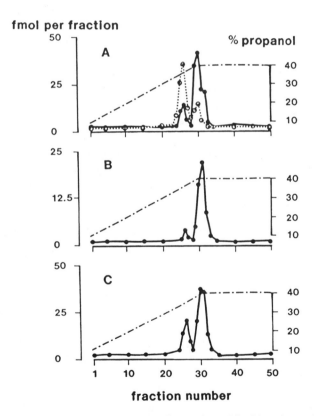

Figure 3. Elution profiles of *A* oxidised (broken line) and unoxidised (continuous line) standard synthetic substance P, *B* substance P-immunoreactivity (SP-IR) extracted from rat brain and *C* SP-IR extracted from human colon, chromatographed on a reverse-phase HPLC column eluted with a linear gradient (broken line) of propanol (10–40%) with 5% acetic acid.

chromatography provides a powerful means of identifying small chemical changes in peptide structure (fig. 3). Oxidation appears to be the most regularly occurring problem of this nature and addition of an anti-oxidant such as dithiothreitol to the extraction medium can be beneficial.

Analytical procedures

Immunoassays may be highly specific for an antigenic determinant, which in the case of a peptide or protein may be a comparatively short amino acid sequence in a particular conformation. With single-site-specific displacement assays, any molecule bearing the antigenic determinant will be registered as 'immunoreactivity', and, in addition, any substances that interfere with label-binding in other ways, e.g. by altering the conformation of the antibody or label or by digesting the label, will also appear as 'immunoreactivity'. A special problem is presented by antibodies raised against oligopeptide haptens, which may be specific to one or two amino acid residues only, and therefore cross-react with proteins or peptides whose only relationship to the hapten is their chance possession of the same residues in an exposed position. For these reasons immunoassay data (and especially conventional competitive binding assay data) on plasma and tissue extracts must be checked by further analytical methods, to characterise the apparent immunoreactivity with respect to other chemical properties.

The easiest and most generally applicable method of further characterisation is by gel filtration (molecular exclusion chromatography). This will separate the immunoreactants and interfering substances according to their molecular size, which provides information of fundamental value for their identification. Most studies that have identified larger molecular forms of peptide immunoreactivities have employed conventional gel filtration as their basic technique. When combined with the identification of immunoreactants within peaks by means of further region-specific antibodies (raised against synthetic fragments of the known peptide, or against synthetic segments of a known precursor sequence), this can very rapidly provide a thumb-nail sketch of the naturally occurring fragments (resulting from post-translational enzymatic processing or subsequent degradation) of a peptide prohormone. However, because gel filtration cannot discriminate between small differences in molecular size, a degree of uncertainty must remain about the exact length of peptide fragments or extended forms. Sometimes this can be resolved by the use of antibodies that have near-absolute specificities for a free N- or C-terminus of an expected peptide fragment. In other cases, precise identification will necessitate isolation and sequencing of the peptide fragment. There is a danger in equating molecular size as estimated by gel filtration in conventional aqueous buffers with molecular weight. The elution position of a peptide or protein on gel filtration (when adsorptive effects are excluded) will depend on its hydrodynamic properties. If peptides (often of widely different fractional ratios) are compared with protein markers of low fractional ratio (e.g. horse, heart, cytochrome *c*), the molecular weight of any peptide of extended conformation will be greatly over-estimated. There is thus an extensive literature on glucagon-like immunoreactivity (GLI) of mol.wt 12,000 daltons as estimated by conventional gel filtration. Isolation and sequencing of the principal GLI 12,000 from the pig showed that its true mol.wt was 8,128. This problem can be overcome to some extent by performing gel filtration in denaturing solutions such as 8 M urea or 6 M guanidine hydrochloride, in which single chain proteins and peptides (any disulphide bridges must be reduced) would adopt similar conformations. However, this needs large quantities of the re-crystallised de-naturing agents, providing solutions that are inconvenient to handle, and the fractions obtained must be further processed to remove the denaturing agent before immunoassay (except in those few cases where dilution is adequate). The alternative is to use sodium dodecylsulphate polyacrylamide gel electrophoresis, but slicing and extraction of immunoreactivities from such gels has proved laborious, and the technique is not widely used in relation to radioimmunoassay.

To give readily interpretable results for analytical use, gel filtration should ideally separate according to molecular size only; however, other interactions between solutes and the gel occur, especially with the more densely cross-linked gels. These are ionic interactions (e.g. a small member of negatively charged groups are found on densely cross-linked dextran gels leading to retardation of basic or early elution of acidic substances in solutions of low ionic strength) and other adsorptive interactions (leading to retardation of peptides with aromatic residues). These interactions can be minimised by using buffer of high ionic strength or containing a moderate admixture of organic solvents, but can also be exploited by specific separations. However, the existence of such interactions must be borne in mind before equating gel filtration results with molecular size.

Other types of chromatography, such as ion exchange or reversed phase chromatography offer the possibility of much greater resolution than gel filtration, but offer less useful information because they separate according to molecular properties that are in general not fundamental for our current understanding of the biosynthesis or degradation of regulatory peptides. With these techniques, probable identification of an immunoreactive peak, short of isolation and sequence determination, depends on showing its exact coelution with a peptide marker of known structure. Non-coelution with the marker will not give much help towards identifying the structural difference from the marker peptide; one can only say whether it is more basic or acidic, or more or less hydrophobic than the marker.

Conventional ion-exchange chromatography is slow and immunoreactivity recoveries may be low, 60–70% being typical. Nevertheless, ion-exchange chromatography has been used very successfully to separate peptide degradation fragments, e.g. circulating gastrin metabolites, and was, until the advent of reversed phase high pressure liquid chromatography (HPLC), the standard method of separating monoradioiodinated peptides from unlabelled and di-iodinated peptides in the preparation of radioimmunoassay tracers.

Reversed phase (hydrophobic) separation of peptides is the method that has worked best with the adoption of HPLC technology. This is because it has been technically possible to prepare stationary phases of the high physical stability and capacity necessary for chromatography of a high efficiency on a finely divided matrix. In general, HPLC has worked less satisfactorily for molecular exclusion chromatography (there may be ionic interactions with inorganically based matrices, organic matrices may be insufficiently rigid, and the size separation range may be sub-optimal for most regulatory peptides) or ion-exchange chromatography (again, the problems of achieving adequate physical stability and chemical capacity of the matrix have not been quite satisfactorily overcome). Reversed phase HPLC with a correctly chosen solvent system can be a very powerful separation method for regulatory peptides. Peptide peaks must be prevented from tailing by the addition of small amounts of inorganic salts, e.g. sodium hydrogen phosphate buffers (inconvenient, because they may have a low solubility in the organic solvent and cannot be removed by evaporation) or organic ion-pairing reagents such as tri-fluoracetic acid or heptafluorobutyric acid. The latter create acidic conditions suppressing the ionisation of carboxyl groups, while pairing with amino and guanidinium groups on the peptide and so increasing their hydrophobicity. However, care must be taken to keep within the pH range for stability of the stationary phase (pH 2–8 for silica-based matrices).

Chromatography of a wide range of regulatory peptides on analytical octadecylsilysilica columns with acetonitrile/water gradients of 1% per min may, however, give a separation that is little better than that obtained on gel filtration. Many brain-gut peptides emerge between 30% and 40% acetonitrile and will not necessarily be well separated in such a standardised system. Further, whereas the extended forms of peptide immunoreactivities will always be separable on gel filtration if the size-difference is appreciable, such extended forms may not necessarily differ sufficiently in hydrophobicity to be well separated on a simple acetonitrile gradient. The resolving power of reversed phase HPLC will only be realised if an effort is made to find conditions appropriate for a particular separation. Once this is done, however, the speed of HPLC will make possible a large number of analyses in a short time, especially with automated sample injection. Sometimes the particular radioimmunoassay employed for quantitating the immunoreactivities will be sensitive and robust enough to allow direct assay of samples from the fractions, any interfering effect of the organic solvent being diluted out in the assay buffer. In other cases, the solvent must be removed by evaporation under reduced pressure or in a stream of nitrogen.

Reversed phase HPLC is now the method of choice for separating radioiodinated from un-iodinated peptides. The mono-iodinated peptide is eluted later than the native peptide and is usually well separated from this under the appropriate isocratic conditions. However, during oxidation, methionyl residues will be oxidised to the sulphoxide and the resulting increase in polarity will lead to earlier elution of the peptide. This illustrates the way in which reversed phase HPLC separates almost too well for some purposes. It will certainly pick up as separate peaks sulphoxide and deamidated forms of peptide immunoreactivities, detecting heterogeneity that has in fact only been created by the preceding handling procedure. This then needs further analysis to identify the origin of the heterogeneity, so that artifact can be distinguished from heterogeneity of biosynthetic or biodegradative origin.

Conclusions

The measurement and analysis of regulatory peptides has undergone a slow evolution over the last decade but is still recognisably similar to that employed by Yalow and Berson in the early sixties. The next decade, however, promises a much faster rate of change. The goals of reliable measurement, sensitive measurement, specific measurement of each peptide form and finally quantitation which is meaningful between laboratories will be achieved. It will then be possible for numerical assessment of regulatory peptide concentrations to become diagnostically useful. It will be possible to report that a particular peptide amount in a particular tissue is abnormal. At the present time this cannot be done.

1 Adrian, T. E., Measurement in plasma, in: Radioimmunoassay of Gut Regulatory Peptides, chap. 5, pp. 28–35. Eds S. R. Bloom and R. G. Long. W. B. Saunders Company Ltd, London 1982.
2 Blake, C., and Gould, B. J., Uses of enzymes in immunoassay techniques. A review. Analyst 109 (1984) 533–547.
3 Bird, E. D., Problems of peptide analysis in human post-mortem brain, in: Neurosecretion and Brain Peptides, pp. 657–672. Eds J. B. Martin, S. Reichlin and K. L. Bick. Raven Press, New York 1981.
4 Cambiaso, C. L., Leek, A. E., de Steenwinkel, F., Billen, J., and Masson, P. L., Particle counting immunoassay (PACIA). A general method for the determination of antibodies, antigens and haptens. J. immun. Meth. 18 (1977) 33–44.
5 Curtiss, E. G., and Patel, J. A., Enzyme multiplied immunoassay techniques: a review. CRC Crit. Rev. clin. Lab. Sci. 9 (1978) 303–318.
6 Dwenger, A., Radioimmunoassay: an overview. J. clin. Chem. clin. Biochem. 22 (1984) 883–894.
7 Edwards, R., Immunoassay. William Heinemann Medical Books, London 1985.
8 Emson, P. C., Rossor, N. M., Hunt, S. P., Marley, P. D., Clement-Jones, V., Rehfeld, J. F., and Fahrenkrug, J., Distribution and post-mortem stability of substance P, met enkephalin, vasoactive intestinal polypeptide and cholecystokinin in normal brain and in Huntington's disease, in: Metabolic Disorders of the Nervous System, pp. 312–321. Ed. F. C. Rose. Pitman, London 1981.

9 Ekins, R. P., The estimation of thyroxine in human plasma by an electrophoretic technique. Clinica chim. Acta 5 (1960) 453–459.

10 Ekins, R. P., Current concepts and future developments, in: Alternative Immunoassays, pp. 219–237. Ed. W. P. Collins, John Wiley & Sons Ltd, Chichester 1985.

11 Goodfriend, T. L., Levine, L., and Fasman, G. D., Antibodies to bradykinin and angiotensin: a use of carbodiimide in immunology. Science 144 (1964) 1344–1346.

12 Ferrier, I. N., Crow, T. J., Roberts, G. W., Johnstone, E. C., Owens, D. G. C., Lee, Y., Bacarese-Hamilton, A. J., McGregor, G. P., O'Shaughnessy, D. J., Polak, J. M., and Bloom, S. R., Alterations in neuropeptides in the limbic lobe in schizophrenia, in: Psychopharmacology of the Limbic System, pp. 244–254. Eds M. R. Trimble and E. Zarifian. Br. Ass. Pharmac. Monogr. No. 5, 1984.

13 Henkel, E., Marker-free immunological analytical methods. J. clin. Chem. clin. Biochem. 22 (1984) 919–926.

14 Johansson, A., Stanley, C. J., and Self, C. H., A fast highly sensitive colorimetric enzyme immunoassay system demonstrating benefits of enzyme amplification. Clinica chim. Acta 148 (1985) 119–124.

15 Kricka, L. J., and Carter, T. J. N., Luminescent immunoassays, in: Clinical and Biochemical Luminescence, pp. 153–178. Eds L. J. Kricka and T. J. N. Carter, Marcel Dekker Inc., New York 1982.

16 Lovgren, T., Hemmila, I., Pettersson, K., and Halonen, P., Time resolved fluorometry in immunoassay, in: Alternative Immunoassays, pp. 203–217. Ed. W. P. Collins, John Wiley & Sons Ltd, Chichester 1985.

17 McGregor, G. P., and Bloom, S. R., Radioimmunoassay and stability of substance P in tissues. Lift Sci. 32 (1983) 655–622.

18 Miles, L. E. M., and Hales, C. N., An immunoradiometric assay of insulin, in: Protein and Polypeptide Hormones, Part I, pp. 61–70. Ed. M. Margoulies. Excerpta Medica, Amsterdam.

19 Mutt, V., Chemistry of the gastrointestinal hormones and hormone-like peptides and a sketch of their physiology and pharmacology. Vitamins Horm. 39 (1982) 231–427.

20 Oellerich, M., Enzyme immunoassay: a review. J. clin. Chem. clin. Biochem. 22 (1984) 895–904.

21 O'Shaughnessy, D. J., Antibodies, in: Radioimmunoassay of Gut Regulatory Peptides, chap. 3, pp. 11–20. Eds S. R. Bloom and R. G. Long. Saunders Co. Ltd, London 1982.

22 Parker, C. W., Nature of the immunological responses and antigen-antibody interaction, in: Principles of Competitive Protein Binding Assays, pp. 25–56. Eds W. D. Odell and W. H. Daughaday. Lippincott, Philadelphia 1971.

23 Radioimmunoassay and saturation analysis. Br. med. Bull. 30 (entire issue devoted to the subject) 1974.

24 Roddis, M. J., Burrin, J. M., Johanssen, A., Ellis, D. H., and Self, C. H., Serum thyrotropin: a first-line discriminatory test of thyroid function. Lancet 1 (1985) 277–278.

25 Rossor, M. N., and Emson, P. C., Neuropeptides in degenerative disease of the central nervous system. Trends Neurosci. 5 (1982) 399–401.

26 Smith, D. S., Enhancement fluorimmunoassay of thyroxine. FEBS Lett. 77 (1977) 25–31.

27 Smith, D. S., Al-Hakiem, M. H. H., and London, J., A review of fluorimmunoassay and immunofluorometric assay. Ann. clin. Biochem. 18 (1981) 253–274.

28 Ullman, E. F., Schwarzberg, M., and Rubenstein, K. E., Fluorescence excitation transfer immunoassay, a general method for the determination of antigens, J. biol. Chem. 251 (1976) 4172–4177.

29 Van Weemen, B. K., and Schuurs, A. H. W. M., Immunoassay using hapten-enzyme conjugates. FEBS Lett. 24 (1972) 77–81.

30 Voller, A., and Bidwell, D. E., Enzyme immunoassays, in: Alternative Immunoassays, pp. 77–86. Ed. W. P. Collins, John Wiley & Sons Ltd, Chichester 1985.

31 Waldmann, H., and Milstein, C., Monoclonal antibodies, in: Clinical Aspects of Immunology, 4th edn. Eds P. J. Lachmann and D. K. Peters. Blackwall Scient. Publ. CRC Press, New York 1982.

32 Weeks, I., Sturgess, M., Siddle, K., Jones, M. K., and Woodhead, J. S., A highly sensitive immunochemiluminometric assay for human thyrotrophin. Clin. Endocr. 20 (1984) 489–495.

33 Wood, W. G., Luminescence immunoassays: problems and possibilities. J. clin. Chem. Biochem. 22 (1984) 905–918.

34 Yalow, R. S., and Berson, S. A., Immunoassay of endogenous plasma insulin in man. J. clin. Invest. 39 (1960) 1157–1175.

Hybridisation histochemistry

J. D. Penschow, J. Haralambidis, P. E. Darling, I. A. Darby,
E. M. Wintour, G. W. Tregear and J. P. Coghlan

Summary. The location of gene expression by hybridisation histochemistry is being applied in many areas of research and diagnosis. The aim of this technique is to detect specific mRNA in cells and tissues by hybridisation with a complementary DNA or RNA probe. Requirements for optimal specificity, sensitivity, resolution and speed of detection may not all be encompassed in one simple technique suitable for all applications, thus appropriate procedures should be selected for specific objectives. With reference to published procedures and our own extensive experience, we have evaluated fixatives, probes, labels and other aspects of the technique critical to the preservation and hybridisation *in situ* of mRNA and detection and quantitation of hybrids.

Hybridisation histochemistry is a technique for the location of gene expression in histological sections[54]. Intracellular messenger RNA (mRNA) is hybridised with a labelled specific complementary DNA (or RNA) probe and the RNA-DNA hybrids in the tissue located by the probe label. The principle is analogous to the location of peptides or proteins by specific labelled antibodies, a technique which has proven extremely useful in diagnosis and research. However, the presence of an intracellular protein antigen does not necessarily indicate whether the gene is being expressed or whether the antigen is a product of the cell in which it is located. A major area of current interest is the regulation of gene expression. For many such studies it is important to identify the cell types in which particular genes are expressed and to observe any cellular changes in the expressing cell population which may occur under varying physiological circumstances and which may affect the level of expression per cell, though not necessarily the level for the tissue as a whole.

The historical time frame for development of hybridisation histochemistry was influenced directly by the availability of specific probes for recombinant DNA technology[42,46,54,89,108]. Particular mRNA populations could be located only if the probe was known to have a specific complementary nucleotide sequence.

The pioneers of nucleic acid hybridizations[50] annealed in solution denatured chromosomal DNA, usually with labelled ribosome fractions and recovered the hybrids by density gradient centrifugation. Immobilization of the DNA component on a nitrocellulose membrane[43] paved the way for many subsequent advances including the first hybridisation *in situ.* In these studies denatured intracellular DNA in cell squashes[38], cell cultures[60], paraffin sections[20] and ultrathin glycol methacrylate sections[57] was hybridised with radio-labelled ribosomal RNA (rRNA) fractions to determine the nuclear locations of rRNA genes. Further progress was hindered in most fields by inadequate probe purification procedures resulting in probes of very limited specificity. However, purified viral nucleic acids were recognised as useful probes for the location of specific virus[80] and poly (U) as a probe for total mRNA[61], which was used mainly in developmental biology.

Although hybridisation histochemistry accompanied the technology explosion occasioned by gene cloning, application of the technique was restricted by the limited number and availability of cloned cDNA probes. Even now, access to the numerous probes for growth factors, peptide hormones, enzymes, oncogene products, cell surface receptors, viruses and other cell proteins for which genes have now been cloned is limited. The successful utilisation of synthetic oligodeoxyribonucleotide probes in the hybridisation histochemistry technique[22] removes the difficulties arising from cDNA probe usage, as functional probes may be synthesised easily, quickly and accurately from published nucleic acid sequences. The other major advance in probe design is the single-stranded RNA probe generated by the SP6 vector system[75]. There are several different labelling systems which incorporate a variety of isotopically or nonisotopically labelled nucleotides into the probe which should be matched to the probe type and to the relevant detection system.

The recent proliferation of *in situ* hybridisation techniques addresses many questions in a diversity of areas. The choice of technique depends ultimately on the particular application however, in our experience, the simplest procedures with the minimum manipulations are most likely to succeed. We have developed a simple, reliable, reproducible method for the location of specific intracellular mRNA populations which is applicable to a wide range of tissues. Frozen sections of whole small animals[22,86], human tumours and biopsies[114], sheep brains[21,24,25,82], plants[7] and numerous other sheep[31,112], rodent[23,25,26,30,58] and human tissues[24,26] have been hybridised successfully. Cell cultures[26], smears and paraffin embedded tissues[24] can be also used routinely. Tissue preparation methods may be adapted to enable immunohistochemistry[24] or receptor localisation to be performed on adjacent sections.

Following a brief outline of the method, which is described in detail elsewhere[82], we shall discuss our experiences and those of others in optimising procedures for specific applications.

General procedure

Fresh tissue is freeze-embedded in OCT compound by immersion in hexane cooled to $-70°C$ with dry ice. 5-μm cryostat sections are cut at

$-15°C$ to $-20°C$, thaw-mounted onto gelatinised slides, laid immediately on dry ice and left for 15–30 min. Sections are fixed at $4°C$ for 5 min in 4% glutaraldehyde in 0.1 M phosphate buffer pH 7.2 with 20% ethylene glycol, rinsed twice in hybridisation buffer [600 mM sodium chloride, 50 mM sodium phosphate pH 7, 5 mM EDTA, 0.02% Ficoll, 0.02% bovine serum albumin, 0.02% polyvinylpyrrolidone, 0.1% DNA (degraded free acid) and 40% deionised formamide], left to soak for 10 min–2 h at $40°C$ in a fresh change of buffer, rinsed in ethanol and left to dry.

The ^{32}P-end-labelled oligodeoxyribonucleotide probe[82] is diluted to 0.4 ng/μl in hybridization buffer, heated to $90°C$ for 1 min then a drop placed on a coverslip (20 μl for 22×22 mm) which is applied to the sections on the slide. Slides are incubated at $30-40°C$ in a humidified chamber for 1–3 days (the temperature is variable depending on the probe length and homology with target mRNA)[82].

After hybridisation slides are immersed in $2 \times$ SSC until coverslips dislodge ($2 \times$ SSC is 0.3 M sodium chloride and 0.03 M sodium citrate), rinsed in fresh $2 \times$ SSC, washed at $40°C$ for 30 min in $1 \times$ SSC then rinsed in ethanol and allowed to dry. Slides are taped to blotting paper in an X-ray film cassette and overlaid with a sheet of X-Omat AR-5 (Kodak) which is left to expose for 24 h (for ^{32}P). The film is developed and fixed and the results evaluated. Slides for liquid emulsion autoradiography are selected, dipped at $40°C$ in K5 (Ilford) diluted 1 : 2 with distilled water and left for 1–14 days (for ^{32}P) over silica gel to expose.

Autoradiographs are developed for 2 min at $15°C$ in D19 (Kodak), rinsed in distilled water, fixed in Hypam (Ilford) diluted 1 : 4, washed and stained.

Methodology

Tissue preparation. The choice of tissue preparation methods for hybridisation histochemistry is limited by the need to preserve and retain intracellular mRNA such that it is accessible to the probe and available for base-pairing. Whilst good morphological preservation is a high priority, many fixation and processing regimes are incompatible with these requirements.

Glutaraldehyde or paraformaldehyde-fixed cryostat sections of fresh frozen tissues give the strongest hybridisation signals[99], which in our experience are approximately equivalent, although glutaraldehyde provides superior morphological preservation. Receptor localisation studies may be performed on adjacent unfixed sections which have been treated appropriately[76]. Paraformaldehyde-perfused, frozen tissues are used widely for hybridisation, although perfusion is not always applicable and the hybridisation signal is generally lower and less consistent than with fresh, post-fixed cryostat sections[99]. Paraformaldehyde-perfused tissues may be used for immunohistochemistry, which permits the location of mRNA and protein antigens on adjacent sections[40]. Both procedures have been performed sequentially on the same frozen section[98] but with a reduced hybridisation signal resulting from the preceding immunohistochemistry. Similarly viral RNA and protein has been colocalised in the same paraffin sections[18,63].

Parameters for hybridisation of fixed, embedded tissues are difficult to evaluate as the size of the tissue sample, ribonuclease content, type and rate of fixation and subsequent processing and embedding procedures must be considered in relation to the hybridisation signal and the quality of morphological preservation.

Immersion fixation of tissues followed by processing for paraffin[49,70,108,110] or methacrylate[59] embedding instead of freezing[84,107], has been used effectively for small samples where target mRNA's are abundant and some hybridisation signal can be sacrificed for improvements in morphology. As prolonged exposure to fixatives tends to reduce the hybridisation signal[19,111], large samples ideally are frozen and post-fixed[15,82,99] or dissected from perfusion-fixed animals then frozen[98,111] or processed immediately for embedding. Paraffin-embedded tissues perfused with paraformaldehyde, glutaraldehyde or a combination of both have been used successfully for hybridisation of viral RNA[18,77]. In our experience using perfused embedded tissues for location of mRNA, glutaraldehyde gave a stronger signal than Karnovsky's fixative[62] or paraformaldehyde, provided that tissues were rinsed briefly in buffer and processed immediately through absolute ethanol to paraffin.

We favour an alternative method for embedded tissues, which causes little or no reduction in hybridisation signal compared to fresh frozen glutaraldehyde-fixed sections. Tissues up to 2-mm-thick are frozen in liquid propane, freeze dried at $-45°C$ and 10^{-2} Torr, fixed under vacuum at $37°C$ for 1 h in paraformaldehyde and vacuum-embedded in paraffin. This is an ideal technique for the location of peptide hormone mRNAs[24] as intracellular storage granules are retained, preservation of antigenicity is excellent for immunohistochemistry and tissue morphology is good.

Several studies using cell culture systems where various types of fixatives were compared[8,41,44,68,77] have shown that extensive cross-linking of proteins is advantageous for retention of mRNA but can cause poor penetration of probes through the fixed tissue matrix. Glutaraldehyde was found to give the best morphological preservation and RNA retention but probe penetration was poor, especially for longer probes. mRNA is more accessible to short probes and proteolytic treatment of fixed cells may be necessary for effective hybridisation with longer probes, especially where vigorously cross-linking fixatives are employed[8,68]. A loss of cellular RNA may result from proteoloytic treatments which are not carefully monitored[8,19,41,68] or in tissues where proteins are not extensively cross-linked[44,68]. Probe penetration with non-cross-linking ethanol-based fixatives is less of a problem[77] but morphological preservation is poor and low hybridisation signals may result from loss of intracellular RNA[8,44,68].

Tissue preparation procedures for the location of genomic viral nucleic acids may differ markedly from those appropriate for hybridisation of mRNA, as virus particles are stable for prolonged periods in post-mortem[39,46,47] and immersion-fixed[19] tissues where mRNA survival is poor. Deproteination of the viral capsid is necessary for probe penetration and denaturation of double-stranded viral nucleic acids to permit hybridisation[14,19,46].

Probes

The selection of the type of probe and label depends largely on the specificity, sensitivity and resolution required by the experiment. The original and most commonly used specific probes for hybridisation histochemistry are recombinant cDNA, which are double-stranded, generally long (unless reduced in size by restriction enzymes) and usually labelled by nick translation[88], although the random primer method[104] is simpler and more efficient. By these methods, labelled copies of both strands are produced and exist with the unlabelled template in the hybridisation mixture. Some or all of the four deoxyribonucleotide components of the probe may be labelled with ^{3}H[14,49,98,108,111], ^{35}S[33,77,115], ^{32}P[24,37,54,58] or less commonly ^{125}I[70,83]. Incorporation into the probe of a biotin molecule[19,72,100] has been the main approach in developing non-radioactive labels and other types of cDNA derivatives[36,53,66] have been designed as probes but the sensitivity of detection methods has not yet been proven superior to autoradiography for hybridisation histochemistry. The effective probe concentration of double-stranded probes is reduced by self-annealing and competition with unlabelled strands, thus single-stranded probes of equivalent specific activity are more efficient.

Oligodeoxyribonucleotides up to approximately 100mer can be prepared in the current generation of DNA synthesisers[82] using the solid phase phosphoramidite technology[2] developed by Beaucage et al.[13]. These have proven very efficient as probes for hybridisation histochemistry[9,21,22,24,97,114] and have an advantage in specificity, as they may be designed for a particular application to exploit minor differences in nucleotide sequences and may thus discriminate closely homologous mRNA's[24,82,106,112]. These single-stranded probes are labelled enzymically by attachment of a single ^{32}P-labelled phosphate from γ-[^{32}PATP] at the 5′ end using T4 kinase[74] or a number of ^{32}P, ^{35}S or ^{3}H-labelled nucleotides at the 3′ end using terminal transferase[105]. A variety of other labelling methods for addition of enzymes[56], fluorophores[4,101] or biotin[56,78] have been reported but not yet demonstrated as effective for hybridisation histochemistry. Preparation of overlapping, complementary oligodeoxyribonucleotides to produce a primed, double-stranded probe enables a greater number of labelled nucleotides to be incorporated.

RNA probes prepared in the SP6 vector system are now being used extensively for hybridisation histochemistry[27,51,63,93]. These are single-stranded copies of a DNA insert which must be in the correct orientation to ensure transcription of the required sequence. The SP6 promotor in the presence of SP6 polymerase can generate multiple copies of the insert, the limitation being the concentration of each ribonucleotide in the reaction mixture. At present ^{32}P, ^{35}S and ^{3}H-labelled ribonucleotides are available, and although expensive, probes of very high specific activity can be produced by using sufficient concentrations of each labelled base[75]. The use of non-optimal concentrations of labelled ribonucleotides often results in short transcripts. Furthermore, ^{3}H probes of high specific activity are liable to degradation by radiolysis.

Other types of probes include single-stranded cDNA which has been used for hybridisation histochemistry whilst incorporated in a biotin-labelled M13 vector[107] and probes prepared from purified viral nucleic acids[17,19,80], or mRNA[73,85] which were developed in the prerecombinant era and have largely been supplanted by cloned sequences.

In our experience there is no single probe and label which is optimal for all requirements of specificity, sensitivity, resolution and speed of detection.

Hybridisation

There is a general belief that parameters which apply to hybridisation of denatured mRNA deposited and dried on nitrocellulose are relevant for mRNA in fixed tissues, which has a secondary structure and is trapped with associated ribosomes in a matrix of cross-linked proteins. Although there are some parallels, recovery and hybridisation of specific mRNA from tissue homogenates is subject to a different set of problems and conditions than preservation of the same mRNA in a hybridisable form in whole tissue samples. Problems which may be encountered include loss of RNA by enzymic degradation, which is somewhat tissue-specific and is more rapid in homogenates, reduced penetration of long probes through fixed tissues[8,17,41,68,77] or loss of RNA from cells which may occur after certain fixation protocols[8,44,68]. Probes may fail to hybridise in tissues where directed against regions of the mRNA molecules which may be inaccessible because of their secondary structure[35] or masked by association with proteins[3].

The similar relationship to probe specificity of salt, formamide concentration and hybridisation temperature applies equally to hybridisation of mRNA *in situ*, on filters or in solution, although optimal hybridisation and washing conditions may vary considerably for a given interaction[17,27].

Decreasing salt and increasing formamide concentration and hybridisation temperature decreases the stability of hybrids and may thus be manipulated to increase the specificity of hybridisation. These variables must be adjusted according to the experiment, with regard to the probe length and degree of homology between the probe and target mRNA, and with some consideration of the effects on tissue morphology. To ensure specificity of hybridisation with low formamide concentration temperatures above 60°C may be necessary. As this can be detrimental to morphology and cause sections to dislodge from slides, for our method of hybridisation histochemistry formamide is maintained at 40% and the hybridisation temperature varied accordingly[82] whilst maintaining salt concentration at 0.6 M. With formamide and salt constant the appropriate temperature depends on the type and length of probe as well as specificity for the target mRNA (fig. 1), 40°C is preferred for specific interactions with DNA probes longer than 27 nucleotides, whereas 30°C

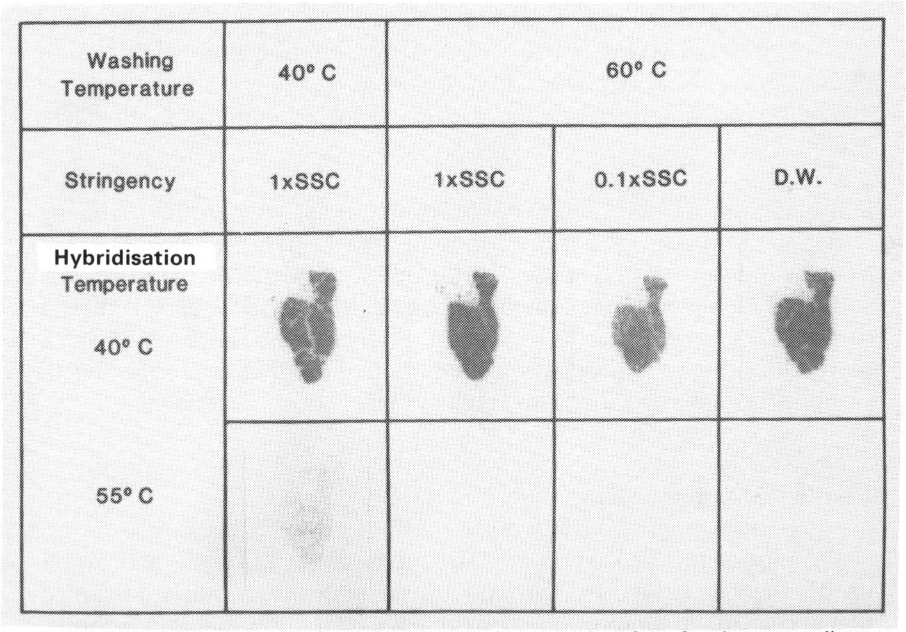

Washing Temperature	40° C	60° C		
Stringency	1xSSC	1xSSC	0.1xSSC	D.W.
Hybridisation Temperature 40° C				
55° C				

Figure 1. Autoradiographs on X-ray film of 5-μm frozen sections from female mouse salivary glands after hybridisation with a ^{32}P-labelled 30mer oligodeoxyribonucleotide probe specific for mouse glandular kallikrein mRNA[87]. For hybridisation, salt concentration was 0.6 M and formamide 40%. The temperature for hybridisation and the temperature and stringency of post-hybridisation washing was varied as illustrated. The remaining procedures were as described previously[82].

or less may be necessary for short probes or for interactions of limited specificity. Higher temperatures may be used as a test of specificity, or for optimal hybridisation with SP6 RNA probes[27] as RNA-RNA hybrids are more stable than DNA-RNA.

In a detailed study of hybridisation kinetics, Cox et al.[27] have shown that the highest signal to background ratio is obtained where probe concentration is optimal for maximum saturation of target sites. Probe is always in great excess and the optimal concentration for double-stranded probes is higher than for single-stranded due to self-reassociation which leads to a continuous decline in effective probe concentration. Short probes have been shown to give a higher signal than long probes of 500 nucleotides at equivalent concentrations[8,17]. This is suggested to be due to differences in diffusion, only short duplexes being formed *in situ*[27]. Formation of networks[109] has been shown for probes longer than 700 base pairs to contribute significantly to the signal[68]. Maximum signal to background ratio will not be achieved by an increase in the duration of hybridisation where the probe is at a sub-optimal concentration as the hybridisation reaction terminates prior to completion, even when probe is in excess[27].

Dextran sulphate has been reported to increase the hybridisation signal, presumably by increasing the effective probe DNA concentration[109], and is often included in hybridisation mixtures[14,19,49,68,70,108]. 5–10% dextran sulphate has been shown to amplify the hybridisation signal with cDNA probes[49,68]. However, for 30mer oligodeoxyribonucleotides at 400 ng/ml we obtained no increase in signal but an increased background.

For prehybridisation, Denhart's solution[32] which contains PVP, BSA and Ficoll is a standard buffer component and has background-reducing effects by saturating non-specific binding sites in tissue. Sonicated DNA from salmon or herring sperm is included frequently for the same reasons[19,69,82,99] but has been shown to reduce the hybridisation signal in some cDNA probe systems[111]. Slides or sections may be acetylated[52] prior to hybridisation to reduce background binding of DNA to charged particles but this has been found in some systems[98] to be ineffective.

Post-hybridisation washing

The aim of this procedure is first to remove the coverslip and excess probe, then to reduce background by washing off as much unwanted non-specifically bound probe as possible whilst retaining specific hybrids. For filter hybridisation, hybrids may be destabilised by increasing washing stringency (lower salt/higher temperature/longer wash), thus hybrids of desired specificity may be retained by repeated washes at increasing stringency. For hybridisation histochemistry we have, amongst others[68], found extensive washing procedures to be irrelevant (fig. 1). Hybrids formed *in situ* are extremely stable to low salt/high temperature washes and non-specific interactions which may have occurred during hybridisation remain through the most stringent of washes. It is thus essential for minimum background, to select appropriate hybridisation conditions and to have probes uncontaminated by vector DNA, very short sequences or unincorporated label. Removal of unbound probe by post-hybridisation treatment with RNase for single-stranded RNA probes[27,93,111] has been shown to greatly reduce background and similar treatment with S1 nuclease has been reported for cDNA[44]. We have encountered few problems of excessive background using specific oligodeoxyribonucleotide probes 24mer and longer.

Hybrid detection and quantitation

Methods for locating hybrids are clearly dependent upon the type of probe label. Autoradiography is the most widely used approach, appropriate methods depending on the energy, half-life and specific activity of the isotope. We have found autoradiography with ^{32}P-labelled probes a very convenient method for hybrid detection as short exposure on fast X-ray film[7,9,15,16,22,26,30,54,82,86,114] provides a rapid screening method for numerous samples. Subsequent liquid emulsion autoradiographs, with resolution to single cells in some cases[15,82], can be obtained in a matter of days following the initial X-ray exposure[15,21–26,31,58,86,114].

Tritium has higher resolution but of course requires longer exposure times due to the longer half-life and lower specific activity than ^{32}P. Addition of enhancers to the photographic emulsion tends to offset the gain in resolution sought from the use of tritium in the first place. X-ray film improved for ^{3}H autoradiography is available (L.K.B. Stockholm; Amersham, Buckinghamshire) but it is comparatively slow, thus X-ray exposures are not often used as a screening procedure. Tritium labels were used to develop in situ hybridisation procedures[20,38,57,60,61,80] and remain popular due to the high resolution which can be obtained in light microscope autoradiographs[18,27,34,40,46,49,59,69,84,99,108].

^{125}I has been used extensively as a probe label for chromosome mapping and occasionally for hybridisation histochemistry[70,83,94] but its effectiveness for the location of mRNA in tissues is reduced by the high non-specific binding to proteins[52]. ^{35}S autoradiography is being used increasingly for hybrid detection[33,47,51,64,77,115] with the advantages of good resolution, somewhere between ^{3}H and ^{32}P, and shorter exposure times than ^{3}H of comparable specific activity. The difference in β particle energy of ^{3}H and ^{35}S has been exploited for the simultaneous location of two viral genomes within the same cell[48]. High non-specific binding to tissues has been reported with this isotope[64].

A variety of approaches to probe detection which avoid radioactivity have been developed and used mainly for chromosome mapping. The first of these employed labelled antibodies to DNA-RNA hybrids[92,102] or direct detection of fluorochrome-labelled probes[10,11]. The lack of sensitivity of these methods led to the signal enhancement approach using an antibody-sandwich to locate hapten-modified probes[12,19,55,67,72]. Further increases in sensitivity were needed, thus reflection contrast microscopy was developed[66] to amplify immunoprecipitates and avidin-peroxidase conjugates for biotin-labelled probes[6,71,100,107]. These methods have the advantage of rapid detection, but have yet to be proven to have greater sensitivity than optimised autoradiography.

It is tempting to use hybridisation histochemistry to measure relative levels of mRNA in different cell populations related to various physiological parameters. Data obtained from extracted mRNA can only provide an average level of gene expression for any tissue sample but cannot resolve other changes in the cell population where a specific gene is being expressed such as proliferation, differentiation or atrophy, which may occur whilst a constant mRNA level is maintained by transcriptional regulation. Conversely, for the same reasons, in comparisons of hybridisation histochemistry from different tissues or animals, relative mRNA levels per cell in any section do not necessarily relate to the total tissue mRNA level. Where possible, a combination of both methods should provide the most informative measure of gene expression.

Methods for quantitation by autoradiography of hybrids in cell cultures[17,68], tissue sections[8,24,26,27,111] and chromosome preparations[103] have been described, however potential sources of error should be recognised. These include variations in probe penetration, efficiency of hybridisation[27,103], cell shape[41] or section thickness, loss of mRNA from cells[44] and formation of probe networks with cDNA probes[68], as well as errors in grain-counting of autoradiographs, which vary for different isotopes[1,90]. We have found quantitation of hybridisation histochemistry by competition experiments with ^{32}P-labelled oligodeoxyribonucleotide probes[24,26] to be a simple, reproducible, means for determining relative mRNA levels in tissue sections. Analysis of X-ray films by computer-assisted densitometry may be undertaken to provide colour-coded images which illustrate differences in signal intensity[45].

Applications

One of the major applications of hybridisation histochemistry, in conjunction with other cell-specific techniques, lies in the reclassification of cells and tissues on a functional basis. This is particularly applicable to embryogenesis and other areas of cell differentiation where the presence in a cell of specific mRNA may be the first detectable criterion of function. Elegant studies of developing *Drosophila*[49,64,72], *Aplysia*[70,94], sea urchins[27,108] and nematodes[34] illustrate the value of this technique in developmental biology. Gene expression for some familiar and newly discovered peptide hormones, enzymes, substrates and growth factors has been located at multiple sites in functionally heterogenous cell populations[33,40,84,86,97] and tissues. This allows further analysis of the hypothesis that 'everything is made everywhere' and recent findings suggest multiple roles for some genes and transcripts which may be differentially processed to heterogenous products which can be cell specific[84,91,95]. For these and other applications of hybridisation histochemistry, specific oligodeoxyribonucleotide probes can be designed to discriminate between closely homologous mRNA sequences which are encoding multigene families, or which share common sequences[5,82,106,114]. These probes are also invaluable for the location in transgenic animals of expression of introduced gene constructs by sectioning and hybridisation of whole animals or isolated tissues. There are clearly many other uses as a research tool and a major role for this technique is becoming increasingly apparent for clinical diagnosis of tumours, infectious diseases and endocrine disorders, or in fact in any situation where a specific recognised mRNA is associated with, or at the root of the problem.

We have selected some examples which show a variety of applications of our method using ^{32}P-labelled probes of different types. Oligodeoxyribonucleotide probes specific for α, β or γ globins, have been used to study globin switching in the ovine foetus by a series of techniques, including

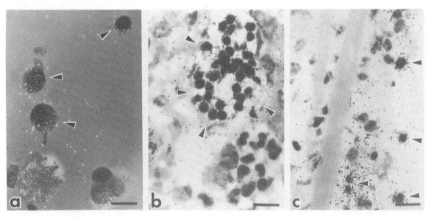

Figure 2. Autoradiograph after hybridisation with ^{32}P-labelled 30mer oligodeoxyribonucleotide probes designed to discriminate mRNA's for ovine β and γ globin[65]. *a* mRNA for β globin shown in erythropoeitic cell (arrowed) from a 3-day methylcellulose culture of ovine 129-day foetal bone marrow. Cytocentrifuged preparation, stained with Wright's and photographed by polarised epi-illumination. Bar = 13 μm. *b* A 5-μm frozen section showing mRNA for γ globin in cells of a haemopoietic focus (arrowed) within liver parenchyma of an ovine 124-day foetus. Stain: H & E. Bar = 30 μm. *c* A 5-μm frozen section of ovine 99-day foetal bone marrow showing mRNA for γ globin in erythropoietic cells (arrowed). Stain H & E. Bar = 35 μm.

62

colony culture of erythropoietic cells and location of gene expression in haemopoietic tissues[112,113] (fig. 2, a-c).

Synthetic probes coding for the E2 region of Ross River virus were used for diagnosis in a human skin biopsy of infection with this virus (fig. 3a), innoculated Vero cell cultures acting as hybridisation controls (fig. 3b). mRNA for a member of the human growth hormone gene family, choriosomatomammotropin, is demonstrated by a 30mer probe in syncytiotrophoblasts of human placenta (fig. 3c). In human primary medullary thyroid carcinoma, mRNA for calcitonin (fig. 3d) was found to coexist in the primary tumour and lymph node metastasis with mRNA for calcitonin-gene-related peptide (CGRP)[114].

mRNA for the enzyme renin is demonstrated with a cDNA probe[81] in the afferent arteriole of the mouse kidney (fig. 4a) and its substrate angiotensinogen with a 36mer probe, including the angiotensin coding region, in neurones of the arcuate nucleus of rat brain (fig. 4b). Attempts

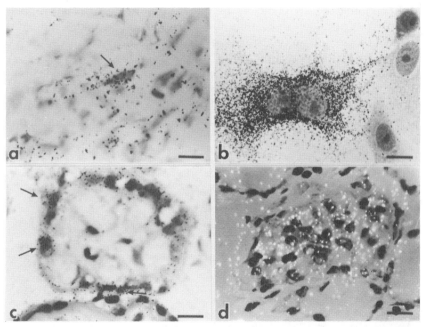

Figure 3. Autoradiographs after hybridisation with ^{32}P-labelled 30mer (a, b, c) or 40mer (d) oligodeoxyribonucleotide probes. a A 10-μm frozen section of human skin biopsy taken 4 days after onset of a rash, mRNA coding for Ross River virus (RRV) envelope protein is demonstrated in an infected cell (arrowed) by a probe corresponding to the E2 region of the RRV genome[29]. Bar = 35 μm. b Vero cells from a culture infected with RRV and hybridised with the RRV E2 probe used in 3(a). Bar = 10 μm. c A 5-μm frozen section of human placenta showing mRNA for choriosomatomammotropin[96] in syncytiotrophoblasts (some arrowed) of foetal cotyledons. Bar = 30 μm. d mRNA for calcitonin in neoplastic C-cells of a human primary medullary thyroid carcinoma, demonstrated in a 5-μm frozen section with a 40mer oligodeoxyribonucleotide probe complementary to the calcitonin coding region. Photomicrograph by polarised epi-illumination. Bar = 25 μm.

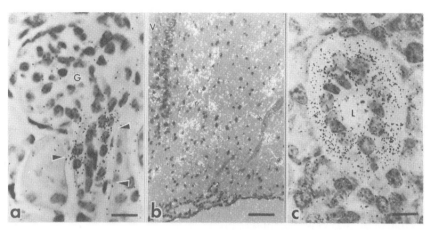

Figure 4. *a* mRNA for renin demonstrated in cells of the afferent arteriole (arrowed) of a renal glomerulus (G) by a ^{32}P-labelled m. renin cDNA probe of 1400 base pairs[81]. Autoradiograph of a 5-μm paraffin section of freeze-dried mouse kidney. Stain: H & E. Bar = 25 μm. *b*. Coronal 8-μm frozen section of rat brain after hybridisation with a ^{32}P-labelled 36mer oligodeoxyribonucleotide complementary to the angiotensin coding region of rat angiotensinogen mRNA[79]. mRNA-DNA hybrids are located by autoradiography in neurones of the arcuate nucleus, adjacent to the third ventricle (V). Photomicrograph by transmitted light dark-field illumination. Stain: cresyl violet. Bar = 300 μm. *c* Autoradiograph of a 4-μm paraffin section of freeze-dried mouse salivary gland after hybridisation with a ^{32}P-labelled RNA probe generated in the SP6 vector system[75] from the 550 base pair mouse kallikrein cDNA pMK-1[87]. Kallikrein-RNA is demonstrated in striated duct cells of the parotid gland. (L) lumen of duct. Stain: H & E. Bar = 20 μm.

by ourselves and others to locate renin mRNA in the brain[33] have been unsuccessful and the mechanism of a brain renin-angiotensin system is still being investigated. Genes of the kallikrein family of serine protease enzymes have been shown to be expressed in a diversity of tissues although the major site of expression of most of the functional genes of this multigene family is in granular convoluted tubule cells of the submandibular gland[106]. Kallikrein mRNA is demonstrated by an SP6 RNA probe containing the mouse kallikrein insert pMK-1[87] in striated ducts of the murine parotid gland (fig. 4c), where products of the renal kallikrein gene, a kininogenase, may have a role in local regulation of blood flow.

Hybridisation histochemistry—*in situ* tissue hybridisation using cDNA, RNA or synthetic oligodeoxynucleotide probes with various labelled compounds is beyond the countdown stage as an emergency technique and is in the early stage of lift-off. Recent contributions from many laboratories have accelerated the transition through generation one, two and three. The technique is user-friendly apart from the autoradiographic end point which is not too difficult. The technique offers valuable, unique information which is not obtainable by application of other procedures. The specific cellular address of gene expression as in the developing embryo during narrow time windows offers very significant advances and insight into differentiation and development. The ability to detect viral, oncogene, growth factor, or hormonal gene expression in carcinomas offers the possibility of a new taxonomy of tumours with the likelihood of an appropriately structured chemotherapy or direction of the most relevant specific cytotoxins; furthermore, the profile of gene expression in secondary metastases will provide vital information about the primary tumour and appropriate treatment. Identification of viruses by this technique will have eventually a permanent niche in the area of diagnosis. Location of gene expression in specific cells is a crucial piece of information in many on-going studies of hormones, enzymes and proteins. This information profile includes transgenic animals, chimeric embryos and immortal cell lines, where knowledge of promoter regions and their interaction with enhancers and other transactive agents will be the bases of the 'new pharmacology'.

Acknowledgments. We are grateful to Professor T. J. Martin and Dr R. J. Fraser from the University of Melbourne Department of Medicine at the Repatriation Hospital and Royal Melbourne Hospital respectively, for their significant contributions to the clinical studies.—This work was supported by grants-in-aid from the National Health and Medical Research Council of Australia, the Myer Family Trusts, the Ian Potter Foundation, and the Howard Florey Biomedical Foundation, and by Grant HD11908 from the National Institutes of Health.—The oligodeoxyribonucleotide probes used in the study were supplied by Syngene Ltd, 225 Queensberry Street, Melbourne, 3000 Australia.

1 Ada, G. L., Humphrey, J. H., Askonas, B. A., McDevitt, H. O., and Nossal, G. J. V., Correlation of grain counts with radioactivity (^{125}I and tritium) in autoradiography. Exp. Cell Res. *41* (1966) 557–572.

2 Adams, S. P., Kavka, K. S., Wykes, E. J., Holder, S. B., and Galluppi, G. R., Hindered dialkylamino nucleoside phosphite reagents in the synthesis of two DNA 51mers. J. Am. chem. Soc. *105* (1983) 661–663.

3 Adams, P. S., Noonan. D., and Jeffery, W. R., A model for the organization of the poly (A) protein complex in messenger ribonucleoprotein. FEBS Lett. *114* (1980) 115–118.

4 Agrawal, S., Christodoulu, C., and Gait, M. J., Efficient methods for attaching non-radioactive labels to the 5'ends of synthetic oligodeoxyribonucleotides. Nucl. Acids Res. *14* (1986) 6226–6245.

5 Amara, S. G., Arriza, J. L., Leff, S. E., Swanson, L. W., Evans. R. M., and Rosenfeld, M. G., Expression in brain of a messenger RNA encoding a novel neuropeptide homologous to calcitonin gene-related peptide. Science *229* (1985) 1094–1097.

6 Ambros, P. F., Matzke, A. J. M., and Matzke, M. A., Localization of agrobacterium rhizogenes T-DNA in plant chromosomes by in-situ hybridization. EMBO *5* (1986) 2073–2077.

7 Anderson, M. A., Cornish, E. C., Mau, S.-L., Williams, E. G., Hoggart, R., Atkinson, A., Bonig, I., Grego, B., Simpson, R., Roche, P. J., Haley, J. D., Penschow, J. D., Niall, H. D., Tregear, G. W., Coghlan, J. P., Crawford, R. J., and Clarke, A. E., Cloning of cDNA for a stylar glycoprotein associated with expression of self-incompatibility in *Nicotiniana alata*. Nature *321* (1986) 38–44.

8 Angerer, L. M., and Angerer, R. C., Detection of poly (A)$^+$ RNA in sea urchin eggs and embryos by quantitative in-situ hybridization. Nucl. Acids Res. *9* (1981) 2819–2840.

9 Arentzen, R., Baldino, F. Jr, Davis. L. G., Higgins, G. A., Lin, Y., Manning. W., and Wolfson, B., In-situ hybridization of putative somatostatin mRNA within hypothalamus of the rat using synthetic oligonucleotide probes. J. Cell Biochem. *27* (1985) 415–422.

10 Bauman, J. G. J., Wiegant. J., Borst, P., and van Duijn. P., A new method of fluorescence microscopical localisation of specific DNA sequences by in-situ hybridization of fluorochrome labelled RNA. Exp. Cell Res. *128* (1980) 485–490.

11 Bauman. J. G. J., Wiegant, J., and van Duijn, P., Cytochemical hybridization with fluorochrome-labelled RNA II. Applications. J. Histochem. Cytochem. *29* (1981) 238–246.

12 Bauman. J. G. J., Wiegant, J., and van Duijn. P., Cytochemical hybridization with fluorochrome-labelled RNA III. Increased sensitivity by the use of anti-flourescein antibodies. Histochemistry *73* (1981) 181–185.

13 Beaucage. S. L., and Caruthers, M. H., Deoxynucleoside phosphoramidites—A new class of key intermediates for deoxypolynucleotide synthesis. Tetrahedron Lett. *22* (1981) 1859–1862.

14 Blum, H. E., Stowring, L., Figus, A., Montgomery, C. K., Haase, A. and Vyas, G. N., Detection of hepatitis B virus DNA in hepatocytes, bile duct epithelium, and vascular elements by in-situ hybridization. Proc. natl Acad. Sci. USA *80* (1983) 6685–6688.

15 Bloch, B., LeGuellec, D., and de Keyzer, Y., Detection of the messenger RNAs coding for the opioid peptide precursors in pituitary and adrenal by in situ hybridization: study in several mammalian species. Neurosci. Lett. *53* (1985) 141–148.

16 Bloch, B., Milner, R. J., Baird, A., Gubler, E., Reymond, C., Bohlen, P., le Guellec, D., and Bloom, F. E., Detection of the messenger RNA coding for preproenkephalin A in bovine adrenal in situ hybridization. Reg. Peptides *8* (1984) 345–354.

17 Brahic, M., and Haase, A. T., Detection of viral sequences of low reiteration frequency by in-situ hybridization. Proc. natl Acad. Sci. USA *75* (1978) 6125–6129.

18 Brahic, M., Haase, A. T., and Cash, E., Simultaneous in-situ detection of viral RNA and proteins. Proc. natl Acad. Sci. USA *81* (1984) 5445–5448.

19 Brigati, D. J., Myerson, D., Leary, J. J., Spalholz, B., Travis, S. Z., Fong, C. K. Y., Hsiung, G. D., and Ward, D. C., Detection of viral genomes in cultured cells and paraffin-embedded tissue sections using biotin-labelled hybridization probes. Virology *126* (1983) 32–50.

20 Buongiorno-Nardelli, S., and Amaldi, F., Autoradiographic detection of molecular hybrids between RNA and DNA in tissue sections. Nature *225* (1970) 946–948.

21 Coghlan, J. P., Aldred, P., Butkus, A., Crawford, R. J., Darby, I. A., Fernley, R. T., Haralambidis, J., Hudson, P. J., Mitri, R., Niall, H. D., Penschow, J. D., Roche, R. J., Scanlon, D. B., Tregear, G. W., Richards, R., van Leeuwen, B. L., Rall, L., Scott, J., and Bell, G., Hybridization histochemistry, in: Endocrinology, pp. 18–24. Eds R. Labrie and L. Proulx. Excerpta Medica, Amsterdam 1984.

22 Coghlan, J. P., Penschow, J. D., Tregear, G. W., and Niall, H. D., Hybridization histochemistry: Use of complementary DNA for tissue localization of specific mRNA populations. in: Receptors, Membranes and Transport Mechanisms in Medicine, pp. 1–11. Eds A. Doyle and F. Mendelson. Excerpta Medica, Amsterdam 1984.

23 Coghlan, J. P., Penschow, J. D., Hudson, P. J., and Niall, H. D., Hybridization histochemistry: Use of recombinant DNA for tissue localization of specific mRNA populations. Clin. exp. Hypert. *A6* (1984) 63–78.

24 Coghlan, J. P., Aldred, P., Haralambidis, J., Niall, H. D., Penschow, J. D., and Tregear, G. W., Hybridization histochemistry. Analyt. Biochem. *149* (1985) 1–28.

25 Coghlan, J. P., Aldred, P., Butkus, A., Crawford, R. J., Darby, I. A., Haralambidis, J., Penschow, J. D., Roche, P. J., Troiani, C., and Tregear, G. W., Hybridization histochemistry—locating gene expression, in: Neuroendocrine Molecular Biology, pp. 33–39. Eds G. Fink, A. J. Harmar and K. W. McKerns. Plenum Publ. Corp., 1986.

26 Coghlan, J. P., Penschow, J. D., Fraser, J. R., Aldred, P., Haralambidis, J., and Tregear, G. W., Location of gene expression in mammalian cells, in: In-Situ Hybridization—Applications to Neurobiology, pp. 24–41. Eds K. Valentino, J. Eberwine and J. D. Barchas. Oxford University Press 1987.

27 Cox, K. H., DeLeon, D. V., Angerer, L. M., and Angerer, R. C., Detection of mRNA's in sea urchin embryos by in-situ hybridization using asymmetric RNA probes. Devl Biol. *101* (1984) 485–502.

28 Craig, R. K., Hall. L., Edbrooke, M. R., Allison, J., and MacIntyre, I., Partial nucleotide sequence of human calcitonin precursor mRNA identifies flanking cryptic peptides. Nature *295* (1982) 345–347.

29 Dalgarno, L., Rice, C. M., and Strauss, J. H., Ross River Virus 26S RNA: complete nucleotide sequence and deduced sequence of the encoded structural proteins. Virology *129* (1983) 170–187.

30 Darby, I. A., Aldred, G. P., Coghlan, J. P., Fernley, R. T., Penschow, J. D., and Ryan, G. B., Use of synthetic oligonucleotide and recombinant DNA probes to study renin gene expression. Clin. exp. Pharm. Physiol. *12* (1985) 199–203.

66

31 Darby, I. A., Aldred, P., Crawford, R. J., Fernley, R. T., Niall, H. D., Penschow, J. D., Ryan, G. B., and Coghlan, J. P., Renin gene expression in vessels of the ovine renal cortex. J. Hypert. *3* (1985) 9–11.

32 Denhart, D. T., A membrane-filter technique for the detection of complementary DNA. Biochem. biophys. Res. Commun. *23* (1966) 641–646.

33 Deschepper, C. F., Mellon, S. H., Cumin, F., Baxter, J. D., and Ganong, W. F., Analysis by immunocytochemistry and in-situ hybridization of renin and its mRNA in kidney, testis, adrenal and pituitary of rat. Proc. natl Acad. Sci. USA *83* (1986) 7552–7556.

34 Edwards, M. K., and Wood, W. B., Location of specific messenger RNA's in *Caenorhabditis elegans* by cytological hybridization. Devl Biol. *97* (1983) 375–390.

35 Favre, A., Morel, C., and Scherrer, K., The secondary structure and poly (A) content of globin messenger RNA as a pure RNA and in polyribosome-derived ribonucleoprotein complexes. Eur. J. Biochem. *57* (1975) 147–157.

36 Forster, A. C., McInnes, J. L., Skingle, D. C., and Symons, R. H., Non-radioactive hybridization probes prepared by the chemical labelling of DNA and RNA with a novel reagent, photobiotin. Nucl. Acids Res. *13* (1985) 746–761.

37 Fuller, P. J., Clements, J. A., Whitfield, P. L., and Funder, J. W., Kallikrein gene expression in the rat anterior pituitary. Molec. cell. Endocr. *39* (1985) 99–105.

38 Gall, J., and Pardue, M., Formation and detection of RNA-DNA hybrid molecules in cytological preparations. Proc. natl Acad. Sci. USA *63* (1969) 378–383.

39 Galloway, D. A., Fenoglio, C. M., Shevchuk, M., and McDougall, J. K., Detection of *Herpes simplex* RNA in human sensory ganglia. Virology *95* (1979) 265–268.

40 Gee, C. E., Chen, C.-L. C., and Roberts, J. L., Identification of proopiomelanocortin neurones in rat hypothalamus by in-situ cDNA-mRNA hybridization. Nature *306* (1983) 374–376.

41 Gee, C. E., and Roberts, J. L., In-situ hybridization histochemistry: a technique for the study of gene expression in single cells. DNA *2* (1983) 155–161.

42 Gerhard, D. S., Kawasaki, E. S., Carter Bancroft, F., and Szabo, P., Localization of a unique gene by direct hybridization in-situ. Proc. natl Acad. Sci. USA *78* (1981) 3755–3759.

43 Gillespie, D., and Spiegelman, S., A quantitative assay for DNA-RNA hybrids with DNA immobilized on a membrane. J. molec. Biol. *12* (1965) 829–842.

44 Godard, C. M., and Jones, K. W., Improved method for detection of cellular transcripts by in-situ hybridization. Detection of poly (A) sequences in individual cells. Histochemistry *65* (1980) 291–300.

45 Goochee, C., Rasband. W., and Sokoloff, L., Computerized densitometry and colour coding of [^{14}C] deoxyglucose autoradiographs. Ann. Neurol. *7* (1980) 359–370.

46 Gowans, E. J., Burrell, C. J., Jilbert, A. R., and Marmion, B. P., Detection of hepatitis B virus sequences in infected hepatocytes by in-situ cytohybridization. J. med. Virol. *8* (1981) 67–78.

47 Haase, A. T., Gantz, G., Eble, B., Walker, D., Stowring, L., Ventura, P., Blum, H., Wietgrefe, S., Zupancic, M., Tourtellotte, W., Gibbs, C. J. Jr. Norrby. E., and Rozenblatt, S., Natural history of restricted synthesis and expression of measles virus genes in subacute sclerosing panencephalitis, Proc. natl Acad. Sci. USA *82* (1985) 3020–3024.

48 Haase, A. T., Walker, D., Stowring, L., Ventura, P., Geballe, A., Blum, H., Brahic, M., Goldberg, R., and O'Brien, K., Detection of two viral genomes in single cells by double label hybridization in-situ and colour microradioautography. Science *227* (1985) 189–191.

49 Hafen, E., Levine, M., Garber, R. L., and Gehring, W. J., An improved in-situ hybridization method for the detection of cellular RNA's in *Drosophila* tissue sections and its application for localising transcripts of the homeotic Antennapedia gene complex. EMBO *2* (1983) 617–623.

50 Hall, B. D., and Spiegelman, S., Sequence complementarity of T-2 DNA and T-2 specific RNA. Proc. natl Acad. Sci. USA *47* (1961) 137–163.

51 Harper, M. E., Marselle, L. M., Gallo, R. C., and Wong-Staal, F., Detection of lymphocytes expressing human T-lymphotropic virus type III in lymph nodes and peripheral blood from infected individuals by in-situ hybridization. Proc. natl Acad. Sci. USA *83* (1986) 772–776.

52 Hayashi, S., Gillam, I. C., Delaney, A. D., and Tener, G. M., Acetylation of chromosome squashes of *Drosophila melanogaster* decreases the background in autoradiographs from hybridization with ^{125}I-labelled RNA. J. Histochem. Cytochem. *36* (1978) 677–679.

53 Hopman, A. H. N., Wiegant, J., Tesser, G. I., and van Duijn, P., A nonradioactive in-situ hybridization method based on mercurated nucleic acid probes and sulfhydryl-hapten ligands. Nucl. Acids Res. *14* (1986) 6471–6488.

54 Hudson, P., Penschow, J., Shine, J., Ryan, G., Niall, H., and Coghlan, J., Hybridization histochemistry: Use of recombinant DNA as a 'homing probe' for tissue localisation of specific mRNA populations. Endocrinology *108* (1981) 353–356.

55 Hutchinson, N. J., Langer-Sofer, P. R., Ward, D. C., and Hamkalo, B. A., In situ hybridization at the electron microscope level: hybrid detection by autoradiography and colloidal gold. J. Cell Biol. *95* (1982) 609–618.

56 Jablonski, E., Moomaw, E. W., Tullis, R. H., and Ruth, J. L., Preparation of oligodeoxynucleotide-alkaline phosphatase conjugates and their use as hybridization probes. Nucl. Acids Res. *14* (1986) 6114–6128.

57 Jacobs, J., Todd, K., Birnstiel, M. L., and Bird, A., Molecular hybridization of ^{3}H-labelled ribosomal RNA and DNA in ultrathin sections prepared for electron microscopy. Biochem. biophys. Acta *228* (1971) 761–766.

58 Jacobs, J. W., Simpson, E., Penschow, J., Hudson, P., Coghlan, J., and Niall, H., Characterisation and localisation of calcitonin messenger ribonucleic acid in rat thyroid. Endocrinology *113* (1983) 1616–1622.

59 Jamrich, M., Mahon, K. A., Gavis, E. R., and Gall, J. G., Histone RNA in amphibian oocytes visualized by in-situ hybridization to methacrylate-embedded tissue sections. EMBO *3* (1984) 1939–1943.

60 John, H. A., Birnstiel, M. L., and Jones, K. W., RNA-DNA hybrids at the cytological level. Nature *223* (1969) 582–587.

61 Jones, K. W., Bishop, J. O., and Brito-da-Cuhna, A., A complex formation between Poly-r-(U) and various chromosomal loci in *Rhyncosciara*. Chromosoma *43* (1973) 375–390.

62 Karnovsky, M. J., A formaldehyde-glutaraldehyde fixative of high osmolarity for use in electron microscopy. J. Cell Biol. *27* (1965) 137A-140A.

63 Koenig, S., Gendelman, H. E., Orenstein, J. M., Dal Canto, M. C., Pezeshkpour, G. H., Yungbluth, M., Janotta, F., Askamit. A., Martin, M. A., and Fauci, A. S., Detection of AIDS virus in macrophages in brain tissue from AIDS patients with encephalopathy. Science *223* (1986) 1089–1093.

64 Kornberg, T., Sidén, I., O'Farrell, P., and Simon, M., The engrailed locus of *Drosophila*: in-situ localization of transcripts reveal compartment-specific expression. Cell *40* (1985) 45–53.

65 Kretschner, P. J., Coon, H. C., Davis, A., Harrison, M., and Nienhuis, A. W., Haemoglobin switching in sheep. J. biol. Chem. *256* (1981) 1975–1982.

66 Landegent, J. E., Jansen in de Wal, N., Ploem, J. S., and van der Ploeg. M., Sensitive detection of hybridocytochemical results by means of reflection-contrast microscopy. J. Histochem. Cytochem. *33* (1985) 1241–1246.

67 Langer-Sofer, P. R., Levine, M., and Ward, D. C., Immunological method for mapping genes on Drosophila polytene chromosomes. Proc. natl Acad. Sci. USA *79* (1982) 4381–4385.

68 Lawrence, J. B., and Singer, R. H., Quantitative analysis of in-situ hybridization methods for the detection of actin gene expression. Nucl. Acids Res. *13* (1985) 1777–1799.

69 Lawrence, J. B., and Singer, R. H., Intracellular localization of messenger RNA's for cytoskeletal proteins. Cell *45* (1986) 407–415.

70 McAllister, L. B., Scheller, R. H., Kandel, E. R., and Axel, R., In-situ hybridization to study the origin and fate of identified neurones. Science *222* (1983) 800–808.

71 McDougall, J. K., Myerson, D., and Beckmann, A., Detection of viral DNA and RNA by in-situ hybridization. J. Histochem. Cytochem. *34* (1986) 33–38.

72 McGinnis, W., Levine, M. S., Hafen, E., Kuroiwa, A., and Gehring, W. J., A conserved DNA sequence in homeotic genes of the *Drosophila* Antennapedia and bithorax complexes. Nature *308* (1984) 428–433.

73 McWilliams, D., and Boime, I., Cytological localisation of placental lactogen messenger ribonucleic acid in syncytiotrophoblast layers of human placenta. Endocrinology *107* (1980) 761–765.

74 Maxam, A. M., and Gilbert, W., Sequencing end-labelled DNA with base-specific chemical cleavages. Meth. Enzymol. *65* (1980) 499–560.

68

75 Melton, D. A., Krieg, P. A., Rebagliati, M. R., Maniatis, T., Zinn, K., and Green, M. R., Efficient in vitro synthesis of biologically active RNA and RNA hybridization probes from plasmids containing a bacteriophage SP-6 promoter. Nucl. Acids Res. *12* (1984) 7035–7056.

76 Mendelsohn, F. A. O., Quirion, R., Saavedra, J. M., Aguilera, G., and Catt, K. J., Autoradiographic localization of angiotensin II receptors in rat brain. Proc. natl Acad. Sci. USA *81* (1984) 1575–1579.

77 Moench, T. R., Gendelman, H. E., Clements, J. E., Narayan, O., and Griffin, D. E., Efficiency of in-situ hybridization as a function of probe size and fixation technique. J. Virol. Meth. *11* (1985) 119–130.

78 Muragagi, A., and Wallace, R. B., Biotin-labelled oligonucleotides: enzymatic synthesis and use as hybridization probes. DNA *3* (1984) 269–277.

79 Ohkubo, H., Kageyama, R., Ujihara, M., Hirose, T., Inayama, S., and Nakanishi, S., Cloning and sequence analysis of cDNA for rat angiotensinogen. Proc. natl Acad. Sci. USA *80* (1983) 2196–2200.

80 Orth, G., Jeantur, P., and Croissant, O., Evidence for and localisation of vegetative viral DNA replication by autoradiographic detection of RNA-DNA hybrids in sections of tumours induced by Shope papilloma virus. Proc. natl Acad. Sci. USA *68* (1971) 1876–1880.

81 Panthier, J. J., Foote, S., Chambrand, B., Strosberg, A. D., Corvol, P., and Rougeon, F., Complete amino acid sequence and maturation of the mouse submaxillary gland renin precursor. Nature *298* (1982) 90–92.

82 Penschow, J. D., Haralambidis, J., Aldred, P., Tregear. G. W., and Coghlan, J. P., Location of gene expression in CNS using hybridization histochemistry. Meth. Enzymol. *124* (1986) 534–548.

83 Pfeifer-Ohlsson, S., Goustin, A. S., Rydnert, J., Bjersing, L., Wahlström. T., Stehelin, D., and Ohlsson, R., Spatial and temporal pattern of cellular myc oncogene expression in developing human placenta: implications for embryonic cell proliferation. Cell *38* (1984) 585–596.

84 Pintar, J. E., Schachter, B. S., Herman, A. B., Durgerian, S., and Krieger, D. T., Characterization and localization of proopiomelanocortin messenger RNA in the adult rat testis. Science *225* (1984) 632–634.

85 Pochet, R., Brocas, H., Vassart, G., Toubeau, G., Seo, H., Refetoff, S., Dumont, J. E., and Pasteels, J. L., Radioisotopic localisation of prolactin mRNA on histological sections by in-situ hybridization. Brain Res. *211* (1981) 433–438.

86 Rall, L. B., Scott, J., Bell, G. I., Crawford, R. J., Penschow, J. D., Niall, M. D., and Coghlan, J. P., Mouse prepro-epidermal growth factor synthesis by the kidney and other tissues. Nature *313* (1985) 228–231.

87 Richards, R. I., Catanzaro, D. F., Mason, A. J., Morris, B. J., Baxter, J. D., and Shine, J., Mouse glandular kallikrein genes. J. biol. Chem. *257* (1982) 2758–2761.

88 Rigby, P. W. J., Dieckmann, M., Rhodes, C., and Berg, P., Labelling deoxyribonucleic acid to high specific activity in vitro by nick translation with DNA polymerase I. J. molec. Biol. *113* (1977) 237–241.

89 Robins, D. M., Ripley, S., Henderson, A. S., and Axel, R., Transforming DNA integrates into the host chromosome. Cell *23* (1981) 29–39.

90 Rogers, A. W., Techniques of autoradiography, pp. 201–284. Elsevier, North Holland, Amsterdam 1979.

91 Rosenfeld, M. G., Mermod. J.-J., Amara, S. G., Swanson, L. W., Sawchenko, P. E., Rivier, J., Vale, W. W., and Evans, R. M., Production of a novel neuropeptide encoded by the calcitonin gene via tissue-specific RNA processing. Nature *304* (1983) 129–135.

92 Rudkin, G. T., and Stollar, B. D., High resolution detection of DNA-RNA hybrids in-situ by indirect immunofluorescence. Nature *265* (1977) 472–473.

93 Schalling, M., Hökfelt, T., Wallace, B., Goldstein, M., Filer, D., Yamin, C., and Schlesinger, D. H., Tyrosine-3-hydroxylase in rat brain and adrenal medulla: hybridization histochemistry and immunohistochemistry combined with retrograde tracing. Proc. natl Acad. Sci. USA *83* (1986) 6208–6212.

94 Scheller, R. H., Kaldany, R.-R., Kreiner, T., Mahon, A. C., Nambu, J. R., Schaefer, M., and Taussig, R., Neuropeptides: mediators of behaviour in *Aplysia*. Science *225* (1984) 1300–1308.

95 Scott, J., Patterson, S., Rall, L., Bell, G. I., Crawford, R., Penschow, J. D., Niall, H. D., and Coghlan, J., The structure and biosynthesis of epidermal growth factor precursor. J. Cell Sci. Suppl. *3 (1985)* 19–28.

96 Seeburg, P., The human growth hormone gene family: nucleotide sequences show recent divergence and predict a new polypeptide hormone. DNA *1* (1982) 239–249.

97 Shivers, B. D., Harlan, R. E., Hejtmancik, J. F., Conn, P. M., and Pfaff, D. W., Localization of cells containing LHRH-like mRNA in rat forebrain using in-situ hybridization. Endocrinology *118* (1986) 883–885.

98 Shivers, B. D., Harlan, R. E., Pfaff, D. W., and Schachter, B. S., Combination of immunocytochemistry and in-situ hybridization in the same tissue section of rat pituitary. J. Histochem. Cytochem. *34* (1986) 39–43.

99 Shivers, B. D., Schachter, B. S., and Pfaff, D. W., In-situ hybridization for the study of gene expression in the brain. Meth. Enzymol. *124* (1986) 497–510.

100 Singer, R. H., and Ward, D. C., Actin gene expression visualized in chicken muscle tissue culture by using in-situ hybridzation with a biotinated nucleotide analog. Proc. natl Acad. Sci. USA *79* (1982) 7331–7335.

101 Smith, F. M., Fung. S., Hunkapiller, M. W., Hunkapiller T. J., and Hood, L. E., The synthesis of oligonucleotides containing an aliphatic amino group at the 5′ terminus: synthesis of fluorescent DNA primers for use in DNA sequence analysis. Nucl. Acids Res. *13* (1985) 2399–2502.

102 Stuart, W. D., and Porter, D. L., An improved in-situ hybridization method. Exp. Cell Res. *113* (1978) 219–222.

103 Szabo, P., Elder, E., Steffensen, D. M., and Uhlenbeck. O. C., Quantitative in-situ hybridization of ribosomal RNA species to polytene chromosomes of *Drosophila melanogaster*. J. molec. Biol. *115* (1977) 539–563.

104 Taylor, J. M., Illmensee, R., and Summers, J., Efficient transcription of RNA and DNA by avian sarcoma virus polymerase. Biochim. biophys. Acta *442* (1976) 324–330.

105 Tu, C.-P. D., and Cohen, S. N., 3′end-labelling of DNA with [α-^{32}P] cordycepin-5-triphosphate. Gene *10* (1980) 177–181.

106 van Leeuwen, B. H., Evans, B. A., Tregear, G. W., and Richards, R. I., Mouse glandular kallikrein genes. J. biol. Chem. *261* (1986) 5529–5535.

107 Varndell, I. M., Polak, J. M., Sikri, K. L., Minth, C. D., Bloom, S. R., and Dixon, J. E., Visualisation of messenger RNA directing peptide synthesis by in-situ hybridization using a novel single-stranded cDNA probe. Histochemistry *81* (1984) 597–601.

108 Venezky, D. L., Angerer, L. M., and Angerer, R. C., Accumulation of histone repeat transcripts in the sea urchin egg pronucleus. Cell *24* (1981) 385–391.

109 Wahl, G. M., Stern, M., and Stark, G. R., Efficient transfer of large DNA fragments from agarose gels to diazobenzyloxymethyl-paper and rapid hybridization using dextran sulphate. Proc. natl Acad. Sci. USA *76* (1979) 3683–3687.

110 Warembourg, M., Tranchant, O., Perret, C., Desplan, C., and Thomasset, M., In-situ detection of vitamin D-induced calcium binding protein (9-kDa CaBP) messenger RNA in rat duodenum. J., Histochem. Cytochem. *34* (1986) 277–280.

111 Wilcox, J. N., Gee, C. E., and Roberts, J. L., In-situ cDNA:mRNA hybridization: development of a technique to measure mRNA levels in individual cells. Meth. Enzymol. *124* (1986) 510–533.

112 Wintour, E. M., Haralambidis, J., Horvath, A., MacIsaac, R. J., Penschow, J. D., and Pontefract, L., The role of the fetal adrenal in hemiglobin switching in sheep, in: Developmental Control of Globin Gene Expression, pp. 541–543. Eds G. Stamatoyannopoulos and A. W. Nienhnis. Alan R. Liss inc. 1987.

113 Wintour, E. M., Smith, M. B., Bell, R. J., McDougall, J. G., and Cauchi, M. N., The role of fetal adrenal hormones in the switch from fetal to adult globin synthesis in sheep. J. Endocr. *104* (1985) 165–170.

114 Zajac, J. D., Penschow, J. D., Mason, T., Tregear, G. W., Coghlan, J. P., and Martin, T. J., Identification of calcitonin and calcitonin gene-related peptide messenger ribonucleic acid in medullary thyroid carcinomas by hybridization histochemistry. J. clin. Endocr. Metab. *62* (1986) 1037–1043.

115 Zawatsky, R., De Maeyer, E., and De Maeyer-Guignard, J. Identification of individual interferon-producing cells by in-situ hybridization, Proc. natl Acad. Sci. USA *82* (1985) 1136–1140.

Regulatory peptide receptors: visualization by autoradiography

J. M. Palacios and M. M. Dietl

Summary. The receptors for regulatory peptides have been extensively characterized using radioligand binding techniques. By combining these binding techniques with autoradiography it is possible to visualize at the light and electron microscopic levels the anatomical and cellular localization of these receptors. In this review we discuss the procedures used to label peptide receptors for autoradiography and the peculiarities of peptides as ligands. The utilization of autoradiography in mapping peptide receptors in brain and peripheral tissues, some of the new insights revealed by these studies, particularly the problem of 'mismatch' between endogenous peptides and receptors, the existence of multiple receptors for a given peptide family and the use of peptide receptor autoradiography in human tissues are also reviewed.

The physiological and pharmacological effects of the regulatory peptides, like those of other hormones and neurotransmitters, are mediated by the interaction of these substances with specific recognition sites named 'receptors'. In parallel with the steadily growing number of peptides being discovered, multiple receptors for these peptides have been characterized[7,50,54,55,116,117]. Pharmacological and biochemical techniques have provided evidence for the existence of multiple receptors for the different peptide families. Historically the characterization of the opioid receptor opened the path for the isolation and characterization of the opiate peptides[12,51,114,118]. The use of the so-called high-affinity radioligand binding techniques was instrumental in beginning the search for endogenous ligands. By using radiolabeled molecules at high specific activity like, for example, the same peptides, it is possible to study the binding of these ligands to membrane preparations. These techniques allow the characterization of sites with high affinity (in the low nanomolar range) and low capacity (fmol/mg of protein) and present selectivity and specificity for a given peptide and its analogues with similar physiological effects. In recent years this relatively simple approach has allowed the detailed study of the structure-activity relationship and the biochemical mechanism involved in peptide action and progress is being made with solubilization and purification and towards the ultimate goal of the elucidation of the molecular characteristics of these receptors[115,123].

One main limitation of these biochemical techniques is, however, the lack of sufficient anatomical resolution to answer questions such as: where are these sites localized at the microscopic level? Are there specific cell populations in a given tissue or organ enriched in a particular peptide receptor? Finally; which subcellular structures bear these receptor sites?

Techniques with high anatomical resolution are required to answer these questions. Because of the wide use of radio-ligand binding assays in the study of peptides and other receptors it is not surprising that *autoradiography* has been one of the techniques most widely used in the localization of peptide receptors at the microscopic level. In particular the *in vitro* autoradiographic procedure originally developed by Young and Kuhar[134] is specially well suited for the study of peptide receptors. Organ and tissue barriers encountered in *in vivo* labeling are overcome, as is the problem of ligand metabolism, a very important problem when dealing with peptide ligand both *in vivo* and *in vitro*[59,60,89]. In this paper we will review the use of this technique in the visualization of peptide receptors both in the central nervous system and peripheral tissues at the light and electron microscopic levels of resolution as well as other alternatives for the visualization of peptide receptors.

Methodology

The procedures for the labeling of peptide receptors for autoradiography are essentially identical with those used for the labeling of other hormone, drug or neurotransmitter receptors[59,60,89]. Some of the particularities encountered in peptide receptor labeling are related to the use, in most cases, of radiolabeled peptides as ligands[123]. These ligands present, in general 1) a high susceptibility to degradation by peptidases present on the tissue, and 2) a tendency to bind in a non-specific way to tissue, glass and gelatine in the histological preparations. An outline of the procedure is summarized in table 1.

Tissue preparation

Different procedures have been used to prepare tissues for peptide receptor localization depending on the method used to label the receptor, i.e., *in vivo* or *in vitro* and the type of resolution desired, i.e., light or electron microscopy.

When *in vivo* labeling procedures are used the ligand is administered systemically, for example, intravenously or subcutaneously and after a given time (chosen to favor a maximal level of specific binding) the animal is sacrificed, generally by perfusion with a fixative. This procedure has been widely used to label brain peptide receptors with non-peptide ligands (such as opiates[1-3]) and peripheral receptors, for example somatostatin receptors[78].

The *in vivo* labeling procedure is limited to the use of metabolically stable analogues or non-peptide ligands, because of the poor ability of the

Table 1. Autoradiographic visualization of regulatory peptide receptors: outline of the procedure

---

1) *Tissue preparation*
   a) Slight *in vivo* fixation followed by *in vitro* prefixation or not depending on desired anatomical resolution or tissue preservation.
   b) Regular *in vivo* fixation after *in vivo* labeling of receptors.
2) *Labeling of receptor*
   a) *In vivo*, systemic administration of ligand followed by perfusion/fixation.
   b) *In vitro*
      I) *Preincubation* with guanylnucleotides and ions, washing of endogenous ligands.
      II) *Incubation* with peptidase-inhibitors, ions, displacers, at different temperatures, addition of chemicals to prevent chemical degradation of ligand and non-specific binding to supports.
      III) *Washing* of unbound ligand.
      IV) *Postfixation*, cross-linking.
3) *Generation of autoradiograms*
   a) 'wet' (liquid) emulsions; dipping of labeled sections.
   b) 'dry' emulsion-coated coverslips, films.
   c) standards.
4) *Quantification of autoradiograms*
   a) Manual counting of autoradiographic grains.
   b) Automatic, computer-assisted microdensitometry or grain-counting.

---

peptide-ligand to penetrate through barriers such as the blood-brain barrier, and the high cost resulting from the use of large amounts of ligand.

These limitations are overcome by using the *in vitro* approach initially developed by Young and Kuhar[134]. In this procedure tissues are obtained postmortem from laboratory animals after perfusion, or not, of the anesthetized animal with buffers containing low concentrations of fixative (routinely 0.04% paraformaldehyde in PBS); higher concentrations might be detrimental to the receptor[134]. The tissues are then sectioned in a cryostat-microtome normally at a 10 to 20 μm thickness and thaw-mounted onto gelatine-coated glass-slides. These procedures have been found to preserve receptors quite adequately: no differences have been found between either the density or the characteristics of receptors in these mounted sections and those in standard membrane preparations from fresh tissues[134]. The stability of peptide receptors is also remarkable even in human postmortem tissues, particularly in brain[18,19,74,91,105].

In some cases a better tissue preservation is desired, as in studies oriented towards the electron microscopic localization of receptors[40,41], or the preservation of antigens for correlative immunohistochemical studies. A compromise has then to be found between the loss of receptor binding following the use of fixatives, and tissue structure and antigen preservation[64].

Ligands

[³H]- and [¹²⁵I]-labeled ligands are generally used in autoradiographic studies of peptide receptors. Many of these ligands are commercially available. Most of the peptides can be iodinated, using the chloramine T method when the peptide has a tyrosine or a histidine residue[37] or in the Bolton-Hunter method[8] when there is a lysine and a terminal NH_2 group. The labeling sites should preferably occur only once in the primary sequence and should represent amino acids of lesser importance in the binding of the ligand to the receptor. In all cases, radiolabeled peptide ligands have to be carefully checked for purity and stability. The biological activity of the radiolabeled peptide, although not always easy to test, is also an important criterion before any binding experiment is attempted. The selection of the isotope is linked to criteria such as stability and purity of the ligand, specific activity requirements, resolution desired, etc. Table 2 lists the most commonly used ligands for the labeling of peptide receptors.

Labeling of receptors for autoradiography

As already mentioned, two procedures can be used to label peptide receptors *in vivo* and *in vitro*. The limitations of the in vivo procedure have already been discussed above, and we will now refer in more detail to the *in vitro* procedure. In labeling peptide receptors in mounted tissue sections (see table 1) the following parameters have to be taken into account 1) the presence of endogenous ligand; 2) the presence of peptidases in the tissue; 3) the possibility of binding of the ligand to sites unrelated to the receptor itself; 4) the nature of the binding of the ligand to the receptor, which is generally reversible and thus presents possibilities of dissociation; 5) the chemical stability of the ligand.

In general endogenous ligands are eliminated by preincubating the tissues for different amounts of time, in conditions in which the released peptide should be accessible to degradation by endogenous peptidases, or is washed out by changing the preincubation buffer. Dissociation of the endogenous ligand is favored by adding to the incubation medium guanylnucleotides or ions[80,123,134].

Routinely a cocktail of several peptidase-inhibitors is included during the incubation period to prevent the degradation of the ligand, if this is a peptide. Anti-oxidants are also generally used to protect the ligand from chemical degradation. Examples of a peptidase inhibitor mixture are those used in the labeling of substance P (SP) receptors or cholecystokinin (CCK) receptors. For SP a 50 mM Tris-HCl buffer at pH 7.4 is used with 200 mg/l BSA to which is added bacitracin 40 mg/l, chymostatin 2 mg/l and leupeptin 4 mg/l, then $MnCl_2$ 5 mM; some investigators like to add polyethylenimine at 0.005%[68,69]. In contrast, for CCK the same buffer is used but at pH 7.7 and with 200 mg/l of bacitracin, $MgCl_2$ 5 mM and dithiothreitol 1 mM[71,126,137].

Table 2. Regulatory peptide receptors: autoradiographic studies

Endogenous peptide	Receptor subtypes	Ligands	References
Opioid peptides			
Met-enkephalin (MET)	μ	[³H]diprenorphine	1, 2, 3, (rat, *in vivo*), 33 (rat),
Leu-enkephalin (LEU)		[³H]etorphine	40, 41 (rat), 46 (rat), 64 (rat),
Dynorphin (DYN)		[³H]bremazocine	74 (human), 76 (rat, *in vivo*),
β-Endorphin (βEND)		[³H]morphine	90 (rat, guinea pig),
		[³H]naloxone	94 (rat, *in vivo*), 130 (monkey)
		[³H]dihydromorphine (DHM)	
		[³H]-[D-Ala², MePhe⁴,	
		Gly-ol⁵]enk. (DAGO)	
		[³H]FK 33824	
	δ	[³H]bremazocine	33 (rat), 36 (rat, guinea pig),
		[³H]-[D-Ala², D-Leu⁵]enk. (DADL)	90 (guinea pig)
		[²⁵I]-[D-Ala², D-Leu⁵]enk. (DADL)	
		[³H]DPDPE	
	κ	[³H]bremazocine	22, 23 (rat, guinea pig), 34 (rat),
		[³H]ethylketocyclazozine	65 (guinea pig), 74 (human),
		[³H]dynorphin A	83 (rat), 90 (guinea pig)
	ε	[³H]-β-endorphin	35 (rat)
	σ	[³H]-3-PPP	39 (rat, guinea pig),
		[³h]phencyclidine	62 (guinea pig), 96 (rat)
		[³H]SKF 10047	
Gut-brain peptides			
Tachykinins			
Substance P (SP)	SP-P (NK1)	[³H]SP	13 (rat), 68 (rat), 79 (rat), 98 (rat)
		[¹²⁵I]BH-SP	5 (rat), 10 (rat gastrointestinal tract),
			70 (rat gastrointestinal tract),
			100 (guinea pig), 108 (rat)
Physalaemin (PHY)	SP-P (NK1)	[¹²⁵I]PHY	133 (rat)
Substance K =	SP-K (NK2)	[¹²⁵I]BH-SK	10 (rat gastrointestinal tract),
Neurokinin A (SK)			69 (rat), 100 (guinea pig).
Eledoisin (ELE)	SP-E (NK3)	[¹²⁵I]BH-ELE	5 (rat), 13 (rat), 109 (rat)
Kassinin (KAS)	SP-E (NK3)	[¹²⁵I]BH-KAS	69 (rat)

Gastrins		
Gastrin		—
Pentagastrin (G-5)	[³H]CCK-5/pentagastrin	26 (rat)
Cholecystokinin (CCK-4)	[³H]CCK-8	126 (rat, pancreas)
Cholecystokinin (CCK-8)	[¹²⁵I]BH-CCK-8	18 (human), 71 (rat)
Cholecystokinin (CCK-33)	[¹²⁵I]CCK-33	42 (rat), 137 (rat, guinea pig)
Vasoactive intestinal polypeptide (VIP)	[¹²⁵I]VIP	17 (rat), 73 (rat)
Motilin (MOT)		—
Pancreatic polypeptide (PPP)		—
Secretin (SECR)		—
Pituitary peptides		
Adrenocorticotropin (ACTH)		—
Growth hormone (GH)		—
Lipotropin		—
α-Melanocyte-stimulating hormone (α-MSH)		—
Arginin-Vasopressin (AVP)	[³H]AVP	4 (rat), 6 (rat), 9 (rat), 15 (rat), 127 (rat)
Oxytocin (OXT)	[³H]OXT	15 (rat), 24 (rat), 128 (rat)
Circulating hormones		
Angiotensin II (AII)	[¹²⁵I]-[Sar¹, Ile⁸]AII	29, 30 (rat, brain, kidney), 43 (rat adrenal gland), 52 (rat), 82 (rat), 119 (dog)
Calcitonin (CT)	[¹²⁵I]CT	21 (human), 44, 45 (rat), 75 (rat), 85 (rat)
Calcitonin-gene-related peptide (CGRP)	[¹²⁵I]CGRP	14 (rat), 45 (rat), 53 (rat, human), 112 (rat)
Insulin (INS)	[¹²⁵I]INS	49 (rat), 135 (rat)
Hypothalamic-releasing hormones		
Thyrotropin-releasing hormone (TRH)	[³H]methyl-TRH	66 (rat), 72 (rat), 92 (rat, guinea pig), 95 (rat)
Corticotropin-releasing factor (CRF)	[NLe²¹, ¹²⁵I-Tyr³²]CRF 103 (rat), 106 (mouse, guinea pig, hamster, rabbit, rat, human)	16 (rat)
Luteinizing-hormone-releasing hormone (LHRH)	[¹²⁵I]LHRH	
Somatostatin (SS)	[¹²⁵I]CGP 23996	38 (rat)
	[¹²⁵I-Tyr¹¹]SS-14	63 (rat), 77 (rat)
	[¹²⁵I-LTT]SS-18	63 (rat)
	SMS 204-0900	104 (rat, pituitary), 105 (human), 122 (rat, human)
	[³H]SS-14	131 (human)

Table 2. (*Continued*)

Endogenous peptide	Receptor subtypes	Ligands	References
Growth-hormone-releasing factor (GRF)			
Others			
Bradykinin (BK)		[³H]BK	67 (guinea pig ileum)
Bombesin (BN)		[¹²⁵I-Tyr⁴]BN	132 (rat)
Neuropeptide tyrosine (NPY)		[¹²⁵I]BH-NPY	see 111 (rat)
Galanin (GAL)		[¹²⁵I]GAL	81 (rat)
Atrial natriuretic factor (ANF)		[¹²⁵I]ANF	99 (rat), 129 (rat, peripheral tissues)
Atriopeptin III (APIII)		[¹²⁵I]APIII	31 (rat, adrenal), 61 (rat, thymus and spleen), 84 (rat kidney)
Neurotensin (NT)		[³H]NT	97 (rat), 86 (rat), 120 (human, rat), 121 (human), 136 (rat)
		[¹²⁵I]NT	110 (human)

The table lists the most important endogenous peptides known until now, the receptors localized and the ligands used for autoradiography in vertebrate tissues (when only the species is mentioned refers to *brain*). Numbers correspond to references; – indicates that no receptor for this peptide has been localized yet.

Other ways to deal with the problem of peptidase activity in the tissue sections are 1) the use of peptide analogues designed to be resistant to catabolic activity, as for example opioid or somatostatin analogues[40,41,77,78] and 2) the use of low incubation temperatures, although this technique also results in slower association kinetics and consequently longer incubation periods. An example is the procedure used to label TRH receptors which decrease peptidase activity[72,92,95] (see however Manaker et al.[66]).

Several procedures have been used to decrease or block the non-specific binding of peptide ligands to a glass surface or to gelatine. One of the most commonly used, which is also postulated to provide protection against peptidase activity, is the addition of bovine serum albumin to the incubation medium. Compounds such as polyethylenimine are reported to decrease the non-specific binding of peptides to glass and have been used in membrane and section labeling of such peptide receptors as those for SP[68,133].

The influence of all these compounds on the characteristics and density of the receptor being studied have to be carefully monitored in preliminary experiments to ensure that no negative influence occurs. The different ligands and receptors behave in a very unpredictable way in response to the parameters mentioned above. As no general rule can be proposed it is advisable to check carefully in each individual situation for all the conditions listed.

Autoradiogram generation

Autoradiograms are generated by putting the labeled tissue sections in close contact with emulsions. This can be done in two ways: 1) dipping the tissues in liquid emulsion[40,41,46]; 2) using a dry emulsion in the form of an emulsion-coated coverslip[134] or a film, such as a tritium-sensitive or X-ray film[87,93,102,125]. The use of the dipping procedure requires that the ligand is irreversibly fixed to the receptor site because emulsions are used in liquid form at relatively high temperature (around 40°C). The use of reversibly bound ligands will lead to the dissociation of the ligand from the receptor, and will produce erroneous autoradiographic pictures. This method has been called by Hamel and Beaudet[40,41] the 'wet' autoradiographic technique in contrast to the 'dry' autoradiographic technique in which emulsion-coated coverslips or films are used (fig. 1). The emulsion-coated coverslip procedure introduced by Young and Kuhar has been extensively used in the mapping of receptors for regulatory peptides. Finally the use of ³H-sensitive films such as the [³H]Ultrofilm from LKB, Sweden or the recently developed ³H-film from Amersham[87,93,102,125] allows an easier handling and quantification of the autoradiograms by computer-assisted image analysis systems. The advantages and

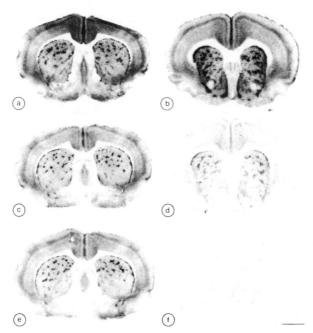

Figure 1. Effects of fixation and dehydration on the topographical distribution of opioid binding sites labeled with [^{125}I]FK 33-824 (left) or [^{125}I]FW 34-569 (right). In *a* and *b* the autoradiographic films were generated before, in *c* and *d* after glutaraldehyde fixation. In *e* and *f*, fixation was followed by dehydration. [^{125}I]FK 33-824 is covalently cross-linked to tissue proteins by aldehydes; no modification in the distribution of this ligand is observed after glutaraldehyde fixation and tissue dehydration (*e*). In contrast, [^{125}I]FW 34-569 does not cross-link and is completely washed out in the course of dehydration (*f*). Bar = 2.0 mm. (Figure kindly provided by A. Beaudet[41], reproduced with permission.)

disadvantages of the use of nuclear tracing emulsions or ^{3}H-sensitive films have been extensively discussed by other authors and the reader is referred to these discussions for more detailed information[59].

Briefly, the main advantage of using the 'wet' procedure or emulsion-coated coverslips is the possibility of visualizing the autoradiographic grains together with the tissue. In the case of the 'wet' emulsion procedure the autoradiographic grains and the tissue are located in the same optical plane, while in the case of the coverslip a slight difference in the optical plane of the autoradiographic grains and the tissue exists. On the other hand, the use of film leads to a loss of anatomical register between the localization of autoradiograms and the actual tissue structure although it allows much easier quantification.

Analysis of autoradiograms

Autoradiograms once generated can be analyzed using many different procedures. In recent times the use of computer-assisted image analysis systems has gained widespread popularity because of the feasibility of performing complete quantitative autoradiographic studies[89]. On the market a large variety of image analysis systems are now available and we will not discuss the differences between the different systems. Basically the autoradiographic films are digitalized using a TV camera and analyzed by microcomputers. The response of the film to radioactivity can be calibrated using appropriate standards and optical densities transformed into receptor densities in terms of fmol/mg protein or area unit[32,58,60,102,125].

One of the most studied problems using receptor autoradiography with tritiated ligands has been 'quenching'[27,28] caused by the differential absorption of β-radiation by different tissue elements[58]. A number of procedures have been designed to overcome the problem in the use of tritium labeled ligands[47]. This difficulty is not encountered with [^{25}I]iodine labeled compounds; this is an important advantage of these ligands, which are the most popular in peptide receptor studies. Regarding the quantification of autoradiograms generated using the 'wet' or 'dry' emulsion procedure there are also some automatic devices for the counting of autoradiographic grains, but in contrast to the devices for the study of films these are expensive, and they are still subject to a number of technical problems. Most commonly the quantification of autoradiographic grains in emulsions is carried out by visual counting of grains[124]. As in the case of the films the use of appropriate standards allows for the full transformation of grain density into receptor density. Of course this procedure is much more time-consuming than fully automatic image analysis.

Uses of peptide receptor localization by autoradiography

Mapping of peptide receptors and the 'mismatch' problem

The most extensive use of peptide receptor autoradiography is the detailed analysis of the regional and anatomical distribution of peptide receptors in different tissues, particularly in the brain. This is the so-called 'mapping of receptors'. Table 2 is a summary of the peptide receptors which have been visualized in different mammalian species and, in some cases, in other vertebrates. Table 2 lists the subtypes of receptors mapped, the ligand used and some key references. The references are, however, not exhaustive because of space limitations; however, the most relevant studies have been cited.

Historically the opioid receptor was the first one to be mapped in detail. Early studies were done in the rat brain after *in vivo* labeling of these receptors using non-peptide opioid ligands[1–3,94]. The results obtained in these studies are an example of the type of analysis and insights provided by peptide receptor mapping in brain and other tissues. The detailed description of the distribution of opioid receptors in the mammalian brain was instrumental in understanding the neuronal mechanisms involved in the physiological actions and pharmacological effects of opioid drugs. Opioid receptors were found to be highly concentrated in some specific areas of the central nervous system, for example, the substantia gelatinosa of the spinal cord, an anatomical locus known to be involved in the processing of sensory information. In addition, other areas of enrichment in opioid receptors were found to be the periaqueductal gray and the medial thalamus. All these localizations provided a clear anatomical substrate for the known analgesic effects opioids exert at both spinal and supra-spinal levels. In addition, the euphoric effects of opiates can now be attributed to the localization of substantial numbers of receptors for these drugs in the limbic system[114,117].

An important consequence of the studies on the localization of opioid receptors in the brain was a search for 'endogenous morphine'[12,51,113]. The development of antibodies against the enkephalins and the immunohistochemical studies of the enkephalins soon provided a detailed mapping of the presynaptic components of the opioid system in the mammalian brain. It was clear from these studies that a relatively good overlap existed between the distribution of endogenous enkephalins and opioid receptors. This led to the proposal that enkephalins were the endogenous ligands for the opioid receptor. More detailed studies have, however, demonstrated that not in every case is there a good overlap between the distribution of endogenous peptide and the distribution of the receptors for this peptide. This has been called the 'mismatch' problem and is the subject of much speculation[48,57,88]. Herkenham and McLean have recently reviewed in detail 'mismatches' between the distribution of the receptors for peptides and their endogenous peptide ligands[48]. Thus, while it was originally proposed that the distribution of endogenous enkephalins and the μ-opioid receptor were in good accordance, an examination of the distribution of opioid receptors in other brain areas such as the diencephalon and forebrain has concluded that 'mismatches' are rather the rule. The hypothalamus is another good example of 'mismatch'. The presence of high densities of opioid peptides as well as opioid-mediated functions on the one hand have to be reconciled with a low density of receptors. This has been explained by the presence of high densities of opioid receptors of the kappa-type. Dynorphine which is postulated to be the endogenous ligand for the kappa-receptor, because of its high affinity for this binding site, has been shown to influence the functions of the hypothalamic system. This explains why κ-receptors are abundant in the hypothalamus while only sparse μ and δ sites are present in this brain areas. On the other hand the opposite situation has been described in the thalamus where very low densities of opioid peptides have been found to correlate with high densities of the μ- and κ-subtypes of the opioid receptors.

Similar mismatches have been observed for many other peptides in the rat brain. A particularly good example is the distribution of SP and that of SP binding sites (fig. 2). The most spectacular example is provided by the substantia nigra where the highest densities of endogenous SP have been found and yet no significant density of SP receptors has been reported. This will lead us to a second problem that we will discuss later, the problem of subtypes of receptors and the relation of peptide receptor subtypes to the presence of multiple endogenous peptides.

Other examples of mismatches are provided by the localization of receptors for the peptides neurotensin, somatostatin, angiotensin II, CCK, thyrotropin-releasing hormone (TRH) and corticotropin-releasing factors (CRF). The problem of the mismatch between endogenous peptides and peptide receptors is further complicated by species differences. Very

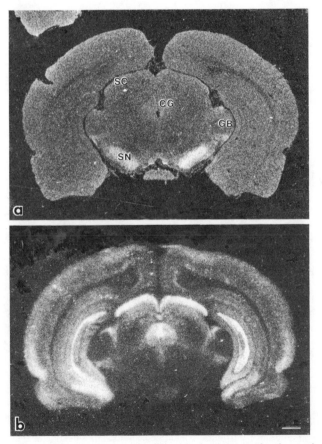

Figure 2. The distribution of substance P localized by immunocytochemistry using internally labeled monoclonal antibodies to substance P (*a*) is compared to substance P binding sites in an equivalent level localized by autoradiography using [^{125}I]Bolton-Hunter substance P (*b*). Bar = 1.0 mm.

82

variable patterns of distribution for peptide receptors are found in different species. Here again the case of the opioid receptors has been a paradigm of the multiplicity of distributions in different species. We have also observed species differences in the distribution of receptors for several regulatory peptides. A particularly striking example of both the

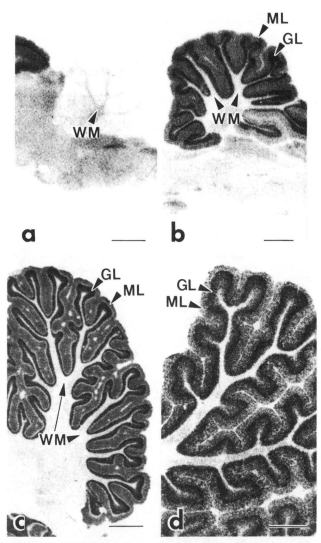

Figure 3. Distribution of [^{125}I]Bolton-Hunter cholecystokinin-8 binding in the cerebellum of 4 different vertebrate species is shown. In the rat (*a*) no cholecystokinin-8 binding sites can be detected in this structure, while guinea pig (*b*), cat (*c*) and human (*d*) show a high density of cholecystokinin-8 binding sites particularly in the granular cell layer. ML, molecular cell layer; GL, granular cell layer; WM, white matter. Bar = 2.0 mm.

mismatch problem and the species differences is illustrated in figure 3 where the localization of CCK receptors in the cerebellar cortex is illustrated. The cerebellum is supposed to contain no measurable density, or a very low level, of the endogenous peptide CCK. However, in the guinea pig and the primates including man, we have found a rich supply of CCK receptors in the cerebellar cortex, while in the rat cerebellum no significant binding was observed.

Several explanations have been proposed for the interpretation of the 'mismatch problem'.

The first classical explanation is that a mismatch is produced by the differences in the refinement of the techniques used. Thus, the lack of resolution of the current immunocytochemical methods being used to visualize endogenous peptides on the one hand and/or the presence of occupied receptors, non-functional spare receptors, and the presence of low affinity receptors which are not visualized using the current high affinity binding techniques on the other hand can lead to an apparent mismatch between neuropeptides and their receptors.

Another explanation is based on the fact that neurotransmitter molecules and receptor molecules are contained often within two completely different neurons which have completely different anatomical distributions[57,60]. In addition, the presence of receptors on non-neuronal elements can also lead to such a mismatch. It has furthermore been suggested that many receptors are not located at synapses and may be sites of action of neuropeptides released at a distance, which would mean a hormone-like transmission of information[40,41,48,80]. Herkenham and McLean[48,80] include the possibility that receptors identified in one region might actually be in transit to another region.

Finally, Schultzberg and Hökfelt[111] have recently proposed a further possibility, based on the finding of the coexistence of neurotransmitters and neuropeptides in the same synapses. They suggest that receptors for two types of co-localized/co-released transmitters may be expressed in a 'coupled' form. Such a coupled expression may occur in regions where only one of the transmitters is present, thus providing a further explanation for the mismatch between transmitters and receptors.

At the present time a full explanation for the mismatch problem is not yet available. The mismatch problem reveals our lack of understanding of the correlation between innervation and receptor density and how this problem relates to the physiological responses of peptides in different tissues and organs.

Multiple peptide receptor subtypes

The existence of more than one pharmacologically characterized receptor type for a given peptide has already been mentioned in the context of the mismatch between opioid peptides and opioid receptors. It appears to be a general rule that multiple subtypes or receptors exist for a given peptide family. Again the historical and foremost example is the case of opioid receptors[117,118]. At least 5–6 putative opioid receptor subtypes (table 2) have been proposed in the literature and attempts have been made to correlate the affinity of different opioid peptides to these receptors in order to establish the most probable endogenous ligand for each receptor subtype. Receptor autoradiography has revealed that these subtypes present very different regional distributions in the mammalian brain and has suggested the most appropriate regions to study the different receptor subtypes. Many other examples of multiplicity of peptide receptors have been reported and have been illustrated by receptor autoradiography.

The most recent example of peptide receptor multiplicity is provided by the tachykinins[101]. At least three subtypes of receptors for the peptides of the family of SP have been proposed and named SP-P, SP-K and SP-E (recently renamed NK-1, NK-2 and NK-3, respectively), depending on the preferential affinity of the different peptides of the tachykinin family for these subtypes. In figure 4 we illustrate the differential distribution of receptors for SP (SP-P) and receptors for eledoisin (SP-E receptors) in the rat brain. The presence of very distinct regional distributions for these ligands clearly suggests that these receptors are different not only in terms of pharmacology but probably in terms of structure. This is an illustration of the power of receptor autoradiography to provide information on the presence of different subtypes of receptors for peptides and information that can be used to explore their different physiological functions.

Peptide receptors in the human brain. Implications for pathology

As mentioned above, the technique of receptor autoradiography can be used in the study of human postmortem material[91]. We have analyzed a large series of neurotransmitter and drug receptors in human postmortem material and found that these molecules survive the chemical changes in the postmortem human brain quite well. We have been able to localize and map receptors for a number of peptides in human postmortem tissues including opioid, neurotensin, somatostatin, TRH, CCK, SP and other receptors. In addition to the already mentioned differences in species distributions and pharmacological characteristics it has been observed that receptors for peptides are changed under some neuropathological conditions[91,120,121]. One of the first reported receptor changes in the human brain was the loss of neurotensin receptors in the substantia nigra of patients dying with Parkinson's disease[121]. This observation followed the observation of high densities of neurotensin receptors in the rat substantia nigra[86] and the loss of these receptors after chemical lesion of the dopaminergic cells using the neurotoxin 6-hydroxydopamine (fig. 5).

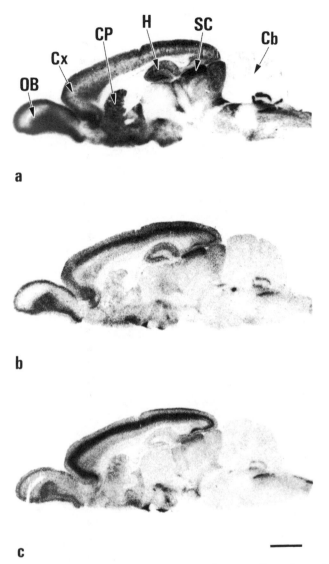

Figure 4. Three sequential sagittal rat sections processed for autoradiography to visualize the three tachykinin receptor subtypes. Three different ligands were used, [^{125}I]Bolton-Hunter substance P (*a*) for the SP-P or NK1 subtype, [^{125}I]Bolton-Hunter substance K (*b*) for the SP-K or NK2 subtype and [^{125}I]Bolton-Hunter eledoisin (*c*) for the SP-E or NK3 subtype were used. Note the differential localization of these receptors in areas such as the neocortex (Cx), caudate-putamen (CP), hippocampus (H) and olfactory bulb (OB). Bar = 1.0 mm.

Other neuropeptide receptors have also been found to be modified in some neuropathological states of the human brain. For example, somatostatin receptors have been found to be decreased in the cortex of patients dying from senile dementia of the Alzheimer type. We have used

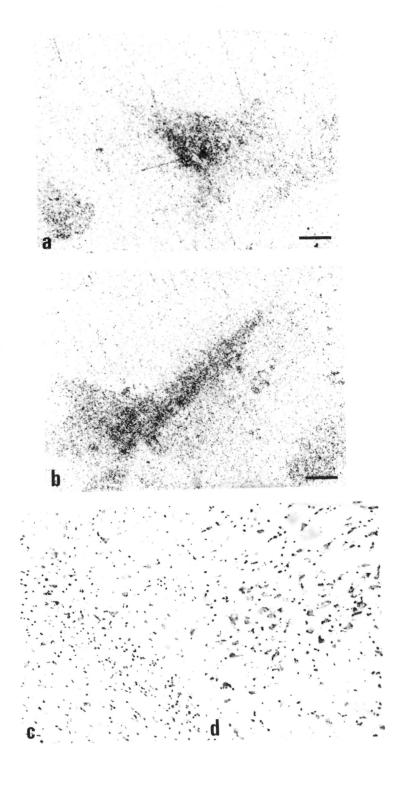

autoradiography to study these changes and observed selective decreases of somatostatin receptors in areas such as the hippocampus and the neocortex of patients with senile dementia while in parkinsonian patients somatostatin receptors appear to be more preferentially decreased in the nuclei caudatus and putamen and the substantia nigra (Cortés et al., unpublished).

The ability of autoradiography to detect subtle changes in pathology offers the possibility of using peptide receptor autoradiography as a tool in neuropathological investigations. This type of investigation will probably be extended to pathologies outside the brain where peptide receptor mechanisms appear to be involved, and in particular diseases of the gut where peptides play an important regulatory role.

Peptide receptors in peripheral tissues

Using *in vitro* autoradiography, peptide receptors have also been localized to peripheral tissues and some of the results in these tissues are listed on table 2. The localization of SP receptors in the guinea pig ileum is illustrated in figure 6. These results clearly demonstrate the possibility of using autoradiography for the study of receptors in peripheral tissues, allowing the microscopic localization of receptors in these tissues and the correlation of the known function of regulatory peptides in the regulation of their functions. Problems similar to those already discussed above, such as the mismatch problem and the presence of receptor subtypes for related peptides have been examined using autoradiography. An example is the presence of the different subtypes of the tachykinin receptors in peripheral organs[10,70].

The technique of receptor autoradiography, however, has not yet been extensively exploited in the study of peripheral tissues and it is very probable that in the future we will be obtaining more and more information about peptide receptors in peripheral tissues.

Cellular localization of peptide receptors and modification of
peptide receptors after drug treatment

One of the main limitations of the most commonly used receptor autoradiography techniques is the lack of cellular and subcellular resolution. In order to compensate for this lack of resolution the technique of

Figure 5. Autoradiographic localization of neurotensin receptors to dopaminergic neurons in rat substantia nigra zona compacta (*a*). Injection of 6-hydroxydopamine results in a loss of neurotensin receptors and dopamine-containing cells in the substantia nigra zona compacta (*b*). Higher power photomicrographs of the same sections stained with cresyl violet (*c*, *d*). Note the loss of cell bodies in the zona compacta in *c* (corresponding to the injected left side, see *a*) compared with the opposite uninjected side in *d*. Bar = 250 μm (*a*, *b*).

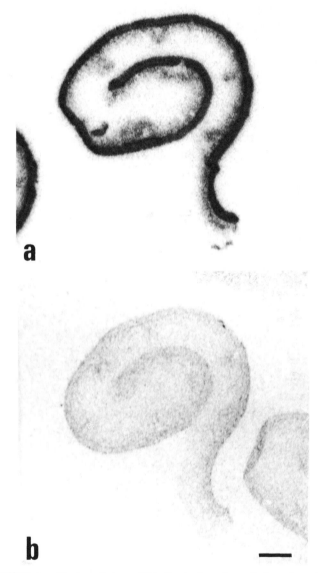

Figure 6. [^{125}I]Bolton-Hunter substance P binding in the guinea pig ileum is mainly localized to the muscle layer (*a*). Non-specific binding in an adjacent section obtained by adding 1 μM of cold substance P to the incubation medium (*b*). Bar = 1.0 mm.

receptor autoradiography has been combined with selective lesions of different cellular types. As mentioned above it was the lesion of the dopaminergic cells in the substantia nigra with the neurotoxin 6-hydroxydopamine which led to the discovery of the presence of high densities of neurotensin receptors in these cells[86] and eventually to the finding of a selective decrease of neurotensin receptors in the parkinsonian substantia nigra[121]. In the central nervous system a large number of studies have addressed the issue of the cellular localization of peptide receptors to different cell populations. In particular, lesions using kainic acid[107], or lesions of selective pathways[13a], have been used to illustrate the preferential localization of receptors in different types of neurotransmitter-identified neurons and pathways in the central nervous system of the rat[107] and the cat[13a].

One of the most important conclusions of all these studies is that peptide receptors appear to be preferentially localized in neurons and selected pathways in the mammalian central nervous system.

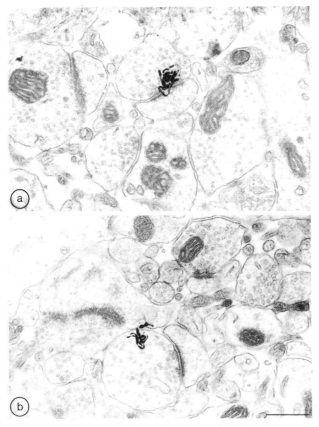

Figure 7. Electron microscopic autoradiographs processed from a vibratome-cut, prefixed section of rat striatum and incubated with [^{125}I]FK 33-824. The radioactivity is detected in the form of isolated silver grains overlying in both instances the apposed plasma membranes of an axon terminal and a dendritic process. Synaptic localization shown in *a*, and non-synaptic localization shown in *b*. Bar = 0.5 μm. (Reproduced with permission from A. Beaudet[41].)

Interestingly, also in the human brain peptide receptors appear to be preferentially localized in neurons. We have analyzed the effects of localized small vascular lesions in the hippocampus of patients dying without a reported neurological disease and found that the loss of neurons was correlated with the loss of binding for peptides, as for example somatostatin[105]. However, the lack of cellular resolution of light microscopic autoradiography is an important issue which has led several groups to attempt to develop procedures for the electron microscopic localization of peptide receptor sites[40,41]. As mentioned in the methodology section this has been facilitated by the fact that peptides can be cross-linked to their receptors. Electron microscopic autoradiography, however, is also limited in its resolution by the fact that in the central nervous system membranes of more than one type co-exist inside the so-called '50% probability circle'[56]. However, the results obtained so far with the μ-subtype[40,41] of the opioid receptor (fig. 7) and neurotensin (Beaudet et al., unpublished) have resulted in a number of interesting conclusions regarding the localization of peptide receptors. The majority of these receptors appear to be extrasynaptic, that is they are localized outside the synaptic terminals as visualized at the electron microscopic level and this has led to the speculation that opioids can act at distance in a hormonal type of transmission.

Conclusions and future trends

The results briefly reviewed in this work illustrate the possibility of using receptor autoradiography in the analysis of the mechanisms of action of regulatory peptides in the brain and other tissues. We have also mentioned a number of limitations of the technique, particularly the relatively low level of subcellular resolution. In the future it is probable that new types of probes will be developed to allow a clearer definition of the cells making peptide receptors. Recently a number of receptors for growth factors, oncogenes and hormones and some neurotransmitters have been cloned and their molecular structure elucidated from the sequence of the genes coding for these proteins. Although until now no receptor for a regulatory peptide has been cloned it is not too speculative to assume that we will soon see published the sequences of DNA coding for opioid, SP and other peptide receptors. This will allow the development of genetic probes for the study of the localization of receptors at the microscopic level, including *in situ* hybridization and the development of synthetic peptides to which antibodies will be raised, and will expand the use of immunohistochemistry in receptor localization.

Advances in medicinal chemistry are also providing us with new non-peptide ligands for peptide receptors. This is an important advance because of the limitations mentioned above on the use of peptides as ligands, particularly for the *in vivo* imaging of receptors in the living man using the technique of positron emission tomography. Opioid receptors have already been visualized using a carfentanil and used in the study of the human opioid system *in vivo*[25]. It is probable that we will see more non-peptide ligands developed for peptide receptors. The most dramatic example is the development of a CCK and non-peptide CCK antagonist by the Merck, Sharp and Dohme Group[11,20].

Acknowledgments. We would like to thank Dr A. Beaudet for providing us with illustrations of electron microscopy localization of peptide receptors, Drs M. Herkenham and S. McLean for their unpublished material and Dr C. R. Jones for the critical reading of the manuscript.

1 Atweh, S. F., and Kuhar, M. J., Autoradiographic localization of opiate receptors in rat brain. I. Spinal cord and lower medulla. Brain Res. *124* (1977) 53–67.

2 Atweh, S. F., and Kuhar, M. J., Autoradiographic localization of opiate receptors in rat brain. II. The brainstem. Brain Res. *129* (1977) 1–12.

3 Atweh, S. F., and Kuhar, M. J., Autoradiographic localization of opiate receptors in rat brain. III. The telencephalon. Brain Res. *134* (1977) 393–405.

4 Baskin, D. G., Petracca, F., and Dorsa, D. M., Autoradiographic localization of specific binding sites for [^{3}H][Arg^8]vasopressin in the septum of the rat brain with tritium-sensitive film. Eur. J. Pharmac. *90* (1983) 155–157.

5 Beaujouan, J. C., Torrens, Y., Saffroy, M., and Glowinski, J., Quantitative autoradiographic analysis of the distribution of binding sites for [^{125}I]Bolton Hunter derivatives of eledoisin and substance P in the rat brain. Neuroscience *18* (1986) 857–875.

6 Biegon, A., Terlou, M., Voorhuis, Th. D., and De Kloet, E. R., Arginine-vasopressin binding sites in rat brain: A quantitative autoradiographic study. Neurosci. Lett. *44* (1984) 229–234.

7 Björklund, A., and Hökfelt, T., Eds, GABA and Neuropeptides in the CNS. Handbook of Chem. Neuroanatomy, vol. 4. Elsevier, Amsterdam 1985.

8 Bolton, A. E., and Hunter, W. M., The labeling of proteins to high specific radioactivities by conjugation to a [^{125}I]-containing acylating agent. Biochem. J. *133* (1973) 529–538.

9 Brinton, R. E., Gee, K. W., Wamsley, J. K., Davis, T. P., and Yamamura, H. I., Regional distribution of putative vasopressin receptors in rat brain and pituitary by quantitative autoradiography. Proc. natn. Acad. Sci. USA *81* (1984) 7248–7252.

10 Burcher, E., Shults, C. W., Buck, S. H., Chase, T. N., and O'Donohue, T. L., Autoradiographic distribution of substance K-binding sites in rat gastrointestinal tract: a comparison with substance P. Eur. J. Pharmac. *102* (1984) 561–562.

11 Chang, R. S. L., and Lotti, V. J., Biochemical and pharmacological characterization of an extremely potent and selective nonpeptide cholecystokinin antagonist. Proc. natn. Acad. Sci. USA *83* (1986) 4923–4926.

12 Cox, B. M., Opheim, K. E., Teschemacher, H., and Goldstein, A., A peptide-like substance from pituitary that acts like morphine. 2. Purification and properties. Life Sci. *16* (1975) 1777–1782.

13 Danks, J. A., Rothman, R. B., Cascieri, M. A., Chicchi, G. G., Liang, T., and Herkenham, M., A comparative autoradiographic study of the distributions of substance P and eledoisin binding sites in rat brain. Brain Res. *385* (1986) 273–281.

13aDashwood, M. R., Gilbey, M. P., and Spyer, K. M., The localization of adrenoceptors and opiate receptors in regions of the cat central nervous system involved in cardiovascular control. Neuroscience *15* (1985) 537–551.

14 Dawbarn, D., Gregory, J., and Emson, P. C., Visualization of calcitonin gene-related peptide receptors in the rat brain. Eur. J. Pharmac. *111* (1985) 407–408.

15 De Kloet, E. R., Rotteveel, F., Voorhuis, Th. A. M., and Terlou, M., Topography of binding sites for neurohypophyseal hormones in rat brain. Eur. J. Pharmac. *110* (1985) 113–119.

16 De Souza, E. B., Perrin, M. H., Insel, T. R., Rivier, J., Vale, W. W., and Kuhar, M. J., Corticotropin-releasing factor receptors in rat forebrain: autoradiographic identification. Science *224* (1984) 1449–1451.

92

17 De Souza, E. B., Seifert, H., and Kuhar, M. J., Vasoactive intestinal peptide receptor localization in rat forebrain by autoradiography. Neurosci. Lett. *56* (1985) 113–120.

18 Dietl, M. M., Probst, A., and Palacios, J. M., On the distribution of cholecystokinin receptor binding sites in the human brain: An autoradiographic study. Synapse *1* (1987) 169–183.

19 Dietl. M. M., Probst, A., and Palacios, J. M., Mapping of substance P receptor sites in the human brain, In preparation.

20 Evans, B. E., Bock, M. G., Rittle, K. E., DiPardo, R. M., Whitter, W. L., Veber, D. F., Anderson, P. S., and Freidinger, R. M., Design of potent, orally effective, nonpeptidal antagonists of the peptide hormone cholecystokinin. Proc. natn. Acad. Sci. USA *83* (1986) 4918–4922.

21 Fischer, J. A., Tobler, P. H., Kaufmann, M., Born, W., Henke, H., Cooper, P. E., Sagar, S. M., and Martin, J. B., Calcitonin: Regional distribution of the hormone and its binding sites in the human brain and pituitary. Proc. natn. Acad. Sci. USA *78* (1981) 7801–7805.

22 Foote, R. W., and Maurer, R., Autoradiographic localization of opiate K-receptors in the guinea-pig brain. Eur. J. Pharmac. *85* (1982) 99–103.

23 Foote, R. W., and Maurer, R., Distribution of opioid binding sites in the guinea-pig hippocampus as compared to the rat: A quantitative analysis. Neuroscience *19* (1986) 847–856.

24 Freund-Mercier, M. J., Stoeckel, M. E., Palacios, J. M., Pazos, A., Reichhart, J. M., Porte, A., and Richard, Ph., Pharmacological characteristics and anatomical distribution of [^{3}H]oxytocin-binding sites in the Wistar rat brain studied by autoradiography. Neuroscience *20* (1987) 599–614.

25 Frost, J. J., Wagner, H. N. Jr., Dannals, R. F., Ravert, H. T., Links, J. M., Wilson, A. A., Burns, H. D., Wong, D. F., McPherson, R. W., Rosenbaum, A. E., Kuhar, M. J., and Snyder, S. H., Imaging opiate receptors in the human brain by positron tomography. J. comp. ass. Tomog. *9* (1985) 231–236.

26 Gaudreau, P., Quirion, R., St. Pierre, S., and Pert, C. B., Tritium-sensitive film autoradiography of [^{3}H]cholecystokinin-5/pentagastrin receptors in rat brain. Eur. J. Pharmac. *87* (1983) 173–174.

27 Geary, W. A., and Wooten, G. F., Regional tritium quenching in quantitative autoradiography of the central nervous system. Brain Res. *336* (1985) 334–336.

28 Geary, W. A., Toga, A. W., and Wooten, G. F., Quantitative film autoradiography for tritium: Methodological consideration. Brain Res. *337* (1985) 99–108.

29 Gehlert, D. R., Speth, R. C., Healy, D. P., and Wamsley, J. K., Autoradiographic localization of angiotensin II receptors in the rat brainstem. Life Sci. *34* (1984) 1565–1571.

30 Gehlert, D. R., Speth, R. C., and Wamsley, J. K., Autoradiographic localization of angiotensin II receptors in the rat brain and kidney. Eur. J. Pharmac. *98* (1984) 145–146.

31 Gibson, R. R., Wildey, G. M., Manaker, S., and Glembotski, Ch. C., Autoradiographic localization and characterization of atrial natriuretic peptide binding sites in the rat central nervous system and adrenal gland J. Neurosci. *6* (1986) 2004–2011.

32 Goochee, C., Rasband, W., and Sokoloff, L., Application of computer-assisted image processing to autoradiographic methods for studying brain functions. Trends Neurosci. *6* (1983) 256–260.

33 Goodman, R. R., Snyder, S. H., Kuhar, M. J., and Young III, W. S., Differentiation of delta and mu opiate receptor localizations by light microscopic autoradiography. Proc. natn. Acad. Sci. USA *77* (1980) 6239–6243.

34 Goodman, R. R., and Snyder, S. H., Kappa opiate receptors localized by autoradiography to deep layers of cerebral cortex: Relation to sedative effects. Proc. natn. Acad. Sci. USA *79* (1982) 5703–5707.

35 Goodman, R. R., Houghten, R. A., and Pasternak, G. W., Autoradiography of [^{3}H]β-endorphin binding in brain. Brain Res. *288* (1983) 334–337.

36 Gouardères, C., Cros, J., and Quirion, R., Autoradiographic localization of mu, delta and kappa opioid receptor binding sites in rat and guinea-pig spinal cord. Neuropeptides *6* (1985) 331–342.

37 Greenwood, F. C., Hunter, W. M., and Glover, J. S., The preparation of [^{131}I]-labeled human growth hormone of high specific radioactivity. Biochem. J. *89* (1963) 114–123.

38 Gulya, K., Wamsley, J. K., Gehlert, D., Pelton, J. T., Duckles, S. P., Hruby, V. J., and Yamamura, H. I., Light microscopic autoradiographic localization of somatostatin receptors in the rat brain. J. Pharmac. exp. Ther. 235 (1985) 254–258.

39 Gundlach, A. L., Largent, B. L., and Snyder, S. H., Autoradiographic localization of sigma receptor binding sites in guinea-pig and rat central nervous system with (+)[³H]3-(-3-hydroxyphenyl)-N-(1-propyl)piperidine. J. Neurosci. 6 (1986) 1757–1770.

40 Hamel, E., and Beaudet, A., Electron microscopic autoradiographic localization of opioid receptors in rat neostriatum. Nature 312 (1984) 155–157.

41 Hamel, E., and Beaudet, A., Localization of opioid binding sites in rat brain by electron microscopic radioautography. J. Electron Microsc. Techn. 1 (1984) 317–329.

42 Hays, S. E., Beinfeld, M. C., Jensen, R. T., Goodwin, F. K., and Paul, S. M., Demonstration of a putative receptor site for cholecystokinin in rat brain. Neuropeptides 1 (1980) 53–62.

43 Healy, D. P., Maciejewski, A. R., and Printz, M. P., Autoradiographic localization of [¹²⁵I]angiotensin II binding sites in the rat adrenal gland. Endocrinology 116 (1984) 1221–1223.

44 Henke, H., Tobler, P. H., and Fischer, J. A., Localization of salmon calcitonin binding sites in rat brain by autoradiography. Brain Res. 272 (1983) 373–377.

45 Henke, H., Tschopp, F. A., and Fischer, J. A., Distinct binding sites for calcitonin gene-related peptide and salmon calcitonin in rat central nervous system. Brain Res. 360 (1985) 165–171.

46 Herkenham, M., and Pert, C. B., In vitro autoradiography of opiate receptors in rat brain suggests loci of 'opiatergic' pathways. Proc. natn. Acad. Sci. USA 77 (1980) 5532–5536.

47 Herkenham, M., and Sokoloff, L., Quantitative receptor autoradiographic tissue defatting eliminates differential self-absorption of tritium radiation in grey and white matter of brain. Brain Res. 321 (1984) 363–368.

48 Herkenham, M., and Mclean, S., Mismatches between receptor and transmitter localization in the brain, in: Quantitative Receptor Autoradiography, pp. 137–171. Eds C. A. Boast, E. W. Snowhill and C. A. Altar. Alan R. Liss, New York 1986.

49 Hill, J. M., Lesniak, M. A., Pert, C. B., and Roth, J., Autoradiographic localization of insulin receptors in rat brain: Prominence in olfactory and limbic areas. Neuroscience 17 (1986) 1127–1138.

50 Hökfelt, T., Johansson, O., Ljungdahl, A., Lundberg, J. M., and Schultzberg, M., Peptidergic neurons. Nature 284 (1980) 515–521.

51 Hughes, J., Smith, T. W., Kosterlitz, H. W., Fothergill, L. A., Morgan, B. A., and Morris, H. R., Identification of two related pentapeptides from the brain with potent opiate agonist activity. Nature 258 (1975) 577–559.

52 Hwang, B. H., Harding, J. W., Liu, D. K., Hibbard, L. S., Wieczorek, C. M., and Wu J.-Y., Quantitative autoradiography of [¹²⁵I]-[Sar¹, Ile⁸]-angiotensin II binding in the brain of spontaneously hypertensive rats, Brain Res. 16 (1986) 75–82.

53 Inagaki, S., Kito, S., Kubota, Y., Girgis, S., Hillyard, C. J., and MacIntyre, I., Autoradiographic localization of calcitonin gene-related peptide binding sites in human and rat brains. Brain Res. 374 (1986) 287–298.

54 Iversen, L. L., Neuropeptides-what next? Trends Neurosci. 4 (1983) 292–294.

55 Krieger, D. T., Brownstein, M. J., and Martin, J. B., Eds, Brain Peptides. John Wiley & Sons, New York 1983.

56 Kuhar, M. J., Taylor, N., Wamsley, J. K., Hulme, E. C., and Birdsall, N. J. M., Muscarinic cholinergic receptor localization in brain by electron microscopic autoradiography. Brain Res. 216 (1981) 1–9.

57 Kuhar, M. J., The mismatch problem in receptor mapping studies. Trends Neurosci. 8 (1985) 190–191.

58 Kuhar, M. J., and Unnerstall, J. R., Quantitative receptor mapping by autoradiography: Some current technical problems. Trends Neurosci. 8 (1985) 49–53.

59 Kuhar, M. J., De Souza, E. B., and Unnerstall, J. R., Neurotransmitter receptor mapping by autoradiography and other methods. A. Rev. Neurosci. 9 (1986) 27–60.

60 Kuhar, M. J., Quantitative receptor autoradiography: An overview, in: Quantitative Receptor Autoradiography, pp. 1–12. Eds C. A. Boast, E. W. Snowhill and C. A. Alter. Alan R. Liss, New York 1986.

94

61 Kurihara, M., Shigematsu, K., and Saavedra, J. M., Localization of atrial natriuretic peptide, ANP-(99-126) binding sites in the rat thymus and spleen with quantitative autoradiography. Reg. Peptides 15 (1986) 341–346.

62 Largent, B. L., Gundlach, A. L., and Snyder, S. H., Pharmacological and autoradiographic discrimination of sigma and phencyclidine receptor binding sites in brain with (+)-[³H]SKF 10,047, (+)-[³H]-3-[3-hydroxyphenyl[-N-(1-propyl)piperidine and [³H]-1-]1(2-thienyl)cyclohexyl]piperidine. J. Pharmac. exp. Ther. 238 (1986) 739.

63 Leroux, P., Quirion, R., and Pelletier, G., Localization and characterization of brain somatostatin receptors as studied with somatostatin-14 and somatostatin-28 receptor radioautography. Brain Res. 347 (1985) 74–84.

64 Lewis, M.E., Khachaturian, H., and Watson, S. J., Visualization of opiate receptors and opioid peptides in sequential brain sections. Life Sci. 31 (1982) 1347–1350.

65 Lewis, M. E., Young, E. A., Houghten, R. A., Akil, H., and Watson, S. J., Binding of [³H]dynorphin a to apparent K-opioid receptors in deep layers of guinea-pig cerebral cortex. Eur. J. Pharmac. 98 (1984) 149–150.

66 Manaker, S., Rainbow, T. C. and Winokur, A., Thyrotropin-releasing hormone (TRH) receptors: Localization in rat and human central nervous system, in: Quantitative Receptor Autoradiography, pp. 103–135. Eds C. A. Boast, E. W. Snowhill and C. A. Altar. Alan R. Liss, New York 1986.

67 Manning, D. C., and Snyder, S. H., Bradykinin receptor-mediated chloride secretion in intestinal function. Nature 299 (1982) 256–259.

68 Mantyh, P. W., Hunt, S. P., and Maggio, J. E., Substance P receptors: Localization by light microscopic autoradiography in rat brain using [³H]SP as the radioligand. Brain Res. 307 (1984) 147–165.

69 Mantyh, P. W., Maggio, J. E., and Hunt, S. P., The autoradiographic distribution of kassinin and substance K binding sites is different from the distribution of substance P binding sites in rat brain. Eur. J. Pharmac. 102 (1984) 361–364.

70 Mantyh, P. W., Goedert, M., and Hunt, S. P., Autoradiographic visualization of receptor binding sites for substance P in the gastrointestinal tract of the guinea-pig. Eur. J. Pharmac. 100 (1984) 133–134.

71 Mantyh, C. R., and Mantyh, P. W., Differential localization of cholecystokinin-8 binding sites in the rat vs. the guinea-pig brain. Eur. J. Pharmac. 113 (1985) 137–139.

72 Mantyh, P. W., and Hunt, S. P., Localization by light microscopic autoradiography in rat brain using [³H][3-Me-His²]TRH as the radioligand. J. Neurosci. 5 (1985) 551–561.

73 Martin, J. L., Dietl, M. M., Hof, P., Palacios, J. M., and Magistretti, P. J., Autoradiographic localization of [Mono(¹²⁵I)Iodo-Tyr¹⁰MetO¹⁷]vasoactive intestinal peptide. Neuroscience 23 (1987) 539–565.

74 Maurer, R., Cortés, R., Probst, A., and Palacios, J. M., Multiple opiate receptor in human brain: an autoradiographic investigation. Life Sci. 33 (1983) 231–234.

75 Maurer, R., Marbach, P., and Mousson, R., Salmon calcitonin binding sites in rat pituitary. Brain Res. 261 (1983) 346–348.

76 Maurer, R., Comparative in vitro/in vivo autoradiography using the opiate ligand ³H-(-)-bremazocine. J. Receptor Res. 4 (1984) 155–163.

77 Maurer, R., and Reubi, J. C., Brain somatostatin receptor subpopulation visualized by autoradiography. Brain Res. 333 (1985) 178–181.

78 Maurer, R., and Reubi, J. C., Somatostatin receptors in the adrenal. Molec. cell. Endocr. 45 (1986) 81–90.

79 Maurin, Y., Buck, S. H., Wamsley, J. K., Burks, T. F., and Yamamura, H. I., Light microscopic autoradiographic localization of [³H]substance P binding sites in rat thoracic spinal cord. Life Sci. 34 (1984) 1713–1716.

80 McLean, S., Rothman, R. B., Jacobson, A. E., Rice, K. C., and Herkenham, M., Distribution of opiate receptor subtypes and enkephalin and dynorphin immunoreactivity in the hippocampus of squirrel, guinea-pig, rat and hamster. J. comp. Neurol. (1987), in press.

81 Melander, R., Hökfelt, T., Nilsson, S., and Brodin, E., Visualization of galanin binding sites in the rat central nervous system. Eur. J. Pharmac. 124 (1986) 381–382.

82 Mendelsohn, F. A. O., Quirion, R., Saavedra, J. M., Aguilera, G., and Catt, K.J., Autoradiographic localization of angiotensin II receptors in rat brain. Proc. natn. Acad. Sci. USA 81 (1984) 1575–1579.

83 Morris, B. J., and Herz, A., Autoradiographic localization in rat brain of kappa opiate binding sites labeled by [³H]bremazocine. Neuroscience *19* (1986) 839–846.

84 Murphy, K. M. M., McLaughlin, L. L., Michener, M. L., and Needleman, P., Autoradiographic localization of atriopeptin III receptors in rat kidney. Eur. J. Pharmac. *111* (1985) 291–292.

85 Olgiati, V. R., Guidobonon, F., Netti, C., and Pecile, A., Localization of calcitonin binding sites in rat central nervous system: evidence of its neuroactivity. Brain Res. *265* (1983) 209–215.

86 Palacios, J. M., and Kuhar, M. J., Neurotensin receptors are located on dopamine-containing neurons in rat midbrain. Nature *294* (1981) 587–589.

87 Palacios, J. M., Niehoff, D. H., and Kuhar, M. J., Receptor autoradiography with tritium-sensitive film: Potential for computerized densitometry. Neurosci. Lett. *25* (1981) 101–105.

88 Palacios, J. M., and Wamsley, J. K., Receptor for amines, amino acids and peptides: Biochemical characterization and microscopic localization, in: Chemical Transmission in the Brain, Progress in Brain Research, vol. 55, pp. 265–278. Eds R. M. Buijs, P. Pévet and D. F. Swaab. Elsevier Biomedical Press 1982.

89 Palacios, J. M., Receptor autoradiography, 10 years later. J. Receptor Res. *4* (1984) 633–644.

90 Palacios, J. M., and Maurer, R., Autoradiographic localization of drug and neurotransmitter receptors: focus on the opiate receptor. Acta histochem. Suppl. *29* (1984) 41–50.

91 Palacios, J. M., Probst, A., and Cortés, R., Mapping receptors in the human brain. Trends Neurosci. *9* (1986) 284–289.

92 Pazos, A., Cortés, R., and Palacios, J. M., Thyrotropin-releasing hormone receptor binding sites: autoradiographic distribution in the rat and guinea-pig brain. J. Neurochem. *45* (1985) 1448–1463.

93 Penney, J. B., Pan, H. S., Young, A. B., Frey, K. A., and Dauth, G. W., Quantitative autoradiography of [³H]muscimol binding in rat brain. Science *214* (1981) 1036–1038.

94 Pert, C. B., Kuhar, M. J., and Snyder, S. H., Opiate receptor: Autoradiographic localization in rat brain. Proc. natn. Acad. Sci. USA *73* (1976) 3729–3733.

95 Pilotte, N. S., Sharif, N. A., and Burt, D. R., Characterization and autoradiographic localization of TRH receptors in sections of rat brain. Brain Res. *293* (1984) 372–376.

96 Quirion, R., Hammer, R. P., Herkenham, M., and Pert, C. B., Phencyclidine (angel dust), 'opiate' receptors: Visualization by tritium-sensitive film. Proc. natn. Acad. Sci. USA *78* (1981) 5881–5885.

97 Quirion, R., Gaudreau, P., St. Pierre, S., Rioux, F., and Pert, C. B., Autoradiographic distribution of [³H]neurotensin receptors in rat brain: visualization by tritium-sensitive film. Peptides *3* (1982) 757–763.

98 Quirion, R., Shults, C. W., Moody, T. W., Pert, C. B., Chase, T. N., and O'Donohue, T. L., Autoradiographic distribution of substance P receptors in rat central nervous system. Nature *303* (1983) 714–716.

99 Quirion, R., Dalpe, M., DeLean, A., Gutkowska, J., Cantin, M., and Genest, J., Atrial natriuretic factor (ANF) binding sites in brain and related structures. Peptides *5* (1984) 1167–1172.

100 Quirion, R., and Dam, T.-V., Multiple tachykinin receptors in guinea-pig brain. High densities of substance K (neurokinin A) binding sites in the substantia nigra. Neuropeptides *6* (1985) 191–204.

101 Quirion, R., Multiple tachykinin receptors. Trends Neurosci. *8* (1985) 183–185.

102 Rainbow, T. C., Bleisch, W., Biegon, A., and McEwen, B. S., Quantitative densitometry of neurotransmitter receptors. J. Neurosci. Meth. *5* (1982) 127–138.

103 Reubi, J. C., and Maurer, R., Visualization of LHRH receptors in the rat brain. Eur. J. Pharmac. *106* (1985) 453–454.

104 Reubi, J. C., and Maurer, R., Autoradiographic mapping of somatostatin receptors in the rat central nervous system and pituitary. Neuroscience *15* (1985) 1183–1193.

105 Reubi, J. C., Cortés, R., Maurer, R., Probst, A., and Palacios, J. M., Distribution of somatostatin receptors in the human brain: an autoradiographic study. Neuroscience *18* (1986) 329–346.

106 Reubi, J. C., Palacios, J. M., and Maurer, R., Specific LHRH receptors in the rat hippocampus and pituitary: an autoradiographical study. Neuroscience *21* (1987) 847–858.

96

107 Ritter, J. K., Gehlert, D. R., Ribb, J. W., Wamsley, J. K., and Hanson, G. R., Neuronal localization of substance P receptors in rat neostriatum. Eur. J. Pharmac. *109* (1985) 431–432.

108 Rothman, R. B., Herkenham, M., Pert, C. B., Liang, T., and Cascieri, M. A., Visualization of rat brain receptors for the neuropeptide, substance P. Brain Res. *309* (1984) 47–54.

109 Rothman, R. B., Danks, J. A., Herkenham, M., Cascieri, M. A., Chicchi, G. G., Liang, T., and Pert, C. B., Autoradiographic localization of a novel peptide binding site in rat brain using the substance P analog, eledoisin. Neuropeptides *4* (1984) 343–349.

110 Sarrieau, A., Javoy-Agid, F., Kitabgi, P., Dussaillant, M., Vial, M., Vincent, J. P., Agid, Y., and Rostène, W. H., Characterization and autoradiographic distribution of neurotensin binding sites in the human brain. Brain Res. *348* (1985) 375–380.

111 Schultzberg, M., and Hökfelt, T., The mismatch problem in receptor autoradiography and the coexistence of multiple messengers. Trends Neurosci. *9* (1986) 109–110.

112 Seifert, H., Chesnut, J., De Souza, E., Rivier, J., and Vale, W., Binding sites for calcitonin gene-related peptide in distinct areas of rat brain. Brain Res. *346* (1985) 195–198.

113 Simantov, R., and Snyder, S. H., Morphine-like factors in mammalian brain: Structure-elucidation and interactions with the opiate receptor. Proc. natn. Acad. Sci. USA *73* (1976) 2515–2519.

114 Snyder, S. H., The opiate receptor in normal and drug altered brain function. Nature *257* (1975) 185–189.

115 Snyder, S. H., and Bennett, J. P. Jr, Neurotransmitter receptors in the brain: Biochemical identification. A. Rev. Physiol. *38* (1976) 153–175.

116 Snyder, S. H., and Innis, R. B., Peptide neurotransmitters. A. Rev. Biochem. *48* (1979) 755–782.

117 Snyder, S. H., Brain peptides and neurotransmitters. Science *209* (1980) 967–983.

118 Snyder, S. H., and Goodman, R. R., Multiple neurotransmitter receptors. J. Neurochem. *35* (1980) 5–15.

119 Speth, R. C., Wamsley, J. K., Gehlert, D. R., Chernicky, C. L., Barnes, K. L., and Ferrario, C. M., Angiotensin II receptor localization in the canine CNS. Brain Res. *326* (1985) 137–143.

120 Uhl, G. R., and Kuhar, M. J., Chronic neuroleptic treatment enhances neurotensin receptor binding in human and rat substantia nigra. Nature *309* (1984) 350.

121 Uhl, G. R., Whitehouse, P. J., Price, D. L., Tourtelotte, W. W., and Kuhar, M. J., Parkinson's disease: Depletion to substantia nigra neurotension receptors. Brain Res. *308* (1984) 186–190.

122 Uhl, G. R., Tran, V., Snyder, S. H., and Martin, J. B., Somatostatin receptors: Distribution in rat central nervous system and human frontal cortex. J. comp. Neurol. *240* (1985) 288–304.

123 Undén, A., Peterson, L.-L., and Bartfai, T., Somatostatin, substance P, vasoactive intestinal polypeptide, and neuropeptide Y receptors: Critical assessment of biochemical methodology and results, in: International Review of Neurobiology, vol. 27. Eds J. R. Smythies and R. J. Bradley. Academic Press, Inc., Orlando, New York 1985.

124 Unnerstall, J. R., Kuhar, M. J., Niehoff, D. L., and Palacios, J. M., Benzodiazepine receptors are coupled to a subpopulation of aminobutyric acid (GABA) receptors: Evidence from a quantitative autoradiographic study. J. Pharmac. exp. Ther. *218* (1981) 797–804.

125 Unnerstall, J. R., Niehoff, D. L., Kuhar, M. J., and Palacios, J. M., Quantitative receptor autoradiography using [^{3}H]Ultrofilm: Application to multiple benzodiazepine receptors. J. Neurosci. Meth. *6* (1982) 59–73.

126 Van Dijk, A., Richards, J. G., Trzeciak, A., Gillessen, D., and Möhler, H., Cholecystokinin receptors: biochemical demonstration and autoradiographical localization in rat brain and pancreas using [^{3}H]cholecystokinin as radioligand. J. Neurosci. *4* (1984) 1021–1033.

127 Van Leeuwen, F. W., and Wolters, P., Light microscopic localization of [^{3}H]arginine-vasopressin binding sites in the rat brain and kidney. Neurosci. Lett. *41* (1983) 61–66.

128 Van Leeuwen, F. W., Van Heerikhuize, J., Van Der Meulen, G., and Wolters, P., Light microscopic autoradiographic localization of [^{3}H]oxytocin binding sites in the rat brain, pituitary and mammary gland. Brain Res. *359* (1985) 320–325.

129 Von Schroeder, H. P., Nishimura, E., McIntosh, C. H. S., Buchan, A. M. J., Wilson, N., and Ledsome, J. R., Autoradiographic localization of binding sites for atrial natriuretic factor. Can J. Physiol. Pharmac, *63* (1985) 1373–1377.

130 Wamsley, J. K., Zarbin, M. A., Young, W. S. III, and Kuhar, M. J., Distribution of opiate receptors in the monkey brain: an autoradiography study. Neuroscience *7* (1982) 595–613.

131 Whitford, C. A., Candy, J. M., Bloxham, C. A., Oakley, A. E., and Snell, C. R., Human cerebellar cortex possesses high affinity binding sites for [^{3}H]somatostatin. Eur. J. Pharmac. *113* (1985) 129–132.

132 Wolf, S. S., Moody, T. W., O'Donohue, T. L., Zarbin, M. A., and Kuhar, M. J., Autoradiographic visualization of rat brain binding sites for bombesin-like peptides. Eur. J. Pharmac. *87* (1983) 163–164.

133 Wolf, S. S., Moody, T. W., Quirion, R., Shults, C. W., Chase, T. N., and O'Donohue, T. L., Autoradiographic visualization of substance P receptors in rat brain. Eur. J. Pharmac. *91* (1983) 157–158.

134 Young, W. S. III, and Kuhar, M. J., A new method for receptor autoradiography: [^{3}H]opioid receptors in rat brain. Brain Res. *179* (1979) 255–270.

135 Young, W. S. III, Kuhar, M. J., Roth, J., and Brownstein, M. J., Radiohistochemical localization of insulin receptors in the adult and developing rat brain. Neuropeptides *1* (1980) 15–22.

136 Young, W. S. III, and Kuhar, M. J., Neurotensin receptor localization by light microscopic autoradiography in rat brain. Brain Res. *206* (1981) 273–285.

137 Zarbin, M. A., Innis, R. B., Wamsley, J. K., Snyder, S. H., and Kuhar, M. J., Autoradiographic localization of cholecystokinin receptors in rodent brain. J. Neurosci. *3* (1983) 877–906.

Combined axonal transport tracing and immunocytochemistry for mapping pathways of peptide-containing nerves in the peripheral nervous system

H. C. Su and J. M. Polak

Summary. The various combinations of axonal transport tracing immunocytochemistry used for mapping pathways of peptide-containing nerves, and in particular those of the peripheral nervous system, are reviewed. The advantages and disadvantages of these methods are discussed. The applications and results presented illustrate the future potential value of this approach.

Introduction

In the beginning of the 1970s neuroanatomy experienced a methodological revolution with the development of a new technique for tracing neural pathways based on the axonal transport of compounds such as horseradish peroxidase (HRP)[50,55,56,100], radiolabelled amino acids[14,20,22] and the fluorescent dye Evans blue (bound to albumin)[49]. New tracers, both ingenious and powerful, are still rapidly being added to the tools of the neuroanatomist[1,2,23,28,31,44,48,52–54,62,63,76,80,92,93,101]. Tracers, such as the fluorescent dyes[1,2,53,54], are taken up by the axonal processes of neurones and retrogradely transported to the parent cell bodies after being administered into their peripheral terminal fields. Others, for example radioactively labelled amino acids[21] and *Phaseolis vulgaris*-leucoagglutinin (PHA-L)[28] are taken up by neuronal cell bodies and subsequently anterogradely transported to their terminals. Compounds, such as HRP[65], wheat germ agglutinin (WGA) or WGA-HRP[57,81] and cholera toxin[92,104], are transported efficiently in both anterograde and retrograde directions simultaneously. Retrograde and anterograde tracing techniques now offer convenient methods for mapping nerve pathways and tracing connections of central neurones, peripheral ganglia and nerves supplying muscles, skin and viscera. However, the use of axonal transport techniques alone does not determine the transmitter content of defined neuronal pathways.

Radioimmunoassay[5,106] and immunocytochemistry[12,13,67,90] were developed at about the same time as these transport techniques and have been used to investigate neuropeptides; these techniques have yielded extensive valuable information about neuropeptides. More than 40 peptides have been localised in the central and peripheral nervous system[3,25,29,34,36,45,70,71,73,79]. Therefore, in addition to classical neurotransmitters, neuropeptides in growing number are now strong candidates for neurotransmitter status.

In order to trace neurotransmitter-specific pathways, especially peptide-containing neuropathways, axonal transport combined with immunocytochemistry has been successfully applied[6–11,32,33,35,38–40,60,66, 75,77,82–84,94–96,102]. In principle, all the combination methods are two-step procedures[37]. First, tracers are non-specifically taken up and transported by neurones regardless of their transmitter content. In the second step neurotransmitters or neurotransmitter candidates in the tracer-labelled neurones are demonstrated by immunocytochemistry. The interpretation of the results is based on the assumption that the neurotransmitter present in the cell body is the same as the one stored and released at its terminals. Combination of axonal transport and immunocytochemistry seems to be the most elegant approach for tracing neurotransmitter-specific pathways. The various combinations used for tracing peptide-containing neurones are reviewed here with emphasis on investigation of the innervation of peripheral organs.

Combination of fluorescent retrograde tracing and immunofluorescence

Many fluorescent substances for tracing neuronal connections have been described by Kuypers and his colleagues[1,2,53,54]. For details of general properties of fluorescent tracers see Steward[91]. When combining fluorescent retrograde tracing with immunofluorescence, it was noticed that many of the fluorescent dyes tend to diffuse out of the cells during the immunofluorescence procedure. Only a few dyes, e.g. True Blue, Fast Blue, Primuline and propidium iodide, resist diffusion during immunohistochemistry and are reliable for the visualisation concomitantly with the positive immunofluorescence staining[37,91].

Using retrograde tracing with True Blue fluorescence combined with the indirect immunofluorescence technique for visualising enkephalin-like immunoreactivity, Hökfelt et al.[38] demonstrated immunoreactive neurones in the medulla oblongata projecting to the spinal cord. A simple method for simultaneous localisation of an antigen and retrogradely transported fluorescent dye (True Blue) within single neurones was described in more detail by Sawchenko and Swanson[75]. The method is based on 1) the efficiency of retrograde neuronal labelling with the fluorescent marker True Blue; 2) the near quantitative persistence of retrogradely transported True Blue localisation after subsequent processing of the tissue for immunohistochemistry: 3) the possibility to distinguish clearly between True Blue and immunocytochemically stained cells by simply using appropriate excitation wave-lengths for each. This method has been widely used to trace peptide-containing nerves in both the central nervous system and peripheral organs[30,32,38,40,75,82,84,94-96]. In brief, the procedure is as follows: 1) True blue (2–5% w/v aqueous suspension) is injected with a Hamilton microsyringe into the terminal field to be investigated. Multiple injection sites are suggested in peripheral organs. 2) After optimal survival time (see below) the injected animals are anaesthetised and transcardially perfused with ice-cold 4% paraformaldehyde. 3) Relevant tissues are dissected out, placed in the same fixative for 1 h or overnight and washed in 0.1 M PBS (pH 7.4) containing 15% sucrose for at least 24 h. 4) Frozen sections are cut and processed for indirect immunofluorescence using fluorescein-labelled second antibodies. 5) Sections are examined by fluorescence microscopy using 340–360 nm excitation for True Blue and 450–490 nm for fluorescein. Co-localisation of True Blue and fluorescein may be observed by switching the filter systems without altering the plane of focus (fig.).

True Blue and fluorescein thus provide sensitive, stable and clearly distinguishable labels for retrograde tracing combined with immunofluorescence in properly fixed tissue[75]. If fluorescent retrograde tracing is combined with immunostaining of thin adjacent sections with different antisera or with elution of the first antiserum after photography and restaining with a second antiserum[99], it can be used for tracing neurones with multiple putative transmitters[59].

When interpreting the results of combined retrograde tracing and immunocytochemistry, some facts require careful consideration.

The degree of specificity of retrograde labelling

False-positive labelling may result from leakage of tracers from the injection sites, the dye being transported to other regions by the vascular system, or spreading to contaminate adjacent organs. It is therefore important that care is taken to avoid any leakage of dye during microinjection. The organs to be investigated are isolated with parafilm during the course of injections. After each injection the needle is left in place for at least 1 min, the organs are then rinsed with saline and thoroughly swabbed and the injection sites inspected. Recently a barrier formed from a plastic wound spray (pyroxylin solution, New Skin) was suggested for application to the surface of injected viscera[24].

Fluorescent dye is avidly taken up not only by nerve terminals, but also by damaged and undamaged fibres-of-passage[75]. Although there are obvious problems associated with interpreting the origins of terminal fields, this feature can be used to advantage in some instances, for example, when labelling a peripheral nerve or a fibre tract in the central nervous system[97]. Recently, the fluorescent retrograde tracer SITS (4-acetamido-4'-isothiocyanostilbene-2,2'-disulfonic acid) has been introduced as a cytoplasmic label, which seems to be taken up primarily from axon terminals[41,78].

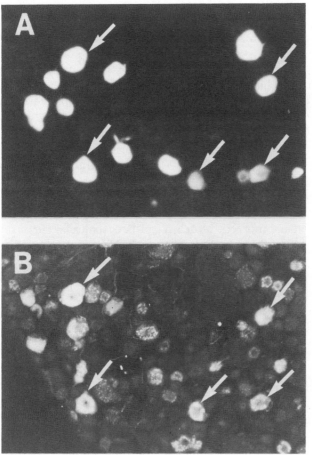

Section from a rat T_9 dorsal root ganglion showing True Blue fluorescence after injection of the dye into the stomach (A). The same field viewed for fluorescein fluorescence showing cells immunostained for CGRP (B). Arrows indicate True Blue retrogradely labelled neuronal cell bodies which display CGRP immunoreactivity. × 157.

Survival time

Proper postinjection survival time is important for optimal neuronal labelling. Survival times must be varied to suit the different tracers used and the system being studied, but they may be divided into three groups, short (less than 48 h), intermediate (2–5 days) and long survival times (5 days to several weeks)[37]. Among the fluorescent tracers which are good for combining with immunocytochemistry, Fast Blue and propidium iodide belong to the group with intermediate survival times, whereas True Blue and Primuline have longer survival times. Although True Blue can be used over a relatively wide range of survival times, it is advisable to determine empirically an optimal time for each system under study.

Visualisation and evaluation of immunostaining

Although experiments have proved that True Blue labelling does not affect the immunostaining, it is possible that some apparently non-immunoreactive cell bodies may contain very low concentrations of neuropeptides which are immunocytochemically undetectable. Immuno-cytochemical visualisation of some peptides requires prior treatment of the animals with colchicine to inhibit axonal transport and cause accumulation of peptides in the cell bodies[15,16,19]. Since colchicine treatment also arrests transport of the retrogradely transported fluorescent dye, these procedures have to be carried out 2–3 days after tracer injection or 24–48 h before sacrifice.

A significant limitation of immunocytochemistry in the study of regulatory peptides as for all other substances is the fact that only material stored in the cell at the moment of death can be visualised. The results do not reflect the biosynthetic activity of the cell. More recently, *in situ* hybridisation has been employed to demonstrate various types of mRNA in cells, including those coding for peptides in neurones[4,26,61,68,86,87,105]. Hybridisation has also been combined with retrograde axonal transport of fluorescent dyes to identify the projections of transmitter-specific neurone populations[74,77].

Combined horseradish peroxidase (HRP) conjugated HRP or other protein tracing with immunocytochemistry

Ljungdahl and his coworkers[60] established a combination of HRP retrograde tracing and immunofluorescence staining of neural pathways. After the optimal postinjection survival time, animals are perfused with fixative and tissues processed. Since the peroxidase-diaminobenzidine reaction product at least partly obscures the immunofluorescence, indirect immunofluorescence is carried out first. The sections are examined under a fluorescence microscope and the stained cells photographed. The coverslips are then removed and the section treated with diaminobenzidine for visualisation of peroxidase. The fluorescence and light micrographs of identified region are then compared. When sections are examined by fluorescence microscopy, it is important to keep the exposure time as short as possible, because the ultraviolet light seems to inactivate the peroxidase.

Successful combination of HRP or WGA-HRP retrograde tracing with peroxidase anti-peroxidase (PAP) immunocytochemistry techniques has been reported by several groups[7-9,57,72,89,103]. Both steps involve a development procedure based on the presence of HRP, but the retrogradely transported HRP appears as punctate granules, whereas the HRP demonstrating immunoreactivity appears as a diffuse homogeneous stain in the cytoplasm[8,103]. Using cobalt chloride to intensify the diaminobenzidine

Anatomically defined peripheral peptide-containing neuropathways

Injection site	Peptide	Localisation of cells containing both tracer and peptides	
Digestive system			
Stomach	Substance P	DRG[82,96]	
	CGRP	DRG[95,96]	NG[95,96]
	NPY	CG[59,95,96]	
Pancreas	Substance P	DRG[82]	
	CGRP	DRG[96]	NG[96]
	NPY	CG[96]	
Colon	Substance P	DRG[96]	
	CGRP	DRG[96]	
	NPY	IMG[96]	
Parotid gland	Substance P	OG[83]	TG[83]
Tongue	CGRP	TG[98]	
Urinary system			
Kidney	Substance P	DRG[51]	
	CGRP	DRG[94]	
Ureter	CGRP	DRG[94]	
Bladder	Substance P	DRG[84]	
	CGRP	DRG[94,107]	
	VIP	PG[30]	DRG[30]
	NPY	PG*	
Urethra	CGRP	DRG[94]	
Female genitalia			
Uterus	CGRP	DRG[42]	
	NPY	PG*	
	VIP	PG[32]	
Vagina	CGRP	DRG[42]	
Preputial gland	CGRP	DRG[42]	
Peripheral ganglia			
IMG	Substance P	DRG (Thoracolumbar)[17]	
	Enkephalin	IML[18]	
Peripheral nerve			
Pelvic nerve	CGRP	DRG[30]	
	VIP	DRG[30,47]	PG[30]
	Substance P	DRG[30,47]	
	Enkephalin	DRG[30,47]	
Pudendal nerve	CGRP	DRG[43]	Motoneurones[43]
	Substance P	DRG[43,47]	
	Enkephalin	DRG[47]	
	VIP	DRG[47]	
Sciatic	CGRP	DRG[43]	Motoneurones[43]
	Substance P	DRG[43]	

CG, coeliac ganglion; DRG, dorsal root ganglia; IMG, inferior mesenteric ganglion; IML, intermediolateral column; NG, nodose ganglia; OG, otic ganglion; PG, pelvic ganglion; TG trigeminal ganglion; *unpublished observations.

product of the first peroxidase reaction, it is possible to increase the colour difference between the retrogradely transported HRP (black) and the antibody-coupled HRP (brown)[6].

However, a disadvantage of these methods is that the granular black reaction product can sometimes obscure a lightly immunostained cell. In contrast, if little HRP has been taken up in the neurones, the reaction product may not be detectable. The most important shortcoming is due to the fact that the appearance of most peptide positive staining does not produce the same homogeneous brown as the conventional transmitters, serotonin and choline acetyltransferase, for which the method was developed[8,103]. Instead, the appearance is granular. This gives rise to much more difficulty in distinguishing retrogradely labelled HRP and the enzyme marking of the immunoreaction. Lechan et al.[57] also pointed out that cells containing the WGA-PAP complex exhibited diminished immunostaining. This was interpreted as being due to steric hindrance to antibody penetration by the WGA accumulation. Furthermore, the fixatives suitable for immunocytochemistry tend to inhibit HRP activity, which is used as an index of the tracer[88]. Thus, the sensitivity of retrograde HRP technique used in this way is less than that in the usual HRP technique. The tetramethyl benzidine (TMB) method is more sensitive than diaminobenzidine (DAB) chromogen[64] and it would be ideal for such combination studies. However, the low pH required to maintain stability of the chromogen and the reaction product is not always compatible with immunocytochemistry. Peschanski and Ralston[69] adapted the chromogen benzidine dihydrochloride (BDHC) for light and electron microscopic tracing studies. They found that excellent sensitivity could be obtained even when the pH was raised above 6.0. At the light microscopical level, the benzidine dihydrochloride reaction product is bluish-green and crystalline. Therefore, using the benzidine dihydrochloride reaction it may be possible to combine HRP/or wheat germ agglutinin HRP retrograde tracing with PAP immunocytochemistry.

More recently, Shiosaka et al.[85] established a sensitive double-labelling technique of retrograde biotinised tracer (biotin-WGA) and immunocytochemistry, which overcame the above disadvantages. Wheat-germ agglutinin conjugated with biotin is used as a tracer. In principle, biotin-WGA labelled cells are visualised with a streptavidin-Texas red conjugate and the antigen localised in neurones is visualised with immunofluorescene. In this method, the detection systems for immunoreactivity and for the tracer (the streptavidin-biotin reaction) are different. There is no interference between the two reactions and the two labels fluoresce with different colours, red and green. Colloidal gold can also be used combined with a tracer. When PAP and the diaminobenzidine reaction are used for immunocytochemistry and colloidal gold particles for detection of the tracer, it is possible to identify double labelled cells, dendrites and axons under the electron microscope. This method may also be used in the peripheral nervous system.

Combined anterograde tracing and immunocytochemistry

A sensitive anterograde tracing method using the lectine, *Phaseolus vulgaris*-leucoagglutinin (PHA-L) as a tracer has been reported and was recently successfully combined with immunocytochemistry to trace chemically specified circuits in the central nervous system by Gerfen and Sawchenko[27,28].

PHA-L is obtained from the red kidney bean. When it is delivered into the central nervous system by iontophoresis, the lectin provides by far the most complete labelling of neurones at the site of injection and is almost exclusively transported in the anterograde direction. Both the transported lectin and the neurotransmitter or related enzyme are visualised in the axons and terminals by a double immunostaining method, which provides a powerful way of tracing antigen-specific projections in the central nervous system. Coexistence of transported PHA-L and neuroactive substance could be demonstrated, for instance, by sequential immunostaining of the same tissue section, after elution of the first reaction product, or by simultaneous double immunofluorescence using primary antisera raised in different species, and localised by non-cross-reacting second antibodies labelled with different fluorochromes. For further details of the procedures see Gerfen and Sawchenko[27,28].

Major disadvantages of the method are: firstly, pressure injections or direct applications of the tracer do not always result in effective anterograde labelling; secondly, it may have only limited use in developing systems, since in early postnatal rats the tracer is invariably taken up by astroglia as well as by neurones. We also do not know yet whether this combination method is able to trace the chemical-specific pathways in the peripheral organs.

Applications and recent discoveries

Retrograde tracing combined with immunocytochemistry has been extensively used to study the projections of peptide-containing neurones in both the central and peripheral nervous system. It is not possible to present all the discoveries here, instead, the principal findings on peripheral peptide-containing neuropathways are reviewed briefly (table).

Sensory substance P and calcitonin gene-related peptide (CGRP) of visceral structures in the rat have been identified and quantified[51,82,84,94,95]. After injections of True Blue into both stomach and pancreas, labelled afferent neurones were visualised in spinal ganglia (mainly at levels T_8–T_{11}) and nodose ganglia. Substance P immunoreactivity was found in 35–50% of gastric spinal afferent and in about 15% of pancreatic spinal afferent neurones[82,96]. Nearly all gastric and pancreatic spinal afferent neurones (about 95%) contain CGRP immunoreactivity[95,96]. Although the nodose ganglia harbour substance P- and CGRP-immunoreactive cell bodies, no True Blue and substance P co-localisation was seen and only a few True Blue-labelled cells contained CGRP immunoreactivity.

The substance P innervation of the rat parotid was also investigated by Sharkey and Templeton[83]. Direct evidence for a dual origin of substance P from the otic and trigeminal ganglia was provided.

Using retrograde tracing and immunocytochemistry it was demonstrated that two segmental groups of dorsal root ganglia at T_{12}-L_2 and L_6-S_1 innervating the bladder contain substance P (10–16%)[84]. Similarly, Kuo et al.[51] demonstrated substance P in renal afferent perikarya. Recently, CGRP immunoreactivity has been shown in a major proportion of afferent neurones supplying the urinary system of the rat[94,107]. Following injection of True Blue into the kidney hilum, ureter or bladder, more than 90% of the True Blue-labelled cells in dorsal root ganglia T_{10}-T_3 contained CGRP-immunoreactive material. Similarly, some 90% of the labelled cells in levels L_6 and S_1 resulting from injections in the ureter and urethra were CGRP-immunoreactive. However, after injection into the wall of the bladder base and dome only 60% of the retrogradely labelled cell bodies in L_6 and S_1 dorsal root ganglia displayed CGRP immunostaining.

It has also been reported that CGRP-immunoreactive nerves in the genitalia of the female rat originate from dorsal root ganglia T_{11}-L_3 and L_6-S_1[42].

A vasoactive intestinal polypeptide (VIP)-containing pathway linking pelvic viscera and the sacral spinal cord has been established by retrograde tracing combined with immunocytochemistry[30,46]. Pelvic visceral afferents with cell bodies in the dorsal root ganglia are a significant source of VIP-containing fibres in the dorsal horn of the sacral spinal cord.

The projection of the postganglionic sympathetic neurones was studied with special reference to the pylorus using a combination of retrograde axonal tracing and indirect immunofluorescence technique[58]. After injection of True Blue into the pyloric sphincter, labelled neurones were found in the coeliac ganglion which also contained NPY- or somatostatin-like immunoreactivity. In elution-restaining experiments, it was established that the majority of these cells were also immunoreactive for tyrosine hydroxylase. Our observations have also proved that extrinsic neuropeptide tyrosine (NPY)-immunoreactive nerves supplying the stomach, pancreas and colon represent postganglionic sympathetic neurones[96].

The origins of the substance P and enkephalin fibres in the inferior mesenteric ganglion have also been characterised by the use of retrograde tracing in combination with indirect immunohistochemistry. Substance P-containing primary sensory neurones[17] and enkephalin-containing sympathetic preganglionic neurones[18] have both been found to project to the inferior mesenteric ganglion.

Retrograde tracing in combination with immunocytochemistry has also been used to analyse the peptide composition of peripheral nerves. For example, among the population of afferent neurones projecting to the pelvic nerve L-enkephalin immunoreactivity was present in 60%, substance P in 30%, and VIP in 15%. Among the population of afferent neurones projecting to the pudendal nerve, L-enkephalin was present in 50%, substance P in 30%, VIP in 3%[47].

Concluding remarks

The combinations of retrograde or anterograde tracing techniques and immunocytochemistry, which have been developed in recent years, have proven their value in defining neuropathways on the basis of their transmitters. However, each of the individual techniques mentioned in this review article has its advantages and drawbacks. The more recently established methods have yet to be applied to the investigation of peripheral organs. Wider use of these methods should provide more data and give guidelines as to which of the newly developed techniques should be chosen for the study of peptide-containing neuropathways.

1 Bentivoglio, M., Kuypers, H. G. J. M., Catsman-Berrevoets, C., and Dann, O., Fluorescent retrograde neuronal labelling in rat by means of substances binding specifically to adenine-thymine rich DNA. Neurosci. Let. *12* (1979) 235–240.

2 Bentivoglio, M., Kuypers, H. G. J. M., and Catsman-Berrevoets, C., Retrograde neuronal labelling by means of hisbenzimide and nuclear yellow (Hoechst, S. 769121): measures to prevent diffusion of the tracers out of retrogradely labelled cells. Neurosci. Lett. *18* (1980) 19–24.

3 Bishop, A. E., Ferri, G. L., Probert, L., Bloom, S. R., and Polak, J. M., Peptidergic nerves, in: Structure of the Gut, pp. 221–237. Eds J. M. Polak, S. R. Bloom, N. A. Wright, and M. J. Daly. Glaxo Group Research Limited, Ware, U.K., 1982.

4 Bloch, B., Le Guellec, D., and de Keyzer, Y., Detection of messenger RNAs coding for the opioid peptide precursors in pituitary and adrenal by 'in situ' hybridisation: study in several mammal species. Neurosci. Lett. *53* (1985) 141–148.

5 Bloom, S. R., and Long, R. G., Radioimmunoassay of Gut Regulatory Peptides. Saunders, London 1982.

6 Bowker, R. M., Westlund, K. N., Sullivan, M. C., and Coulter, J. D., A combined retrograde transport and immunocytochemical staining method for demonstrating the origins of serotonergic projections. J. Histochem. Cytochem. *30* (1982) 805–810.

7 Bowker, R. M., Steinbusch, H., and Coulter, J. D., Serotonin and peptidergic projections to the spinal cord demonstrated by a combined retrograde HRP histochemical and immunocytochemical staining method. Brain Res. *211* (1981) 412–417.

8 Bowker, R. M., Westlund, K. N., and Coulter, J. D., Serotonergic projections to the spinal cord from midbrain of the rat: an immunocytochemical and retrograde transport study. Neurosci. Lett. *24* (1981) 221–226.

9 Bowker, R. M., Westlund, K. N., and Coulter, J. D., Origins of serotonergic projections to the spinal cord in rat: an immunocytochemical-retrograde transport study. Brain Res. *226* (1981) 187–199.

10 Bowker, R. M., Westlund, K. N., and Coulter, J. D., Origins of serotonergic projections in the lumbal spinal cord in the monkey using a combined retrograde transport and immunocytochemical technique. Brain Res. Bull. *9* (1982) 271–278.

11 Brann, M. R., and Emson, P. C., Microiontophoretic injection of fluorescent tracer combined with simultaneous immunofluorescence histochemistry for the demonstration of

108

efferents from the caudate putamen projecting to the globus pallidus. Neurosci. Lett. *16* (1980) 61–66.

12 Coons, A. H., Fluorescent antibody methods, in: General Cytochemical Methods. pp. 399–422. Ed. J. F. Danielli. Academic Press, New York 1958.

13 Coons, A. H., Leduc, E. H., and Connolly, J. M., Studies on antibody production. I. A method for the histochemical demonstration of specific antibody and its application to a study of the hyperimmune rabbit. J. exp. Med. *102* (1955) 42–60.

14 Cowan, W. M., Gottlieb, D. L., Hendrickson, A. E., Price, J. L., and Woolsey, T. A., The autoradiographic demonstration of axonal connections in the central nervous system. Brain Res. *37* (1972) 21–51.

15 Dahlström, A., Effect of vinblastine and colchicine on monoamine-containing neurons of the rat, with special regard to axoplasmic transport of amine granules. Acta neuropath. *5* (1971) 226–237.

16 Dahlström, A., Heiwall, P. O., and Häggendal, J., Effect of antimitotic drugs on the intraaxonal transport of neurotransmitters in rat adrenergic and cholinergic nerve. Ann. N.Y. Acad. Sci. *253* (1975) 507–516.

17 Dalsgaard, C.-J., Hökfelt, T., Elfvin, L.-G., Skirboll, L., and Emson, P., Substance P-containing primary sensory neurons projecting to the inferior mesenteric ganglion: evidence from combined retrograde tracing and immunohistochemistry. Neuroscience *7* (1982) 647–654.

18 Dalsgaard, C.-J., Hökfelt, T., Elfvin, L.-G., and Terenius, L., Enkephalin-containing sympathetic preganglionic neurons projecting to the inferior mesenteric ganglia: evidence from combined retrograde tracing and immunohistochemistry. Neuroscience *7* (1982) 2039–2050.

19 Dube, D., and Pelletier, G., Effect of colchicine on the immunohistochemical localisation of somatostatin in the rat brain: light and electron microscopic studies, J. Histochem. Cytochem. *27* (1979) 1577–1582.

20 Edwards, S. B., and De Olmos, J. S., Autoradiographic studies of the midbrain reticular formation: ascending projections of nucleus cuneiformis. J. comp. Neurol. *165* (1976) 417–432.

21 Edwards, S. B., and Hendrickson, A., The autoradiographic tracing of axonal connections in the central nervous system, in: Neuroanatomical Tract Tracing Methods, pp. 171–205. Eds L. Heimer and M. J. RoBards. Plenum Press, New York 1983.

22 Edwards, S. B., The ascending and descending projections of the red nucleus in the cat: An experimental study using an autoradiographic tracing method. Brain Res. *48* (1972) 45–63.

23 Fillenz, M., Gagnon, C., Stockel, K., and Thoenen, H., Selective uptake and retrograde axonal transport of dopamine-β-hydroxylase antibodies in peripheral adrenergic neurons. Brain Res. *14* (1976) 293–303.

24 Fox, E. A., and Powley, T. L., Tracer diffusion has exaggerated CNS maps of direct preganglionic innervation of pancreas. J. auton. nerv. Syst. *15* (1986) 55–69.

25 Furness, J. B., and Costa, M., Types of nerves in the enteric nervous system. Neuroscience *5* (1980) 1–20.

26 Gee, C. E., Chen, C. L. C., Roberts, J. L., Thompson, R., and Watson, S. J., Identification of proopiomelanocortin neurons in rat hypothalamus by in situ cDNA-mRNA hybridization. Nature *306* (1983) 374–376.

27 Gerfen, C., and Sawchenko, P. E., A method for anterograde axonal tracing of chemically specified circuits in the central nervous system: combined Phaseolus vulgaris-leucoagglutinin (PHA-L) tract tracing and immunohistochemistry. Brain Res. *343* (1985) 144–150.

28 Gerfen, C. R., and Sawchenko, P., An anterograde neuroanatomical tracing method that shows the detailed morphology of neurons, their axons and terminals: immunohistochemical localisation of an axonally transported plant lectin, Phaseolus vulgaris leucoagglutinin (PHA-L) Brain Res. *290* (1984) 219–126.

29 Gibson, S. J., and Polak, J. M., Neurochemistry of the spinal cord, in: Immunocytochemistry, 2nd edn, pp. 360–389. Eds J. M. Polak and S. Van Noorden. Wright, Bristol 1986.

30 Gibson, S. J., Polak, J. M., Anand, P., Blank, M. A., Yiangou, Y., Su, H. C., Terenghi, G., Katagiri, T., Morrison, J. F. B., Lumb, B. M., Inyama, C., and Bloom, S. R., A VIP/PHI-containing pathway links urinary bladder and sacral spinal cord. Peptides *7* (1986) 205–219.

31 Gonatas, N. K., Harper, C., Mizutani, T., and Gonatas, J. O., Superior sensitivity of conjugates of horseradish peroxidase with wheat germ agglutinin for studies of retrograde axonal transport. J. Histochem. Cytochem, *27* (1979) 728–734.

32 Gu, J., Polak, J. M., Su, H. C., Blank, M. A., Morrison, J. F. B., and Bloom, S. R., Demonstration of paracervical ganglion origin for the vasoactive intestinal peptide-containing nerves of the rat uterus using retrograde tracing techniques combined with the immunocytochemistry and denervation procedures. Neurosci. Lett. *51* (1984) 377–382.

33 Hökfelt, T., Phillipson, O., and Goldstein, M., Evidence for a dopaminergic pathway in the rat descending from the AII cell group to the spinal cord. Acta physiol. scand. *107* (1979) 393–395.

34 Hökfelt, T., Johansson, O., Ljungdahl, A., Lundberg, J. M., and Schultzberg, M., Peptidergic neurons. Nature *284* (1980) 515–521.

35 Hökfelt, T., Skirboll, J., and Rehfeld, J. F., A subpopulation of mesencephalic dopamine neurons projecting to limbic areas contains a cholecystokinin-like peptide: Evidence from immunohistochemistry combined with retrograde tracing. Neuroscience *5* (1980) 2093–2124.

36 Hökfelt, T., Lundberg, J. M., Schultzberg, M., Johansson, O., Skirboll, L., Änggard, A., Fredholm, B., Hamberger, B., Pernow, B., Rehfeld, J., and Goldstein, M., Cellular localization of peptides in neural structures. Proc. R. Soc. Lond. B210 (1980) 63–77.

37 Hökfelt, T., Skagerberg, G., Skirboll, L., and Björklund, A., Combination of retrograde tracing and neurotransmitter histochemistry, in: Handbook of Chemical Neuroanatomy, vol. 1: Methods in Chemical Neuroanatomy, pp. 228–285. Eds A. Björklund and T. Hökfelt. Elsevier Science Publisher, B. V., 1983.

38 Hökfelt, T., Terenius, L., Kuypers, H. G. J. M., and Dann, O., Evidence for enkephalin immunoreactive neurons in the medulla oblongata projecting to the spinal cord. Neurosci. Lett. *14* (1979) 55–60.

39 Holets, V. R., and Elde, R., The differential distribution and relationship of serotoninergic and paptidergic fibres of sympathoadrenal neurons in the intermediolateral cell column of the rat: A combined retrograde axonal transport and immunofluorescence study. Neuroscience *7* (1982) 1155–1174.

40 Holets, V. R., Mullett, T. J., and Elde, R. P., Simultaneous use of the retrograde tracer True Blue and immunohistochemistry: innervation of the rat adrenal medulla. Soc. Neurosci., Abstr. *6* (1980) 336.

41 Huisman, A. M., Kuypers, H. G. J. M., and Ververs, B., Retrograde neuronal labelling of cells of origin of descending brainstem pathways in rat using SITS as a retrograde tracer. Brain Res. *289* (1983) 305–310.

42 Inyama, C. O., Wharton, J., Su, H. C., and Polak, J. M., CGRP-immunoreactive nerves in the genitalia of the female rat originate from dorsal root ganglia T_{11}-L_3 and L_6-S_1: a combined immunocytochemical and retrograde tracing study. Neurosci. Lett. *69* (1986) 13–18.

43 Katagiri, T., Gibson, S. J., Su, H. C., and Polak, J. M., Composition and central projections of the pudendal nerve in the rat investigated by combined peptide immunocytochemistry and retrograde fluorescent labelling. Brain Res *372* (1986) 313–322.

44 Katz, L. C., Burkhatter, A., and Dreyer, W. J., Fluorescent latex microspheres as a retrograde neuronal marker for the in vivo and in vitro studies of visual cortex. Nature *310* (1984) 498–500.

45 Kawatani, M., Lowe, I., Mossy, J., Martinez, J., Nadelhaft, I., Eskay, R., and de Groat, W. C., Vasoactive intestinal polypeptide (VIP) is localised to the lumbosacral segments of the human spinal cord. Soc. Neurosci. Abstr. *9* (1983) 294.

46 Kawatani, M., Lowe, I. P., Nadelhaft, I., Morgan, C., and de Groat, W. C., Vasoactive intestinal polypeptide in visceral afferent pathways to the sacral spinal cord of the cat. Neurosci. Lett. *42* (1983) 311–316.

47 Kawatani, M., Nagel, J., Houston, M. B., Eskay, R., Lowe, I. P., and de Groat, W. C., Identification of leucine-enkephalin and other neuropeptides in pelvic and pudendal afferent pathways to the spinal cord of the cat. Soc. Neurosci. Abstr. *10* (1984) 589.

48 Keizer, K., Kuypers, H. G. J. M., Huisman, A. M., and Dann, O., Diamidino Yellow Dihydrochloride (DY 2HCl); a new fluorescent retrograde neuronal tracer which migrates only very slowly out of the cell. Exp. Brain Res. *51* (1983) 179–191.

49 Kristensson, K., Transport of fluorescent protein tracer in peripheral nerves. Acta neuropath. *16* (1970) 293–300.

50 Kristensson, K., and Olsson, Y., Retrograde axonal transport of protein. Brain Res. *27* (1971) 363–365.

110

51 Kuo, D. C., Oravitz, J. J., Eskay, R., and de Groat, W., Substance P in renal afferent perikarya identified by retrograde transport of fluorescent dye. Brain Res. *323* (1984) 167–171.

52 Kuypers, H. G. J. M., Bentivoglio, M., Catsman-Berrevoets, C. E., and Bharos, A. T., Double retrograde neuronal labelling through divergent axon collaterals, using two fluorescent tracers with the same excitation wavelength which label different features of the cell. Exp. Brain Res. *40* (1980) 383–392.

53 Kupers, H. G. J. M., Bentivoglio, M., Van der Kooy, D., and Catsman-Berrevoets, C. E., Retrograde transport of bisbenzimide and propidium iodide through axons to their parent cell bodies. Neurosci. Lett. *12* (1977) 1–7.

54 Kuypers, H. G. J. M., Bentivoglio, M., Van der Kooy, D., and Catsman-Berrevoets, C. E., Retrograde axonal transport of fluorescent substances in the rat's forebrain. Neurosci. Lett. *6* (1979) 127–135.

55 La Vail, J. H., and La Vail, M. M., Retrograde axonal transport in the central nervous system. Science *176* (1972) 1416–1417.

56 La Vail, J. H., and La Vail, M. M., The retrograde intraaxonal transport of horseradish peroxidase in the chick visual system: A light and electron microscopic study. J. comp. Neurol. *157* (1974) 303–358.

57 Lechan, D. M., Nestler, J., and Jacobsen, S. J., Immunohistochemical localisation of retrogradely and anterogradely transported wheat germ agglutinin (WGA) within the central nervous system of the rat: application to immunostaining of a second antigen within the same neuron. J. Histochem. Cytochem. *29* (1981) 255–262.

58 Lindh, B., Dalsgaard, C.-J., Elfvin, L.-G., Hökfelt, T., and Cuello, A. C., Evidence of substance P immunoreactive neurons in dorsal root ganglia and vagal ganglia projecting to the guinea pig pylorus. Brain Res. *269* (1983) 365–369.

59 Lindh, B., Hökfelt, T., Elfvin, L.-G., Terenius, L., Fahrenkrug, J., Elde, R., and Goldstein, M., Topography of NPY-, somatostatin-, and VIP-immunoreactive; neuronal subpopulations in the guinea pig celiac-superior mesenteric ganglion and their projection to the pylorus. J. Neurosci. *6* (1986) 2371–2383.

60 Ljungdahl, A., Hökfelt, T., Goldstein, M., and Park, D., Retrograde peroxidase tracing of neurons combined with transmitter histochemistry. Brain Res. *84* (1975) 313–319.

61 McCabe, J. T., Morrell, J. I., Ivell, R., Schmale, H., Richter, D., and Pfaff, D. W., In situ hybridization technique to localize rRNA and mRNA in mammalian neurons. J. Histochem. Cytochem. *34* (1986) 45–50.

62 Menetrey, D., Retrograde tracing of neural pathways with a protein-gold complex. I. Light microscopic detection after silver intensification. Histochemistry *83* (1985) 391–395.

63 Menetrey, D., and Lee, C. L., Retrograde tracing of neural pathways with a protein gold complex. II. Electron microscopic demonstration of projections and collaterals. Histochemistry *83* (1985) 525–530.

64 Mesulam, M.-M., Tetramethyl benzidine for horseradish peroxidase neurohistochemistry: a non-carcinogenic blue reaction-product with superior sensitivity for visualizing neural afferents and efferents. J. Histochem. Cytochem. *26* (1978) 106–117.

65 Mesulam, M.-M., Principles of horseradish peroxidase neurohistochemistry and their applications for tracing neural pathways—axonal transport, enzyme histochemistry and light microscopic analysis, in: Tracing Neural Connections with Horseradish Peroxidase, pp. 1–151. Ed. M.-M. Mesulam. John Wiley and Sons, New York 1982.

66 Nahin, R. L., and Micevych, P. E., A long ascending pathway of enkephalin-like immunoreactive spinoreticular neurons in the rat. Neurosci. Lett. *65* (1986) 271–276.

67 Nakane, P. K., and Pierce, G. B. Jr, Enzyme labelled antibodies for the light and electron microscopic localisation of tissue antigens. J. Cell Biol. *33* (1967) 307–318.

68 Nojiri, H., Sato, M., and Urano, A., In situ hybridization of the vasopressin mRNA in the rat hypothalamus by use of a synthetic oligonucleotide probe. Neurosci. Lett. *58* (1985) 101–105.

69 Peschanski, M., and Ralston, H. J. III. Light and electron microscopic evidence of transneuronal labelling with WGA-HRP to trace somatosensory pathways to the thalamus. J. comp. Neurol. *236* (1985) 29–41.

70 Polak, J. M., and Bloom, S. R., Immunocytochemistry of the diffuse neuroendocrine system, in: Immunocytochemistry, 2nd edn, pp. 328–348. Eds J. M. Polak and S. Van Noorden. Wright, Bristol 1986.

71 Polak, J. M., and Bloom, S. R., Peptidergic nerves of the gastrointestinal tract. Invest. Cell Path. *1* (1978) 301–326.

72 Priestley, J. V., Somogyi, P., and Cuello. C., Neurotransmitter specific projection neurons revealed by combined immunohistochemistry with retrograde transport of HRP. Brain Res. *320* (1981) 231–240.

73 Roberts, G. W., and Allen, Y. S., Immunocytochemistry of brain neuropeptides, in: Immunocytochemistry. 2nd edn. pp. 349–359. Eds J. M. Polak and S. Van Noorden. Wright, Bristol 1986.

74 Roberts, J. L., and Wilcox, J. N., Hybridization histochemistry: identification of specific mRNAs in individual cells. Immunocytochemistry, in: 2nd edn. pp. 198–204. Eds J. M. Polak and S. Van Noorden. Wright, Bristol 1986.

75 Sawchenko, P. E., and Swanson, L. W., A method for tracing biochemically defined pathways in the central nervous system using combined fluorescence retrograde transport and immunohistochemical techniques. Brain Res. *210* (1981) 31–41.

76 Sawchenko, P. E., and Gerfen, C. R., Plant lectins and bacterial toxins as tools for tracing neuronal connections. TINS *8* (1985) 378–384.

77 Schalling, M., Hökfelt, T., Wallace, B., Goldstein, M., Filer, D., Yamin, C., and Schlesinger, D. H., Tyrosine 3-hydroxylase in rat brain and adrenal medulla: Hybridization histochemistry and immunohistochemistry combined with retrograde tracing. Proc. natn. Acad. Sci. USA *83* (1986) 6208–6212.

78 Schmued, L. C., and Swanson, L. W., SITS: a covalently bound fluorescent retrograde tracer does not appear to be taken up by fibres of passage. Brain Res. *249* (1982) 137–141.

79 Schultzberg, M., Hökfelt, T., Nilsson, G., Terenius, L., Rehfeld, J. F., Brown, M., Elde, R., Goldstein, M., and Said, S., Distribution of peptide- and catecholamine-containing neurons in the gastrointestinal tract of rat and guinea pig: immunohistochemical studies with antisera to substance P, vasoactive intestinal polypeptide, enkephalins, somatostatin, gastrin/cholecystokinin, neurotensin and dopamine β-hydroxylase. Neuroscience *5* (1980) 689–744.

80 Schwab, M. E., and Thoenen, H., Selective binding, uptake, and retrograde transport of tetanus toxin by nerve terminals in the rat iris: An electron microscope study using colloidal gold as a tracer. J. Cell Biol. *77* (1978) 1–13.

81 Schwab, M. E., Javoy-Agid, F., and Agid, Y., Labelled wheat germ agglutinin (WGA) as a new, highly sensitive retrograde tracer in the rat brain hippocampal system. Brain Res. *152* (1978) 145–150.

82 Sharkey, K. A., Williams, R. G., and Dockray, G. J., Sensory substance P innervation of the stomach and pancreas. Demonstration of capsaicin-sensitive sensory neurons in the rat by combined immunohistochemistry and retrograde tracing. Gastroenterology *87* (1984) 814–921.

83 Sharkey, K. A., and Templeton, D., Substance P in the rat parotid gland: evidence for a dual origin from the otic and trigeminal ganglia. Brain Res. *304* (1984) 392–396.

84 Sharkey, K. A., Williams, R. G., Schutzberg, M., and Dockray, G. J., Sensory substance P-innervation of the urinary bladder: possible site of action of capsaicin in rats. Neuroscience *10* (1983) 861–868.

85 Shiosaka, S., Shimada, S., and Tohyama, M., Sensitive double-labelling technique of retrograde biotinized tracer (biotin-WGA) and immunocytochemistry: light and electron microscopic analysis. J. Neurosci. Meth. *16* (1986) 9–18.

86 Shivers, B. D., Harlan, R. E., Pfaff, D. W., and Schachter, B. S., Combination of immunocytochemistry and in situ hybridization in the same tissue section of rat pituitary. J. Histochem. Cytochem. *34* (1986) 39–43.

87 Siegel, R. E., and Young, W. S. III, Detection of preprocholecystokinin and preproenkephalin A mRNAs in rat brain by hybridization histochemistry using complementary RNA probes. Neuropeptides *6* (1985) 573–580.

88 Smolen, A. J., Glazer, E. J., and Ross, L. L., Horseradish peroxidase histochemistry combined with glyoxylic acid-induced fluorescence used to identify brain stem catecholaminergic neurons which project to the chick thoracic spinal cord. Brain Res. *160* (1979) 353–357.

89 Sofroniew, M. V., and Schrell. U., Evidence for a direct projection from oxytocin and vasopressin neurons in the hypothalamic paraventricular nucleus to the medulla oblongata: immunohistochemical visualization of both the horseradish peroxidase transported and the peptide produced by the same neurons. Neurosci. Lett. *22* (1981) 211–217.

112

90 Sternberger, L. A., Hardy, P. H. Jr, Cuentis, J. J., and Meyer, H. G., The unlabelled antibody enzyme method of immunohistochemistry. Preparations and properties of soluble antigen-antibody complex (horseradish peroxidase-antihorseradish peroxidase) and its use in identification of spirochetes. J. Histochem. Cytochem. *18* (1970) 315–324.

91 Steward, O., Horseradish peroxidase and fluorescent substances and their combination with other techniques, in: Neuroanatomical Tract-Tracing Methods, pp. 279–310. Eds L. Heimer and M. J. Robards. Plenum Press, New York and London 1983.

92 Stockel, K., Schwab, M. E., and Thoenen, H., Role of gangliosides in the uptake and retrograde axonal transport of cholera and tetanus toxin as compared to nerve growth factor and wheat germ agglutinin. Brain Res. *132* (1977) 273–285.

93 Stockel, K., Schwab, M., and Thoenen, H., Comparison between the retrograde axonal transport of nerve growth factor and tetanus toxin in motor, sensory and adrenergic neurons. Brain Res. *99* (1975) 1–16.

94 Su, H. C., Wharton, J., Polak, J. M., Mulderry, P. K., Ghatei, M. A., Gibson. S. J., Terenghi, G., Morrison, J. F. B., Ballesta, J., and Bloom, S. R., Calcitonin gene-related peptide-immunoreactivity in afferent neurones supplying the urinary tract: combined retrograde tracing and immunocytochemistry. Neuroscience *18* (1986) 727–747.

95 Su, H. C., Bishop, A. E., Gibson, S. J., and Polak, J. M., The origins and nature of the CGRP- and NPY-immunoreactive nerves of the rat stomach determined by retrograde tracing and denervation procedures combined with immunocytochemistry. Reg. Peptides *13* (1985) 66.

96 Su, H. C., Bishop, A. E., Power, R. F., Hamada, Y., and Polak, J. M., Dual intrinsic and extrinsic origins of CGRP- and NYP-immunoreactive nerves of rat gut and pancreas. J. Neurosci. *7* (1987) 2674–2687.

97 Swanson, L. W., and Sawchenko, P. E., Tracing pathways with combined axonal transport and immunohistochemical methods at the light microscopic level. Society of Neuroscience, Short course I, Modern Neuroanatomical methods, pp. 2–17, 1985.

98 Terenghi, G., Polak, J. M., Rodrigo, J., Mulderry, P. K., and Bloom, S. R., Calcitonin gene-related peptide-immunoreactive nerves in the tongue, epiglottis and pharynx of the rat: Occurrence, distribution and origin. Brain Res. *365* (1986) 1–14.

99 Tramu, G., Pillez. A., and Leonardelli, J., An efficient method of antibody elution for the successive or simultaneous localization of two antigens by immunocytochemistry. J. Histochem. Cytochem. *26* (1978) 322–324.

100 Turner, P. T., and Harris, A. B., Ultrastructure of exogenous peroxidase in cerebral cortex. Brain Res. *74* (1974) 305–326.

101 Van de Kooy, D., The organisation of the thalamic, nigral and raphe afferents to the medial versus lateral caudate putamen in rat. A fluorescent retrograde double labelling study. Brain Res. *169* (1979) 381–387.

102 Van der Kooy, D., and Steinbusch, H. W. M., Simultaneous fluorescent retrograde axonal tracing and immunofluorescent characterization of neurons. J. Neurosci. Res. *5* (1980) 479–584.

103 Wainer, B. H., and Rye, D. B., Retrograde horseradish peroxidase tracing combined with localization of choline acetyltransferase immunoreactivity. J. Histochem. Cytochem. *32* (1984) 439–443.

104 Wan, X. S. T., Trojanowski, J. Q., and Gonatas, K. O., Cholera toxin and wheat germ agglutinin conjugates as neuroanatomical probes: their uptake and clearance, transganglionic and retrograde transport and sensitivity. Brain Res. *243* (1982) 215–224.

105 Wolfson, B., Manning, R. W., David, L. G., Arentzen, R., and Baldino, F. Jr, Colocalization of corticotropin releasing factor and vasopressin mRNA in neurons after adrenalectomy. Nature *315* (1985) 59–61.

106 Yalow, R. S., Radioimmunoassay: a probe for the fine structure of biologic system. Science *200* (1978) 1236–1245.

107 Yokokawa, K., Tohyama, M., Shiosaka, S., Shiotani, Y., Sonada, T., Emson, P. C., Hillyard, C. V., Girgis, S., and MacIntyre, I., Distribution of calcitonin gene-related peptide-containing fibers in the urinary bladder of the rat and their origin. Cell Tiss. Res. *244* (1986) 271–278.

The use of cell and tissue culture techniques in the study of regulatory peptides

C. J. S. Hassall, T. G. J. Allen, B. S. Pittam and G. Burnstock

Summary. Cell and tissue culture preparations have a number of general advantages for the study of biological processes: cells are more accessible for study, diffusion delays and barriers to applied substances are minimised, the humoral and cellular components of the culture environment can be controlled and progressive changes in intracellular and intercellular events can be directly monitored. These significant advantages mean that culture preparations can provide unique opportunities for investigation of the properties and functions of regulatory peptides. Culture preparations also have disadvantages and not all cultures are suitable for use in all types of experiments; therefore, the choice of preparation must be made accordingly. Here we describe different types of culture preparation and give examples where cultures have been used to examine peptide synthesis, storage, secretion and receptor localisation, as well as the short-term and trophic actions of regulatory peptides.

The isolation and growth of cells in culture preparations has been an invaluable way of answering questions that would be difficult to resolve *in situ*, either because of the complexity of whole animals or the inaccessibility of many cells to experimental techniques. Tissue and cell cultures have been used in a wide variety of experiments to determine the fundamental properties of different cell types and the factors involved in the regulation of their development and differentiation.

The first use of tissue culture techniques was made by Harrison[59], who maintained frog neural tube cells *in vitro* for long enough to observe that individual neurones extend neurites, thereby disproving an alternative hypothesis that nerve fibres originated from the fusion of many cells. Subsequently, cell and tissue culture techniques and their applications have become very diverse[1,38,41,42,111], but some fundamental principles still apply. Firstly, culture preparations generally enable more direct study of specific cell types, their interactions and their properties. Secondly, most culture preparations permit experiments to be carried out under conditions where any diffusion delays or barriers to drugs are minimised. Thirdly, and perhaps most importantly, both the humoral and cellular components of the culture conditions may be controlled so that they are subject to less variation and are far more easily manipulated than the environment of cells *in vivo*. However, on this point it should be noted that most conventional growth media are supplemented with serum and/ or tissue extracts, the constituents of which are often variable, undefined and may, for example, act as attachment factors, mitogens or affect cellular differentiation[6,7,8,13,76,121,137]. Therefore, to permit reproducible control of a culture medium, it should be chemically defined, but it must support cell growth and it should also be permissive for the expression of all cellular properties without exerting any influence to alter normal, developmental changes. Unfortunately, this type of medium has yet to be developed for most types of culture preparations.

Although appropriate culture preparations can be used as model systems to give results that may be directly related to the equivalent, more complex and inaccessible *in vivo* preparations, a valid argument against the use of cultures is that the observations made may be the result of the artificial conditions of culture. For instance, the culture environment may only be suitable for some of the cell types present in the tissue of origin, resulting in the selective death of the others; or it may instruct changes in the phenotype of an individual subpopulation of cells. Consequently, the homology between culture and in vivo preparations should be determined where possible. On the other hand, as discussed later, the findings of experiments in which the cells either display specific requirements for growth, or phenotypic alterations, are very interesting since they could lead to the discovery of the environmental factors regulating particular aspects of cellular development and differentiation *in vivo*.

The types of culture preparations available for study of regulatory peptides are described in the next section. The rest of this paper is concerned with the specific applications of tissue and cell culture techniques to investigations of regulatory peptides. The literature on peptides and cultured cells is extensive and will not be reviewed in full. Instead, we have chosen examples to illustrate the potential and the problems of the use of culture preparations in the study of regulatory peptides. The sites of regulatory peptide synthesis, storage and receptors, as well as the short-term, electrophysiological and long-term, trophic effects of peptides will be discussed.

Types of culture preparation

There are a number of ways of maintaining cells in culture: they can be grown in suspension in a liquid medium, allowed to attach to a substrate which is then bathed in growth medium, or embedded in a gel containing the necessary nutrients for growth[41,42,111]. Each of these conditions for the maintenance of cells in culture, as well as the various types of culture preparation that have been developed, have inherent advantages and disadvantages. Therefore it is necessary to carefully select the most suitable for use in each investigation. The main characteristics of the general types of culture preparation are briefly summarised here.

Organ cultures, as the name suggests, consist of the entire organ maintained in isolation, accordingly the relationships of the organ with other tissues is disrupted, but its own anatomy and functional integrity may be retained[38,41,42,106]. The disadvantages associated with such preparations include poor gaseous and nutrient exchange in the centre of the organ which may lead to cell death and prevent long-term growth. Also, cells may not be particularly accessible or visible, so that studies of individual cell types or cell relationships are difficult to carry out.

Organotypic culture preparations contain selected regions of the tissues typically present in an organ[35,36,99,124]. The advantages and disadvantages of this type of culture are similar to those of organ cultures; however, because smaller pieces of tissue are used, there is less cell death at the centre of the culture.

Explant cultures are made up of fragments of tissue or whole peripheral ganglia[24,26,27,30,74] which can be maintained in culture for extended periods of time because their small size facilitates nutrient and gaseous exchange. In addition, many of the cells in the explant tend to migrate outwards, causing the explant to flatten with time and this facilitates the study of cell interactions in particular. Although the migration disrupts the organisation of the tissue, several investigations have shown that the cells retain many of their differentiated properties[3,26,27,75]. However, organ, organotypic and explant culture preparations all consist of mixed cell types so they would not be suitable for studies of the precise requirements or characteristics of an individual cell type.

Dispersed cell culture preparations consist of single, or small groups of cells dissociated from organs or tissues which can be employed in a variety of different types of investigation[30,38,105,139]. Using such preparations, biochemical studies can be carried out on relatively homogeneous, purified populations of cells. Also, it is possible to identify and observe individual cells directly in dissociated cultures to study their morphology, to monitor their behaviour, what influences them and the way that they interact with each other over an extended period of time. However, the main disadvantages connected with the use of this type of culture preparation are that the relatively vigorous dissociation procedures required to disperse the tissue, and which often involve mechanical and/or enzymatic treatments, not only disrupt tissue organisation totally, but may also result in a low or inconsistent yield of cells and variable proportions of the different cell types present. Furthermore, these cells may have been damaged or altered by the preparative procedures alone.

Continuous cell lines have similar advantages to dissociated primary cell cultures and can provide large numbers of cells for experimentation, but their properties in culture may be more difficult to relate to the tissue of origin because they consist of transformed cells capable of permanent growth and division in culture[38,41,42,47,49].

**Localisation of the sites of regulatory peptide synthesis,
storage and receptors on cultured cells**

The ability to maintain a large variety of different cell types in culture,
together with recent technical advances (see this volume) has facilitated
many detailed studies of regulatory peptides. For example: organotypic
culture preparations of sections of the gut wall, explants of myenteric
ganglia and dissociated gut preparations containing myenteric neurones
have all been used to localise a number of neuropeptides in enteric neurones in culture[66,73,75,105,122,124]. Dissociated cell cultures have also been
used to demonstrate that some of the neuropeptides found in nerve fibres
in the heart, bladder and trachea *in situ* are also present in cultured
intramural neurones, which indicates that a proportion of the immunoreactive neuropeptides observed *in situ* are of intrinsic, rather than
extrinsic origin in these tissues[20,21,34,60,61] (fig. 1). A culture preparation
has been used to show that calcitonin gene-related peptide is released
from trigeminal ganglion cells[97]; while the coexistence of a neuropeptide
with another neuroactive molecule in the same neurone has been demonstrated using both explant and dissociated cell cultures[11,62] (fig. 1). Some
of the particular uses of culture preparations, together with their advantages and disadvantages, for the study of regulatory peptide synthesis,
storage and receptors are considered below.

The improved accessibility of cells in culture to experimental techniques

Since a number of techniques can be applied concurrently to the same
culture preparation, several related questions can be asked about a particular peptide or cell type at the same time. This approach has enabled
the unequivocal demonstration of the synthesis, storage and secretion of
atrial natriuretic peptide (ANP) by ventricular myocytes, as well as atrial
myocytes from newborn rat heart maintained in dissociated cell culture
preparations[10,23,48,51,65,126,129,140]. Hamid et al.[51] and Hassall et al.[65] employed several techniques: *in situ* hybridisation was used to localise ANP
mRNA to both atrial and ventricular myocytes in culture[51]; ANP was
visualised in the cultured myocytes (fig. 2) and was found to be associated
with the specific granules present in the cells by immunocytochemistry at
the light and electron microscopic levels respectively; the amount of ANP
stored in the myocytes in culture and secreted into their growth medium
was measured by a specific radioimmunoassay, while the molecular form
of the stored and secreted peptide was determined by gel column chromatography[65]. The results of these studies on the cell culture preparations
from the two different areas of the heart have shown that the gene for
ANP is expressed in the same cells that store and release ANP under the
conditions of culture. They also provide evidence for the

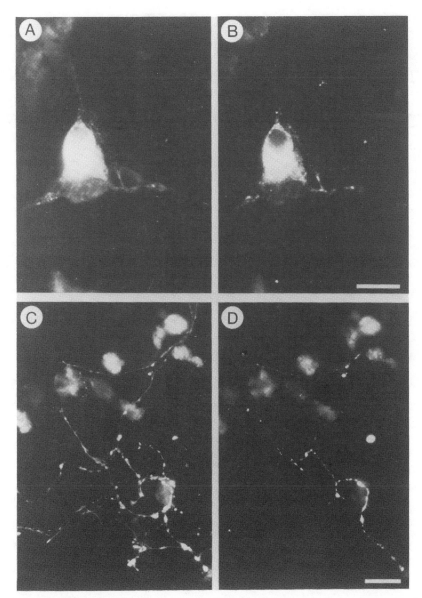

Figure 1. Eight-day-old culture containing newborn guinea-pig intracardiac neurones grown in serum-free, defined medium, supplemented with 100 μM 5-HT for 2 h prior to fixation and double immunostaining for 5-HT (*A*, *C*) and NPY (*B*, *D*). *A*, *B*: the colocalisation of 5-HT-immunoreactivity (*A*) and NPY-immunoreactivity (*B*) in the same mononucleate neuronal cell body. *C*, *D*: a different field of the same culture showing varicose 5-HT-immunoreactive neurites (*C*), only one of which contains NPY-immunoreactivity (*D*). Bars = 25 μm. (From Hassall and Burnstock[63]. With permission.)

118

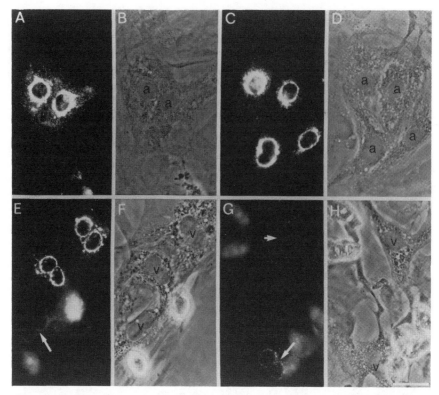

Figure 2. Fluorescence and phase-contrast micrographs showing ANP-like immunoreactive rat atrial (*A–D*) and ventricular (*E–H*) myocytes in 10-day cultures maintained in either serum-supplemented medium (199-FCS; *A, B, E, F*) or in serum-free, hormone-supplemented medium (199-N1; *C, D, G, H*). *A* Fluorescence and *B* phase-contrast micrographs of the same field showing two intensely ANP-like immunoreactive mononucleate atrial myocytes (a) grown in medium 199-FCS. *C* Fluorescence and *D* phase-contrast micrographs of four bright ANP-like-immunoreactive atrial myocytes (a) maintained in medium 199-N1. *E* Fluorescence and *F* phase-contrast micrographs of the same field in a culture grown in medium 199-FCS showing bright ANP-like immunoreactivity in two binucleate ventricular myocytes (v) and weak, perinuclear immunofluorescence in a third binucleate ventricular myocyte (arrow). *G* Fluorescence and *H* phase-contrast micrographs of two binucleate ventricular myocytes (v), one of which displays weak ANP-like immunoreactivity (long arrow), whereas the other is negative (short arrow), in a culture maintained in medium 199-N1. Scale bar = 15 μm. (From Hassall et al.[65]. With permission.)

breakdown of gamma-ANP (the precursor molecule) in the blood after, rather than before, secretion from the cardiac myocytes. In addition, by the use of a serum-free, defined growth medium for part of this work, the cleavage of gamma- to alpha-ANP was shown to be serum-independent and therefore likely to be due to factors present on, or released by, the cultured heart cells. Furthermore, the expression of ANP by the ventricular myocytes was influenced by the length of time in culture and by the presence of serum. Since changes in ventricular ANP have been described during development and in disease *in vivo*[10,37,48,70,86,87,128,131], the possibility that some of the factors regulating ANP expression in culture[65] are equivalent to those acting in the heart under these conditions *in vivo* is worth investigation. Future studies employing the same, or similar, preparations and techniques in parallel should prove to be valuable by facilitating long-term, direct and comprehensive analysis of the factors influencing different aspects of ANP metabolism in both cultured atrial and ventricular myocytes. Culture preparations could also be used to differentiate between those agents that directly affect ANP *in vivo* and those that have indirect effects, for example, by causing an alteration in blood volume which then changes ANP release into the blood[2].

Another example of work carried out on a culture preparation combining different techniques to give a more complete characterisation of cell subtypes has recently been reported[16,17]. In this case a differential distribution of receptors on neurones identified by their peptide content was revealed[17]. Living myenteric plexus cultures explanted from the taenia coli of newborn guinea-pig caecum[3,74,75] were first labelled using the irreversible muscarinic acetylcholine receptor antagonist, ³H-propylbenzilylcholine mustard; the cultures were then fixed to localise vasoactive intestinal polypeptide (VIP), enkephalin (ENK), somatostatin (SOM) or substance P (SP) and subsequently processed for immunocytochemistry and autoradiography. Thus, the morphology, neurochemical content and presence of muscarinic receptors were visualised simultaneously on myenteric neurones in the same preparation. It was found that while most ENK-, SOM- or SP-immunoreactive neurones did not possess muscarinic acetylcholine receptors, many VIP-immunoreactive neurones were uniformly covered in these receptors (fig. 3). However, the majority of neurones that expressed muscarinic receptors were not VIP-immunoreactive and their neurochemical identity has yet to be determined.

This type of study could be extended by the incorporation of immunocytochemical techniques to colocalise neuroactive substances[11,62] (fig. 1) and/or electrophysiological characterisation of the properties of single neurones in culture prior to the labelling techniques. Such experiments should be very useful because they would enable detailed classification of neuronal subpopulations that would be very difficult, although not impossible to achieve *in situ*[12]. Detailed studies of the gut have led Furness, Costa and colleagues to propose that a principle of chemical coding can be applied and that this should enable the circuitry of the intrinsic and extrinsic neural connections in the gut to be determined[31]. This is based on the premise that the neurochemical profiles of subpopulations of neurones in the gut can be employed to unequivocally identify them and their projections. If the neurochemical identities of the neurones with different functions in the gut are shown to be unique, such an approach would provide an important means of analysis of the gastrointestinal tract and many other tissues. In addition, if the properties of the cultured cells can

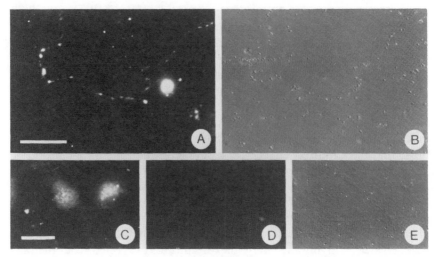

Figure 3. Concurrent localisation of muscarinic receptors and VIP-immunopositive fibres in cultures of the myenteric plexus. The fluorescence micrograph *A* shows a VIP-immunopositive neurite that is not part of a fibre bundle, whilst the autoradiograph *B* shows the same field viewed with Nomarski interference optics. Comparison of the two micrographs reveals autoradiograph grains overlying the VIP-immunopositive neurite. Fluorescent micrographs *C* and *D* show the appearance of control cultures incubated with a non-immune rabbit serum (1:100, *C*; 1:1000, *D*). Autoradiograph *E* shows a control culture incubated in the presence of 1 μM atropine. Scale bars = 20 μm. (From Buckley et al.[17]. With permission.)

be correlated with those of the same cells in the intact animal, then culture preparations should be valuable for further, direct study of the neurochemical properties characterising particular cell types. However, the possibility that alterations in the properties of cells occur in culture should always be considered.

Changes in neuronal phenotype observed under the conditions of culture

There are several reports of the expression of particular properties by neurones in culture that are not consistent with those of neurones at the same stage of development *in vivo*[44,63,98,107]. For example, together with the well-characterised changes in the adrenergic/cholinergic transmitter status of dissociated, cultured sympathetic neurones that results from their co-culture with certain types of non-neuronal cells or their growth in medium conditioned by these cells[43,108–110,115], levels of SP and SOM in both sympathetic and sensory neurones are also influenced by various aspects of the culture environment, including the presence of non-neuronal cells, pineal gland conditioned medium, neuronal density and membrane depolarisation[77–79,82]. Kessler[79] has shown that the levels of SP and SOM and the activity of choline acetyltransferase and tyrosine hydroxylase respectively appear to be linked to some extent in dissociated sympathetic neurones in culture, but the changes seen were not always coupled in this way and therefore these features of the neuronal phenotype are independently regulated.

Although the properties of cultured neurones may not always be expressed by neurones at the equivalent stage *in vivo*, they could represent a normal change in differentiation that occurs in development or under certain physiological conditions in the whole animal. For example, the types of changes that take place in sympathetic neurones in culture have also been seen *in situ*: the adrenergic/cholinergic transition observed in culture occurs during the normal development of sympathetic neurones innervating sweat glands[84,85]. Also, Kessler and Black[80] have shown that the expression of SP in the adult superior cervical ganglion may be trans-synaptically regulated. Furthermore, there is evidence that the development and maintenance of many types of enteric neuronal phenotype may be regulated by factors present in the target tissue of the gut[45,46,88]. Hence, the study of neurones in culture preparations, such as explants of myenteric ganglia[3,74,75,122], in which the neurones are both extrinsically denervated and have little or no association with their target cells, may be very useful for investigation of these regulatory factors since mechanisms causing alterations in phenotype may be the same in culture as in vivo. Potentially, this type of experiment is of great interest and may well lead to important advances in our understanding of the changes seen in cells during development and in disease.

A problem fundamental to many types of experiments, including culture studies, is that major differences in the expression of particular properties of cells tend to be much easier to detect than less extreme, but nevertheless important alterations, such as a change in the distribution of a particular peptide, or a slight modification in its form. Although these changes may be very difficult to demonstrate, they should always be considered in the analysis of results from cultured cells, otherwise negative findings in particular may be misinterpreted since they could be the result of a problem in the detection of the substance. However, if caution is used, culture preparations provide a unique opportunity to distinguish the sites of peptide synthesis, storage, secretion and receptor sites.

Localisation of peptide binding sites

The general advantages afforded by use of culture preparations also hold true for radioligand binding studies on a single cell type in culture, or for autoradiographic techniques when more than one cell type is present in the preparation. These techniques have been employed in a wide variety of culture preparations using a number of different ligands, but the study of peptide receptors or binding sites has been particularly limited by difficulties associated with the reversibility of binding of many, if not all peptide receptor ligands, which are generally the peptide itself, or an analogue (apart from opiate receptor ligands in particular[5]). This reversibility in binding results in diffusion of the ligand which may decrease the resolution of the technique, thereby counteracting many of the advantages gained by the autoradiographic study of cells in culture. Other problems include: the necessity to prevent adsorption of the ligand to the cells, coverslips, etc.; in general, the ligand should be incubated with unfixed cells in culture because fixation may well change the conformation of the receptor molecule and prevent binding; and the experimental conditions necessary to preserve the morphological and physiological integrity of the cells (such as temperature) may result in the breakdown of the ligand and influence the properties of the receptor-ligand complex[57]. Therefore, several different peptidase inhibitors may be required in the incubation medium, but these, in turn, may badly damage the cultured cells. Thus, optimum conditions for ligand binding and particularly autoradiography may be difficult to achieve, but these problems can be overcome. ANP binding sites have been demonstrated on cultures of aortic smooth muscle cells[68] and SOM receptors on a rat pituitary tumour cell line have been characterised[123]. Furthermore, autoradiography has been used to visualise SP binding sites on spinal cord neurones in organotypic culture[69], as well as on a subpopulation of neurones dissociated from the superior cervical ganglion in culture[71] (fig. 4). However, the development of new, irreversible peptide receptor ligands should greatly facilitate further studies of the distribution and characteristics of regulatory peptide receptor sites on cells in culture.

It should be noted that radioligand binding techniques are of limited use in mixed cell cultures since the binding sites cannot be attributed to any of the constituent cells and possible variation in the relative numbers of the different cell types present means that accurate quantitation of results is impossible. In contrast, autoradiographic techniques, combined with the use of culture preparations, enable binding sites to be visualised on identified cell types with far greater resolution than can be achieved when tissue sections are studied (compare refs 16 and 64 with 15 and 56 for example), although quantitation is still a problem. However, the inability to demonstrate the exact cellular sites of binding on cells *in situ* means that it is very difficult to establish whether the localisation of receptors on particular cell types, or their distribution on the cell membrane, changes under the conditions of culture. Use of additional techniques, such as electrophysiology, may be necessary to resolve this problem.

Electrophysiological analysis of the effects of regulatory peptides on cultured neurones

Culture preparations can offer some distinct advantages over *in situ* preparations in a number of electrophysiological experiments. Whilst the

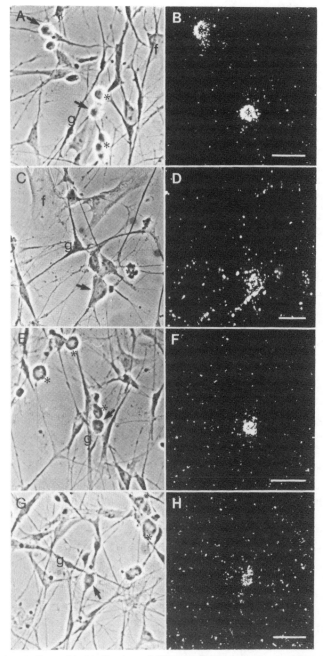

Figure 4. Autoradiographic localisation of [^{125}I]-SP binding sites over a subpopulation of neurones dissociated from rat superior cervical ganglion and grown in culture. *A, C, E, G* Phase-contrast photomicrographs of selected regions of the cultures; *B, D, F, H* show the corresponding dark field views of the same regions. A subpopulation of neurones present in culture are heavily labelled (arrows), while others are devoid of silver grains (asterisks). Non-neuronal cells such as fibroblasts (f) and glia (g) that are also present in these cultures are unlabelled. Bars = 50 μm. (Micrograph courtesy of Sharon James; see also James and Burnstock[71])

emphasis of this section will be confined to studies of the actions of neuropeptides on neurones, it must be stressed that many of the points raised will be equally valid for experiments performed on other cell types.

To a large extent, the advantages afforded by culture preparations are technical ones. Many neurones, especially those within parasympathetic intramural ganglia, are covered by a tough fibrous capsule or are embedded within muscle. This produces a variety of different problems in the employment of most electrophysiological techniques. Firstly, penetration of the tough capsule surrounding the ganglia requires the use of very fine and therefore by necessity, high impedance microelectrodes, with all their inherent drawbacks with regard to noise, rectification and limited current passing. Secondly, where ganglia lie within muscle, the problems can be further exacerbated by either spontaneous rhythmic contractions or contractions brought about by exogenously applied drugs, both of which can cause an electrode to be dislodged. In a number of cases, these problems have been reduced by direct chemical blockade of muscle contractions and by the use of enzymes or mechanical means to soften or cut the ganglion capsule; however all these actions have their drawbacks. The use of tissue culture techniques can offer a number of alternative solutions. For example, in the enteric nervous system, where the ganglia lie between layers of smooth muscle, some of the problems of *in situ* preparations have been overcome by carefully dissecting out sections of the myenteric plexus and growing them as explant cultures[72]. These cultures retain the integrity of the ganglionated plexus, thus allowing synaptic interactions to be studied[54]. Initially, immediately after placing the explant in culture, the only real advantage to be gained by this procedure is the avoidance of problems associated with muscle contraction. However, as in a number of other preparations, the explant develops through a number of stages until single neurones can be clearly identified using conventional phase contrast optics[72,74].

Where ganglia are more diffusely localised in a tissue and explant culture techniques cannot be employed, or when isolated neurones are required, then dissociated cell culture preparations can often be the best solution[61,105,114,117]. The obvious disadvantage of completely dissociated cell culture preparations compared with explants is the loss of all synaptic interactions, although some reinnervation can occur with time, as will be discussed later. The main advantage they afford is that the cells can be easily visualised and are free from overlying tissues. This allows easy placement of electrodes as well as greater access to exogenously applied drugs. Unlike *in situ* preparations, where drug onset and washout times are often protracted requiring impalements to be maintained for considerable periods, the virtual monolayer of cells in a dissociated cell culture can enable similar experiments to be performed in a fraction of the time. In addition, discrete application of drugs is more easily facilitated. Pressure application of neuropeptides through blunt microelectrodes is generally preferred to iontophoretic techniques as many peptides are difficult to eject in ionised solutions. With the clear access to the cell soma and neurites afforded by dissociated cell culture preparations, neuropeptides may be applied at known concentrations with little or no diffusional delay. Studies performed in this manner using cultured neurones suggest that in some cases there may be fast peptidergic actions, as well as the traditional slower ones previously found *in situ*[53,133,135] (fig. 5). The fact that fast peptidergic responses have not been reported using *in situ* preparations could of course indicate that some alteration has taken place due to the conditions of culture. On the other hand, it may simply be that in dissociated culture the ability to apply neuropeptides extremely rapidly and focally allows the detection of fast events before any desensitisation occurs.

A further advantage of cultured cells for electrophysiological investigations is that they are ideally suited for patch-clamp recording. The patch-clamp technique relies on the ability of the experimenter to obtain a very high resistance seal (giga-seal) between the electrode and the cell membrane. To do this, it is vital that the patch electrode should not become

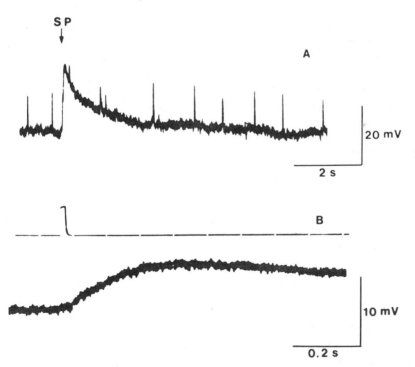

Figure 5. Fast response to SP. *A* a 50 ms pulse of SP (50 μM) evoked a fast response. This cell showed spontaneous activity of subthreshold potentials. *B* recording from another cell at a faster time scale. The top trace indicates the duration and timing of injection of SP (50 μM). (From Hanani and Burnstock[53]. With permission.)

contaminated by passing it through surrounding tissues and that the cell surface should be clean and free from overlying cells[52,102]. In this respect dissociated cell cultures are ideal. The use of patch-clamp techniques to study peptidergic responses, however, does have its problems. This is because most peptide responses are mediated via second messenger systems that can be disrupted using patch-clamp techniques. This is true even with whole cell patch recordings[96] because dilution of intracellular contents with those of the patch pipette occurs. A new technique known as 'slow whole cell patch' that is specifically designed to overcome these problems has recently been reported[89]. In this technique, the membrane inside the electrode is permeabilised to give electrical continuity; however the holes created in the membrane do not permit the passage of large molecules out of the cell.

There is now considerable evidence for the establishment of functional synaptic interactions between neurones and between neurones and muscle cells in culture[32,33,39,104,116]. For example, chick ciliary ganglion neurones synthesise and release acetylcholine when maintained in culture[104,130] and form functional cholinergic synapses with one another[9,94]. Although ciliary ganglion neurones have not been shown to form synapses with each other *in vivo*, about 70% of all the neurones display spontaneous cholinergic excitatory postsynaptic potentials (EPSPs) after 4 days in culture[94,125]. Subsequently a number of neuropeptides have been found to modulate these synaptic properties. For instance, application of ENK to these cells causes a reversible reduction in EPSP amplitude that is antagonised by naloxone. This is thought to result from a decrease in calcium entry into the presynaptic terminal which reduces the amount of acetylcholine released[95]. Additional support for this mechanism of action was provided by the observation that although ENK had no effect on resting membrane potential or conductance, it did reduce the duration of the calcium-dependent component of the somal action potential. Similar actions of ENK on calcium entry have also been reported in for example, cultured chick dorsal root ganglion neurones[40,101], mouse sensory neurones[134], and myenteric neurones dissociated from the rat enteric nervous system[135]. As well as modifying cholinergic transmission, ENK has also been shown to have multiple membrane actions and modulate the actions of iontophoretically applied GABA, glycine and glutamate in spinal neurones[4].

Other peptides have also been shown to have modulatory roles in synaptic transmission. For example, by using tight-seal whole cell patch recording techniques on cultured sympathetic and parasympathetic ganglion cells, Role[120] was able to demonstrate a modulatory role for SP, whereby it produced a marked increase in the rate of receptor desensitisation to acetylcholine. This work was further supported by intracellular studies on cultured ciliary ganglion neurones[95]. Using a multidisciplinary approach, the correlation between neuropeptide content and synaptic events has also been examined in cultured enteric neurones[136]. In this study, SOM-like immunoreactive neurones evoked larger amplitude cholinergic fast EPSPs than non-SOM-like immunoreactive neurones, whilst neurones displaying VIP-like immunoreactivity induced non-cholinergic depolarisations in addition to fast nicotinic EPSPs. Exogenously applied SOM did not produce an increase in EPSP amplitude so it could not be considered a co-transmitter/neuromodulator; however, VIP did mimic the slow EPSP indicating that in VIP-like immunoreactive neurones it may act as a co-transmitter with acetylcholine.

In addition to these complex pre- and postsynaptic regulatory roles for neuropeptides, marked alterations in membrane excitability have been recorded in response to other peptides for example VIP, SOM, SP and ENK[53,55,135].

So far, discussion has been confined to the benefits of culture preparations for electrophysiological studies of the short-term effects of peptides, but experiments need not be restricted to this. With culture preparations it is technically feasible, although difficult, to record from the same cell on two or more occasions and in this way establish whether the properties of a cell are fixed or labile. This technique, referred to as serial physiological assay by Potter and his colleagues[115], has been used to show that an individual neurone can change its transmitter and that, during a transition period, both transmitters are released at functional synapses on to a target tissue[115]. This technique could be used to establish whether the release and/or the responses to peptides change with time or in different culture conditions, and whether peptides acting as long-term trophic factors (see next section) alter the functional characteristics of a cell.

In summary therefore, cell culture preparations have helped to answer a number of questions concerning the roles played by neuropeptides in neurotransmission and neuromodulation. However, to date, they have not been utilised to their full potential.

Long-term, trophic actions of regulatory peptides on cultured cells

The term trophic factor is usually applied to a substance that acts over a period of days, weeks or even longer to influence the development, function or regeneration of its target cells[18,19,132]. Therefore, culture preparations are particularly suitable for the study of interactions between trophic factors and cells because the cells can be maintained and observed under controlled conditions for long periods of time. This advantage was recognised from the start of research in this field and culture preparations have been used in many studies, including some of the original work on nerve growth factor[50].

Although most research has been focussed on the examination of possible transmitter, modulatory and hormonal roles of regulatory peptides, there is also evidence to suggest that some of these peptides may well have trophic actions *in situ*. For example, insulin and the insulin-like growth factors, which are structurally related, have been shown to have effects *in situ*[67] and, in addition, further work has been carried out on a number of different cell types in culture. Insulin appears to be an essential supplement for cell survival in many different defined, serum-free culture media[7]; it enhances rat and human vascular smooth muscle cell proliferation and endothelial cell DNA synthesis[112,113,127]. Neuronal survival and neurite formation in cultured sympathetic and sensory neurones are also influenced by insulin and insulin-like growth factor II[119]. Together with these changes in the survival and growth of cells in culture, insulin affects the phenotype of cultured sympathetic neurones by inducing electrical coupling between them[83,138].

Other studies of culture preparations have indicated that regulatory peptides may act as trophic factors. VIP increases the survival of spinal cord neurones in culture[14], while exogenous SP and SOM have been shown to influence the activity of tyrosine hydroxylase in cultured superior cervical ganglion neurones; furthermore SP appears to act on specific receptors on these cells to mediate this effect[81]. SP also causes increases in neuronal cell body size, neurite extension and their regeneration in explant cultures of the trigeminal ganglion[90-92]. Angiotensin II, vasopressin, SP and substance K all act as mitogens in cultures of vascular smooth muscle, as well as endothelial cells.[22,58,103]

Even when the molecule responsible for particular trophic effects has not been identified or isolated, preparations in which two pieces of tissue were grown together in co-culture have been used to demonstrate the trophic actions of one upon the other[25,28,29,100,118] (fig. 6). In addition, the effects of exposure of a population of cells to growth medium conditioned by a particular tissue or cell type may also be investigated with comparative ease in culture[43,44,93,98,115,118]. Thus, culture preparations are not only of use in direct investigations of the consequences of adding a known molecule to the culture medium, they can also be employed in the more general study of unknown trophic factors since the causes of changes in the properties of cells, under different conditions in culture, may be identified. Furthermore, examination of what mimics or blocks the observed effects of trophic factor(s) in culture can provide valuable insight into the nature and roles of these substances *in vivo*.

Future directions

Culture preparations are valuable for the study of several aspects of regulatory peptides: firstly, the properties of the peptide molecules can be analysed; secondly, the factors influencing the differentiation of peptide-containing cells can be investigated; and thirdly, the characteristics of

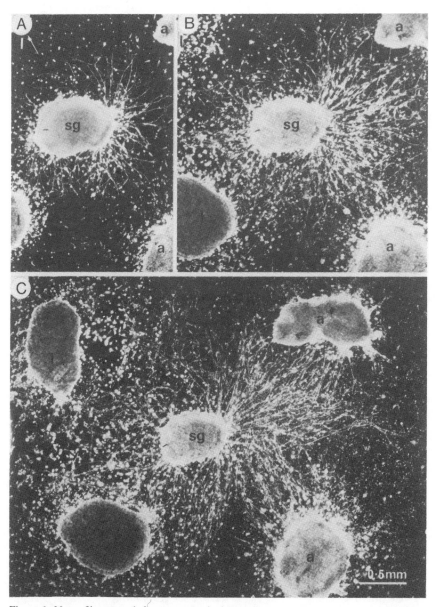

Figure 6. Nerve fibre growth from a sympathetic ganglion explant (sg) to explants of lung (l) and atrium (a). 5-day-old rat. *A* 2 days *in vitro*. *B* 3 days *in vitro*. *C* 5 days *in vitro*. At all times a greater number of nerve fibres grew on the side of the sympathetic ganglion explant nearer the atrium than lung explants. The nerve fibre growth to the atrium explants appeared to be directed, many fibres changing course to penetrate atrium explants. Nerve fibre growth to lung explants always appeared random. (Adapted from Chamley et al.[28]. With permission.)

cells that are responsive to peptides and the nature of these responses can be determined. In a few studies, a combination of techniques have been applied to the same culture, and these multidisciplinary studies hold out great promise for the future. Cultures provide the opportunity to correlate the morphology, the peptide content, and the receptor expression of a single cell with its functional properties. Furthermore, molecular biological techniques that enable the demonstration of mRNA transcripts of regulatory peptide genes may be used in conjunction with culture preparations to provide a unique means of studying the factors which regulate peptide expression in development and disease.

Acknowledgments. This work was funded in part by grants from The British Heart Foundation, Asthma Research Council and Medical Research Council.

1 Acton, R. T., and Lynn, J. D., Eukaryotic cell cultures: Basics and applications, in: Advances in Experimental Medicine and Biology, vol. 172. Plenum Press, New York/London 1984.

2 Ballermann, B. J., and Brenner, B. M., Role of atrial peptides in body fluid homeostasis. Circ. Res. 58 (1986) 619–630.

3 Baluk, P., Jessen, K. R., Saffrey, M. J., and Burnstock, G., The enteric nervous system in tissue culture. II. Ultrastructural studies of cell types and their relationships. Brain Res. 262 (1983) 37–47.

4 Barker, J. L., Smith, T. G., and Neale, J. N., Multiple membrane actions of enkephalin revealed using cultured spinal neurons. Brain Res. 154 (1978) 153–158.

5 Barnard, E. A., and Demoliou-Mason, C., Molecular properties of opioid receptors. Br. med. Bull. 39 (1983) 37–45.

6 Barnes, D., and Sato, G. H., Methods for growth of cultured cells in serum-free medium. Analyt. Biochem. 102 (1980) 255–270.

7 Barnes, D., and Sato, G., Serum-free cell culture: a unifying approach. Cell 22 (1980) 649–655.

8 Barnes, D. W., Sirbasku, D. A., and Sato, G. H., Cell Culture Methods for Molecular and Cell Biology, vols 1–4. Alan R. Liss Inc., New York 1984.

9 Berg, D. K., Jacob, M. H., Margiotta, J. F., Nishi, R., Stollberg, J. M., Smith, M. A., and Lindstrom, J., Cholinergic development and identification of synaptic components for chick ciliary ganglion neurons in cell culture, in: Molecular Bases of Neural Development, pp. 363–383. Eds G. Edelman. W. E. Gall and W. M. Cowan. Neurosciences Research Foundation, New York 1985.

10 Bloch, K. D., Seidman, J. G., Naftilan, J. D., Fallon, J. T., and Seidman, C. E., Neonatal atria and ventricles secrete atrial natriuretic factor via tissue-specific pathways. Cell 47 (1986) 695–702.

11 Bohn, M. C., Kessler, J. A., Adler, J. E., Markey, K., Goldstein, M., and Black, I. B., Simultaneous expression of the SP-peptidergic and noradrenergic phenotypes in rat sympathetic neurons. Brain Res. 298 (1984) 378–381.

12 Bornstein, J. C., Furness, J. B., and Costa, M., Sources of excitatory synaptic inputs to neurochemically identified submucous neurons of guinea-pig small intestine, J. auton. nerv. Syst. 18 (1987) 83–91.

13 Bottenstein, J. E., Skaper, S. D., Varon, S. S., and Sato, G. H., Selective survival of neurons from chick embryo sensory ganglionic dissociates utilizing serum-free supplemented medium. Exp. Cell Res. 125 (1980) 183–190.

14 Brenneman, D. E., and Foster, G. A., Structural specificity of peptides influencing neuronal survival during development. Peptides 8 (1987) 687–694.

15 Buckley, N., and Burnstock, G., Autoradiographic localisation of muscarinic receptors in guinea-pig intestine: distribution of high and low affinity agonist binding sites. Brain Res. 294 (1984) 15–22.

16 Buckley, N. J., and Burnstock, G., Localization of muscarinic receptors on cultured myenteric neurons: a combined autoradiographic and immunocytochemical approach. J. Neurosci. *6* (1986) 531–540.

17 Buckley, N. J., Saffrey, M. J., Hassall, C. J. S., and Burnstock, G., Localization of muscarinic receptors on peptide-containing neurones of the guinea-pig myenteric plexus in tissue culture. Brain Res. *445* (1988) 152–156.

18 Burnstock, G., Neuropeptides as trophic factors, in: Systemic Role of Regulatory Peptides, pp. 423–441. Eds S. R. Bloom, J. M. Polak and E. Lindenlaub. F. K. Schattauer Verlag, Stuttgart/New York 1982.

19 Burnstock, G., Recent concepts of chemical communication between excitable cells, in: Dale's Principle and Communication Between Neurones, pp. 7–35. Ed. N. N. Osborne. Pergamon Press, Oxford/New York 1983.

20 Burnstock, G., Allen, T. G. J., Hassall, C. J. S., and Pittam, B. S., Properties of intramural neurones cultured from the heart and bladder, in: Histochemistry and Cell Biology of Autonomic Neurons and Paraganglia, Experimental Brain Research, Series 16, pp. 323–328. Ed. C. Heym. Springer-Verlag, Heidelberg 1987.

21 Burnstock, G., Allen, T. G. J., and Hassall, C. J. S., The electrophysiologic and neurochemical properties of paratracheal neurones in situ and in dissociated cell culture. Am. Rev. resp. Dis. *136* (1987) 523–526.

22 Campbell-Boswell, M., and Robertson, A. L., Effects of angiotensin II and vasopressin on human smooth muscle cells in vitro. Exp. molec. Path. *35* (1981) 265–276.

23 Cantin, M., Dagenais, N., Salmi, L., Gutkowska, J., Ballak, M., Thibault, G., Garcia, R., and Genest, J., Secretory patterns of atrial natriuretic factor (ANF) by cultured cardiocytes of right and left atrium from newborn and adult rats. Clin. exp. Hypertens. [A] *7* (1985) 685–705.

24 Carbonetto, S., Evans, D., and Cochard, P., Nerve fiber growth in culture on tissue substrata from central and peripheral nervous systems. J. Neurosci. *7* (1987) 610–620.

25 Chamley, J. H., and Campbell, G. R., Trophic influences of sympathetic nerves and cyclic AMP on differentiation and proliferation of isolated smooth muscle cells in culture. Cell Tiss. Res. *161* (1975) 497–510.

26 Chamley, J. H., Mark, G. E., and Burnstock, G., Sympathetic ganglia in culture. II. Accessory cells. Z. Zellforsch. *135* (1972) 315–327.

27 Chamley, J. H., Mark, G. E., Campbell, G. R., and Burnstock, G., Sympathetic ganglia in culture. I. Neurons. Z. Zellforsch. *135* (1972) 287–314.

28 Chamley, J. H., Goller, I., and Burnstock, G., Selective growth of sympathetic nerve fibers to explants of normally densely innervated autonomic effector organs in tissue culture. Devl Biol. *31* (1973) 362–379.

29 Chamley, J. H., Campbell, G. R., and Burnstock, G., Dedifferentiation, redifferentiation and bundle formation of smooth muscle cells in tissue culture: influence of cell number and nerve fibres. J. Embryol. exp. Morph. *32* (1974) 297–323.

30 Chamley-Campbell, J., Campbell, G. R., and Ross, R., The smooth muscle cell in culture. Physiol. Rev. *59* (1979) 1–61.

31 Costa, M., Furness, J. B., and Gibbins, I. L., Chemical coding of enteric neurons, in: Progress in Brain Research, vol. 68, pp. 217–239. Eds T. Hökfelt, K. Fuxe and B. Pernow. Elsevier Science Publishers, Amsterdam 1986.

32 Crain, S. M., Development of functional neuromuscular connections between separate explants in fetal mammalian tissues after maturation in culture. Anat. Rec. *160* (1968) 466.

33 Crain, S. M., Neurophysiologic Studies in Tissue Culture. Raven Press, New York 1976.

34 Crowe, R., Haven, A. J., and Burnstock, G., Intramural neurones of the guinea-pig urinary bladder: histochemical localization of putative neurotransmitters in cultures and newborn animals. J. auton. nerv. Syst. *15* (1986) 319–339.

35 Dreyfus, C. F., Bornstein, M. B., and Gershon, M. D., Synthesis of serotonin by neurons of the myenteric plexus in situ and in organotypic tissue culture. Brain Res. *128* (1977) 125–139.

36 Dreyfus, C. F., Sherman, D. L., and Gershon, M. D., Uptake of serotonin by intrinsic neurons of the myenteric plexus grown in organotypic tissue culture. Brain Res. *128* (1977) 109–123.

37 Edwards, B. S., Ackerman, D. M., Wold, L. E., and Burnett, J. C., Ventricular occurrence

of 'atrial' natriuretic factor: a marker for the failing ventricle. J. Am. coll. Cardiol. *9* (1987) 1A.

38 Fedoroff, S., and Hertz, L., Cell, Tissue, and Organ Cultures in Neurobiology. Academic Press, New York 1977.

39 Fischbach, G. D., Synapse formation between dissociated nerve and muscle cells in low density cell culture. Devl Biol. *28* (1972) 407–429.

40 Fischbach, G. D., Dunlap, K., Mudge, A., and Leeman, S., Peptide and amine transmitter effects on embryonic chick sensory neurons in vitro, in: Neurosecretion and Brain Peptides, pp. 175–188. Eds J. B. Martin, S. Reichlin and K. I. Bick. Raven Press, New York 1981.

41 Freshney, R. I., Culture of Animal Cells. A Manual of Basic Technique. Alan R. Liss Inc., New York 1983.

42 Freshney, R. I., Animal Cell Culture. A Practical Approach. IRL Press, Oxford 1986.

43 Furshpan, E. J., Landis, S. C., Matsumoto, S. G., and Potter, D. D., Synaptic functions in rat sympathetic neurons in microcultures. I. Secretion of norepinephrine and acetylcholine. J. Neurosci. *6* (1986) 1061–1079.

44 Furshpan, E. J., Potter, D. D., and Matsumoto, S. G., Synaptic functions in rat sympathetic neurons in microcultures. III. A purinergic effect on cardiac myocytes. J. Neurosci. *6* (1986) 1099–1107.

45 Gershon, M. D., and Rothman, T. P., Experimental and genetic approaches to the study of the development of the enteric nervous system. Trends Neurosci. *7* (1984) 150–155.

46 Gershon, M. D., Payette, R. F., and Rothman, T. P., Development of the enteric nervous system. Fedn Proc. *42* (1983) 1620–1625.

47 Giacobini, E., Vernadakis, A., and Shahar, A., Tissue Culture in Neurobiology. Raven Press, New York 1980.

48 Glembotski, C. C., Oronzi, M. E., Li, X., Shields, P. P., Johnston, J. F., Kallen, R. G., and Gibson, T. R., The characterization of atrial natriuretic peptide (ANP) expression by primary cultures of atrial myocytes using an ANP-specific monoclonal antibody and an ANP messenger ribonucleic acid probe. Endocrinology *121* (1987) 843–852.

49 Greene, L. A., and Tischler, A. S., Establishment of a noradrenergic clonal line of rat adrenal pheochromocytoma cells which respond to nerve growth factor. Proc. natl Acad. Sci. USA *73* (1976) 2424–2428.

50 Greene, L. A., and Shooter, E. M., The nerve growth factor: biochemistry, synthesis, and mechanism of action. A. Rev. Neurosci. *3* (1980) 353–402.

51 Hamid, Q., Wharton, J., Terenghi, G., Hassall, C. J. S., Aimi, J., Taylor, K. M., Nakazato, H., Dixon, J. E., Burnstock, G., and Polak, J. M., Localization of atrial natriuretic peptide mRNA and immunoreactivity in the rat heart and human atrial appendage. Proc. natl Acad. Sci. USA *84* (1987) 6760–6764.

52 Hamill, O. P., Marty, A., Neher, E., Sakmann, B., and Sigworth, F., Improved patch-clamp techniques for high resolution current recording from cells and cell free membrane patches. Pflügers Arch. *391* (1981) 85–100.

53 Hanani, M., and Burnstock, G., Substance P evokes slow and fast responses in cultured myenteric neurones of the guinea-pig. Neurosci. Lett. *48* (1984) 19–23.

54 Hanani, M., and Burnstock, G., Synaptic activity in myenteric neurons in tissue culture. J. auton. nerv. Syst. *14* (1985) 49–60.

55 Hanani, M., and Burnstock, G., The actions of substance P and serotonin on myenteric neurons in tissue culture. Brain Res. *358* (1985) 276–281.

56 Hancock, J. C., Hoover, D. B., and Houghland, M. W., Distribution of muscarinic receptors and acetylcholinesterase in the rat heart. J. auton. nerv. syst. *19* (1987) 59–66.

57 Hanley, M., Peptide binding assays, in: Neurotransmitter Receptor Binding, pp. 91–102. Eds H. I. Yamamura, S. J. Enna and M. J. Kuhar. Raven Press, New York 1985.

58 Hanley, M. R., Neuropeptides as mitogens. Nature *315* (1985) 14–15.

59 Harrison, R. G., Observations on the living developing nerve fiber. Proc. Soc. exp. Biol. *4* (1907) 140–143.

60 Hassall, C. J. S., and Burnstock, G., Neuropeptide Y-like immunoreactivity in cultured intrinsic neurones of the heart. Neurosci. Lett. *52* (1984) 111–115.

61 Hassall, C. J. S., and Burnstock, G., Intrinsic neurones and associated cells of the guinea-pig heart in culture. Brain Res. *364* (1986) 102–113.

62 Hassall, C. J. S., and Burnstock, G., Evidence for uptake and synthesis of 5-hydroxytryptamine by a subpopulation of intrinsic neurones in the guinea-pig heart. Neuroscience 22 (1987) 413–423.

63 Hassall, C. J. S., and Burnstock, G., Immunocytochemical localisation of neuropeptide Y and 5-hydroxytryptamine in a subpopulation of amine-handling intracardiac neurones that do not contain dopamine-β-hydroxylase in tissue culture. Brain Res. 422 (1987) 74–82.

64 Hassall, C. J. S., Buckley, N. J., and Burnstock, G., Autoradiographic localisation of muscarinic receptors on guinea-pig intracardiac neurones and atrial myocytes in culture. Neurosci. Lett. 74 (1987) 145–150.

65 Hassall, C. J. S., Wharton, J., Gulbenkian, S., Anderson, J. V., Frater, J., Bailey, D. J., Merighi, A., Bloom, S. R., Polak, J. M., and Burnstock, G., Ventricular and atrial myocytes of newborn rats synthesise and secrete atrial natriuretic peptide in culture: Light- and electron-microscopical localisation and chromatographic examination of stored and secreted molecular forms. Cell Tiss. Res. 251 (1988) 161–169.

66 Haynes, L. W., and Zakarian, S., Development of enkephalin interneurones from two regions of the nervous system in tissue culture. Reg. Pep. 3 (1982) 73.

67 Herschman, H. R., Polypeptide growth factors and the CNS. Trends Neurosci. 9 (1986) 53–57.

68 Hirata, Y., Tomita, M., Yoshimi, H., and Ikeda, M., Specific receptors for atrial natriuretic factor (ANF) in cultured vascular smooth muscle cells of rat aorta. Biochem. biophys. Res. Commun. 125 (1984) 562–568.

69 Hösli, E., and Hösli, L., Binding sites for [^{3}H]substance P on neurons of cultured rat spinal cord and brain stem: an autoradiographic study. Neurosci. Lett. 56 (1985) 199–203.

70 Izumo, S., Lompre, A-M., Nadal-Ginard, B., and Mahdavi, V., Expression of the genes encoding the fetal isoforms of contractile proteins and atrial natriuretic factor (ANF) in the hypertrophied adult rat ventricles. J. Am. coll. Cardiol. 9 (1987) 1A.

71 James, S., and Burnstock, G., Autoradiographic localization of binding sites for ^{125}I-substance P on neurones from cultured rat superior cervical ganglion. Brain Res. (1988) in press.

72 Jessen, K. R., McConnell, J. D., Purves, R. D., Burnstock, G., and Chamley-Campbell, J. H., Tissue culture of mammalian enteric neurons. Brain Res. 152 (1978) 573–579.

73 Jessen, K. R., Saffrey, M. J., Van Noorden, S., Bloom, S. R., Polak, J. M., and Burnstock, G., Immunohistochemical studies of the enteric nervous system in tissue culture and in situ: localization of vasoactive intestinal polypeptide (VIP), substance-P and enkephalin immunoreactive nerves in the guinea-pig gut. Neuroscience 5 (1980) 1717–1735.

74 Jessen, K. R., Saffrey, M. J., and Burnstock, G., The enteric nervous system in tissue culture. I. Cell types and their interactions in explants of the myenteric and submucous plexuses from guinea-pig, rabbit and rat. Brain Res. 262 (1983) 17–35.

75 Jessen, K. R., Saffrey, M. J., Baluk, P., Hanani, M., and Burnstock, G., The enteric nervous system in tissue culture. III. Studies on neuronal survival and the retention of biochemical and morphological differentiation. Brain Res. 262 (1983) 49–62.

76 Kaufman, L. M., and Barrett, J. N., Serum factor supporting long-term survival of rat central neurons in culture. Science 220 (1983) 1394–1396.

77 Kessler, J. A., Non-neuronal cell conditioned medium stimulates peptidergic expression in sympathetic and sensory neurons in vitro. Devl Biol. 106 (1984) 61–69.

78 Kessler, J. A., Environmental co-regulation of substance P, somatostatin and neurotransmitter synthesizing enzymes in cultured sympathetic neurons. Brain Res. 321 (1984) 155–159.

79 Kessler, J. A., Differential regulation of peptide and catecholamine characters in cultured sympathetic neurons. Neuroscience 15 (1985) 827–839.

80 Kessler, J. A., and Black, I. B., Regulation of substance P in adult rat sympathetic ganglia. Brain Res. 234 (1982) 182–187.

81 Kessler, J. A., Adler, J. E., and Black, I. B., Substance P and somatostatin regulate sympathetic noradrenergic function. Science 221 (1983) 1059–1061.

82 Kessler, J. A., Adler, J. E., Bell, W. O., and Black, I. B., Substance P and somatostatin metabolism in sympathetic and special sensory ganglia in vitro. Neuroscience 9 (1983) 309–318.

134

83 Kessler, J. A., Spray, D. C., Saez, J. C., and Bennett, M. V., Determination of synaptic phenotype: insulin and cAMP independently initiate development of electrotonic coupling between cultured sympathetic neurons. Proc. natl Acad. Sci. USA *81* (1984) 6235–6239.

84 Landis, S. C., Neurotransmitter plasticity during the development of cholinergic sympathetic neurons in culture and in vivo, in: Developing and Regenerating Vertebrate Nervous Systems, pp. 107–120. Eds P. W. Coates, R. R. Markwald and A. D. Kenny. Alan R. Liss Inc., New York 1983.

85 Landis, S. C., and Keefe, D., Evidence for neurotransmitter plasticity in vivo: developmental changes in properties of cholinergic sympathetic neurons. Devl Biol. *98* (1983) 349–372.

86 Lattion, A.-L., Michel, J.-B., Arnauld, E., Corvol, P., and Soubrier, F., Myocardial recruitment during ANF mRNA increase with volume overload in the rat. Am. J. Physiol. *251* (1986) H890–H896.

87 Lattion, A.-L., Nussberger, J., Waeber, B., Flückiger, J. P., Aubert, J. F., and Brunner, H. R., The effect of short- and long-term sodium load on the cardiac gene expression of atrial natriuretic peptide in normotensive rats. Fedn Proc. *46* (1987) 671.

88 Le Douarin, N. M., The Neural Crest, Developmental and Cell Biology 12. Cambridge University Press 1982.

89 Lindau, M., and Fernandez, J. M., IgE-mediated degranulation of mast cells does not require opening of ion channels. Nature *319* (1986) 150–153.

90 Lindner, G., and Grosse, G., The effect of substance P on the regeneration of nerve fibres in vitro. Z. mikrosk.-anat. Forsch., Leipzig *95* (1981) 390–394.

91 Lindner, G., Grosse, G., Oehme, P., and Jentzsch, K. D., Effect of substance P on the ganglion trigeminale in tissue culture. Z. mikrosk.-anat. Forsch., Leipzig *95* (1981) 607–616.

92 Lindner, G., Grosse, G., Oehme, P., Jentzsch, K. D., and Neubert, K., Effect of substance P (SP) and SP partial sequences on the neurite extension in tissue culture. Z. mikrosk.-anat. Forsch., Leipzig *96* (1982) 643–655.

93 Manthorpe, M., Luyten, W., Longo, F. M., and Varon, S., Endogenous and exogenous factors support neuronal survival and choline acetyltransferase activity in embryonic spinal cord cultures. Brain Res. *267* (1983) 57–66.

94 Margiotta, J. F., and Berg, D. K., Functional synapses are established between ciliary ganglion neurones in dissociated cell cultures. Nature *296* (1982) 152–154.

95 Margiotta, J. F., and Berg, D. K., Enkephalins and substance P modulate synaptic properties of chick ciliary ganglion neurones in cell culture. J. Neurosci. *18* (1986) 175–182.

96 Marty, A., and Neher, E., Tight seal whole cell recording, in: Single Channel Recording. Eds B. Sakmann and E. Neher. Plenum Press, New York 1983.

97 Mason, R. T., Peterfreund, R. A., Sawchenko, P. E., Corrigan, A. Z., Rivier, J. E., and Vale, W. W., Release of the predicted calcitonin gene-related peptide from cultured rat trigeminal ganglion cells. Nature *308* (1984) 653–655.

98 Matsumoto, S. G., Sah, D., Potter, D. D., and Furshpan, E. J., Synaptic functions in rat sympathetic neurons in microcultures. IV. Nonadrenergic excitation of cardiac myocytes and the variety of multiple transmitter states. J. Neurosci. *7* (1987) 380–390.

99 Meyer, T., Burkart, W., and Jockusch, H., Choline acetyltransferase induction in cultured neurons: dissociated spinal cord cells are dependent on muscle cells, organotypic explants are not. Neurosci. Lett. *11* (1979) 59–62.

100 Mudge, A. W., Schwann cells induce morphological transformation of sensory neurones in vitro. Nature *309* (1984) 367–369.

101 Mudge, A. W., Leeman, S. E., and Fischbach, G., Enkephalin inhibits release of substance P from sensory neurones in culture and decreases action potential duration. Proc. natl Acad. Sci. USA *76* (1979) 526–530.

102 Neher, E., Sakmann, B., and Steinbach, J. H., The intracellular patch clamp: A method of resolving currents through individual open channels in biological membranes. Pflügers Arch. *375* (1978) 219–228.

103 Nilsson, J., von Euler, A. M., and Dalsgaard, C.-J, Stimulation of connective tissue cell growth by substance P and substance K. Nature *315* (1985) 61–63.

104 Nishi, R., and Berg, D. K., Dissociated ciliary ganglion cells in vitro: survival and synapse formation. Proc. natl Acad. Sci. USA *74* (1977) 5171–5175.

135

105 Nishi, R., and Willard, A. L., Neurons dissociated from rat myenteric plexus retain differentiated properties when grown in cell culture. I. Morphological properties and immunocytochemical localization of transmitter candidates. Neuroscience *16* (1985) 187–199.

106 Parsons, R. L., Neel, D. S., McKeon, T. W., and Carraway, R. E., Organization of a vertebrate cardiac ganglion: A correlated biochemical and histocemical study. J. Neurosci. *7* (1987) 837–846.

107 Patterson, P. H., Environmental determination of autonomic neurotransmitter functions. A. Rev. Neurosci. *1* (1978) 1–17.

108 Patterson, P. H., and Chun, L. L. Y., The influence of non-neuronal cells on catecholamine and acetylcholine synthesis and accumulation in cultures of dissociated sympathetic neurons. Proc. natl Acad. Sci. USA *71* (1974) 3607–3610.

109 Patterson, P. H., and Chun, L. L. Y., The induction of acetylcholine synthesis in primary cultures of dissociated rat sympathetic neurons. I. Effects of conditioned medium. Devl Biol. *56* (1977) 263–280.

110 Patterson, P. H., and Chun, L. L. Y., The induction of acetylcholine synthesis in primary cultures of dissociated rat sympathetic neurons. II. Developmental aspects. Devl Biol. *60* (1977) 473–481.

111 Paul, J., Cell and Tissue Culture. Churchill Livingstone, London 1975.

112 Pfeifle, B., and Ditschuneit, H., Effect of insulin on growth of cultured human arterial smooth muscle cells. Diabetologia *20* (1981) 155–158.

113 Pfeifle, B., Ditschuneit, H. H., and Ditschuneit, H., Insulin as a cellular growth regulator of rat arterial smooth muscle cells in vitro. Horm. Metab. Res. *12* (1980) 381–385.

114 Pittam, B. S., Burnstock, G., and Purves, R. D., Urinary bladder intramural neurones: an electrophysiological study utilising a tissue culture preparation. Brain Res. *403* (1987) 267–278.

115 Potter, D. D., Landis, S. C., Matsumoto, S. G., and Furshpan, E. J., Synaptic functions in rat sympathetic neurons in microcultures. II. Adrenergic/cholinergic dual status and plasticity. J. Neurosci. *6* (1986) 1080–1098.

116 Purves, R. D., Hill, C. E., Chamley, J. H., Mark, G. E., Fey, D. M., and Burnstock, G. Functional autonomic neuromuscular junctions in tissue culture. Pflügers Arch. *350* (1974) 1–7.

117 Ransom, B. R., Neale, E., Henkart, M., Bullock, P. N., and Nelson, P. G., Mouse spinal cord in cell culture: I. Morphology and intrinsic neuronal electrophysiologic properties. J. Neurophys. *40* (1977) 1132–1150.

118 Rawdon, B. B., and Dockray, G. J., Trophic factors from the gut influence growth of sympathetic and sensory neurons in culture. Reg. Pep. suppl. *2* (1983) S64.

119 Recio-Pinto, E., Rechler, M. M., and Ishii, D. N., Effects of insulin, insulin-like growth factor II, and nerve growth factor on neurite formation and survival in cultured sympathetic and sensory neurons. J. Neurosci. *6* (1986) 1211–1219.

120 Role, L. W., Substance P modulation of acetylcholine-induced currents in embryonic chicken sympathetic and ciliary ganglion neurons. Proc. natl Acad. Sci. USA *81* (1984) 2924–2928.

121 Sato, G. H., Pardee, A. B., and Sirbasku, D. A., Growth of Cells in Hormonally Defined Media, Books A and B, Cold Spring Harbor Conferences on Cell Proliferation, vol. 9. Cold Spring Harbor Laboratory 1982.

122 Saffrey, M. J., and Burnstock, G., Peptide-containing neurons in explant cultures of guinea-pig myenteric plexus during development in vitro: gross morphology and growth patterns. Cell Tiss. Res. *293* (1988) 105–114.

123 Schonbrunn, A., and Tashjian, A. H., Characterization of functional receptors for somatostatin in rat pituitary cells in culture. J. Biol. Chem. *253* (1978) 6473–6483.

124 Schultzberg, M., Dreyfus, C. F., Gershon, M. D., Hökfelt, T., Elde, R. P., Nilsson, G., Said, S., and Goldstein, M., VIP-, enkephalin-, substance P- and somatostatin-like immunoreactivity in neurons intrinsic to the intestine: immunohistochemical evidence from organotypic tissue cultures. Brain Res. *155* (1978) 239–248.

125 Smith, M. A., Margiotta, J. F., and Berg, D. K., Differential regulation of acetylcholine sensitivity and alpha-bungarotoxin-binding sites of ciliary ganglion neurons in cell culture. J. Neurosci. *3* (1983) 365–370.

126 Sylvestre, D. L., Zisfein, J. B., Graham, R. M., and Homcy, C. J., Serum-mediated enhancement of ANF accumulation in the culture medium of cardiac myocytes. Biochem. biophys. Res. Commun. *140* (1986) 151–159.

136

127 Taggart, H., and Stout, R. W., Control of DNA synthesis in cultured vascular endothelial and smooth muscle cells. Response to serum, platelet-deficient serum, lipid-free serum, insulin and oestrogens. Atherosclerosis *37* (1980) 549–557.

128 Takayanagi, R., Imada, T., and Inagami, T., Synthesis and presence of atrial natriuretic factor in rat ventricle. Biochem. biophys. Res. Commun. *142* (1987) 483–488.

129 Thibault, G., Benjamet, S., Gutkowska, J., Garcia, R., Chrétien, M., and Cantin, M., Atrial natriuretic factor (ANF) secreted by cultured atrial and ventricular cardiocytes. Fedn Proc. *46* (1987) 1074.

130 Tuttle, J. B., Suszkiw, J., and Ard, M., Long term survival and development of dissociated parasympathetic neurons in culture. Brain Res. *183* (1980) 161–180.

131 Thompson, R. P., Simson, J. A. V., and Currie, M. G., Atriopeptin distribution in the developing rat heart. Anat. Embryol. *175* (1986) 227–233.

132 Varon, S. S., and Bunge, R. P., Trophic mechanisms in the peripheral nervous system. A. Rev. Neurosci. *1* (1978) 327–361.

133 Vincent, J.-D, and Barker, J. L., Substance P: evidence for diverse roles in neuronal function from cultured mouse spinal neurons. Science *205* (1979) 1409–1412.

134 Werz, M. A., and MacDonald, R. L., Opioid peptides decrease calcium-dependent action potential duration of mouse dorsal root ganglion neurons in cell culture. Brain Res. *239* (1982) 315–321.

135 Willard, A. L., and Nishi, R., Neurons dissociated from the rat myenteric plexus retain differentiated properties when grown in cell culture. II. Electrophysiological properties and responses to neurotransmitter candidates. J. Neurosci. *16* (1985) 201–211.

136 Willard, A. L., and Nishi, R., Neuropeptides mark functionally distinguishable cholinergic enteric neurons. Brain Res. *422* (1987) 163–167.

137 Wolinsky, E. J., and Patterson, P. H., Rat serum contains a developmentally regulated cholinergic inducing activity. J. Neurosci. *5* (1985) 1509–1512.

138 Wolinsky, E. J., Patterson, P. H., and Willard, A. L., Insulin promotes electrical coupling between cultured sympathetic neurons. J. Neurosci. *5* (1985) 1675–1679.

139 Wollenberger, A., Seventy-five years of cardiac tissue and cell culture. Trends pharmac. Sci. *6* (1985) 383–387.

140 Zisfein, J. B., Sylvestre, D., Homcy, C. J., and Graham, R. M., Analysis of atrial natriuretic factor biosynthesis and secretion in adult and neonatal rat atrial cardiocytes. Life Sci. *41* (1987) 1953–1959.

Quantitative analysis of autoradiograms

A. P. Davenport, R. G. Hill and J. Hughes

Summary. Quantitative autoradiography of macroscopic specimens using computer-assisted image analysis is now widely used for studying the distribution of peptide receptors in the brain and peripheral tissues and more recently has been used to measure mRNA in tissue sections by *in situ* hybridisation. The spatial distribution of radiolabelled substances in tissue can be detected by the blackening of the emulsion in sheets of radiation-sensitive film and the resulting pattern of optical densities within the autoradiogram can be quantified by comparison with a calibrated radioactive scale. In this review, the technique of computer-assisted densitometry is described together with guidelines for the selection and preparation of radioactive standards. Strategies are discussed for ensuring that radioactivity can be measured and quantified using film-based emulsions with a precision approaching that of conventional counting techniques.

Image analysis is the process of quantifying identifiable parts of an image by size, shape, position and optical density. The speed and precision by which this information can be obtained have advanced rapidly with the development of computer-assisted image analysis systems[1,15,45]. In the study of regulatory peptides a major application of this technique has been the analysis of autoradiograms which reveal the anatomical localisation and density of receptors in the brain and peripheral tissues[24]. In macro-autoradiography, the spatial distribution of radiolabelled compounds within thin (10–25 μm) sections of tissue can be detected by the blackening of the emulsion in sheets of radiation-sensitive films apposed directly to the specimen[1,11,17,21,23,28,42]. On developing the film, the pattern of optical densities within the resulting autoradiogram reflects the amount of radioactivity and can be measured by comparison with a calibrated radioactive scale co-exposed with the sections of radiolabelled tissue. In complex structures such as the brain, it is theoretically possible to resolve over seventy discrete anatomical regions in a single coronal section without pre-selection of the tissue to be analysed. Resolution is however limited to groups of cells and the tissue must be separated from the film in order to make optical density measurements[1,35]. An alternative approach is to use micro-autoradiography in which radiolabelled sections are coated with a thin layer of nuclear emulsion or apposed to emulsion-coated cover slips[43] (Palacios and Dietl, this volume). On developing the emulsion, radioactivity is detected using an optical microscope as the pattern of individual silver grains lying above the specimen. Using this method with [³H] labelled compounds, it is possible to measure the amount of radioactivity associated with a single cell[35].

Autoradiography has been used to visualise the binding of radio-labelled ligands to receptors for over forty different peptides (see Palacios and Dietl, this volume). This review will describe the methods used in computer-assisted densitometry and will discuss potential errors in quantitative autoradiography of macroscopic specimens using film based emulsions as the technique has been widely used for receptor studies[7,11,18,21,24,32] and is now being employed to measure mRNA coding for regulatory peptides in tissue sections by *in situ* hybridisation[3,8,9,29,37,38,40,44]. Further information concerning quantitative autoradiography and image analysis can be found in a number of articles and reviews[1,2,15,22–24,28,30,33,34,42,43,45].

Selection and preparation of standards

Ligands labelled with [¹²⁵I] and [³H] are mainly used for receptor studies (Palacios and Dietl, this volume). Probes for *in situ* hybridisation are also labelled with [³⁵S] and [³²P] (Penschow et al., this volume).

The majority of workers have adopted the simplest approach to the generation of a calibrated radioactive scale for the quantification of receptor autoradiograms which is to mix increasing quantities of the radiolabel to be measured with tissue paste derived from the same target tissues as those used in the assay (see for example Unnerstall et al.[42] for [³H] and Davenport and Hall[12,13] for [¹²⁵I] standards). The concentration of radioactivity that should be added to the tissue paste to obtain a suitable range of optical density values in the resulting image on radiation-sensitive film must be determined empirically and will depend on a number of factors including the desired thickness of the section, the sensitivity of the film used and the anticipated length of exposure. Typically, for standard sections cut to a thickness of 10 μm which are used for the quantification of [¹²⁵I] receptor autoradiograms produced by exposing labelled neural tissue to radiation-sensitive film for 1–14 days, radiolabel is added to give concentrations ranging from 0.5 to 9 μCi/g tissue[13]. After extensive mixing to ensure a homogeneous distribution of the label, sections are cut to the same thickness as the experimental tissue. The amount of radioactivity in representative sections is measured by liquid scintillation or gamma counting with appropriate corrections for background and counting efficiency[20]. It is not necessary to prepare standards with the radiolabelled ligand used in a receptor assay. Any compound labelled with the same isotope can be used, provided it is not volatile and produces a homogeneous distribution when mixed with tissue paste[13]. The major errors associated with the production of tissue paste standards is failing to produce a uniform distribution of the label but this will become apparent from the observation of poor reproducibility when they are apposed to radiation-sensitive film.

Alternatively, commercially prepared standards are available for [³H] and [¹²⁵I][12,13] (Amersham International plc) in which radiolabel is incorporated at the molecular level in a methacrylate co-polymer separated by inert coloured layers. The polymer has a density of 1.11, similar to tissue[13]. The concentration of radioactivity in each activity level doubles at each step to give a uniform spacing on a logarithmic scale. Standards are available in pre-cut strips or as a multilayer block. Sections are cut to the same thickness as the experimental tissue using a standard rotary microtome, expanded on water at 60°C and brushed flat onto gelatin-subbed slides to remove creases. Representative sections are sub-divided into the individual activity levels and counted as previously described. The main advantage of these standards is that the dimensions of the microscale are suitable for mounting on microscope slides together with sections of tissue[13].

Selection of radiation-sensitive films

In macro-autoradiography, radiolabelled tissue is placed in direct contact with an emulsion containing silver halide crystals in gelatin on a polyester film base[1,28]. It is thought that one electron emitted from the radioisotope in the tissue is sufficient to activate one silver halide crystal[14] which on development is converted to a stable image consisting of silver grains. Ultrofilm (supplied by Cambridge Instruments or Amersham International as Hyperfilm [³H]) is now almost universally used for the detection of ligands labelled with [³H] and [¹²⁵I] owing to the ability of the emulsion to detect low energy electrons[1,25] (table). The film consists of a single coat of emulsion (4 μm thick) with a high silver content and large crystal size[13]. The emulsion characteristics of Hyperfilm βmax (Amersham International) is similar except the emulsion layer is three times thicker than [³H] and is protected by an anti-scratch layer[13]. As a result, βmax film detects a different spectrum of radioactive emissions to [³H] film and is recommended for the detection of [³⁵S], [¹²⁵I] and [³²P] (table).

Computer-assisted image analysis

Many different methods have been used to measure the response of emulsions to radioactivity[35]. For film-based macro-autoradiography,

140

Physical data on isotopes used in quantitative autoradiography

	Half-life	Radiation	Energy (KeV_{max})	Recommended film type
[3H]	12y	β-	19	[3H]*
[^{35}S]	5370y	β-	156	βmax$^+$
[^{32}P]	88d	β-	1710	βmax
[^{125}I]	60d	e-	30	[^{3}H] βmax

*Ultrofilm (Cambridge Instruments) or Hyperfilm [^{3}H] (Amersham International).$^+$ Hyperfilm βmax (Amersham International).

optical density measurements are simple and easy to make using a videocamera in conjunction with an image-analyser although measurements must be made at comparatively low magnifications to avoid resolving individual silver grains[35].

A wide range of computer-assisted image analysers are available ranging from those which are dedicated to a specific task such as morphometric analysis to multi-user instruments with a wide range of image-analysis applications[1,28]. The major components of the latter type of system are shown in figure 1, which is based on a Quantimet 920 (Cambridge Instruments, Cambridge, U.K.) but shares certain features with other image-analysers.

Image acquisition

The image of an autoradiogram is captured by a television type videocamera with high resolution (896 pixels horizontally by 704 pixels

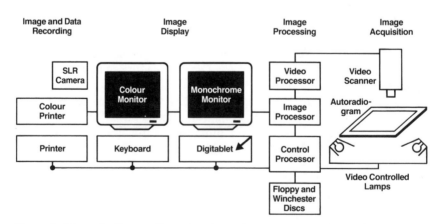

Figure 1. Schematic diagram illustrating the major components of a modern computer-assisted image analysis system.

vertically) and good signal to noise ratio which is achieved by a slow scan speed of 11 Hz. For macro-autoradiography, the camera is equipped with a zoom lens allowing magnification in the range × 0.7–4. The autoradiograms are illuminated by reflected white light from four tungsten lamps controlled by the computer to ensure optimum illumination within the dynamic range of the camera.

The videosignal must be transformed by the videoprocessor from an analogue to digital format before it can be processed by the computer. The image is quantified spatially by digitising into an array of 630 000 square pixels and by energy content of each pixel in the range 0–255 grey values where 0 is black and 255 is white.

Interacting with the image

The image processor performs digital signal processing tasks on images presented to the operator on a high-resolution monochrome television monitor. These functions are controlled either from a digi-tablet employing a hand-operated cursor or an alpha-numeric keyboard. At this stage, parts of the image are separated from spurious information or the 'background' by procedures such as detection, erosion, dilation, contrast enhancement and image editing. Colours can be assigned to particular grey values on the basis of a previously measured standard curve and are displayed on a colour monitor. Colour-coded images are recorded directly from the screen onto 35-mm colour film or to a colour printer for documentation. In this system the control processor is a Dec LSI-11/23 with a 512 Kbyte memory. Programmes and data are stored on dual floppy diskettes and Winchester drives with a combined capacity of 22 Mbytes.

Calibration of the image-analyser for densitometry

In order to make reproducible optical density measurements of areas of interest throughout the autoradiogram, the image-analyser must be calibrated for each film so that pixel areas can be correlated with physical dimensions and the grey value of each pixel correlated with optical density[1,28]. Magnification is altered by means of the zoom lens to adjust the size of the measuring field according to the dimensions of the autoradiogram to be imaged. The number of pixels per unit area is calibrated using a measuring box in conjunction with vernier calipers. For macro-autoradiography of coronal sections of rodent brains magnification is adjusted to give a spatial resolution of 20–40 μm/pixel.

142

The shading corrector is set by imaging a blank but homogeneous area of the film which reveals the variation in grey values across an apparently uniform field of view (fig. 2A). The shading corrector compensates for this variation introduced by the illumination system and light transmission properties of the optical system, to give a uniform tone over the measuring field (fig. 2B). If the image-analyser is not equipped to make shading corrections, measurements must be restricted to a small field preferably central.

The system is calibrated against neutral density filters to convert the grey values of each pixel into optical density. The theoretical maximum optical density that can be measured by this system is 2.41 but values above 1.0 become increasingly inaccurate. In practice, radiation sensitive films rapidly approach saturation above an optical density of 1.0[1] and determinations made below an optical density of 0.01 are close to the detection limits of the image-analyser giving a useful range for making accurate measurements of 0.1–0.8[13].

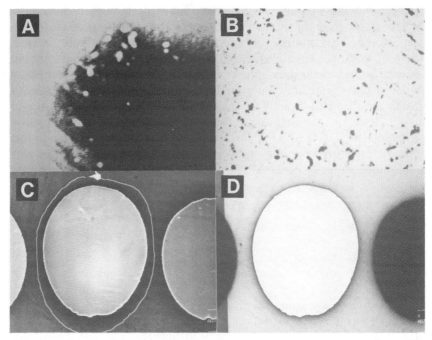

Figure 2. Calibration of a computer-assisted image analysis system for densitometry. The shading corrector is set by imaging a blank but homogeneous area of the film which reveals the variation in grey values across an apparently uniform field of view (A). The shading corrector compensates for this variation to give a uniform tone over the measuring area (B). The optical density of each tissue paste standard is measured by using a cursor to draw around the detected image (C). The image-analyser isolates this region and measures the integrated optical density and the area of the standard (D).

Measuring the optical density of calibrated standards

The procedure for measuring the optical densities produced by exposing a series of standards to radiation-sensitive film is shown in figure 2C–D. The threshold is set by altering the illumination system to detect each standard in turn and the cursor used to draw around the detected image (fig. 2C). The image-analyser isolates this region from the background and measures the integrated diffuse optical density which is the sum of the individual optical densities of each pixel within the area occupied by the standard. Thus slight variation in the distribution of radiolabelled material within the tissue will be averaged and the dimensions of the standard are not critical since this is always accurately measured.

The amount of radioactivity previously estimated by gamma or liquid scintillation counting is then divided by the area to calculate radioactivity in DPM/mm^2. Typically, eight to ten standards are used to generate a calibration curve consisting of the natural log plot of optical densities versus radioactivity.

Measuring the optical density within autoradiograms

An example of a film-based autoradiogram is shown in figure 3A which shows the total binding of the sulphated octapeptide of cholecystokinin to a coronal section of rat brain produced by incubating a cryostat section with 0.25 ± 0.01 nM of [^{125}I] CCK-8S for 120 min at 20°C[4–6] following the general method for *in vitro* receptor autoradiography described by Palacios and Dietl (this volume). After washing and drying, sections were apposed to Ultrofilm for 7 days. An adjacent section to that used to produce figure 3A was co-incubated with a competitive displacer (1 μm unlabelled CCK-8S) to produce the autoradiogram for non-specific [^{125}I] CCK-8S binding (fig. 3B).

The image-analyser is used to measure optical densities in discrete areas within the autoradiogram as shown in figure 4A–F. A cursor is used to draw within an anatomical region defined by reference to a stereotaxic atlas[31] (fig. 4A). The computer isolates the area of interest and measures the integrated optical density (fig. 4B). Other regions within the autoradiographical image can be selected and separate measurements made. When all of these have been completed, the threshold for detecting the image is increased (fig. 4C) to produce a template of the total binding image (fig. 4D), which is used to align the autoradiographical image of non-specific binding to an adjacent section (fig. 4E). The second image is digitally subtracted from the first to obtain a new image (fig. 4F) of the amount of specific CCK-8S bound in attomoles (1×10^{-18} moles) per square millimetre (amol/mm^2), calculated by interpolating from the standard curve.

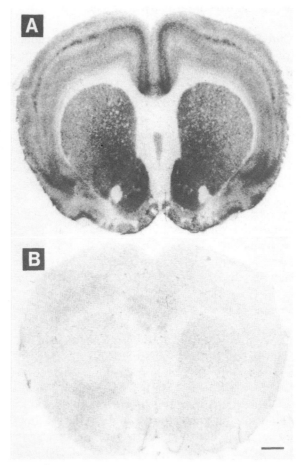

Figure 3. *A* An example of a receptor autoradiogram for the binding of the neuropeptide cholecystokinin (CCK-8S) to a coronal section of rat brain. The section was incubated with 0.25 nM of [^{125}I] CCK-8S for 120 min at 20°C. After washing and drying, sections were apposed to Ultrofilm for 7 days at 20°C. *B* An adjacent section to that used to produce (*A*) was co-incubated with unlabelled CCK-8S (1 μM) to produce the autoradiogram for non-specific [^{125}I] CCK-8S binding[4–6]. Scale bar = 1 mm.

Potential errors in quantitative autoradiography

Many factors can affect the response of the emulsion to radioactivity and therefore the resulting optical densities within the autoradiogram[1,28,35]. In order to minimise slide-to-slide variation, sections from control and experimental animals should ideally be mounted on the same slide. Slides are normally placed in contact with films in light-tight X-ray cassettes where a maximum of 64 slides including one standard can be exposed to a large sheet of film (35 × 43 cm) at the same time.

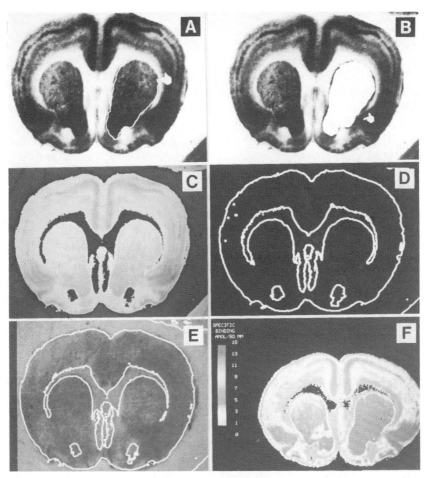

Figure 4. Procedure for measuring the optical density in an autoradiographical image of the total binding of CCK-8S to a coronal section of rat brain. A cursor is used to draw within an anatomical region (*A*). The computer isolates the area of interest and measures the integrated optical density (*B*). The threshold for detecting the image is then increased (*C*) to produce a template (*D*), which is used to align the autoradiographical image of non-specific binding to an adjacent section (*E*). The second image is digitally subtracted from the first to obtain a new image (*F*) of the amount of specific CCK-8S bound in attomoles (1×10^{-18} moles) per square millimetre (amol/mm^2), calculated by interpolating from the standard curve.

Slides bearing replicate sections should be positioned throughout the area occupied by the film, to minimise possible variation in the thickness of the emulsion. Uneven pressure which may be applied by the cassette locking mechanism can result in blurred images giving 50% lower optical density values compared to autoradiographical images in sharp focus[1]. The use of intensifying screens to increase the autoradiographical efficiency by converting ionising radiations to light should be avoided as resolution is frequently decreased, the low intensities of light produce disproportionately faint images and the response of the emulsion to radioactivity is complex[35] although this can be overcome in part, by pre-exposing film to a hypersensitising light flash and cooling the film to $-70°C$[26].

The conditions of developing the film at the end of the exposure period are particularly important and are influenced by the composition, concentration, temperature and age of the developer[1,35]. In order to minimise these variables, it is essential that standards of the same radio-isotope used to label the tissue sections are co-exposed with each film[1,13]. A calibration curve should be constructed for each film as it is unsafe to assume that optical density values produced by a set of standards will be the same for a different film even when processed under similar conditions[1].

Non-linear response of the emulsion to radioactivity

The relationship between optical density and radioactivity is not linear; optical density does not increase as rapidly as tissue radioactivity increases and the response is affected by the characteristics of the radiation source, the type of emulsion, conditions of exposure and method of development[1,23,35]. A number of mathematical transformations have been used to describe this relationship[23]; we (see Davenport and Hall[13]) have chosen to plot the natural log of optical density versus the natural log of radioactivity in order to give a long central linear portion. In figure 5, the response of two types of radiation sensitive film, Hyperfilm [^{3}H] and βmax have been compared following exposure to [^{125}I] standards incorporated either with polymer (Amersham International) (shown as squares) or tissue paste derived from areas of the brain containing mainly grey matter (circles) or white matter (triangles). For both grey and white tissue paste standards there was a linear relationship between the natural log of optical density and radioactivity when exposed to either type of film from 1 to 14 days and there was no significant difference between the slopes and intercepts of these lines. The linear relationship between optical density and radioactivity was lost for the higher activity levels of the polymer standards (indicated by asterisks on fig. 5) and should be removed, depending on the anticipated length of exposure and age of the product. However, when these data were truncated to the domain of the tissue paste standards, there was no significant difference between either the slopes or intercepts of brain paste or polymer standards. Thus polymer standards can be used to quantify the binding of [^{125}I] labelled peptide to tissue sections[13].

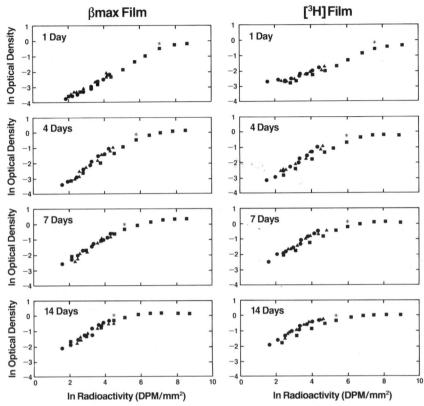

Figure 5. Natural log plot of optical density versus radioactivity/mm² for standards produced by mixing [¹²⁵I] label with tissue paste containing mainly grey (●) or white matter (▲) and with polymer (■) following direct exposure for 1–14 days to βmax or [³H] films. For polymer standards, the linear relationship between ln optical density and radioactivity is lost at higher levels of radioactivity as indicated by * but there was no difference between slopes and intercepts of tissue and the lowest five activity levels of the polymer standards[13].

Saturation of the film optical density

A major potential error in the quantification of autoradiograms is illustrated by figure 5. Both types of Hyperfilm approach saturation at an optical density of 1 (ln 0)[13] and similar values have been obtained with Ultrofilm[23]. At this point changes in tissue radioactivity produce little or no change in optical density. It is important to ensure that when using [¹²⁵I], [³⁵S] or [³²P] labelled compounds that no areas of the autoradiogram are so heavily irradiated so as to produce images approaching saturation of the film. This could occur with iodinated peptides where discrete areas of the brain may contain high concentrations of binding sites and an example of film saturation produced by using [^{125}I] insulin which could lead to erroneous results is described by Baskin and Dorsa[1]. It is therefore important to determine the upper and lower limits for making accurate measurements for a particular type of film and image analyser. Tissue sections should be exposed so as to produce autoradiograms with optical densities falling well within the linear range of the standards[1,13].

Comparison of [^{3}H] tissue paste and commercial polymer standards

Although [^{125}I] polymer standards can be used for the quantification of iodinated compounds in a variety of tissues[13], equivalent amounts of tritium incorporated into tissue paste or polymer will not necessarily produce the same optical densities in the film, owing to the different self-absorption properties of the two media. The response of Hyperfilm [^{3}H] exposed to tritium standards incorporated in polymer (Amersham International) and in grey tissue paste standards cut to the same thickness is shown in figure 6. There was no difference in the slopes of these lines, but there was a significant difference in the intercept. For sections containing tritium, infinite thickness is reached at 5 μm[35]. Any

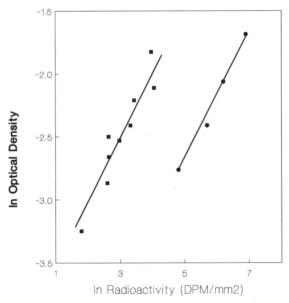

Figure 6. Natural log plot of optical density versus radioactivity/mm^2 for standards produced by mixing [^{3}H] label with tissue paste containing mainly grey matter (■) and with polymer (●) and cut to a thickness of 30 μm following direct exposure for 3 months to Hyperfilm [^{3}H]. There was no difference in the slopes of these lines, but there was a significant difference in the intercept and a correction factor must be applied[16] before the polymer standards can be used to quantify the binding of tritium-labelled ligands to tissue section.

increase in the thickness of the section beyond this point produces no further increase in the optical density of the emulsion and only a proportion of the radioactivity incorporated into 30-μm thick polymer standards will contribute to the blackening of the film. However, cryostat sections of tissue dehydrate and the thickness can be reduced by about 70%[41] so that a higher proportion of the tritium emissions will be detected. Ideally polymer-based standards should be calibrated against tissue paste standards and a correction factor applied[16] before the former can be used to quantify the binding of tritium-labelled ligands to tissue sections.

Regional variation in tissue absorption with [^{3}H]

Previous studies mapping receptors in the brain with [^{3}H] labelled ligands have suggested that white matter absorbed more beta particles than grey matter, resulting in varying autoradiographical efficiency over a tissue section[16,17,19] (Palacios and Dietl, this volume). This effect is shown in figure 7. The ability of two areas of the brain containing mainly grey or white matter to absorb beta particles was measured by thaw-mounting unlabelled sections of tissue onto microscope slides previously coated with tritium. When the thickness of the section was increased to 20 μm, white matter absorbed significantly more beta particles than grey matter.

A number of techniques for coping with regional differences in absorption have been developed[23]. The magnitude of the self-absorption factor for tritium has been determined for a number of brain regions and correction factors have been calculated[16,19]. Self-absorption appears related to the lipid content of tissue. Some ligands can be covalently linked to tissue with gaseous formaldehyde, so that lipid can be removed from the sections of brain prior to autoradiography, thus eliminating differential absorption between grey and white matter[19]. This method has been applied for some radiolabelled peptides[27] but not all ligands can be covalently linked in this way and the extraction of lipids may remove significant quantities of ligand from the tissue[23]. Ideally, absorption of tritium should be determined for each type of tissue rather than relying on published correction factors. The results shown in figure 7 and studies by Camus et al.[10] and Savastava et al.[36] suggest that provided receptor binding assays are performed on tissue sections which are less than 20-μm thick regional absorption of beta particles may not be significant.

A major advantage in the use of iodinated peptides is that no regional differences in the absorption [^{125}I] emissions have been detected in standards prepared from grey or white matter[13,23] and different areas of the brain appear to absorb [^{125}I] emissions to the same extent[23].

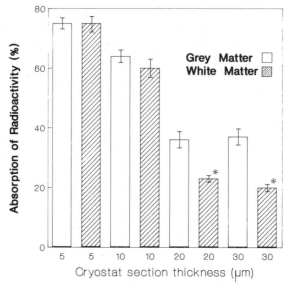

Figure 7. Differential absorption of emissions from tritium by areas of the brain containing predominantly grey or white matter. Cryostat sections of unlabelled rat brain were cut at various thicknesses and thaw-mounted onto microscope slides previously coated with tritium and apposed to Hyperfilm [³H] for two months. Optical density measurements were made in two areas containing grey or white matter. White matter absorbed significantly more beta emissions in sections cut to a thickness of 20 μm.

However, at the higher levels of resolution obtained with micro-autoradiography regional differences in silver grain densities may occur at the cellular level[39].

No differences in self-absorption by soft tissue have been reported for detecting compounds labelled with [³⁵S] or [³²P] by macro-autoradiography, owing to the higher energies of beta particles emitted by these isotopes[35,41].

Reporting the results

Calibrated radioactive scales are co-exposed with tissue sections so that optical densities can be converted into molar quantities of receptor bound radioligand. At present there is no generally accepted convention for expressing the results of autoradiographical studies. It is common to express the results of conventional homogenate binding assays as molar quantities of receptor bound peptide per unit weight of protein[1,11,18] since the protein content of the membrane preparation can be readily measured. A number of investigators have adopted this convention for receptor autoradiography by relating the amount of ligand in tissue samples to the protein content of their standards[1,11,18]. However, in complex structures such as the brain protein content varies considerably between nuclei (A. P. Davenport, unpublished observations) whereas protein in standards is homogeneous. A number of investigators[1,11,13,18] have argued that making assumptions about the protein content of small areas of tissue can be avoided by expressing the amount of ligand bound per unit area actually measured by the image-analyser. We (see Davenport and Hall[13]) therefore measure radioactivity detected by the emulsion in DPM/mm^2 because three measurements can be easily made namely, optical density, surface area and radioactivity content which can be readily converted into molar quantities per surface area by dividing by the specific activity of the ligand.

Conclusion

Many factors affect the response of film-based emulsions to radio-activity but some of these can be avoided or controlled by co-exposing a set of calibrated radioactive standards with the sections of labelled tissue. As a result of developments in computer-assisted analysis systems and provided care is taken to avoid these errors, autoradiograms can be quantified with a speed and accuracy approaching other methods of measuring radioactivity. Autoradiography has however three major advantages; decay rates of less than one disintegration per day can be measured, the morphological features of the tissue are maintained and a permanent record of the spatial distribution of radio-labelled compounds is produced which can be re-analysed. The technique is ideally suited to measuring subtle changes in the density of peptide receptors within complex structures such as the brain following drug administration, during disease or during development which would be difficult or impossible to detect by conventional homogenate-based assays.

1 Baskin, D. G., and Dorsa D. M., Quantitative autoradiography and in vitro radioligand binding, in: Functional Mapping in Biology and Medicine: Computer-assisted Autoradiography, pp. 204–234. Ed. D. L. McEachron. Karger, Basel 1986.

2 Baskin, D. G., Davidson, D., Corp, E. S., Lewellen, T., and Graham, M., An inexpensive microcomputer digital imaging system for densitometry: quantitative autoradiography of insulin receptors with [^{125}I] and LKB Ultrofilm. J. Neurosci. Meth. 16 (1986) 119–129.

3 Baldino, F., and Davis, L. G., Gluccocorticoid regulation of vasopressin messenger RNA, in: In Situ Hybridization in Brain, pp. 97–116. Ed. G. R. Uhl. Plenum Press, New York 1986.

4 Beresford, I. J. M., Davenport, A. P., Hall, M. D., Hill, R. G., and Hughes, J., Autoradiographic localization of CCK receptors on intrinsic striatal neurones. Neuroscience 22 (1987a) S803.

5 Beresford, I. J. M., Hall, M. D., Clark, C. R., Hill, R. G., Hughes, J., and Sirinathsinghji, D. J. S., Striatal lesions and transplants demonstrate that cholecystokinin receptors are localized on intrinsic striatal neurones: a quantitative autoradiographic study. Neuropeptides 10 (1987b) 109–136.

152

6 Beresford, I. J. M., Davenport, A. P., Hall, M. D., Hill, R. G., Hughes, J., and Sirinathsinghji, D. J. S., Experimental hemiparkinsonism in the rat following chronic unilateral infusion of MPP+ into the nigrostriatal dopamine pathway: II. Differential localization of dopamine and cholecystokinin receptors. Neuroscience 27 (1988) in press.

7 Boast, C. A., Snowhill, E. W., and Altar, C. A., Quantitative Receptor Autoradiography. Alan R. Liss, New York 1986.

8 Brown, M. J., Davenport, A. P., Emson, P. C., Hall, M. D., and Nunez, D. J., Adrenal cells express atrial natriuretic mRNA (ANP mRNA). An 'in-situ' hybridization study. J. Physiol. 398 (1988a) 76P.

9 Brown, M. J., Davenport, A. P., Emson, P. C., Hall, M. D., and Nunez, D. J., Measurement of atrial natriuretic peptide messenger RNA (ANP mRNA) in the neonatal rat heart by computer-assisted image analysis of 'in-situ' hybridization (ISH). J. Physiol. 398 (1988b) 75P.

10 Camus, A., Javoy-agid, F., Dubois, A., and Scatton, B., Autoradiographic localization and quantification of Dopamine D2 receptors in normal human brain with [^{3}H]N-n-propylnorapomorphine. Brain Res. 375 (1986) 135–149.

11 Clark, C. R., and Hall, M. D., Hormone receptor autoradiography: Recent developments. TIBS 11 (1986) 195–199.

12 Davenport, A. P., Hall, M. D., Hill, R. G., and Hughes, J., Evaluation of [^{125}I] reference sources for quantitative receptor autoradiography: comparison between brain paste and polymer standards. Neuroscience 22 (1987) S371.

13 Davenport, A. P., and Hall, M. D., Comparison between brain paste and polymer [^{125}I] standards for quantitative receptor autoradiography. J. Neurosci. Meth. 25 (1988) 75–82.

14 Ehn, E., and Larsson, B., Properties of an antiscratch-layer-free X-ray film for the autoradiographic registration of tritium. Sci. Tools 26 (1979) 24–29.

15 Eilbert, J. L., Quantitative analysis of autoradiographs, in Functional Mapping in Biology and Medicine: Computer assisted Autoradiography, pp. 122–142. Ed. D. L. McEachron. Karger, Basel 1986.

16 Geary, W. A., and Wooten, G. F., Regional tritium quenching in quantitative autoradiography in the central nervous system. Brain Res. 336 (1985) 334–336.

17 Geary, W. A., Toga, A. W., and Wooten, G. F., Quantitative film autoradiography for tritium: Methodological considerations. Brain Res. 337 (1985) 99–108.

18 Hall, M. D., Davenport, A. P., and Clark, C. R., Quantitative receptor autoradiography. Nature 324 (1986) 493–494.

19 Herkenham, M., and Sokoloff, L., Quantitative receptor autoradiography: Tissue defatting eliminates differential self-absorption of tritium radiation in grey and white matter of brain. Brain Res. 321 (1984) 363–368.

20 Horrocks, D. L., Standardizing ^{125}I sources and determining ^{125}I counting efficiencies of well-type gamma counting systems. Clin. Chem. 21 (1975) 370–375.

21 Kuhar, M. J., Autoradiographic localization of drugs and neurotransmitter receptors, in: Handbook of Chemical Neuroanatomy: Methods in Chemical Neuroanatomy, vol. 1, pp. 398–415. A. Eds A. Bjorklund and T. Hökfelt. Elsevier, Amsterdam.

22 Kuhar, M. J., Recent progress in receptor mapping: which neurons contain the receptors. TINS 10 (1987) 308–310.

23 Kuhar, M. J., and Unnerstall, J. R., Quantitative receptor mapping by autoradiography: Some current technical problems. TINS 8 (1985) 49–53.

24 Kuhar, M. J., De Souza, E. B., and Unnerstall, J. R., Neurotransmitter receptor mapping by autoradiography and other methods. A. Rev. Neurosci. 9 (1986) 27–59.

25 Larsson, B., and Ullberg, S., A film for the registration of tritium. Sci. Tools, Special Issue on Whole Body Autoradiography (1977) 30–33.

26 Laskey, R. A., Radioisotope detection by fluorography and intensifying screens. Amersham International plc, Amersham, UK 1984.

27 Lewis, M., Pert, A., Pert, C., and Herkenham, M., Opiate receptor localization in rat cerebral cortex. J. comp. Neurol. 216 (1983) 339–358.

28 McEachron, D. L., Adler, N. T., and Tretiak, O. J., Two views of functional mapping and autoradiography, in: Functional Mapping in Biology and Medicine: Computer assisted Autoradiography, pp. 1–46. Ed. D. L. McEachron. Karger, Basel 1986.

153

29 Nunez, D. J., Davenport, A. P., Emson, P. C., Hall, M. D., and Brown, M. J., In-situ hybridization to localize artriuetic peptide messenger RNA (ANP mRNA). Quantitation using computer-assisted (Quantimet 920) image analysis. Clin. Sci. molec. Med. Suppl. 17 (1987) 1p.

30 Pan, H. S., Frey, K. A., Young, A. B., and Penney, J. B., Changes in [^{3}H]-muscimol binding in the substantia nigra, endopeduncular nucleus, globus pallidus and thalamus after striatal lesions as demonstrated by quantitative receptor autoradiography. J. Neurosci. 3 (1983) 1189–1198.

31 Paxinos, G., and Watson, C., The rat brain in stereotaxic coordinates. Academic Press, Sydney 1986.

32 Palacios, J. M., Probst, A., and Cortes R., Mapping receptors in human brain. TINS 9 (1986) 284–289.

33 Rainbow, T. C., Bleisch, W. V., Biegon, A., and McEwen, B. S., Quantitative densitometry of neurotransmiter receptors. J. Neuorsci. Meth. 5 (1982) 127–138.

34 Rainbow, T. C., Biegon, A., and Berck, D. J., Quantitative receptor autoradiography with tritium-labelled ligands: comparison of biochemical and densitometric measurements. J. Neurosci. Meth. 11 (1984) 231–241.

35 Rogers, A. W., Techniques of Autoradiography, 3rd edn. Elsevier, Amsterdam 1979.

36 Savasta, M., Dubois, A., and Scatton, B., Autoradiographic localization of D1 Dopamine receptors in rat brain with [^{3}H]SCH 23390. Brain Res. 375 (1986) 291–301.

37 Sherman, T. G., Kelsey, J. E., Khachaturian, H., Burke, S., Akil, H., and Watson, S. J., Opioid peptides and vasopressin: Application of in situ hybridization to studies of the hypothalamus and pituitary, in: In Situ Hybridization in Brain, pp. 49–62. Ed. G. R. Uhl. Plenum Press, New York 1986.

38 Shivers, B. D., Harlan, R. E., Romano, G. J., Howells, R. D., and Pfaff, D. W., Cellular location and regulation of proenkephalin mRNA in rat brain, in: In Situ Hybridization in Brain, pp. 3–20. Ed. G. R. Uhl. Plenum Press, New York 1986.

39 Thiel, E., Dormer, P., and Ruppelt, W., Thierfelder, S., Quantitative immunoautoradiography at the cellular level II. Absolute measurements using labelled standard cells as a source of reference. J. Immun. Meth. 12 (1986) 237–251.

40 Uhl, G. R., Evans, J., Parta, M., Walworth, C., Hill, K., Sasek, C., Voigt, M., and Reppert, S., Vasopressin and somatostatin mRNA in situ hybridization, in: In Situ Hybridization in Brain, pp. 21–47. Ed. G. R. Uhl. Plenum Press, New York 1986.

41 Ullberg, S. A., The technique of whole body autoradiography. Sci. Tools, Special Issue on Whole Body Autoradiography (1987) 2–29.

42 Unnerstall, J. R., Niehoff, D. L., Kuhar, M. J., and Palacios, J. M., Quantitative receptor autoradiography using [^{3}H]Ultrofilm: application to multiple benzodiazepine receptors, J. Neurosci. Meth. 6 (1982) 59–73.

43 Young, W. S., and Kuhar, M. J., A new method for receptor autoradiography: [^{3}H]opioid receptors in rat brain. Brain Res. 179 (1979) 255–270.

44 Young, W. S., and Kuhar, M. J., Quantitative in situ hybridization and determination of mRNA content, in: In Situ Hybridization in Brain, pp. 245–248. G. R. Uhl. Plenum Press, New York 1986.

45 Zilles, K., Schleicher, A., Rath, M., Glaser, T., and Traber, J., Quantitative autoradiography of transmitter binding sites with an image analyzer. J. Neurosci. Meth. 18 (1986) 207–220.

Coexistence of peptides with classical neurotransmitters

T. Hökfelt, D. Millhorn, K. Seroogy, Y. Tsuruo, S. Ceccatelli, B. Lindh, B. Meister, T. Melander, M. Schalling, T. Bartfai and L. Terenius

Summary. In the present article the fact is emphasized that neuropeptides are often located in the same neurons as classical transmitters such as acetylcholine, 5-hydroxytryptamine, catecholamines, γ-aminobutyric acid (GABA) etc. This raises the possibility that neurons produce, store and release more than one messenger molecule. The exact functional role of such coexisting peptides is often difficult to evaluate, especially in the central nervous system. In the periphery some studies indicate apparently meaningful interactions of different types with the classical transmitter, but other types of actions including trophic effects have been observed. More recently it has been shown that some neurons contain more than one classical transmitter, e.g., 5-HT plus GABA, further underlining the view that transfer of information across synapses may be more complex than perhaps hitherto assumed.

Biochemical and modern molecular biological techniques have defined a large number of bioactive substances in the central and peripheral nervous system (CNS and PNS). In addition to earlier described low molecular weight compounds such as acetylcholine (ACh), catecholamines and certain amino acids (γ-aminobutyric acid (GABA), glycine), which are considered to act as neurotransmitters, an increasing number of peptides ranging in size from a few up to 40 amino acids and more have been identified in neurons[135]. The biochemical[79,121] and immunohistochemical[12] demonstration of peptides in well-defined neuronal systems in widespread areas of the nervous system, taken together with physiological investigations, indicate that some peptides, at least in some systems, may have a transmitter role[116,135]. For example, several lines of evidence suggested a transmitter function for substance P (SP) in primary sensory neurons[117,124]. Important clues for alternate roles for peptides have also been presented[146]. Subsequent analyses have indicated that peptides exert a wide range of effects.

When the interest in neuronal peptides became manifest 15 or so years ago, and their presence in distinct subsets of peripheral and central neuron populations had been demonstrated, it seemed possible to assume that they might have a transmitter role. For instance, classical transmitters identified at that time, such as ACh, catecholamines and 5-hydroxytryptamine (5-HT) had been found to be present in only a small population of neurons in the central nervous system[34,71,120]. The addition of numerous peptides thus appeared to represent a meaningful way to 'fill up' neuronal systems, i.e. those cells that did not contain a classical transmitter produced a peptide. However, immunohistochemical analysis of the distribution of various peptides in comparison to, for example, catecholamine and 5-HT systems revealed that in many cases peptides could be observed in the same neuron that also contained a classical transmitter. This has been documented in many articles and reviews[5,18,55,62,66,86,109,113,134] as well as discussed at meetings[21,30,65,114]. The co-localization of classical neurotransmitter and peptide in the same neuron represents a logical continuation of earlier demonstrations showing the presence of biogenic amines and peptide hormones in the same endocrine cells[119,123] and possible cellular coexistence of transmitters in invertebrate neurons (e.g. Brownstein et al.[16], for discussion, see Osborne[113]). These findings, in a general sense, could be interpreted to mean that neurons either contain more than one transmitter substance (classical transmitter + peptide) or that the peptide in these neurons may be responsible for other types of functions, for example, they could exert long-term trophic effects.

Many examples of neurons that contain more than one peptide but no classical transmitter have been reported. However, it is unclear whether these neurons really lack a classical transmitter or whether the proper marker for a classical transmitter in these neurons is simply 'missing'. There is also increasing evidence that neurons may contain more than one classical transmitter. For example, 5-HT and GABA appear to coexist in the same cells of both pontine and medullary raphe nuclei as first reported by Pujol and collaborators[8,9,103]. It does not seem unlikely that these neurons in addition contain one or more peptides. Finally, increasing evidence suggests that adenosine nucleotides may participate as co-messengers in neurotransmission, as early advanced by Burnstock[18].

It is important to note that coexistence of several types of compounds with possible messenger function still largely represents a histochemical concept based on immunohistochemical demonstration of these substances, using antisera raised against various transmitters, transmitter synthesizing enzymes and peptides. Apart from problems concerning the specificity and sensitivity of these techniques, the most important questions are, of course, to what extent are these compounds actually released from the nerve endings and how do they participate in the transmission process? This article represents an initial account of an emerging view that transmission of messages across synapses is a more complicated event than perhaps previously assumed.

How to define coexistence

The nervous system is an extremely heterogeneous tissue, and it is therefore not possible to study coexistence with biochemical techniques with the present status of sensitivity. An exception may be some invertebrate neurons, which are so large that they can be isolated individually and that their content of neuroactive compounds possibly can be determined biochemically[113]. Biochemistry can be used however to demonstrate coexistence indirectly. For example, there are 'specific' neurotoxins such as 6-hydroxydopamine[142] and 5,6-dihydroxytryptamine[6] which destroy catecholamine and concomitant depletion of 5-HT neurons, respectively. With the latter compounds, a concomitant depletion of 5-HT, SP and TRH has been shown and was interpreted to indicate coexistence of these compounds in single neurons[48].

Because neurons can be visualized individually in the microscope and because antibodies can identify substances within a single cell or a slice of a cell, immunohistochemistry[27,31,105,126,138] offers the most accurate method for determining coexistence. Various immunohistochemical approaches can be used to study multiple antigens in a neuron. They include the *'adjacent section method'*, where consecutive sections are incubated with different primary antisera. No cross-reaction between antisera can occur and consequently there are no problems of specificity due to interference between antibodies. Only large objects such as cell bodies can be studied but with sufficiently thin sections a cell body can often be identified in two or even more consecutive sections. When epoxy resin-embedded material is used, sections can be cut at 1 μm or thinner, and then numerous sections through a single cell body can be analyzed[10]. *'Elution-restaining methods'*[105,143] have been extensively utilized. After photography of the first staining pattern, the antibodies are eluted with acid solutions, and the sections are then reincubated with a new antiserum, and the new staining patterns are compared with the previously taken photographs. This method can, in our experience, not be used with all antisera, since the elution procedure seems to damage some antigens. The third approach is *'direct double-staining'*, which is based on availability of antisera raised in different species (fig. 1, *a–h*). Secondary antibodies labelled with different chromogens (e.g. green fluorescent fluorescein isothiocyanate, FITC, and red fluorescent tetramethyl rhodamine isothiocyanate, TRITC) and directed against IgG from the two respective species then allow visualization of the two antigens in the same section by switching between appropriate filter combinations (fig. 1, *a–h*; see ref. 105). In fact, it has recently been shown that three antigens can be visualized in a single section using a third, blue fluorescent dye conjugated to an appropriate secondary antibody[137]. By combining this triple staining technique with elution-restaining, it should be possible to visualize four or even more antigens in a section. The final analysis of coexistence will, however, include electron microscopic studies. It has, for example, been shown that 5-HT and SP are stored in the same vesicles in some nerve endings in the spinal cord[122], and also at the ultrastructural level there are now methods to demonstrate three antigens in one section[37].

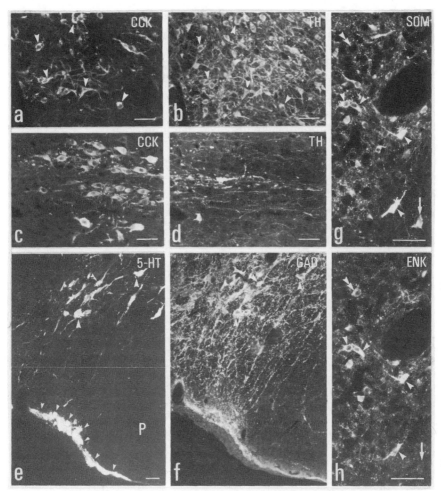

Figure 1. Immunofluorescence micrographs of the ventral tegmental area (VTA) (*a, b*), the periaqueductal central grey (PAG) (*c, d*), the ventral medulla oblongata (*f, g*) and the nucleus tractus solitarii (NTS) (*g, h*), after incubation with antibodies to cholecystokinin (CCK) (*a, c*), tyrosine hydroxylase (TH), a marker of dopamine (DA) neurons (*b, d*), 5-hydroxytryptamine (5-HT) (*e*), glutamic acid decarboxylase (GAD), a marker for GABA neurons (*f*), somatostatin (SOM) (*g*), and methionine-enkephalin (ENK) (*h*). *a* and *b, c* and *d, e* and *f* as well as *g* and *h* show, respectively, the same sections which have been processed according to double staining technique using primary antisera raised in different species and secondary antibodies labelled with green fluorescent FITC and red fluorescent TRITC, respectively. This series of micrographs are meant to illustrate coexistence of classical transmitter and peptide (DA plus CCK) (*a, b*), two classical transmitters (5-HT plus GABA) and two peptides (SOM plus ENK). *a–d* Numerous cell bodies (arrow heads) in the VTA contain both CCK-and TH-like immunoreactivity (LI), whereas in the PAG no double-labelled cells can be seen. *e, f* In the area lateral to the pyramidal tract (P) numerous cell bodies (big arrow heads) contain both 5-HT- and GAD-LI. Note numerous 5-HT cells (small arrow heads) along the ventral surface of the brain which seem to lack GAD-LI. *g, h* In the NTS numerous cell bodies (arrow heads) contain both peptides, but there are also cells containing only one of the peptides (double arrow heads point to SOM-negative, ENK-positive cell and arrows point to SOM-positive, ENK-negative cell). Bars indicate 50 μm. (From refs 103, 104.)

Immunohistochemical methods should, in spite of their power and usefulness, be considered with some caution both with regard to specificity and their sensitivity. Thus, it cannot be excluded that the antisera cross-react with compounds which are structurally similar to the immunogens. Recently evidence has been presented that one single amidated amino acid in the C-terminal position may be sufficient to cause cross-reactivity[11,74]. Therefore, expressions such as 'somatostatin(SOM)-like immunoreactivity'. 'SOM-immunoreactive', etc. should be used.

It is also important to emphasize the sensitivity problem and that negative results should be interpreted with great caution. It has been demonstrated repeatedly that improvement of the fixation technique and/or production of antibodies with a higher affinity and/or higher avidity reveal a certain antigen in places where it had not been demonstrated earlier. Also, peptide levels in cell bodies are often too low to be visualized in central neurons, but can be increased by pretreatment of experimental animals with a mitosis inhibitor, colchicine[33], and in this way visualized.

Coexistence—overview

During the last years an increasing number of neurons containing co-existing messenger molecules have been described both in the central and peripheral nervous systems. Limited space does not allow a complete account of this work, therefore only selected cases are included. As indicated above, different types of combinations have been encountered: 1) classical transmitter + peptide(s) (fig. 1, *a–d*), 2) more than one classical transmitter (fig. 1, *e, f*), and 3) more than one peptide (fig. 1, *g, h*). In table 1 coexistence of classical transmitters and peptides in the mammalian CNS are listed, which have been described in papers in 1985 and earlier. The purpose of this table is to demonstrate that for each of the classical transmitters there is at least one example of coexistence with one or more peptides. In table 2, examples of the recent evidence that neurons may contain more than one classical transmitter are summarized. The first evidence concerned coexistence of a biogenic amine (5-HT) and an amino acid (GABA)[8,9], and it seemed possible to argue that 5-HT and GABA indeed belong to different classes of compounds and thus that they might complement each other in some unknown way. More recently, however, there is evidence for occurrence of two inhibitory amino acids (GABA and glycine) in the same Golgi neurons in the rat cerebellum[118]. Also indirect evidence suggests such a coexistence, since GABA nerve endings have been shown to be located opposite to postsynaptic membranes that contain glycine receptors[144].

Table 1. Coexistence of classical transmitters and peptides in the mammalian CNS[a] (selected cases)

Classical transmitter	Peptide[b]	Brain region (species)	References
Dopamine	CCK	Ventral mesencephalon (rat, cat, mouse, monkey, man?)	57, 58, 64, 66
	Neurotensin	Ventral mesencephalon (rat)	59
		Hypothalamic arcuate nucleus (rat)	59, 70
Norepinephrine	Enkephalin	Locus coeruleus (cat)	24, 80
	NPY	Medulla oblongata (man, rat)	40, 60, 130
		Locus coeruleus (rat)	40
	Vasopressin	Locus coeruleus (rat)	19
Epinephrine	Neurotensin	Medulla oblongata (rat)	59
	NPY	Medulla oblongata (rat)	40, 130
	Substance P	Medulla oblongata (rat)	84
	Neurotensin	Solitary tract nucleus (rat)	59
5-HT	Substance P	Medulla oblongata (rat, cat)	20, 22, 54, 73, 85
	TRH	Medulla oblongata (rat)	56, 73
	Substance P + TRH	Medulla oblongata (rat)	73
	CCK	Medulla oblongata (rat)	94
	Enkephalin	Medulla oblongata, pons (cat)	49, 68
		Area postrema (rat)	4
ACh	Enkephalin	Superior olive (guinea pig)	2
		Spinal cord (rat)	76
	Substance P	Pons (rat)	145
	VIP	Cortex (rat)	38
	Galanin	Basal forebrain (rat, monkey)	100, 101
	CGRP	Medullary motor nuclei (rat)	140
GABA	Motilin (?)	Cerebellum (rat)	23
	Somatostatin	Thalamus (cat)	111
		Cortex, hippocampus (rat, cat, monkey)	52, 72, 131, 136
	CCK	Cortex, hippocampal formation (cat, monkey, rat)	52, 78, 136
	NPY	Cortex (cat, monkey)	52
	Enkephalin	Retina (chicken)	147
		Ventral pallidum, hypothalamus (rat)	75, 150
	Opioid peptide	Basal ganglia (rat)	110
	Galanin	Hypothalamus (rat)	102
	Substance P	Hypothalamus (rat)	75
	VIP	Hippocampal formation (rat)	78
Glycine	Neurotensin	Retina (turtle)	148

[a] The coexistence situations have been defined mainly by immunohistochemistry. Only papers published in 1985 or earlier are included. [b] This column contains the peptide against which the antiserum used for immunohistochemistry was raised. The exact structure of the peptide coexisting with the classical transmitter has for the most part not been defined.

Table 2. Coexistence of two classical transmitters in the CNS[a]

Classical transmitter 1	Classical transmitter 2	Brain region (species)	References
GABA	5-HT	Nucleus raphe dorsalis (rat)	8, 9
		Medullary raphe nuclei and adjacent areas (rat)	8, 9, 103
		Retina (rabbit)	115
GABA	DA	Arcuate nucleus (rat)	41
		Olfactory bulb (rat)	46, 77
GABA	Histamine	Hypothalamus (rat)	133
GABA	ACh	Medial septum/diagonal band (rat)	15
GABA	Glycine	Cerebellum (rat)	118

[a] The coexistence situations have been defined mainly by immunohistochemistry using antisera raised against the transmitter itself and/or a transmitter synthesizing enzyme.

Coexistence situations in the CNS

As shown in table 1, peptides can be found in virtually all types of classical transmitter neurons in many parts of the central nervous system. It does not seem unlikely that further research will find more and more such examples and that coexistence is a rule rather than an exception. It is of particular interest that many GABA neurons in cortical areas including hippocampus contain one or more peptides[52,78,131,136]. It should be noted that many types of coexistence combinations seem to occur in an unpredictable way and that often only subpopulations of neurons seem to contain a certain peptide. For example, 5-HT neurons in the lower medulla oblongata contain a SP-[22,54] and also a thyrotropin-releasing hormone(TRH)-like peptide[48,56,73], whereas so far no such coexistence has been reported in pontine and mesencephalic 5-HT cells. Moreover, the proportion of 5-HT neurons that contain the two peptides vary within the medullary raphe nuclei[73]. Differential coexistence is also observed in the catecholamine neurons. Thus of the multiple groups originally described and defined by Dahlström and Fuxe[34] (A1–A12; see refs 61 and 63), the A1 and A6 noradrenergic and the C1–C3 adrenergic neurons contain a neuropeptide Y(NPY)-like peptide[13,30,40], the parvocellular C2 adrenaline group a neurotensin[59]—and a cholecystokinin (CCK)[64]-like peptide. Some A1 and most A6 neurons express a galanin like peptide[102], and many mesencephalic and some hypothalamic dopamine (DA) neurons exhibit neurotensin-like immunoreactivity (LI)[42,59,70], whereby the A12 TH-positive neurons contain galanin-, neurotensin- and growth hormone releasing factor (GRF)-LI[97,98,112]. Finally the caudal part of the A13 DA cell group has a SOM-like peptide[99]. The distribution of CCK-LI in the mesencephalic DA neurons is particularly intricate and is illustrated in figure 2 showing e.g. that the A10 cells in the ventral tegmental area (fig. 1, a, b) have an increasing proportion of coexistence

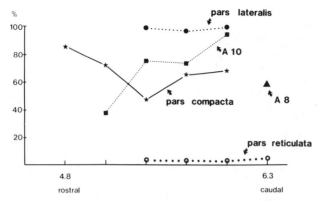

Figure 2. Schematic illustration of the percentage of dopamine (DA) neurons containing cholecystokinin (CCK)-LI in various subregions of the ventral mesencephalon at different rostral-caudal levels. (The most rostral level is approximately 4.8 mm behind the Bregma, the most caudal point 6.3 mm behind the Bregma and sections have been analyzed at 0.3-mm intervals.) Areas analyzed are pars compacta, pars lateralis and pars reticulata of the substantia nigra as well as the ventral tegmental area (A10 DA cell group) and the A8 DA cell group in the mesencephalic reticular formation. Note high proportion of DA/CCK coexistence in pars lateralis and low percentage in pars reticulata. In the ventral tegmental area there is an increasing incidence of coexistence in the caudal direction. (From Staines, Hökfelt, Goldstein et al., in ref. 66.)

in caudal direction, whereas neurons of the pars lateralis have almost 100% coexistence and hardly any cells containing both DA and CCK are found in pars reticulata. A galanin-like peptide has recently been observed in the basal forebrain cholinergic neurons both in rat and monkey; these neurons project to the hippocampal formation[100,101].

Coexistence in the PNS

In the peripheral nervous system coexistence is frequently encountered; in fact, it can be observed in most systems. Particularly complicated patterns have been observed in the gastrointestinal tract with up to four peptides in presumably cholinergic neurons (see ref. 29). The sympathetic and parasympathetic systems are also rich in peptides (see ref. 87). Originally SOM-LI was found in a population of sympathetic noradrenergic neurons[53]. Further study has extended these findings. For example, in immunohistochemical analysis the coeliac-superior mesenteric ganglion in guinea pig has shown at least three distinct populations of neurons (figs 3, a–c; 4)[82,91,93]:

1) noradrenergic ganglion cells containing an NPY-like peptide (approximately 65% of all neurons), 2) noradrenergic cells containing SOM-LI (25%), and 3) a small population of vasoactive intestinal polypeptide (VIP)/peptide histidine isoleucine(PHI)-positive cell bodies. Some of the latter ones contained NPY-LI and sometimes also noradrenaline (NA).

162

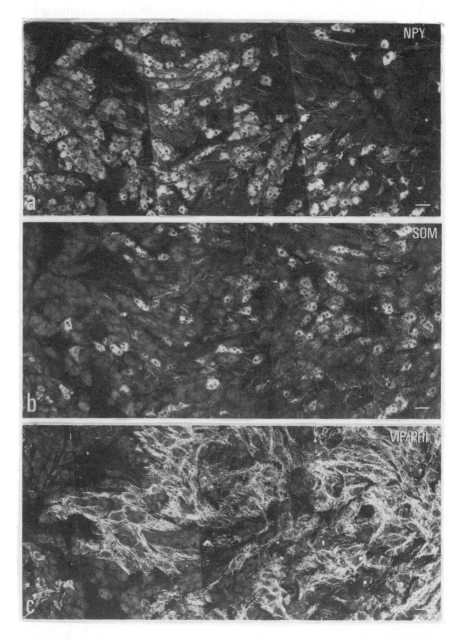

Figure 3. Immunofluorescence micrographs (montages) of the coeliac-superior mesenteric ganglion of guinea pig after incubation with antiserum to neuropeptide Y(NPY) (*a*), somatostatin (SOM) (*b*) and peptide histidine isoleucine (PHI) (*c*). The montages show semiadjacent sections in the border zone between the NPY and SOM domains. On the left-hand side the NPY-positive cell bodies dominate, whereas SOM cell bodies are seen mainly to the right, but in this particular area a considerable intermingling takes place. Note that PHI-positive fibers originating in the gastro-intestinal wall preferentially innervate SOM-positive cell bodies. Bar indicates 50 μm. (From ref. 82.)

These neurons have specific domains within the ganglion and have been shown to project to different targets in the gastrointestinal wall (fig. 4)[28,45]. Moreover, they seem to be controlled by different afferent inputs (fig. 4). Thus, whereas the afferents from the intestine containing e.g.

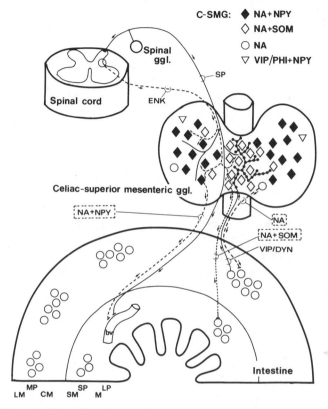

Figure 4. Schematic illustration of the coeliac-superior mesenteric ganglion (C-SMG) and its connection with the spinal cord and intestine with special reference to peptides and coexistence systems. In the C-SMG two main populations of ganglion cells are seen characterized by the presence of noradrenaline (NA) plus neuropeptide Y(NPY)-like immunoreactivity (LI) and NA plus somatostatin(SOM)-LI, respectively. The former are located in the lateral parts of the ganglion, whereas the NA + SOM ganglion cells occupy its mid portion. Small population of cells contain vasoactive intestinal peptide (VIP)/peptide histidine isoleucine (PHI) plus NPY, and they are located in the lateral aspects of the 'NPY domaine'. Some NA cells seem to lack a peptide. The NA + SOM neurons project to the submucous ganglion, whereas the NA + NPY neurons innervate blood vessels in the intestinal wall. Cell bodies containing NA alone project to the myenteric ganglia. Projections from the intestine to the C-SMG arise from the myenteric ganglia and contain multiple peptides including VIP, PHI and dynorphin (DYN). They seem to innervate exclusively NA plus SOM cell bodies in the midline areas of the ganglion. The fibers from the spinal cord contain an opioid peptide, possibly an enkephalin(ENK)like peptide, and they distribute diffusely over the ganglion. To what extent these fibers also contain acetylcholine has not been established. Finally, primary sensory neurons containing i.a. substance P give rise to a diffuse plexus within the ganglion, and these fibers represent collaterals of axons continuing on to the gastrointestinal wall and innervating blood vessels. bv, blood vessel; CM, circular muscle layer; LM, longitudinal muscle layer; LP, lamina propria; M, mucosa; MP, myenteric plexus; SM, submucosa; SP, submucous plexus. (This schematic drawing is based mainly on work in refs 28, 35, 36, 42, 82, 91, 93, 96.)

VIP/PHI and DYN exclusively terminate around the SOM-positive cell bodies (cf. fig. 3, *b* and *c*), afferents from the spinal cord and spinal ganglia have a more widespread distribution[35,36,82,93,96]. These findings suggest that chemical coding of neurons by a particular peptide in the peripheral nervous system may reflect its participation in a well-defined physiological event.

Are coexistence combinations preserved during phylogeny?

This question has been studied only to a limited extent. However, there are examples both of variation and preservation of certain coexistence situations among different species. For example, NA and SOM coexist in sympathetic neurons in guinea pig and rat but not in cat[90,91]. Coexistence of DA and CCK-LI has been observed in mouse, rat, cat, monkey and probably man, but the proportions and exact distribution of coexistence neurons in the ventral mesencephalon seem to vary among the different species[64]. In contrast, no CCK-LI has so far been observed in mesencephalic DA cell bodies in guinea pig. We have analyzed one of the most primitive vertebrates, the lamprey fish, and although coexistence situations have been encountered, there is so far no evidence for any major coexistence of those compounds which have been described in mammals[17].

Functional significance of coexistence

The functional significance of coexistence of multiple putative messenger molecules is not very well understood. A key issue is whether or not it is meaningful to have numerous compounds simultaneously conveying messages between neurons or a neuron and an effector cell; i.e. whether these messengers can produce selective and differential responses. There are several models which explain how multiple messengers might work. One shows that the neuron under all conditions releases all types of messenger molecules at the same time, and that the distribution and type of receptors provide selectivity and specificity, i.e. post-synaptic selectivity. An alternative model would be the ability to release the messenger differentially i.e., presynaptic selectivity. Both, of course, may operate together and other types of mechanisms should also be considered. In the following we shall present some morphological evidence for the view that differential release can be obtained and that this is related to differential storage of the transmitter substance.

In general, nerve endings contain at least two types of vesicles, the synaptic vesicle (diameter about 500 Å) and a large type of vesicle (diameter about 1000 Å), often containing an electron-dense core and termed 'large dense-core' or 'granular' vesicle. Immunohistochemical studies at the ultrastructural level have revealed that peptides seem to be located in the large dense-core vesicles. For example, Pelletier et al.[122] demonstrated that SP is present in large dense-core vesicles in nerve endings in the ventral horn of the spinal cord. This general idea is supported by subcellular fractionation studies demonstrating that VIP in the cat salivary gland[89] and NPY in rat vas deferens[44] seem to appear exclusively in a heavy fraction characterized by the presence of large dense-core vesicles. These fractions also contained overlapping peaks with the coexisting classical transmitters, i.e. ACh and NA, respectively. In contrast, the lighter fractions, presumably characterized by content of small synaptic vesicles, only contain classical transmitters. These findings suggest that peptides at least in some tissues are stored exclusively in large vesicles, whereas classical transmitters are found in both dense-core and synaptic vesicles. Thus, if a mechanism would exist allowing selective activation and release from the two types of vesicles, it should be possible to obtain differential release of transmitter substance from the nerve terminal. There is evidence that the classical transmitter and peptide can be released differentially and that this release is dependent on the frequency of action potentials[39,90]. According to this hypothesis, a low impulse frequency selectively activates small vesicles resulting in the release of the classical transmitter, whereas at higher frequencies or by bursts of impulses the large vesicles also release their content in addition. In this way the classical transmitter is released selectively or in combination with a peptide(s)[86].

Interaction of coexisting messengers

Some experimental models have yielded interesting and perhaps meaningful results concerning possible interaction among transmitter substances. For example, the cat salivary gland receives a parasympathetic innervation containing ACh together with VIP and PHI and noradrenergic sympathetic, perivascular fibers containing NPY[90]. ACh induces both secretion and an increase in blood flow and these effects are both atropine sensitive[90]. VIP alone has no apparent effect on secretion but causes increased blood flow, thus co-operating with ACh in the regulation of blood flow[90]. Moreover, VIP potentiates ACh-induced secretion, and additive effects on blood flow are seen when ACh and VIP are infused together[90]. With regard to the sympathetic control of blood flow, NA and NPY cooperate in causing vasoconstriction, whereby NPY alone exhibits a slowly developing, long lasting effect[88,90]. A different type of interaction has been observed in rat vas deferens which is innervated by noradrenergic fibers containing NPY[1a,92,139], since here the peptide inhibits release of NA. Thus, the peptide seems to exert an antagonistic action at the presynaptic level.

A second model was tested in the autonomic nervous system of the bullfrog by Jan and collaborators[14]. The frog sympathetic ganglion allows a thorough analysis of the coexistence concept, since it is possible to define the roles of the coexistence messengers also with electrophysiological techniques. Neurons in some ganglia of the lumbar chain contain ACh and a luteinizing hormone-releasing hormone (LHRH)-like peptide, and physiological experiments indicate that the preganglionic C-fibers release both ACh and the peptide[14]. However, the targets for the two compounds are not identical. Acetylcholine exerts its actions only on the so-called C-cells, which are in synaptic contact with the preganglionic C-fiber from which ACh is released. The LHRH-like peptide causes responses only in some C-cells but does in addition activate B-cells which are many μm apart and thus not in synaptic contact with the preganglionic C-fibers[14]. The physiological analysis reveals that the LHRH-peptide causes slow excitatory postsynaptic potentials (EPSP), whereas it is known that ACh causes a fast EPSP[14]. In conclusion, in this particular model the two compounds both induce excitatory postsynaptic potentials but, whereas ACh induces a fast potential, the LHRH-like peptide is responsible for the slow EPSP. Moreover, whereas ACh acts synaptically on C-cells, the LHRH-like peptide activates only a proportion of these cells but can in addition induce slow EPSP in B-cells located up to 10 μm away. As pointed out by Branton et al.[14], the distribution of receptors and the ability of the messenger molecule to 'survive' long diffusion distances represent important factors for deciding upon what effects are evoked. These and the previous examples indicate that coexistence of messenger molecules is not a 'homogeneous' phenomenon and that different types of interaction may take place.

Are extracellular enzymes targets for neuropeptides?

Finally, in the discussion of functional significance of multiple messengers, we would like to focus on a recently discovered peptide, calcitonin gene-related peptide (CGRP)[3,127]. Using antibodies raised against this peptide, immunohistochemical studies have revealed characteristic and unique distribution patterns within the nervous system, including its presence in primary sensory neurons[127]. In fact, these CGRP-positive primary sensory neurons seem in part to be identical to previously described SP-immunoreactive neurons[47,149]. It therefore seems likely that CGRP and SP are released from the same nerve endings both in peripheral tissues as well as in the superficial layers of the dorsal horn of the spinal cord[95,129]. Possible interactions between CGRP and SP have been studied in the spinal cord after intrathecal administration of the two peptides, separately or in combination[149]. After intrathecal injection of SP at the lumbar level, rats exhibit a characteristic behavior with caudally-directed biting and scratching[69,125], and this could be confirmed in our study on rats, exhibiting a fairly short-lasting behavior (2–4 min)[149]. CGRP alone in doses up to 20 μg did not cause any observable effects. However, if SP and CGRP were injected together, a marked increase in the duration of this behavior was seen, lasting for 30 min or more[149]. A partial explanation for the prolongation of SP-induced behavior by CGRP has been forwarded by Terenius and collaborators[81]. They observed that CGRP is a potent inhibitor of a SP endopeptidase isolated from human CSF[81], suggesting that CGRP may prolong transmission at SP 'synapses' by inhibiting a degrading enzyme. This may represent a new type of interaction of two compounds released from the same nerve endings and raised some general questions concerning chemical transmission, indicating that messenger molecules may not always interact with membrane-bound receptors but perhaps, as in this case, with an enzyme located in the extracellular space.

It has been suggested[67] that such a hypothetical action of a messenger molecule on an extracellular enzyme may be a more general principle. For example, in the substantia nigra it has been reported that nerve cells can secrete acetylcholinesterase[25,51] (see Greenfield[50]), and it is known that this enzyme can hydrolyze SP[26], which is present in very dense fiber networks in the zona reticulata of the substantia nigra[32,83]. In spite of this, several groups have failed to demonstrate binding sites for tachykinins in the rat substantia nigra with receptor autoradiography[7,128,132]. One hypothetical explanation could thus be that SP released from nerve endings in the zona reticulata primarily interacts with an extracellular enzyme[67].

Studies on CGRP have also suggested another role for a coexisting peptide. It was early observed that CGRP-LI in the spinal cord is present not only in central branches of primary sensory neurons, but also in motoneurons and it therefore seems likely that CGRP coexists with ACh[47,127,140,141]. Recently evidence has been obtained by two groups that CGRP may be involved in regulation of receptor density. Thus, CGRP added to cultured chicken myotubes causes an increase in the number of surface ACh receptors, probably by acting as a long-term anterograde factor in the biogenesis and maturation of the endplate postsynaptic membrane[43,107]. These findings further underline the view that coexisting messengers may interact and act in a wide variety of ways, characteristic of the particular system in which the coexistence occurs.

Conclusions and speculations

The functional significance of the histochemical demonstration of coexistence of multiple messengers is at present difficult to evaluate, but evidence has been obtained from studies in the PNS that classical transmitters and peptides are co-released and interact in a cooperative way on effector cells. Other types of interaction may also occur, however, since peptides have been shown to inhibit the release of the coexisting classical transmitter. In the CNS, the situation is even less clear but similar mechanisms may also operate. Indirect evidence suggest that peptides may in some cases strengthen transmission at synaptic (or non-synaptic) sites and in other cases inhibit release of the coexisting classical transmitter. Thus, multiple messengers may provide a mechanism for relaying differential responses and for increasing the amount of information transmitted at synapses.

It is emphasized that coexisting messengers may not necessarily be involved directly in the transmission process at synapses but could also exert other types of actions, for example have trophic effects or induce other types of long-term events in neurons and effector cells. For instance, it has been shown that SP exerts growth-stimulatory effects on smooth muscle cells[108], and as discussed above CGRP may be involved in regulation of expression of transmitter receptors[43,107]. In fact, it may be argued that the coexistence phenomenon as such, i.e. that neurons in addition to classical transmitter(s) contain other compounds, suggests that peptides are involved in other functions, since the neurons already have a classical transmitter at their disposal for accomplishing the task of fast cell-to-cell communication.

Redundancy of neurons and neuronal systems is an important feature of the central nervous system and one may ask the question, why should it not be sufficient to have one transmitter at each synapse when there are so many nerve cells? An answer could be that redundancy is also present at the level of the individual synapse. A highly differentiated transmission process may also be necessary to achieve the enormous operational capacity of our brain, which also includes transfer of messages for long-term effects.

At this point it is, however, also wise to look upon the coexistence phenomenon with a critical eye, in view of the fact that physiological implications so far are very little elucidated. It cannot be excluded that coexistence of multiple messengers is a paraphenomenon representing a consequence of evolution. It is possible that peptides have been important messengers in lower species, and that they have been replaced by the more efficient, small-molecule transmitters, especially in phylogenetically young areas of the brain such as cortex, and that peptides at least in some places are carried along more or less as 'silent passengers'. It will be an important task to establish in the future whether or not, in fact, peptides and classical transmitters are released from the same nerve endings and under which conditions this occurs. Furthermore, it will be important to determine the models of action and interaction of the different transmitter substances.

Finally, the question may be raised whether or not the coexistence phenomenon is of interest in relation to pathological processes. So far, little evidence for such an involvement has been presented. It is obvious that, for example, the presence of a CCK-like peptide in certain mesencephalic dopamine neurons (see above) could be discussed in relation to schizophrenia, since this disease according to one of the hypothesis is related to hyperactivity of mesolimbic dopamine systems (see book edited by Matthysse and Kety[96a]), but this issue has so far not been sufficiently penetrated.

It may, however, be relevant in this overview to speculate how coexisting messengers in a general way could interact in the development of a pathologic process. As an example we have chosen coexistence of acetylcholine and the newly discovered peptide galanin[141a] in forebrain neurons of rat[100] and monkey[101] projecting to the hippocampus[100]. As shown by many groups, these cholinergic neurons may be important for higher brain functions such as memory and learning[104a,131a]. Their cholinergic nature and projections to cortical areas have been established in many studies (see Fibiger[42a] and Wainer et al.[145a] for review), and there is strong evidence that they are degenerated in Alzheimer's disease and senile dementia[148a]. It is therefore not unreasonable to consider if and how a possibly coexisting peptide, galanin, could be involved in the development of this disease.

In the hypothalamus galanin may inhibit the release of dopamine in a system where galanin and dopamine coexist[108a]. If galanin inhibits the release of acetylcholine also in the human cholinergic forebrain system, this peptide could be of importance for the development of Alzheimer's disease (fig. 5). Our reasoning is based on the hypothesis described above that a coexisting peptide is stored in the large dense-core vesicles (fig. 5) and is preferentially released when neurons are firing at a high rate or with a certain frequency pattern[39,90]. In the case of galanin, it may have the purpose to prevent excessive release of the coexisting transmitter (fig. 5, a–c). A further basis for our discussion is that the degeneration of the cholinergic forebrain neurons is a sequential process and that partial destruction of a system causes a marked hyperactivity in the remaining neurons, as demonstrated experimentally on the nigrostriatal dopamine system by Agid et al.[1]. It may be speculated that such changes occur during progressive degeneration of the cholinergic forebrain system in Alzheimer's disease. As shown in figure 5b, hyperactivity, i.e. increased firing in the remaining, non-lesioned neurons would lead to a substantial release of galanin and consequently decreased acetylcholine release. If, as one may anticipate, feedback mechanisms operate, low acetylcholine levels in the synaptic space would lead to further increase in impulse activity, further release of galanin and stronger suppression of acetylcholine release. Thus, provided that galanin biosynthesis can be maintained, the more the activity increases, the less acetylcholine is released, taking the system into a vicious circle. Such an increased strain on the neurons could also lead to accelerated degeneration and thus faster development of the

170

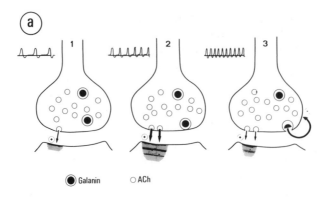

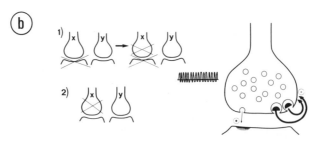

Figure 5. Schematic illustration of a cholinergic nerve ending in the hippocampus containing galanin and originating in basal forebrain. *a*) Under normal conditions acetylcholine (ACh) is released in increasing amounts with increasing impulse frequency, causing an increased postsynaptic response (1, 2). However, with very high activity (3), galanin is also released causing inhibition of ACh release. *b*) If a proportion of the cholinergic forebrain neurons is damaged (x), either as a consequence of degeneration of postsynaptic neurons (1)[135a], or by presynaptic degeneration (2), the remaining neurons (y) may exhibit hyperactivity, leading to increased galanin release, increased inhibition of ACh release and a diminished postsynaptic response. This in turn, via feed-back mechanisms, may further activate the forebrain neurons, leading to a vicious circle causing accelerated cell death.

disease. It would be interesting to know if one could counteract this process by a compound which blocks galanin binding sites. Such a galanin antagonist could be a potential drug for the treatment of Alzheimer's disease, should one become available in the future.

Acknowledgments. This study was supported by the Swedish MRC (04X-2887;04X-3776), Konung Gustafis och Drottning Victorias Stiftelse and Alice och Knut Wallenbergs Stiftelse. We thank Ms W. Hiort, Ms S. Nilsson, Ms S. Soltesz-Mattisson, Ms K. Åman and Ms E. Björklund for their assistance. D. M. was on leave from the Department of Physiology, University of North Carolina, Chapel Hill, N.C., USA. He is supported by the Swedish MRC and an Established Investigatorship of the American Heart Association and NIH Grant HL 33831. K. S. was supported by a NATO Postdoctoral Fellowship in Science. Y. T. was supported by the Swedish MRC. S. C. was supported by the Wenner-Gren Foundation and the Swedish Institute.
    We thank Drs J. Brown, Vancouver, Canada, J. Fahrenkrug, Copenhagen, Denmark, M. Goldstein, New York, USA, W. Oertel, Munich, FRG, L. Terenius, Uppsala, Sweden,

A. Verhofstad, Nijmegen, The Netherlands and J. Walsh, Los Angeles, USA, for supplying antisera used in the illustrations of this paper. For details, we refer to the original papers cited in the text and the reference list.

1 Agid, Y., Javoy, F., and Glowinski, J., Hyperactivity of remaining dopaminergic neurones after partial destruction of the nigro-striatal dopaminergic system in the rat. Nature, New Biol. 245 (1973) 150–151.

1a Allen, J. M., Tatemoto, K., Polak, J. M., Hughes, J., and Bloom, S. R., Two novel related peptides, neuropeptide Y (NPY) and peptide YY (PYY) inhibit the contraction of the electrically stimulated mouse vas deferens. Neuropeptides 3 (1982) 71–77.

2 Altschuler, R. A., Parakkal, M. H., and Fex, J., Localization of enkephalin-like immunoreactivity in acetylcholinesterase-positive cells in the guinea-pig lateral superior olivary complex that project to the cochlea. Neuroscience 9 (1983) 621–630.

3 Amara, S. G., Jonas, V., Rosenfeld, M. G., Ong, E. S., and Evans, R. M., Alternative RNA-processing in calcitonin gene expression generates mRNAs encoding different polypeptide products. Nature 298 (1982) 240–244.

4 Armstrong, D. M., Miller, R. J., Beaudet, A., and Pickel, V. M., Enkephalin-like immunoreactivity in rat area postrema: Ultrastructural localization and coexistence with serotonin. Brain Res. 310 (1984) 269–278.

5 Bartfai, T., Presynaptic aspects of the coexistence of classical neurotransmitters and peptides. TIPS 8 (1985) 331–334.

6 Baumgarten, H. G., Björklund, A., Lachenmayer, L., Nobin, A., and Stenevi, U., Longlasting selective depletion of brain serotonin by 5,6-dihydroxytryptamine. Acta physiol. scand., Suppl. 373 (1971) 1–16.

7 Beaujouan, J. C., Torrens, Y., Saffroy, M., and Glowinski, J., Quantitative autoradiographic analysis of the distribution of binding sites for (^{125}I) Bolton Hunter derivatives of eledoisin and substance P in the rat brain. Neuroscience 18 (1984) 857–875.

8 Belin, M. F., Weisman-Nanopoulos, D., Steinbusch, H., Verhofstad, A., Maitre, M., Jouvet, M., and Pujol, J. F., Mise en évidence de glutamate décarboxylase et de sérotonine dans un même neurone au niveau du noyau du raphé dorsalis du rat par des méthodes de double marquage immunocytochimique. C.r. Acad. Sci. 293 (1981) 337–341.

9 Belin, M. F., Nanopoulos, D., Didier, M., Aguera, M., Steinbusch, H., Verhofstad, A., Maitre, M., and Pujol, J. F., Immunohistochemical evidence for the presence of γ-aminobutyric acid and serotonin in one nerve cell. A study on the raphe nuclei of the rat using antibodies to glutamate decarboxylase and serotonin. Brain Res. 275 (1983) 329–339.

10 Berod, A., Chat, M., Paut, L., and Tappaz, M., Catecholaminergic and GABAergic anatomical relationship in the rat substantia nigra, locus coeruleus, and hypothalamic median eminence: Immunocytochemical visualization of biosynthetic enzymes on serial semithin plastic-embedded sections. J. Histochem. Cytochem. 32 (1984) 1331–1338.

11 Berkenboesch, F., Linton, E. A., and Tilders, F. J. H., Colocalization of PHI- and CRF-immunoreactivity in neurons of the rat hypothalamus: a surprising artefact. Neuroendocrinology 44 (1986) 338–346.

12 Björklund, A., and Hökfelt, T., Eds, Handbook of Chemical Neuroanatomy, vol. 4: GABA and Neuropeptides in the CNS, Part I. Elsevier, Amsterdam 1985.

13 Blessing, W. W., Howe, P. R. C., Joh, T. H., Oliver, J. R., and Willoughby, J. O., Distribution of tyrosine hydroxylase and neuropeptide Y-like immunoreactive neurons in rabbit medulla oblongata, with attention to colocalization studies, presumptive adrenaline-synthesizing perikarya, and vagal preganglionic cells. J. comp. Neurol. 248 (1986) 285–300.

14 Branton, W. D., Phillips, H. S., and Jan, Y. N., The LHRH family of peptide messengers in the frog nervous system, in: Progress in Brain Research, pp. 205–215. Eds T. Hökfelt, K., Fuxe and B. Pernow. Elsevier, Amsterdam 1986.

15 Brashear, H. R., Záborszky, L., and Heimer, L., Distribution of GABAergic and cholinergic neurons in the rat diagonal band. Neuroscience 17 (1986) 439–445.

16 Brownstein, M. J., Saavedra, J. M., Axelrod, J., and Carpenter, D. O., Coexistence of several putative neurotransmitters in single identified neurons of aplysia. Proc. natn. Acad. Sci. USA 71 (1974) 4662–4665.

17 Buchanan, J. T., Brodin, L., Hökfelt, T., and Grillner, S., Survey of neuropeptide-like immunoreactivity in the lamprey spinal cord. Brain Res. 408 (1987) 299–302.

172

18 Burnstock, G., Do some nerve cells release more than one transmitter? Neuroscience *1* (1976) 239–248.

19 Caffé, A. R., and van Leeuwen, F. W., Vasopressin-immunoreactive cells in the dorsomedial hypothalamic region, medial amygdaloid nucleus and locus coeruleus of the rat. Cell Tissue Res. *233* (1983) 23–33.

20 Chan-Palay, V., Combined immunocytochemistry and autoradiography after in vivo injection of monoclonal antibody to substance P and ^{3}H-serotonin: Coexistence of two putative transmitters in single raphe cells and fiber plexuses. Anat. Embryol. *156* (1979) 241–254.

21 Chan-Palay, V., and Palay, S. L., Eds, Coexistence of Neuroactive Substances in Neurons. John Wiley & Sons, New York 1984.

22 Chan-Palay, V., Jonsson, G., and Palay, S. L., Serotonin and substance P coexist in neurons of the rat's central nervous system. Proc. natn. Acad. Sci. USA *75* (1978) 1582–1586.

23 Chan-Palay, V., Nilaver, G., Palay, S. L., Beinfeld, M. C., Zimmerman, E. A., Wu, J.-Y., and O'Donohue, T. L., Chemical heterogeneity in cerebellar Purkinje cells: evidence and coexistence of glutamic acid decarboxylase-like and motilin-like immunoreactivities. Proc. natn. Acad. Sci. USA *78* (1981) 7787–7791.

24 Charnay, Y., Léger, L., Dray, F., Bérod, A., Jouvet, M., Pujol, J. F., and Dubois, P. M., Evidence for the presence of enkephalin in catecholaminergic neurons of cat locus coeruleus. Neurosci. Lett. *30* (1982) 147–151.

25 Chubb, I. W., Goodman, S., and Smith, A. D., Is acetylcholinesterase secreted from central neurons into the cerebrospinal fluid? Neuroscience *1* (1976) 57–62.

26 Chubb, I. E., Hodgson, A. J., and White, G. H., Acetylcholinesterase hydrolyzes substance P. Neuroscience *5* (1980) 2065–2072.

27 Coons, A. H., Fluorescent antibody methods, in: General Cytochemical Methods, pp. 399–422. Ed. J. F. Danielli Academic Press, New York 1958.

28 Costa, M., and Furness, J. B., Somatostatin is present in subpopulation of noradrenergic nerve fibres supplying the intestine. Neuroscience *13* (1984) 911–919.

29 Costa, M., Furness, J. B., and Gibbins, I. L., Chemical coding of enteric neurons, in: Progress in Brain Research, vol. 68, pp. 217–239. Eds T. Hökfelt, K. Fuxe and B. Pernow. Elsevier, Amsterdam 1986.

30 Cuello, A. C., Ed., Co-transmission. MacMillan, London and Basingstoke 1982.

31 Cuello, A. C., Ed., Immunohistochemistry; IBRO Handbook Series: Methods in the Neurosciences, vol. 3. John Wiley & Sons, Chichester 1983.

32 Cuello, A. C., and Kanazawa, I., The distribution of substance P immunoreactive fibers in the rat central nervous system. J. comp. Neurol. *178* (1978) 129–156.

33 Dahlström, A., Effects of vinblastine and colchicine on monoamine containing neurons of the rat with special regard to the axoplasmic transport of amine granules. Acta neuropath. Suppl. *5* (1971) 226–237.

34 Dahlström, Å., and Fuxe, K., Evidence of the existence of monoamine-containing neurons in the central nervous system. I. Demonstration of monoamines in the cell bodies of brain stem neurons. Acta physiol. scand. *62*. Suppl. 232 (1984) 1–55.

35 Dalsgaard, C.-J., Hökfelt, T., Elfvin, L.-G., Skirboll, L., and emson, P., Substance P-containing primary sensory neurons projecting to the inferior mesenteric ganglion: Evidence from combined retrograde tracing and immunohistochemistry. Neuroscience *7* (1982) 647–654.

36 Dalsgaard, C.-J., Hökfelt, T., Schultzberg, M., Lundberg, J. M., Terenius, L., Dockray, G. J., and Goldstein, M., Origin of peptide-containing fibers in the inferior mesenteric ganglion of the guinea pig: Immunohistochemical studies with antisera to substance P, enkephalin, vasoactive intestinal polypeptide, cholecystokinin and bombesin. Neuroscience *9* (1983) 191–211.

37 Doerr-Schott, J., Multiple immunocytochemical labelling methods for the simultaneous ultrastructural localization of various hypophysial hormones, in: Pars Distalis of the Pituitary Gland—Structure, Function and Regulation. Excerpta Medica Int. Congr. Ser. 673, pp. 95–106. Eds F. Yoshimura and A. Gorbman. Elsevier Science Publ., Amsterdam 1986.

38 Eckenstein, F., and Baughman, R. W., Two types of cholinergic innervation in cortex, one co-localized with vasoactive intestinal polypeptide. Nature *309* (1984) 153–155.

39 Edwards, A. V., Järhult, J., Andersson, P.-O., and Bloom, S. R., The importance of the pattern of stimulation in relation to the response of autonomic effectors, in: Systemic Role of Regulatory Peptides, pp. 145–148. Eds S. R. Bloom, J. M. Polak and E. Lindenlaub. Schattauer, Stuttgart 1982.

40 Everitt, B. J., Hökfelt, T., Terenius, L., Tatemoto, K., Mutt, V., and Goldstein, M., Differential co-existence of neuropeptide Y (NPY)-like immunoreactivity with catecholamines in the central nervous system of the rat. Neuroscience *11* (1984) 443–462.

41 Everitt, J. E., Hökfelt, T., Wu, J.-Y., and Goldstein, M., Coexistence of tyrosine hydroxylase-like and gamma-aminobutyric acid-like immunoreactivities in neurons of the arcuate nucleus. Neuroendocrinology *39* (1984) 189–191.

42 Everitt, B. J., Meister, B., Hökfelt, T., Melander, T., Terenius, L., Rökaeus, Å., Theodorsson-Norheim, E., Dockray, G., Edwardson, J., Cuello, C., Elde, R., Goldstein, M., Hemmings, H., Ouimet, C., Walaas, I., Greengard, P., Vale, W., Weber, E., Wu, J.-Y., and Chang, K.-J., The hypothalamic arcuate nucleus—median eminence complex: Immunohistochemistry of transmitters, peptides and DARPP-32 with special reference to coexistence in dopamine neurons. Brain Res. Rev. *11* (1986) 97–155.

42a Fibiger, H. C., The organization and some projections of cholinergic neurons of the mammalian forebrain. Brain Res. Rev. *4* (1982) 327–388.

43 Fontaine, B., Klarsfeld, A., Hökfelt, T., and Changeux, J.-P., Calcitonin gene-related peptide, a peptide present in spinal cord motoneurons, increases the number of acetylcholine receptors in primary cultures of chick embryo myotubes. Neurosci. Lett. *71* (1986) 59–65.

44 Fried, G., Terenius, L., Hökfelt, T., and Goldstein, M., Evidence for the differential localization of noradrenaline and neuropeptide Y (NPY) in neuronal storage vesicles isolated from rat vas deferens. J. Neurosci. *5* (1985) 450–458.

45 Furness, J. B., Costa, M., Emson, P. C., Håkanson, R., Moghimzadeh, E., Sundler, F., Taylor, J. L., and Chance, R. E., Distribution, pathways and reactions to drug treatment of nerves with neuropeptide Y- and pancreatic polypeptidelike immunoreactivity in the guinea pig digestive tract. Cell Tissue Res. *234* (1983) 71–92.

46 Gall, C., Henry, S. H. C., Seroogy, K. B., and Jones, E. G., Colocalization of GABA- and tyrosine hydroxylase-like immunoreactivities in neurons of the rat main olfactory bulb. Soc. Neurosci. Abstr. *15* (1985) 89.

47 Gibson, S. J., Polak, J. M., Bloom, S. R., Sabate, I. M., Mulderry, P. M., Ghatei, M. A., McGregor, G. P., Morrison, J. F. B., Kelly, J. S., Evans, R. M., and Rosenfeld, M. G., Calcitonin gene-related peptide immunoreactivity in the spinal cord of man and of eight other species. J. Neurosci. *4* (1984) 3101–3111.

48 Gilbert, R. F. T., Emson, P. C., Hunt, S. P., Bennett, G. W., Marsden, C. A., Sandberg, B. E. B., Steinbusch, H., and Verhofstad, A. A. J., The effects of monoamine neurotoxins on peptides in the rat spinal cord. Neuroscience *7* (1982) 69–88.

49 Glazer, E. J., Steinbusch, H., Verhofstad, A., and Basbaum, A. I., Serotonin neurons in nucleus raphe dorsalis and paragigantocellularis of the cat contain enkephalin, J. Physiol., Paris *77* (1981) 241–245.

50 Greenfield, S. A., The significance of dendritic release of transmitter and protein in the substantia nigra. Neurochem. int. *7* (1985) 887–901.

51 Greenfield, S. A., Cheramy, A., Leviel, V., and Glowinski, J., In vivo release of acetylcholinesterase in the cat substantiae nigrae and caudate nuclei. Nature *284* (1980) 355–357.

52 Hendry, S. H. C., Jones, E. G., DeFelipe, J., Schmechel, D., Brandon, C., and Emson, P. C., Neuropeptide-containing neurons of the cerebral cortex are also GABAergic. Proc. natn. Acad. Sci. USA *81* (1984) 6526–6530.

53 Hökfelt, T., Elfvin, L.-G., Elde, R., Schultzberg, M., Goldstein, M., and Luft, R., Occurrence of somatostatin-like immunoreactivity in some peripheral sympathetic noradrenergic neurons. Proc. natn. Acad. Sci. USA *74* (1977) 3587–3591.

54 Hökfelt, T., Ljungdahl, Å., Steinbusch, H., Verhofstad, A., Nilsson, G., Brodin, E., Pernow, B., and Goldstein, M., Immunohistochemical evidence of substance P-like immunoreactivity in some 5-hydroxytryptamine-containing neurons in the rat central nervous system. Neuroscience *3* (1978) 517–538.

55 Hökfelt, T., Johansson, O., Ljungdahl, Å., Lundberg, J. M., and Schultzberg, M., Peptidergic neurons. Nature *284* (1980) 515–521.

174

56 Hökfelt. T., Lundberg, J. M., Schultzberg, M., Johansson, O., Ljungdahl, Å., and Rehfeld, J., Coexistence of peptides and putative transmitters in neurons, in: Neural Peptides and Neuronal Communication, pp. 1–23. Eds E. Costa and M. Trabucchi. Raven Press, New York 1980.

57 Hökfelt, T., Rehfeld, J. F., Skirboll, L., Ivemark, B., Goldstein, M., and Markey, K., Evidence for coexistence of dopamine and CCK in mesolimbic neurones. Nature 285 (1980) 476–478.

58 Hökfelt, T., Skirboll, L., Rehfeld, J. F., Goldstein, M., Markey, K., and Dann, O., A subpopulation of mesencephalic dopamine neurons projecting to limbic areas contains a cholecystokinin-like peptide: evidence from immunohistochemistry combined with retrograde tracing. Neuroscience 5 (1980) 2093–2124.

59 Hökfelt, T., Everitt, B. J., Theodorsson-Norheim, E., and Goldstein, M., Occurrence of neurotensinlike immunoreactivity in subpopulations of hypothalamic, mesencephalic, and medullary catecholamine neurons. J. comp. Neurol. 222 (1984) 543–559.

60 Hökfelt, T., Lundberg, J. M., Lagercrantz. H., Tatemoto, K., Mutt, V., Lindberg, J., Terenius, L., Everitt, B. J., Fuxe, K., Agnati, L. F., and Goldstein, M., Occurrence of neuropeptide Y (NPY)-like immunoreactivity in catecholamine neurons in the human medulla oblongata. Neurosci. Lett. 36 (1983) 217–222.

61 Hökfelt, T., Johansson, O., and Goldstein, M., Central catecholamine neurons as revealed by immunohistochemistry with special reference to adrenaline neurons, in: Handbook of Chemical Neuroanatomy, vol. 2: Classical Transmitters in the CNS, Part I, pp. 157–276. Eds. A. Björklund and T. Hökfelt. Elsevier, Amsterdam 1984.

62 Hökfelt, T., Johansson, O., and Goldstein, M., Chemical anatomy of the brain. Science 225 (1984) 1326–1334.

63 Hökfelt, T., Mårtensson, R., Björklund, A., Kleinau, S., and Goldstein, M., Distributional maps of tyrosine hydroxylase immunoreactive neurons in the rat brain, in: Handbook of Chemical Neuroanatomy, vol. 2: Classical Transmitters in the CNS, Part I, pp. 277–379. Eds A. Björklund and T. Hökfelt. Elsevier, Amsterdam 1984.

64 Hökfelt, T., Skirboll, L., Everitt, B. J., Meister, B., Brownstein, M., Jacobs, T., Faden, A., Kuga, S., Goldstein, M., Markstein, R., Dockray, G., and Rehfeld, J., Distribution of cholecystokinin-like immunoreactivity in the nervous system with special reference to coexistence with classical neurotransmitters and other neuropeptides, in: Neuronal Cholecystokinin, pp. 255–274. Eds J. J. Vanderhaeghen and J. Crawley. Ann. N.Y. Acad. Sci., New York 1985.

65 Hökfelt, T., Fuxe, K., and Pernow, B., Eds, Coexistence of neuronal messengers: a new principle in chemical transmission, in: Progress in Brain Research, vol. 68. Elsevier, Amsterdam 1986.

66 Hökfelt, T., Holets, V. R., Staines, W., Meister, B., Melander, T., Schalling, M., Schutzberg, M., Freedman, J., Björklund, H., Olson, L., Lindh, B., Elfvin, L.-G., Lundberg, J. M., Lindgren, J. Å., Samuelsson, B., Pernow, B., Terenius, L., Post, C., Everitt, B., and Goldstein, M., Coexistence of neuronal messengers—an overview, in: Progress in Brain Research, vol. 68. pp. 33–70. Eds T. Hökfelt, K. Fuxe and B. Pernow. Elsevier, Amsterdam 1986.

67 Hökfelt, T., and Terenius, L., More on receptor mismatch. TINS 10 (1987) 22.

68 Hunt, S. P., and Lovick, T. A., The distribution of serotonin, metenkephalin and β-lipotropin-like immunoreactivity in neuronal perikarya of the cat brain stem. Neurosci. Lett. 30 (1982) 139–145.

69 Hylden, J. L. K., and Wilcox, G. L., Intrathecal substance P elicits caudally-directed biting and scratching behavior in mice. Brain Res. 217 (1981) 212–215.

70 Ibata, Y., Fukui, K., Okamura, H., Kawakami, T., Tanaka, M., Obata, H. L., Isuto, T., Terubayashi, H., Yanaihara, C., and Yanaihara, N., Coexistence of dopamine and neurotensin in the hypothalamic arcuate and periventricular nucleus. Brain Res. 269 (1983) 177–179.

71 Jacobowitz, D. M., and Palkovits, M., Topographic atlas of catecholamine and acetyl-cholinesterase-containing neurons in the rat brain. I. Forebrain (Telencephalon, Diencephalon). J. comp. Neurol. 157 (1974) 13–28.

72 Jirikowski, G., Reisert, I., Pilgrim, Ch., and Oertel, W. H., Coexistence of glutamate decarboxylase and somatostatin immunoreactivity in cultured hippocampal neurons of the rat. Neurosci. Lett. 46 (1984) 35–39.

175

73 Johansson, O., Hökfelt, T., Pernow, B., Jeffcoate, S. L., White, N., Steinbusch, H. W. M., Verhofstad, A. A. J., Emson, P. C., and Spindel, E., Immunohistochemical support for three putative transmitters in one neuron: coexistence of 5-hydroxytryptamine-. substance P-, and thyrotropin releasing hormone-like immunoreactivity in meduallary neurons projecting to the spinal cord. Neuroscience 6 (1981) 1857–1881.

74 Ju, G., Hökfelt, T., Fischer, J. A., Frey, P., Rehfeld, J. F. and Dockray, G. J., Does cholecystokinin-like immunoreactivity in rat primary sensory neurons represent calcitonin gene related peptide? Neurosci. Lett. 68 (1986) 305–310.

75 Köhler, C., Swanson, L. W., Haglund, L., and Wu, J.-Y., The cytoarchitecture, histochemistry and projections of the tuberomammillary nucleus in the rat. Neuroscience 16 (1985) 85–110.

76 Kondo, H., Kuramoto, H., Wainer, B. H., and Yanaihara, N., Evidence for the coexistence of acetylcholine and enkephalin in the sympathetic preganglionic neurons of rats. Brain Res. 335 (1985) 309–314.

77 Kosaka, T., Hataguchi, Y., Hama, K., Nagatsu, I., and Wu. J.-Y., Coexistence of immunoreactivities for glutamate decarboxylase and tyrosine hydroxylase in some neurons in the periglomerular region of the rat main olfactory bulb: possible coexistence of gamma-aminobutyric acid (GABA) and dopamine. Brain Res. 343 (1985) 166–171.

78 Kosaka, T., Kosaka, K., Tateishi, K., Hamaoka, Y., Yanaihara, N., Wu, J.-Y., and Hama, K., GABAergic neurons containing CCK-8-like and/or VIP-like immunoreactivities in the rat hippocampus and dentate gyrus. J. comp. Neurol. 239 (1985) 420–430.

79 Krieger, D. T., Brownstein, M. J., and Martin, J. B., Eds, Brain Peptides. John Wiley & Sons, New York 1983.

80 Léger, L., Charnay, Y., Chayvialle, J. A., Bérod, A., Dray, F., Pujol, J. F., Jouvet, M., and Dubois, P. M., Localization of substance P- and enkephalin-like immunoreactivity in relation to catecholamine-containing cell bodies in the cat dorsolateral pontine tegmentum: an immunofluorescence study. Neuroscience 8 (1983) 525–546.

81 Le Grevés, P., Nyberg, F., Terenius, L., and Hökfelt, T., Calcitonin gene-related peptide is a potent inhibitor of substance P degradation Eur. J. Pharmac. 115 (1986) 309–311.

82 Lindh, B., Hökfelt, T., Elfvin, G., Terenius, L., Fahrenkrug, J., Elde, R., and Goldstein, M., Topography of NPY-, somatostatin-, and VIP-immunoreactive neuronal subpopulations in the guinea pig celiac-superior mesenteric ganglion and their projection to the pylorus. J. Neurosci. 6 (1986) 2371–2383.

83 Ljungdahl, Å., Hökfelt, T., and Nilsson, G., Distribution of substance P-like immunoreactivity in the central nervous system of the rat. I. Cell bodies and nerve terminals. Neuroscience 3 (1978) 861–943.

84 Lorenz, R. G., Saper, C. B., Wong, D. L., Ciaranello, R. D., and Loewy, A. D., Co-localization of substance P- and phenylethanolamine-N-methyltransferase-like immunoreactivity in neurons of ventrolateral medulla that project to the spinal cord: Potential role in control of vasomotor tone. Neurosci. Lett. 55 (1985) 255–260.

85 Lovick, T. A., and Hunt, S. P., Substance P-immunoreactive and serotonin-containing neurones in the ventral brainstem of the cat, Neurosci. Lett. 36 (1983) 223–228.

86 Lundberg, J. M., and Hökfelt, T., Coexistence of peptides and classical neurotransmitters. TINS 6 (1983) 325–333.

87 Lundberg, J. M., and Hökfelt. T., Multiple co-existence of peptides and classical transmitters in peripheral autonomic and sensory neurons—functional and pharmacological implications, in: Progress in Brain Research, vol. 68, pp. 241–262. Eds T. Hökfelt, K. Fuxe and B. Pernow. Elsevier, Amsterdam 1986.

88 Lundberg, J. M., and Tatemoto, K., Pancreatic polypeptide family (APP, BPP, NPY, and PYY) in relation to sympathetic vasoconstriction resistant to α-adrenoceptor blockade. Acta physiol. scand. 116 (1982) 393–402.

89 Lundberg, J. M., Fried, G., Fahrenkrug, J., Holmstedt, B., Hökfelt, T., Lagercrantz, H., Lundgren, G., and Änggård, A., Subcellular fractionation of cat submandibular gland: comparative studies on the distribution of acetylcholine and vasoactive intestinal polypeptide (VIP). Neuroscience 6 (1981) 1001–1010.

90 Lundberg, J. M., Hedlund, B., Änggård, A., Fahrenkrug, J., Hökfelt, T., Tatemoto, K., and Bartfai, T., Co-storage of peptides and classical transmitters in neurons, in: Systemic Role of Regulatory Peptides, pp. 93–119. Eds S. R. Bloom, J. M., Polak and E. Lindenlaub. Schattauer, Stuttgart and New York 1982.

176

91 Lundberg, J. M., Hökfelt, T., Änggård, A., Terenius, L., Elde, R., Markey, K., Goldstein, M., and Kimmel, J., Organizational principles in the peripheral sympathetic nervous system: subdivision by coexisting peptides (somatostatin-, avian pancreatic polypeptide-, and vasoactive intestinal polypeptide-like immunoreactive materials). Proc. natn. Acad. Sci. USA 79 (1982) 1303-1307.

92 Lundberg, J. M., Terenius, L., Hökfelt, T., Martling, C. R., Tatemoto, K., Mutt, V., Polak, J., Bloom, S., and Goldstein, M., Neuropeptide Y (NPY)-like immunoreactivity in peripheral noradrenergic neurons and effects of NPY on sympathetic function. Acta physiol scand. 116 (1982) 477-480.

93 Macrae, I. M., Furness, J. B., and Costa, M., Distribution of subgroups of noradrenaline neurons in the coeliac ganglion of the guinea-pig. Cell Tissue Res. 244 (1986) 173-180.

94 Mantyh, P. W., and Jung, S. P., Evidence for cholecystokininlike immunoreactive neurons in the rat medulla oblongata which project to the spinal cord. Brain Res. 291 (1984) 49-54.

95 Mason, R. T., Peterfreund, R. A., Sawchenko, P. E., Corrigan, A. Z., Rivier, J. E., and Vale, W. W., Release of the predicted calcitonin gene-related peptide from cultured rat trigeminal ganglion cells. Nature 308 (1984) 653-655.

96 Matthews, M. R., and Cuello, A. C., Substance P-immunoreactive peripheral branches of sensory neurones innervate guinea-pig sympathetic neurons. Proc. natn. Acad. Sci. USA 79 (1982) 1668-1672.

96a Matthysse, S. W., and Kety, S. S., Eds, Catecholamines and Schizophrenia. Pergamon Press, Oxford 1975.

97 Meister, B., Hökfelt, T., Vale, W. W., and Goldstein, M., Growth hormone releasing factor (GRF) and dopamine coexist in hypothalamic arcuate neurons. Acta physiol. scand. 124 (1985) 133-136.

98 Meister, B., Hökfelt, T., Vale, W. W., Sawchenko, P. E., Swanson, L., and Goldstein, M., Coexistence of dopamine and growth hormone releasing factor (GRF) in a subpopulation of tubero-infundibular neurons of the rat. Neuroendocrinology 42 (1986) 237-247.

99 Meister, B., Hökfelt, T., Brown, J., Joh, T., and Goldstein, M., Dopaminergic cells in the caudal A13 cell group express somatostatin-like immunoreactivity. Exp. Brain Res. 67 (1987) 441-444.

100 Melander, T., Staines, W. A., Hökfelt, T., Rökaeus, Å., Eckenstein, F., Salvaterra, P. M., and Wainer, B. H., Galanin-like immunoreactivity in cholinergic neurons of the septum-basal forebrain complex projecting to the hippocampus of the rat. Brain Res. 360 (1985) 130-138.

101 Melander, T., and Staines, W. A., A galanin-like peptide coexists in putative cholinergic somata of the septum-basal forebrain complex and in acetylcholinesterase-containing fibers and varicosities within the hippocampus in the owl monkey (Aotus trivirgatus). Neurosci. Lett. 68 (1986) 17-22.

102 Melander, T., Hökfelt, T., Rökaeus, Å., Cuello, A. C., Oertel, W. H., Verhofstad, A., and Goldstein, M., Coexistence of galanin-like immunoreactivity with catecholamines, 5-hydroxytryptamine, GABA and neuropeptides in the rat CNS. J. Neurosci. 6 (1987) 3640-3654.

103 Millhorn, D. E., Hökfelt, T., Seroogy, K., Oertel, W., Verhofstad, A., and Wu, J.-Y., Immunohistochemical evidence for colocalization of gamma-aminobutyric acid (GABA) and serotonin in neurons of the ventral medulla oblongata projecting to the spinal cord. Brain Res. 410 (1987) 179-185.

104 Millhorn, D., Hökfelt, T., Terenius, L., Buchan, A., and Brown, J. C., Somatostatin- and enkephalin-like immunoreactivities are frequently colocalized in neurons in the caudal brain stem of the rat. Exp. Brain. Res. 67 (1987) 420-428.

104a Mishkin, M., A memory system in the monkey. Phil. Trans. R. Soc. Lond. Biol. Ser. B298 (1982) 85-95.

105 Nairn, R. C., Ed., Fluorescent Protein Tracing, 3rd edition. E. & S. Livingstone Ltd., Edinburgh and London 1968.

106 Nakane, P. K., Simultaneous localization of multiple tissue antigens using the peroxidase-labeled antibody method: a study in pituitary glands of the rat. J. Histochem. Cytochem. 16 (1968) 557-560.

107 New, H. V., and Mudge, A. W., Calcitonin gene-related peptide regulates muscle acetyl-choline receptor synthesis. Nature 323 (1986) 809-811.

108 Nilsson, J., von Euler, A. M., and Dalsgaard, C.-J., Stimulation of connective tissue cell growth by substance P and substance K. Nature *315* (1985) 61–63.

108a Nordström, Ö., Melander, T., Hökfelt, T., Bartfai, T., and Goldstein, M., Evidence for an inhibitory effect of the peptide galanin on dopamine release from the rat median eminence. Neurosci. Lett. *73* (1987) 21–26.

109 O'Donohue, T. L., Millington, W. R., Handelmann, G. E., Contreras, P. C., and Chronwall, B. M., On the 50th anniversary of Dale's law: Multiple neurotransmitter neurons. TIPS *6* (1985) 305–308.

110 Oertel, W. H., and Mugnaini, E., Immunocytochemical studies of GABAergic neurons in rat basal ganglia and their relations to other neuronal systems. Neurosci. Lett. *47* (1984) 233–238.

111 Oertel, W. H., Graybiel, A. M., Mugnaini, E., Elde, R. P., Schmechel, D. E., and Kopin, E. J., Coexistence of glutamic acid decarboxylase- and somatostatin-like immunoreactivity in neurons of the feline nucleus reticularis thalami. J. Neurosci. *3* (1983) 1322–1332.

112 Okamura, H., Murakami, S., Chihara, K., Nagatsu, I., and Ibata, Y., Coexistence of growth hormone releasing factor-like and tyrosine hydroxylase-like immunoreactivities in neurons or the rat arcuate nucleus. Neuroendocrinology *41* (1985) 177–179.

113 Osborne, N. N., Is Dale's principle valid? TINS *2* (1979) 73–75.

114 Osborne, N. N., Ed. Dale's Principle and Communication between Neurones. Pergamon Press, Oxford and New York 1983.

115 Osborne, N. N., and Beaton, D. W., Direct histochemical localisation of 5,7-dihydroxytryptamine and the uptake of serotonin by a subpopulation of GABA neurones in the rabbit retina. Brain Res. *382* (1986) 158–162.

116 Otsuka, M., and Takahashi, T., Putative peptide neurotransmitters. A. Rev. Pharmac. Toxic. *17* (1977) 425–439.

117 Otsuka, M., Konishi, S., Yanagisawa, M., Tsunoo, A., and Akagi, H., Role of substance P as a sensory transmitter in spinal cord and sympathetic ganglia, in: Substance P in the Nervous System, pp. 13–34. Ciba Foundation Symposium 91. Pitman, London 1982.

118 Ottersen, O. P., Storm-Mathisen, J., Laake, J. H., and Madsen, S., Distribution of possible amino acid transmitters and some cellular markers in the cerebellum. Neurosci. Lett. Suppl. *26* (1986) 399.

119 Owman, Ch., Håkanson, R., and Sundler, F., Occurrence and function of amines in polypeptide hormone producing cells. Fedn Proc. Fedn. Am. Soc. exp. Biol. *32* (1973) 1785–1791.

120 Palkovits, M., and Jacobowitz, D. M., Topographic atlas of catecholamine and acetylcholinesterase-containing neurons in the rat brain. II. Hindbrain (Mesencephalon, Rhombencephalon). J. comp. Neurol. *157* (1974) 29–41.

121 Palkovits, M., and Brownstein, M. J., Distribution of neuropeptides in the central nervous system using biochemical micromethods, in: Handbook of Chemical Neuroanatomy, vol. 4: GABA and Neuropeptides in the CNS, Part 1, pp. 1–71. Eds A. Björklund and T. Hökfelt. Elsevier, Amsterdam 1985.

122 Pelletier, G., Steinbusch, H. W., and Verhofstad, A., Immunoreactive substance P and serotonin present in the same dense core vesicles. Nature *293* (1981) 71–72.

123 Pearse, A. G. E., The cytochemistry and ultrastructure of polypeptide hormone producing cells of the APUD series and the embryologic, physiologic and pathologic implications of the concept. J. Histochem. Cytochem. *17* (1969) 303–313.

124 Pernow, B., Substance P. Pharmac. Rev. *35* (1983) 85–141.

125 Piercey, M. F., Dobry, P. J. K., Schroeder, L. A., and Einspahr, F. J., Behavioral evidence that substance P may be a spinal cord sensory neurotransmitter. Brain Res. *210* (1981) 407–412.

126 Polak, J. M., and Van Noorden, S., Ed., Immunocytochemistry. Practical Applications in Pathology and Biology. Wright–PSG, Bristol 1983.

127 Rosenfeld, M. G., Mermod, J.-J., Amara, S. G., Swanson, L. W., Sawchenko, P. E., Rivier, J., Vale, W. W., and Evans, R. M., Production of a novel neuropeptide encoded by the calcitonin gene via tissue-specific RNA processing. Nature *304* (1983) 129–135.

128 Rothman, R. B., Herkenham, M., Pert, C. B., Liang, T., and Cascieri, M. A., Visualization of rat brain receptors for the neuropeptide substance P. Brain Res. *309* (1984) 47–54.

178

129 Saria, A., Gamse, R., Petermann, J., Fischer, J. A., Theodorsson-Norheim, E., and Lundberg, J. M., Simultaneous release of several tachykinins and calcitonin gene-related peptide from rat spinal cord slices, Neurosci, Lett. *63* (1986) 310–314.

130 Sawchenko, P. E., Swanson, L. W., Grzanna, R., Howe, P. R. C., Polak, J. M., and Bloom, S. R., Co-localization of neuropeptide-Y immunoreactivity in brainstem catecholaminergic neurons that project to the paraventricular nucleus of the hypothalamus J. comp. Neurol. *241* (1985) 138–153.

131 Schmechel, D. E., Vickrey, B. G., Fitzpatrick, D., and Elde, R. P., GABAergic neurons of mammalian cerebral cortex: widespread subclass defined by somatostatin content. Neurosci. Lett. *47* (1984) 227–232.

132 Shults, C. W., Quirion, R., Chronwall, B., Chase, T. N., and O'Donohue, T., A comparison of the anatomical distribution of substance P and substance P receptors in the rat central nervous system. Peptides *5* (1984) 1097–1128.

133 Senba, E., Daddona, P. E., Watanabe, T., Wu, J.-Y., and Nagy, J. I., Coexistence of adenosine deaminase, histidine decarboxylase, and glutamate decarboxylase in hypothalamic neurons of the rat. J. Neurosci. *5* (1985) 3393–3402.

134 Smith, A. D., Dale's principle today: Adrenergic tissues, in: Neuron Concept Today. Eds J. Szentágothai, J. Hámori and E. S. Vizi. Symposium, Tihany, Akadémiai Kiado, Budapest 1976.

135 Snyder, S., Brain peptides as neurotransmitters. Science *209* (1980) 976–983.

135a Sofroniew, M. V., Pearson, R. C. A., Eckenstein, F., Cuello, A. C., and Powell, T. P. S., Retrograde changes in cholinergic neurons in the basal forebrain of the rat following cortical damage. Brain Res. *289* (1983) 370–374.

136 Somogyi, P., Hodgson, A. J., Smith, A. D., Nunzi, M. G., Gorio, A., and Wu, J.-Y., Different populations of GABAergic neurons in the visual cortex and hippocampus of cat contain somatostatin- or cholecystokinin-immunoreactive material. J. Neurosci. *4* (1984) 2590–2603.

137 Staines, W. A., Meister, B., Melander, T., Nagy, J. I., and Hökfelt, T., Three-colour immunohistofluorescence allowing triple labelling within a single section. J. Histochem. Cytochem. *36* (1988) 145–151.

138 Sternberger, L. A., Ed., Immunocytochemistry, 2nd edition. John Wiley & Sons. New York 1979.

139 Stjärne, L., and Lundberg, J. M., Neuropeptide Y (NPY) depresses the secretion of ^{3}H-noradrenaline and the contractile response evoked by field stimulation in rat vas deferens. Acta physiol. scand. *120* (1984) 477–479.

140 Takami, K., Kawai, Y., Shiosaka, S., Lee, Y., Girgis, S., Hillyard, C. J., MacIntyre, I., Emson, P. C., and Tohyama, M., Immunohistochemical evidence for the coexistence of calcitonin gene-related peptide- and choline acetyltransferase-like immunoreactivity in neurons of the rat hypoglossal, facial and ambiguous nuclei. Brain Res. *328* (1985) 386–389.

141 Takami, K., Kawai, Y., Uchida, S., Tohyama, M., Shiotani, Y., Yoshida, H., Emson, P. C., Girgis, S., Hillyard, C. J., and MacIntyre, J., Effect of calcitonin gene-related peptide on contraction of striated muscle in the mouse. Neurosci. Lett. *60* (1985) 227–230.

141a Tatemoto, K., Röckaeus, Å., Jörnvall, H., McDonald, T. J., and Mutt, V., Galanin— A novel biologically active peptide from porcine intestine. FEBS Lett *164* (1983) 124–128.

142 Thoenen, H., and Tranzer, J. P., Chemical sympathectomy by selective destruction of adrenergic nerve endings with 6-hydroxydopamine. Arch. Pharmak. exp. Path. *261* (1968) 271–288.

143 Tramu, G., Pillez, A., and Leonardelli, J., An efficient method of antibody elution for the successive or simultaneous location of two antigens by immunocytochemistry. J. Histochem. Cytochem. *26* (1978) 322–324.

144 Triller, A., Cluzeaud, F., and Korn, H., GABA-containing terminals can be opposed to glycine receptors at central synapses. Brain Res. (1987) in press.

145 Vincent, S. R., Satoh, K., Armstrong, D. M., and Fibiger, H. C., Substance P in the ascending cholinergic reticular system. Nature *306* (1983) 688–691.

145a Wainer, B. H., Levey, A. I., Mufson, E. J., and Mesulam, M. M., Cholinergic systems in mammalian brain identified with antibodies against choline acetyltransferase. Neurochem. Int. *6* (1984) 163–182.

146 Wall, P. D., and Fitzgerald, M., If substance P fails to fulfil the criteria as a neurotransmitter in somatosensory afferents, what might be its function?, in: Substance P in the Nervous System, pp. 249–266. Ciba Foundation Symposium 91. Pitman, London 1982.

147 Watt, C. B., Su, Y. T., and Lam. D. M.-K., Interactions between enkephalin and GABA in avian retina. Nature *311* (1984) 761–763.

148 Weiler, R., and Ball, A. K., Co-localization of neurotensin-like immunoreactivity and ^{3}H-glycine uptake system in sustained amacrine cells of turtle retina. Nature *311* (1984) 759–761.

148a Whitehouse, P. J., Price, D. L., Strable, R. G., Clark, A. W., Coyle, J. T., and DeLong, M. R., Alzheimer's disease and senile dementia: loss of neurons in the basal forebrain. Science *215* (1982) 1237–1239.

149 Wiesenfeld-Hallin, Z., Hökfelt, T., Lundberg, J. M., Forssmann, W. G., Reinecke, M., Tschopp, F. A., and Fischer, J. A., Immunoreactive calcitonin gene-related peptide and substance P coexist in sensory neurons to the spinal cord and interact in spinal behavioural responses of the rat. Neurosci. Lett. *52* (1984) 199–204.

150 Zahm, D. S., Zaborszky, L., Alones, V. E., and Heimer, L., Evidence for the coexistence of glutamate decarboxylase and met-enkephalin immunoreactivities in axon terminals of rat ventral pallidum. Brain Res. *325* (1985) 317–321.

Peptides and epithelial growth regulation

R. A. Goodlad and N. A. Wright

Summary. There is now considerable evidence implicating several peptides in the control of gastrointestinal epithelial cell proliferation and cell renewal. While some of these may act directly, many may be involved in regulating the powerful trophic effects of the intake and digestion of food on the gut epithelium. — Several peptides have been associated with the regulation of intestinal cell proliferation. There is little doubt that gastrin is trophic to the stomach, but, its role in the rest of the gastrointestinal tract is debatable. Enteroglucagon has often been associated with increased intestinal epithelial proliferation, but at the moment all the evidence for this is circumstantial. The effects of peptide YY and bombesin warrant further study. The availability of recombinant epidermal growth factor (EGF) has recently enabled us to demonstrate a powerful trophic response to infused EGF throughout the gastrointestinal tract. The increasing availability of peptides will eventually allow the rigorous *in vivo* evaluation of the trophic role of these potentially very important peptides.

In many ways the gastrointestinal epithelium is an ideal model for the study and investigation of the control of epithelial cell proliferation, as it is continuously and rapidly renewed with its cell division restricted to an anatomically discrete zone. It is also capable of adapting its rates of proliferation to a wide variety of physiological and other stimuli. The study of epithelial cell renewal is also of considerable importance since most tumours are of epithelial origin[83]. Three main mechanisms are generally considered to be involved in the control of epithelial renewal in the gut namely, a (local?) negative feedback system from the functional (villus) to the reproductive zone (crypt), the direct or indirect effects of food (luminal nutrition and/or intestinal workload) and the effects of humoral factors[110].

Parabiotic studies in which the blood systems of two animals are linked have indicated that a hormonal factor may cross-circulate from a stimulated animal to its partner.[58,107]. A similar response has also been noted in less extreme models where isolated loops of small intestine still respond to altered food intake[21], and after intestinal resection[7,43].

The study of cell renewal and epithelial growth control necessitates the use of suitable methods, and unfortunately many studies in this field have been bedeviled by the use of totally inappropriate methods. The problems involved have been spelt out in detail elsewhere[5,20,39,73,109,110], and are as follows: 1) The intestine contains a large proportion of non-epithelial cells (muscle, submucosa lymphoid aggregates); thus any gross measure may give a misleading result. Even the mucosa itself is approximately

20% non-epithelial[20]. 2) The choice of a suitable denominator is of vital importance, as many measures, such as labelling index and mitotic index will not detect a general increase in compartment size. These measures also suffer from being 'state' measures, and as such can be misleading if the duration of the DNA synthesis phase or mitosis is altered. 3) Measures based on the gross uptake of tritiated thymidine can be especially misleading, as although usually equated with growth, tritiated thymidine uptake can be affected by a variety of stimuli. Thymidine itself is not a precursor in the *de novo* synthesis of DNA, but is incorporated by a salvage pathway which depends on the activity of several enzymes and transport mechanisms plus the size of the endogenous thymidine pool. All of these factors can be influenced by hormones or growth factors. Thymidine can also be stored and recycled, and it can also be taken up by bacteria[73].

Most of these pitfalls can be avoided if the accumulation of arrested metaphases in microdissected crypts is determined. This 'rate' measure also avoids the several problems involved in the quantification of sectioned material, and expressing the results on a per crypt basis can account for all the factors that may influence epithelial cell production (cell cycle time, size of the growth fraction and size of the crypt itself)[5,20,39,73,109,110]

Gastrin as a trophic hormone in the gastrointestinal tract

There is a considerable body of evidence for a powerful pharmacological and possibly physiological modulation of cell proliferation by gastrin in the stomach[22,70,106], and especially of the enterochromaffin-like cells[59]. There is also evidence, unfortunately mainly based on the gross uptake of tritiated thymidine, that this trophism extends into the small intestine and colon[51,52,54,65]. Claims by Johnson[49,50] for a major trophic role for gastrin were also supported by a study of the effects of gastrin on primary duodenal explants in short-term culture[65]; but this study is especially open to criticism[108].

On the other hand several groups of workers have failed to show any structural or functional changes in the small bowel after a variety of manoeuvers. No proliferative effects were noted after pentagastrin infusion[72], and a large series of experiments designed to give a wide range of gastrin levels failed to show any relationship between plasma gastrin levels and cell proliferation[76,81,82]. The previously observed increases in tritiated thymidine uptake could alternatively be due to gastrin increasing cellular permeability and transport[82,91]. Thymidine kinase activity increases after refeeding, but before the increase in gastrin[68]. The reported increase in tritiated thymidine uptake without any increase in tissue mass or protein or DNA content[95] also argues for alterations in cellular permeability confounding gross thymidine measures.

The use of more robust cell kinetic measures to quantify crypt cell production has failed to show any correlation between gastrin and proliferation in starved and refed rats[40], after intestinal resection[6,88,89] and after a variety of dietary manipulations[34].

Thus although there is good evidence for a trophic effect of gastrin in the stomach (but only in the fundus, not in the antrum[19]), a general trophic role for gastrin is not proven.

Enteroglucagon

Enteroglucagon (also called glicentin) is considered by many to be the prime candidate for the title of 'enterotrophin' and there is a considerable body of evidence to support this.

An enteroglucagon secreting renal tumour was associated with marked mucosal hypertrophy which was reversed on removal of the tumour[9,33]. Plasma enteroglucagon levels rise in a variety of hyperproliferative models such as after intestinal resection[6,12,47,88,89], in lactating and in hypothermic-hyperphagic rats[27,48], and elevated plasma levels are seen in several human pathological conditions associated with intestinal hyperplasia[14].

There is also an excellent correlation between plasma enteroglucagon and crypt cell production rate in a wide range of hypo- and hyperproliferative models of intestinal adaptation. These include starvation and refeeding[40], intestinal resection[88,89], pancreatico-bilary diversion and resection[6] and dietary manipulation[34]. A contraindication of the enteroglucagon hypothesis is nonetheless provided by a pilot immunoneutralization study where an anti-pancreatic glucagon monoclonal antibody failed to prevent proliferative adaptation after intestinal resection (in fact it elevated the augmented mitotic index[42]). Enteroglucagon cells are located throughout the gut, but most are localised in the distal gut, especially the terminal ileum[14], which is the strategic position for monitoring the efficiency of digestion and either delaying intestinal transit or increasing absorptive function (via increased crypt cell output). Enteroglucagon is part of a preproglucagon which contains glicentin related pancreatic polypeptide (GRPP), glucagon, oxyntomodulin and glucagon-like peptides 1 and 2[11]. GRPP, enteroglucagon, GLP1 and GLP2 are all produced and secreted separately in the gut, with enteroglucagon sometimes cleaving to release GRPP and oxyntomodulin (fig.). GLP1 has now been tested in some preliminary studies[11] (Goodlad, unpublished), but with no proliferative effects observed. An initial infusion study of the effects of oxyntomodulin also proved negative (Goodlad, unpublished).

The enteroglucagon hypothesis is still attractive, but as it is entirely based on circumstantial evidence the definitive test of the hypothesis cannot be performed until pure enteroglucagon is purified or isolated.

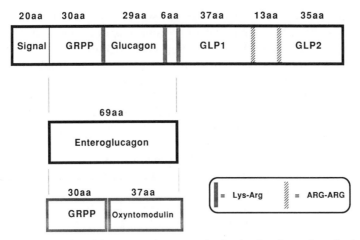

Schematic representation of the pre-pro-glucagon molecule, showing the number of amino acids (aa) in each component. Enteroglucagon, GRPP, oxyntomodulin, GLP1 and GLP2 can all be released by the intestinal mucosa.

Peptide YY (PYY)

PYY is a novel (36 amino acid) candidate hormone which may to be co-localised with enteroglucagon[3,28,29], and like enteroglucagon it can also inhibit gastric acid secretion and emptying[4,97]. It also has a high degree of sequence homology with pancreatic polypeptide and neuropeptide Y. Like enteroglucagon it is found throughout the gastrointestinal tract, with most cells in the distal gut, but the majority of PYY cells are located in the colon[1]. PYY receptors have been located in the small intestine[57]. PYY abnormalities have been reported in various disease states associated with malabsorption[2]. PYY levels rise after intestinal resection and correlate well with intestinal crypt cell production rates, but the direct infusion of PYY via osmotic mini-pumps appeared to have little effect on intestinal crypt cell production[87]: Nonetheless plasma PYY levels also correlate quite well with intestinal cell proliferation after dietary manipulations, and correlate very well with plasma enteroglucagon levels[34]. PYY receptors are mainly found in the small intestine while most PYY containing cells are localised in the colon, thus the possibility of a feedback loop from the hind gut to the small intestine seems attractive.

Bombesin

Bombesin is a tetradecapeptide first isolated from amphibian skin, and bombesin-like immunoreactivity is present along the digestive tract of mammals[84,105]. Bombesin can stimulate gastrin release and gastric acid secretion[102]. It can also effect the release of several gut hormones in man[13]. The mammalian equivalent of bombesin is thought to be gastrin releasing peptide (GRP), a 27 amino acid peptide first isolated from porcine gut[74,75]. *In vitro* administration of bombesin stimulates proliferation of the 3T3 mouse fibroblast line[86]. Bombesin also acts as an autocrine growth factor for some lung tumours[23]. In the suckling rat bombesin stimulates growth of the entire gastrointestinal tract and pancreas[63] while in the adult it has been reported that it can stimulate antral gastrin cell proliferation[61]. Another study has shown that bombesin can also stimulate intestinal crypt cell production in transected rats, but it could not stimulate the already elevated rates of proliferation in animals with intestinal resection[90].

Thus bombesin is another peptide whose role in the control of intestinal epithelial cell proliferation warrants further investigation.

Epidermal growth factor (EGF)

The location of the main sites of production of EGF in the salivary glands and Brunner's glands of the duodenum of man[44] and the rat[78] would imply that EGF may have a role in the maintenance of gastrointestinal homeostasis.

While the growth promoting actions of EGF in vitro are well characterised[18], its rôle *in vivo* is uncertain: EGF stimulates the proliferation and differentiation of the epidermis, maturation of the pulmonary epithelium and accelerates the healing of corneal epithelium in the foetus and newborn[18]. EGF also stimulates the proliferation and maturation of the neonatal intestine[16,67,77], where it increases the activity of ornithine decarboxylase[30], an enzyme associated with the initiation of cell proliferation[66]. The presence of EGF in a variety of body fluids, including saliva, plasma[18] and milk[17], its production by the salivary and Brunner's glands[44,78], the reports of a trophic action of saliva on the intestine[64,79], the demonstration of EGF receptors in intestinal epithelial cells[32,96] and its reported cytoprotective effects on the duodenal mucosa[56] all suggest that it has a rôle in the control of gastrointestinal homeostasis other than the inhibition of gastric acid secretion.

The injection of EGF into rodents has produced conflicting reports, some finding that it can increase the incorporation of tritiated thymidine into DNA throughout the gastrointestinal tract[92,93], others only observing this in the stomach[53] or only in starved animals[24]. EGF may also aid gastrointestinal growth in undernourished young rats[69]. A study of the short-term effects of EGF administration using the crypt cell production method[8] showed a trophic effect in some sites of the intestine.

The ideal model of the hypoplastic intestine is provided by maintaining animals on isocaloric total parenteral nutrition (TPN), which is generally agreed to be the pertinent system for the study of effects of humoral factors on the intestine[85]; since the intestine of the TPN rat is in a steady state and basal level of proliferation.

The TPN model was used to investigate the effects of recombinant EGF (human β-urogastrone) on cell proliferation. EGF-urogastrone infusion increased intestinal crypt cell production throughout the gut[36], especially in the colon. It also progressively increased proliferation with increasing dose, and was equally effective whether given continuously or when given after hypoplasia had become established[37,38]. A proliferative effect on the intestine has also been seen in a human infant maintained on intravenous infusion[104]. EGF was not effective when given intragastrically[37,38,80]. The continuous intra-ileal infusion of EGF has nonetheless been reported as increasing intestinal cell proliferation both in the perfused section and in the jejunum (which did not receive any luminal EGF[101]). EGF can be absorbed from the intestine, at least in the young animal[100], but may be partially degraded as it passes through.

It is thus likely that EGF may have both local and systemic effects on the gut. The evidence for a systemic rôle for EGF in the control of gastrointestinal epithelial cell proliferation is far stronger than that obtained for any other peptide, but the question of whether this is a physiological or a pharmacological effect remains to be seen. A final twist to the EGF story is provided by the discovery that transforming growth factor is both structurally and functionally very similar to EGF[71,98].

Other peptides

Cholecystokinin (CCK) is a peptide closely related to gastrin and while there is some evidence for it having a trophic effect on the gut[45,46], the direct infusion of low and high doses of CCK had no effect on intestinal structure and function (but did markedly stimulate the pancreas)[26]. Neurotensin can induce gastric growth[31]. The levels of gastrointestinal somatostatin increase on starvation[103] and somatostatin can inhibit cell proliferation in the stomach[62] by a mechanism involving, at least in part, the suppression of gastrin release[25]. In the duodenum[15], it can also inhibit the rise in crypt cell proliferation and enteroglucagon normally seen after resection[90]. Somatostatin also inhibits EGF secretion from Brunner's glands[55].

The above list of possible trophic agents cannot be regarded as final, as there are still more peptides to be discovered let alone investigated, for example growth hormone releasing factor (somatocrinin) has recently been reported to stimulate intestinal epithelial cell proliferation in the stomach and duodenum[60].

186

Conclusion

The control of gastrointestinal epithelial cell proliferation is undoubtedly a multifactorial affair, involving local negative feedback, the direct and indirect effects of food (food intake is one of the best predictors of intestinal cell production and intestinal function[35]) and the local and systemic effects of humoral factors. The ultimate action of these factors will involve either the modulation of the 'traditional' second messengers, via calcium, lipids and cyclic nucleotides or highly specific ligand-receptor interactions, which activate very specific transduction pathways[41].

The data presently available suggest that several peptides may play a rôle in the control of epithelial cell renewal. The relative importance of these peptides has yet to be established, and the further investigation of these important factors demands the use of valid techniques, which should be applied to the entire gastrointestinal tract, as the response of the stomach, colon and small intestine has been seen to vary.

Acknowledgment. We thank the Cancer Research Campaign for financial assistance.

1 Adrian, T. E., Ferri, G.-L., Bacarese-Hamilton, A. J., Fuessl, H. S., Polak, J. M., and Bloom, S. R., Human distribution and release of a putative new gut hormone, Peptide YY. Gastroenterology 89 (1985) 1070–1077.
2 Adrian, T. E., Savage, A. P., Bacarese-Hamilton, H. J., Wolfe, K., Besterman, H. S., and Bloom, S. R., Peptide YY abnormalities in gastrointestinal disease. Gastroenterolgy 90 (1986) 379–384.
3 Ali-Rachedi, Varndell, I. M., Adrian, T. E., Gapp, D. A., Van Noorden, S., Bloom, S. R., and Polak, J. M., Peptide YY (PYY) immunoreactivity is co-stored with glucagon-related immunoreactants in endocrine cells of the gut and pancreas. Histochemistry 80 (1984) 487–491.
4 Allen, J. M., Fitzpatrick, M. L., Yeats, J. C., Darcy, K., Adrian, T. E., and Bloom, S. R., Effects of peptide YY and neuropeptide Y on gastric emptying in man. Digestion 30 (1984) 255–262
5 Al-Mukhtar, M. Y. T., Polak, J. M., Bloom, S. R., and Wright, N. A., The search for appropriate measurements of proliferative and morphological status in studies on intestinal adaptation, in: Intestinal Adaptation II, pp. 3–25. Eds J. W. L. Robinson, R. H. Dowling and J. W. L. Reicken. MTP Press Ltd, Lancaster 1982.
6 Al-Mukhtar, M. Y. T., Sagor, G. R., Ghatei, M. A., Bloom, S. R., and Wright, N. A., The role of pancreatico-biliary secretions in intestinal adaptation after resection, and its relationship to plasma enteroglucagon. Br. J. Surg. 70 (1983) 398–400.
7 Al-Mukhtar, M. Y. T., Sagor, G. R., Ghatei, M. A., Polak, J. M., Koopmans, H. S., Bloom, S. R., and Wright, N. A., The relationship between endogenous gastrointestinal hormones and cell proliferation in models of intestinal adaptation, in: Intestinal Adaptation II, pp. 243–255. Eds J. W. L Robinson, R. H. Dowling and J. W. L. Reicken. MTP Press Ltd, Lancaster 1982.
8 Al-Nafussi, A., and Wright, N. A., The effect of epidermal growth factor EGF: on cell proliferation of the gastrointestinal mucosa of rodents. Virchows Arch. (Cell Path) 40 (1982) 63–69.
9 Bloom, S. R., An enteroglucagon tumour. Gut 13 (1972) 520–523.
10 Bloom, S. R., Gut and brain—endocrine connections. J. R. Coll. Physiol. (Lond.) 14 (1980) 51–57.
11 Bloom, S. R., Gut hormones in adaptation. Gut 28 (1987) S1, 31–35.

12 Bloom, S. R., Besterman, H. S., Adrian, T. E., Christophides, N. A., Sarson, D. L. Mallison, C. N., Pero, A., and Modigliani, R., Gut hormone profiles following resection of large and small bowel. Gastroenterology 76 (1978) 1101.

13 Bloom, S. R., Ghatei, M. A., Christofides, N. D., Blackburn, A. M., Adrian, T. E., Lezoche, E., Basso, N., Carlei, F., and Speranza, F., Bombesin infusion in man, pharmokinetics and effect on gastrointestinal and pancreatic hormonal peptides. J. Endocr, 83 (1979) 51.

14 Bloom, S. R., and Polak, J. M., The hormonal pattern of intestinal adaptation. A major role for enteroglucagon. Scand. J. Gastroent. 17 Suppl. 74 (1982) 93–104.

15 Bosshard, A., Pensu, D., Chayvialle, J. A., and Reinberg, A., Tissue related inhibition of jejunal cell renewal and gastrin synthesis by somatostatin. Reg. Pept Suppl. 1 (1980) 814.

16 Calvert, R., Beaulieu, J. F., and Menard, D., Epidermal growth factor EGF: accelerates the maturation of foetal mouse intestinal mucosa in utero. Experientia 38 (1982) 1096.

17 Carpenter, G., Epidermal growth factor is a major growth promoting agent in human milk. Science 210 (1980) 198–199.

18 Carpenter, G., Epidermal growth factor, in: Tissue Growth Factors, pp. 89–123. Ed. R. Baserga. Springer-Verlag, Berlin 1981.

19 Castelyn, P. P., Dubrasquet, M., and Willems, G., Opposite effects of gastrin on cell proliferation in the antrum and other parts of the upper gastrointestinal tract in the rat. Dig. Dis. 22 (1977) 798–804.

20 Clarke, R. M., Progress in measuring epithelial turnover in the villus of the small intestine. Digestion 8 (1973) 161–175.

21 Clarke, R. M., Intestinal function and epithelial replacement. MD thesis, University of Cambridge 1975.

22 Crean, G. P., Gunn, A. A., and Rumsey, R. D. E., Parietal cell hyperplasia of the gastric mucosa induced by the administration of pentagastrin to rats. Gastroenterology 57 (1969) 147–155.

23 Cuttitta, F., Carney, D. N., Mulshine, J., Moody, T. W., Fedorko, J., Fischler, A., and Minna, J. D., Bombesin-like peptides can function as Autocrine growth factors in human small cell cancer. Nature 4 316(6031) (1985) 823–826.

24 Dembinski, A., Gregory, H., Konturek, S. J., and Polanski, M., Trophic action of epidermal growth factor on the pancreas and gastroduodenal mucosa in rats. J. Physiol. (Lond.) 325 (1982) 35–42.

25 Dembinski, A., Warzecha, Z., Konturek, S. J., and Schally, A. V., Effects of somatostatin on the growth of gastrointestinal mucosa and pancreas in rats. Role of endogenous gastrin. Gut 28 (1987) S1, 227–232.

26 Dowling, R. H., Small bowel adaptation and its regulation. Scand. J. Gastroent. 17 Suppl. 74 (1982) 54–74.

27 Elias, E., and Dowling, R. H., The mechanism for small bowel adaptation in lactating rats. Clin. Sci. molec. Med. 51 (1976) 427–433.

28 El-Salhy, M., Grimelius, L., Wilander, E., Ryberg. B., Terenius, L., Lundberg, J. M., and Tatemoto, K., Immunocytochemical identification of polypeptide YY (PYY) cells in the human gastrointestinal tract. Histochemistry 77 (1983) 15–23,

29 El-Salhy, M., Wilander, E., Juntti-Berggren, L., and Grimelius, L., The distribution and ontogeny of polypeptide YY (PYY) and pancreatic polypeptide (PP)-immunoreactive cells in the gastrointestinal tract of rat. Histochemistry 78 (1983) 53–60.

30 Feldman, E. J., Aures, D., and Grossman, M. I., Epidermal growth factor stimulates ornithine decarboxylase activity in the digestive tract of the mouse. Proc. Soc. exp. Biol. Med. 159 (1978) 400–402.

31 Feurle, G. E., Muller, B., and Rix, E., Neurotensin induces hyperplasia of the pancreas and growth of the gastric antrum in rats. Gut 28 (1987) S1, 19–23.

32 Forgue-Lafitte, M. E., Laburthe, M., Chambblier, M. C., Moody, A. J., and Rosselin, G., Demonstration of specific receptors for EGF/urogastrone in isolated rat intestinal epithelial cells. FEBS Lett. 114 (1980) 243–246.

33 Gleeson, M. H., Bloom, S. R., Polak, J. M., Henry, K., and Dowling, R. H., Endocrine tumour in kidney affecting small bowel structure, motility and absorptive function. Gut 12 (1971) 773–782.

34 Goodlad, R. A., Lenton, W., Ghatei, M. A., Adrian, T. E., Bloom, S. R., and Wright, N. A., Effects of an elemental diet, inert bulk and different types of dietary fibre on the

188

response of the intestinal epithelium to refeeding in the rat and relationship to plasma gastrin, enteroglucagon and PYY concentrations. Gut *28* (1987) 171–180.

35 Goodlad, R. A., Plumb, J. A., and Wright, N. A., The relationship between intestinal crypt cell production and intestinal water absorption measured in vitro in the rat. Clin. Sci. *72* (1987) 297–304.

36 Goodlad, R. A., Wilson, T. G., Lenton, W., Wright, N. A., Gregory, H., and McCullagh, K. G., Urogastrone-epidermal growth factor is trophic to the intestinal epithelium of parenterally fed rats. Experientia *41* (1985) 1161–1163.

37 Goodlad, R. A., Wilson, T. G., Lenton, W., Wright, N. A., Gregory, H., and McCullagh, K. G., Intravenous but not intragastric urogastrone-EGF is trophic to the intestine of parenterally fed rats. Gut *28* (1987) 573–578.

38 Goodlad, R. A., Wilson, T. G., Lenton, W., Wright, N. A., Gregory, H., and McCullagh, K. G., Proliferative effects of urogastrone-EGF on the intestinal epithelium. Gut *28* (1987) S1, 37–43.

39 Goodlad, R. A., and Wright, N. A., Quantitative studies on epithelial replacement in the gut, in: Techniques in the Life Sciences. Techniques in Digestive Physiology, vol. P2, pp. 212/1–212/23. Ed. T. A. Titchen. Elsevier Biomedical Press, Ireland 1982.

40 Goodlad, R. A., and Wright, N. A., Cell proliferation, plasma enteroglucagon and plasma gastrin levels in starved and refed rats. Virchows Arch. (Cell Path.) *43* (1983) 55–62.

41 Gorelick, F. S., Second messanger systems and adaptation. Gut *28* (1987) S1, 79–84.

42 Gregor, M., Menge, H., Stossel, R., and Riecken, E. O., Effects of monoclonal antibodies to enteroglucagon on ileal adaptation after proximal small bowel resection. Gut *28* (1987) S1, 9–14.

43 Hanson, W. R., Proliferative and morphological adaptation of the intestine to experimental resection. Scand. J. Gastroent. *17* Suppl. 74 (1982) 11–20.

44 Heitz, P. U., Kasper, M., Van Noorden. S., Polak, J. M., Gregory, H., and Pearse, A. G. E., Immunohistochemical localisation of urogastrone to human duodenal and submandibular glands. Gut *19* (1978) 408–413.

45 Hughes, C. A., Bates, T., and Dowling, R. H., Cholycystokinin and secretin prevent the intestinal mucosal hyperplasia of total parenteral nutrition in the dog. Gastroenterology *75* (1978) 34–41.

46 Hughes, C. A., Hatoff, D. A., Ducker, D. A., and Dowling, R. H., The effect of CCK octapeptide on the pancreas: a study in rats during total parenteral nutrition and oral feeding. Eur. J. clin. Invest. *10* (1980) 16p.

47 Jacobs, L. R., Polak, J. M., Bloom, S. R., and Dowling, R. H., Does enteroglucagon play a trophic role in the intestine. Clin. Sci. molec. Med. *50* (1976) 14–15.

48 Jacobs, L. R., Bloom, S. R., and Dowling, R. H., Response of plasma and tissue levels of enteroglucagon immunoreactivity to intestinal resection, lactation and hyperphagia. Life Sci. *29* (1981) 2003–2007.

49 Johnson, L. R., The trophic action of gastrointestinal hormones. Gastroenterology *70* (1976) 278–288.

50 Johnson, L. R., Regulation of gastrointestinal growth, in: Physiology of the Digestive Tract, pp. 169–196. Ed. L. R. Johnson. Raven Press, New York 1981.

51 Johnson, L. R., and Chandler, A. M., RNA and DNA of gastric and duodenal mucosa in antrectomised and gastrin treated mice. Am. J. Physiol. *224* (1973) 937–940.

52 Johnson, L. R., and Guthrie, P. D., Secretin inhibition of gastrin stimulated DNA synthesis. Gastroenterology *67* (1974) 601–606.

53 Johnson, L. R., and Guthrie, P. D., Stimulation of rat oxyntic gland mucosal growth by epidermal growth factor. Am. J. Physiol. *238* (1980) G45–G49.

54 Johnson, L. R., Lichtenburger, L. M., Copeland, E. M., Dudrick, S. J., and Castro, G. A., Action of gastrin on gastrointestinal structure and function. Gastroenterology *68* (1975) 1184–1192.

55 Kirkegaard, P., Olsen, P. S., Nexo, E., Holst, J. J., and Poulsen, S. S., Effect of vasoactive intestinal polypeptide and somatostatin on secretion of epidermal growth factor and bicarbonate from Brunner's glands. Gut *25* (1984) 1225–1229.

56 Kirkegaard, P., Olsen, P. S., Poulsen, S. S., and Nexo, E., Epidermal growth factor inhibits cysteamine-induced duodenal ulcers. Gastroenterology *85* (1983) 1277–1283.

57 Laburthe, M., Chenut, B., Rouyer-Fessard, C., Tatemoto, K., Couvineau, A., Servin, A.,

and Amiranoff, B., Interaction of Peptide YY with rat intestinal epithelial plasma membranes: binding of the radioiodinated peptide. Endocrinology *118* (1986) 1910–1917.

58 Laplace, J. P., Intestinal resection in chronically blood-crossed twin pigs; blood carried factor(s). Digestion *10* (1974) 229–232.

59 Larssson, H., Carlsson, E., Mattsson, H., Lundell, L., Sunder, F., Sundell, G., Wallmark, B., Watanabe, T., and Hakanson, R., Plasma gastrin and gastric enterochromaffin cell activation and proliferation. Studies with omeprazole and ranitidine in intact and antrectomized rats. Gastroenterology *90* (1986) 391–399.

60 Lehy, T., Accary, J. P., Dubrasquet, M., and Lewin, M. J. M., Growth hormone releasing factor (somatocrinin) stimulates epithelial cell proliferation in the rat digestive tract. Gastroenterology *90* (1986) 646–653.

61 Lehy, T., Accary, J. P., Labeille, D., and Dubrasquet, M., Chronic administration of bombesin stimulated antral gastrin cell proliferation in the rat. Gastroenterology *84* (1983) 914–919.

62 Lehy, T., Dubrasquet, M., and Bonfils, S., Effect of somatostatin on normal and gastrin stimulated cell proliferation in the gastric and intestinal mucosae of the rat. Digestion *9* (1979) 99–109.

63 Lehy, T., Puccio, F., Chariot, J., and Labeille, D., Stimulating effect of bombesin on the growth of gastrointestinal tract and pancreas in suckling rats. Gastroenterology *90* (1986) 1942–1949.

64 Li, A. K. C., Schattenkerk, M. E., Huffman, R. G., Ross, J. S., and Malt, R. A., Hypersecretion of submandibular saliva in male mice: trophic response in small intestine. Gastroenterology *84* (1983) 949–955.

65 Lictenburger, L., Muller, L. R., Erwin, D. W., and Johnson, L. R., Effects of pentagastrin on adult rat duodenal cells in culture. Gastroenterology *65* (1973) 242–251.

66 Luk, G. D., Marten, L. J., and Baylin, S. B., Ornithine decarboxylase is important in intestinal mucosal maturation and recovery from injury in rats. Science *210* (1980) 195–198.

67 Malo, C., and Menard, D., Influence of epidermal growth factor on the development of suckling mouse intestinal mucosa. Gastroenterology *83* (1982) 28–35.

68 Majumdar, A. P. N., Effects of fasting and refeeding on antral, duodenal and serum gastrin levels and on colonic thymidine kinase activity in rats. Hormone Res. *19* (1984) 127–134.

69 Majumdar, A. P. N., Postnatal undernutrition: Effects of epidermal growth factor on growth and function of gastrointestinal tract in rats. J. Pediatr. Gastroent. Nutr *3* (1984) 618–625.

70 Mak, K. M., and Chang, W. W. L., Pentagastrin stimulates epithelial cell proliferation in duodenal and colonic crypts in fasted rats. Gastroenterology *71* (1976) 1117–1120.

71 Marquardt, H., Hunkapiller, M. W., Hood, L. E., and Todara, G. J., Rat transforming growth factor type I: Structure and relation to epidermal growth factor. Science *223* (1984) 1079–1081.

72 Mayston, D. D., Barrowman, J. A., and Dowling, R. H., Effect of pentagastrin on small bowel structure and function in the rats. Digestion *12* (1975) 78–84.

73 Maurer, H. R., Potential pitfalls of 3H thymidine techniques to measure cell proliferation. Cell Tiss. Kinet. *14* (1981) 111–120.

74 McDonald, T. J., Jornvall, H., Nilsson, G., Vagne, M., Ghatei, M. A., Bloom, S. R., and Mutt, V., Bombesin stimulation of DNA synthesis and cell division in cultures of Swiss 3T3 cells. Biochem. biophys. Res. Commun. *90* (1979) 217–233.

75 McDonald, T. J., Nilsson, G., Vagne, M., Ghatei, M. A., Bloom, S. R., and Mutt, V., A gastrin releasing peptide from the porcine non-antral gastric tissue. Gut *19* (1978) 767–774.

76 Morin, C. L., and Ling, V., Effects of pentagastrin on the rat small intestine after resection. Gastroenterology *75* (1978) 224–229.

77 Oka, Y., Ghrisan, F. K., Greene, H. L., and Orth, D. N., Effect of mouse epidermal growth factor/urogastrone on functional maturation of rat intestine. Endocrinology *112* (1983) 940–944.

78 Olsen, P. S., and Nexo, E., Quantitation of epidermal growth factor in the rat: Identification and partial characterisation of duodenal EGF. Scand. J. Gastroent. *18* (1983) 771–776.

190

79 Olsen, P. S., Poulsen, S. S., Kirkegaard, P., and Nexo, E., Role of submandibular saliva and epidermal growth factor in gastric cytoprotection. Gastroenterology *87* (1984) 103–108.

80 O'Loughlin, E. V., Chung, M., Hollenberg, M., Hayden, J., Zahavi, I., and Gall, D. G., Effects of epidermal growth factor on ontogeny of the gastrointestinal tract. Am. J. Physiol. *249* (1985) G674–G678.

81 Oscarson, J., Hakanson, R., Liedberg, G., Lundqvist, G., Sundler, F., and Thorell, J., Varied serum gastrin concentration: Trophic effects on the gastrointestinal tract of rats. Acta physiol. scand. (1979) suppl. 475.

82 Oscarson, J. E. A., Veen, H. F., Williamson, R. C. N., Ross, J. S., and Malt, R. A., Compensatory postresectional hyperplasia and starvation atrophy in small bowel: dissociation from endogenous gastrin levels. Gastroenterology *72* (1977) 890–895.

83 Peto, R., Epidemiology, multistage models and short-term mutagenicity tests. Cold Spring Harbor Laboratories 1977.

84 Polak, J. M., Ghatei, M. A., Wharton, J., Bishop, A. E., Bloom, S. R., Solcia, E., Brown, M. R., and Pearse, A. G. E., Bombesin-like immunoreactivity in the gastrointestinal tract, lung and central nervous system. Scand. J. Gastroenterology *13* Suppl. 49 (1978) 148.

85 Robinson, J. W. L., Dowling, R. H., and Reichen, E. O. (Eds), Intestinal Adaptation II. MTP Press, Lancaster 1982.

86 Rozengurtz, E., and Sinnett-Smith, J., Characterisation of a gastrin releasing peptide from porcine non-antral gastrin tissue. J. Proc. natl Acad. Sci. USA *80* (1983) 2936–2940.

87 Savage, A., Gornacz, G. E., Adrian, T. E., Goodlad, R. A., Wright, N. A., and Bloom, S. R., Elevation of PYY following intestinal resection in the rat is not responsible for the adaptive response. Gut *26* (1985) 1353–1358.

88 Sagor, G. R., Ghatei, M. A., Al-Mukhtar, M. Y. T., Wright, N. A., and Bloom, S. R., The effects of altered luminal nutrition on cellular proliferation and plasma concentrations of enteroglucagon after small bowel resection in the rat. Br. J. Surg. *69* (1982) 14–18.

89 Sagor, G. R., Ghatei, M. A., Al-Mukhtar, M. Y. T., Wright, N. A., and Bloom, S. R., Evidence for a humoral mechanism after intestinal resection, exclusion of gastrin but not enteroglucagon. Gastroenterology *54* (1983) 902–916.

90 Sagor, G. R., Ghatei, M. A., O'Shaughnessy, D. J., Al-Mukhtar, M. Y. T., Wright, N. A., and Bloom, S. R., Influence of somatostatin and bombesin on plasma enteroglucagon and cell proliferation after intestinal resection in the rat. Gut *26* (1985) 89–94.

91 Schwartz, M. A., and Storozuk, R. B., Enhancement of small intestine absorption by intraluminal gastrin. Gastroenterology *88* (1985) 1578.

92 Schieving, L. A., Yeh, Y. C., and Scheiving, L. E., Circadian phase-dependent stimulatory effects of epidermal growth factor on deoxyribonucleic acid synthesis in the tongue, oesophagus, and stomach of the adult male mouse. Endocrinology *105* (1979) 1475–1480.

93 Schieving, L. A., Yeh, Y. C., Tsai, T. H., and Scheiving, L. E., Circadian phase-dependent stimulatory effects of epidermal growth factor on deoxyribonucleic acid synthesis in the duodenum, jejunum, ileum, caecum, colon and rectum of the adult male mouse. Endocrinology *106* (1980) 1498–1503.

94 Smith, J., Cook, E., Fotheringham, I., Pheby, S., Derbyshire, R., Eaton, M. A. W., Doel, M., Lilley, D. M. J., Pardon, J. F., Pator, T., Lewis, H., and Bell, L. D. R., Chemical synthesis and cloning of a gene for human B-urogastrone. Nucl. Acid Res. *10* (1982) 4467–4482.

95 Solomon, T. E., Trophic effects of pentagastrin on gastrointestinal tract in fed and fasted rats. Gastroenterology *91* (1986) 108–116.

96 St Hilaire, R. J., Gospodarowicz, D., and Kim, Y. S., Epidermal growth factor: effect on the growth of a human colon adenocarcinoma cell line. Gastroenterology *78* (1981) 1271.

97 Sukuki, T., Nakaya, M., Itoh, Z., Tatemoto, K., and Mutt, V., Inhibition of interdigestive contractile activity in the stomach by peptide YY in Heidenhain pouch dogs. Gastroenterology *85* (1983) 114–121.

98 Tam, J. P., Physiological effects of transforming growth factor in newborn mouse. Science *229* (1985) 673–675.

99 Tatemoto, K., and Mutt, V., Isolation of two novel candidate hormones using a chemical method for finding naturally occurring polypeptides. Nature *285* (1980) 417–418.

100 Thornburg, W., Matrisian, L., Magun, B., and Koldovsky, O., Gastrointestinal absorption of epidermal growth factor in suckling rats. Am. J. Physiol. *246* (1984) G80–G85.

101 Ulshen, M. H., Lyn-Cook, L. E., and Raasch, R. H., Effects of intraluminal epidermal growth factor on mucosal proliferation in the small intestine of adult rats. Gastroenterology *91* (1986) 1134–1140.

102 Varner, A. A., Modlin, I. M., and Walsh, J. H., High potency of bombesin for stimulation of human gastrin release and gastric acid secretion. Reg. Pep. (1981) 289.

103 Voyles, N. R., Awoke, S., Wade, A., Bhathena, S. J., Smith, S. S., and Recant, L., Starvation increases gastrointestinal somatostatin in normal and obese Zuker rats: a possible regulatory mechanism. Horm. Metab. Res. *14* (1982) 392–395.

104 Walker-Smith, J. A., Phillips, A. D., Walford, N., Gregory, H., Fitzgerald, J. D., MacCullagh, K., and Wright, N. A., Intravenous epidermal growth factor/urogastrone increases small intestinal cell proliferation in congenital microvillous atrophy. Lancet *ii* (1985) 1239–1240.

105 Wharton. J., Polak, J. M., Bloom, S. R., Ghatei, M. A., Solcia, E., Brown, M. R., and Pearse, A. G. E., Bombesin-like immunoreactivity in the lung. Nature *273* (1978) 769.

106 Willems, G., Cell renewal in the gastric mucosa. Digestion *6* (1972) 46–63.

107 Williamson, R., Intestinal adaptation 2. Mechanisms of control. N. Engl. J. Med. *298* (1978) 1444–1450.

108 Wright, N. A., Regulation of growth by peptides, in: Gut Hormones. Eds S. R. Bloom and J. M. Polak. Churchill Livingstone, Edinburgh 1981.

109 Wright, N. A., and Alison, M., The Biology of Epithelial Cell Populations, vol. 1. Clarendon Press, Oxford 1984.

110 Wright, N. A., and Alison, M., The Biology of Epithelial Cell Populations, vol. 2. Clarendon Press, Oxford 1984.

Post-translational proteolytic processing of precursors to regulatory peptides

P. C. Andrews, K. A. Brayton and J. E. Dixon

Summary. Precursors to regulatory peptides undergo maturation processes which include proteolytic processing. The enzymes involved in this process remove the hydrophobic peptide located at the amino-terminus of the precursor. Endoprotease cleavage also occurs at single and two adjacent basic residues, this is followed by a removal of basic residues located at the C-terminus of the peptides by a carboxypeptidase-like enzyme.

Regulatory peptides are diverse in their function and localization; however, they share a common property in that they all are initially synthesized as larger precursors which are processed proteolytically to form biologically active products[1-3]. Figure 1 is a schematic representation of several regulatory peptide precursors showing their processing sites and indicating that these precursors can have a molecular weight greater than 10 times that of the biologically active peptides[4], or they may lose only a few amino acids during their maturation process[5].

Precursors to regulatory peptides may have in common (i) a similar 'route' from their site of synthesis to the ultimate export of their products from the cell, and (ii) they all undergo proteolytic processing events. The proteolytic processing events include removal of the signal sequence, which is necessary for sequestration of the protein into the endoplasmic reticulum as well as subsequent endoproteolytic and exoproteolytic cleavages. Specific regulatory peptides can also undergo other post-translational modifications which include disulphide bond formation, carbohydrate addition, sulphation, phosphorylation, acetylation, and amidation to mention only a few of the numerous modifications which have been described[6].

In this brief review, it is not possible to examine thoroughly all aspects of post-translational modifications of precursors to regulatory peptides. We will, instead, focus on one of the events that regulatory peptides have in common, namely, the proteolytic processing of the precursors to form their corresponding regulatory peptides.

Biosynthesis of precursors to regulatory peptides. The nomenclature used to identify the specific intermediates observed during protein biosynthesis arose from the early work of Steiner and his colleagues[7] who identified a large precursor form of insulin which they referred to as proinsulin. When a precursor to proinsulin was observed in cell-free translation experiments, the larger precursor was referred to as

PRO-OPIOMELANOCORTIN PEPTIDE

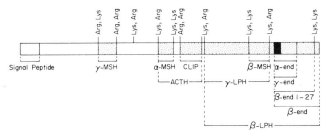

β-NEO-ENDORPHIN/DYNORPHIN PRECURSOR PEPTIDE

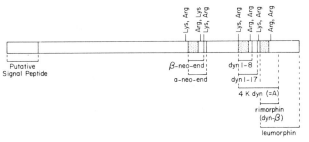

PREPROTACHYKININS

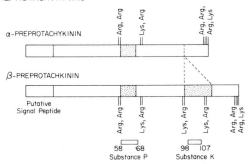

Figure 1. Schematic representation of peptide precursors. Stippled areas indicate biologically important products; Black indicates methionine enkephalin. Data sources are as follows: Proglucagon[98]; Preproinsulin[99]; Corticotropin-releasing factor precursor[100]; Prepro LH-RH precursor[101]; Thyrotropin-releasing hormone[102]; Prepro-enkephalin A[103]; Prepro-opiomelanocortin[104]; Prepro-enkephalin B[105]; Preprotachykinins[106].

194

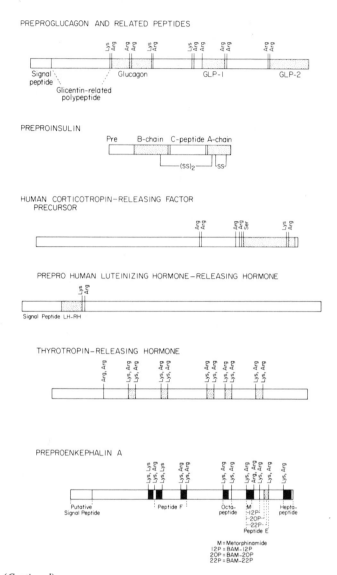

Figure 1. (*Continued*)

preproinsulin[8] (fig. 1). This terminology has generally been applied to the various intermediates of regulatory peptide biosynthesis.

The signal peptide

Proteins destined for secretion from the cell are initially synthesized in the cytoplasm, but become tightly bound to the endoplasmic reticulum as synthesis proceeds. The generally accepted model proposed by Blobel and Dobberstein in 1975 for protein export from the cell has been referred to as the signal hypothesis[9,10]. This model suggests that the initial synthesis of the growing peptide chain in the cytoplasm results in attachment of the amino terminus of the exported protein to membranes of the endoplasmic reticulum[11]. The interactions between the endoplasmic reticulum and growing peptide-ribosomal complex are complicated, but they clearly involve protein complexes such as the signal recognition particle, which is composed of six proteins and a 7S RNA molecule[12-15]. In addition, other proteins appear to play important roles in this interaction[15].

All precursors to regulatory peptides which are transported via this vectorial mechanism into the endoplasmic reticulum, harbor an amino terminal sequence of approximately 16 to 30 amino acid residues which generally includes 4 to 12 hydrophobic residues. This amino acid sequence has been referred to as the 'signal sequence' because it appears to contain the 'code' which designates that this protein is to be routed to the endoplasmic reticulum[9,10].

Signal sequences are not normally observed on newly synthesized proteins found within cells. This suggests that the signal sequence is rapidly removed from the precursor, while the growing peptide is still attached to the ribosome. In order to observe the signal sequence, it is necessary to carry out cell-free translation experiments with isolated messenger RNAs. In the absence of membranes from endoplasmic reticulum, the signal sequence is not removed and can be readily identified following isolation of the regulatory peptide precursor.

An example of a signal sequence identified on a preproregulatory protein was recently described by Minth et al.[16]. RNA isolated from a human phaeochromocytoma was translated in a cell-free translation system with radioactive leucine or methionine. Minth et al.[16], isolated the precursor by immunoprecipitation with antibodies directed towards the regulatory peptide, in this instance an antibody produced against neuropeptide Y. The antibodies recognized the larger precursor (preproneuropeptide Y) and selectively precipitated the product. The results of the cell-free translation and immunoprecipitation of the precursor are illustrated in figure 2.

The immunoprecipitated product was subjected to automated Edman degradation. This results in sequential removal of amino acids from the amino terminus. When the radioactivity in each cycle of the Edman degradation is plotted against the amino acid sequence deduced from the cDNA sequence one can see that there is exact agreement between the [^{3}H] leucine in cycles 2, 7, 9, 12, 14, 16, 18, 19, 22, 25, and the locations of the deduced leucine residues. The only cycle in which [^{35}S] methionine is seen is cycle 1. The results of the sequencing reactions and the deduced amino acid sequence of the signal peptide of prepro-neuropeptide Y are

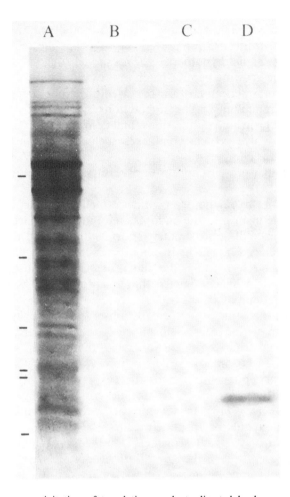

Figure 2. Immunoprecipitation of translation products directed by human pheochromocytoma RNA. Translations were carried out in a wheat germ cell-free translation system and then analyzed on sodium dodecylsulphate/15% polyacrylamide gels. Lanes: A, total translation products (one-tenth of reaction); B, precipitation with preimmune serum; C, precipitation with anti-NPY serum (YN-12) in the presence of 5 μg of porcine NPY; D, precipitation with anti-NPY (YN-12). Protein standards (in Da) from top to bottom are ovalbumin (43,000), α-chymotrypsinogen (25,700), β-lactoglobulin (17,400), lysozyme (14,300), cytochrome c (12,300), and bovine trypsin inhibitor (6200).

shown in figure 3. The signal sequence of prepro-neuropeptide Y is removed as a result of proteolysis between Ala-28 and Tyr-29[17]. Some of the factors which govern the removal of the signal sequence from secretory proteins are now under study.

Cleavage of the signal peptide

Both eukaryotic and prokaryotic cells appear to use similar mechanisms for protein export. Bacterial cells have been shown to be capable of

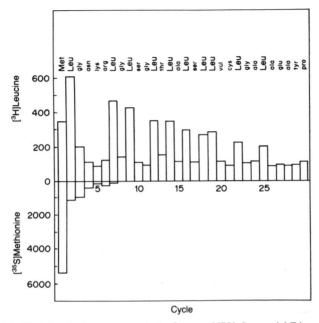

Figure 3. The NH₂-terminal sequence analysis of prepro-NPY. Sequential Edman degradation was carried out on the in vitro cell-free translation product immunoprecipitated with antiserum directed against porcine NPY. The radioactivity corresponding to [³⁵S] methionine and [³H] leucine was determined for 30 cycles. The positions of the radiolabeled amino acids in the immunoprecipitated translation product are shown along with the amino acid sequence determined from nucleotide sequence analysis. The positions of Leu and Met are capitalized.

exporting various eukaryotic proteins[18,19] and conversely, eukaryotic cells will secrete certain bacterial proteins[20]. Much of our knowledge concerning the bacterial protease responsible for cleavage of the signal sequence from the growing nascent protein comes from studies by Wickner and his colleagues on the biogenesis of the M13 procoat protein[21]. The protease which cleaves the signal peptide from the M13 protein has been isolated and characterized. It appears to be an integral membrane protein of 37,000 daltons[22]. This peptidase isolated from bacteria seems to have some similar substrate specificity to that of the analogous signal peptidase found in the eukaryotic endoplasmic reticulum[21]. For example, the bacterial enzyme will cleave preproproteins such as the honeybee pre-promellitin, human-prepro placental lactogen and preproinsulin[21]. The specific site of cleavage of the signal peptide in numerous regulatory peptide precursors has been established. No specific 'sequence' of amino acids would appear to be responsible for recognition by the signal peptidase; rather, higher order structure (i.e. secondary structure) seems to play an important role in determining the exact site of cleavage. Inouye and colleagues[23-26] have examined the structure-function relationship of residues located within the signal sequence. The importance of the central hydrophobic region is readily apparent from a number of reports which describe secretion-defective mutants with substitutions in this region[27-30]. In addition, using site-directed mutagenesis, Inouye et al. showed that selective alterations of amino acids within the hydrophobic segment have dramatic effects upon the secretory potential of the signal sequence[24]. Substitution of amino acids at or near the cleavage site can result in either the absence of processing or alternate processing at new sites. The analysis of the conformation required for the cleavage of prolipoprotein suggests that the residues near the cleavage site are probably in the β-turn structure. Substitution of residues which reduce the β-turn potential seem to alter dramatically the ability of the signal peptidase to carry out effective cleavage of the precursor sequence.

Recently, Evans et al.[31] reported the purification to near homogeneity from canine pancreatic microsomes, of the eukaryotic signal peptidase. In contrast to the prokaryotic protease, the purified canine enzyme consists of a complex of six polypeptides with apparent molecular masses of 25, 33, 22, 18 and 12 kDa. The 22 and 23 kDa proteins appear to be glycoproteins, however, only one of the subunits appears to function in signal peptide cleavage.

Cleavage at two adjacent basic residues

As discussed in the previous section, all regulatory peptides to date appear to be synthesized initially as preproproteins with rapid removal of the signal peptide. The resulting propeptides can be proteolytically processed at either single basic residues or at two adjacent basic residues. For processing at adjacent basic residues, arginine rather than lysine seems to be preferred in the P_1 position[32]. Figure 1 shows a number of examples of proregulatory proteins which harbor these processing sites.

The mammalian enzyme which recognizes and cleaves the prohormone at the two basic residues has not been isolated in pure form, nor characterized. However, a clue to the enzymes characteristics is provided by the unicellular eukaryote *Saccharomyces cerevisiae* which produces precursors for at least two secreted biologically active peptides, prepro-α-factor and preprokiller toxin which contain pairs of basic residues at presumptive processing sites. In KEX 2 mutant yeast cells neither of these proteins is produced in their mature form. The KEX 2 mutation in yeast resides in a gene that encodes a novel endoprotease specific for cleavage of these two substrates on the carboxyl side of pairs of basic residues[33].

The KEX 2 gene has been cloned and introduced into yeast cells on multicopy plasmids such that the KEX 2 protease is overproduced about 30-fold. From these cells the KEX 2 endoprotease has been purified about 100-fold, and the following catalytic properties have been elucidated. The membrane bound enzyme has a neutral pH optimum, and the enzyme displays a marked substrate preference. The relative rates of hydrolysis of peptide bonds are shown below.

$$-ArgArg\downarrow \cong -LysArg\downarrow \gg -LysLys\downarrow \cong -Arg\downarrow \gg Lys\downarrow$$

The enzyme also requires an active site thiol, as determined by its susceptibility to thiol inhibitors and is unaffected by several classical serine protease inhibitors. The most potent enzyme inhibitor found to date is a reagent originally prepared by Kettner and Shaw (1981), alanyl-lysyl-arginyl-chloromethylketone. The KEX 2 endoprotease can be covalently tagged and radioactively labeled by a derivative of this inhibitor, which will do the same for cathepsin B. The activity of this enzyme also appears to be completely dependent on the presence of Ca^{++} ions[34].

The yeast KEX 2 endoprotease resembles the mammalian proteases called calpains. The calpains are Ca^{++}-dependent neutral thiol proteases[35,36]. In experiments using synthetic fluorogenic substrates to determine calpain substrate specificity, it was found that cleavage occurred on the carboxyl side of a Tyr, Met or Arg residue provided that the preceding residue is hydrophobic[37]. Although many proteins have been examined as substrates of calpain cleavage, the true biological substrates of the calpains have not been determined.

It has not been easy to distinguish the 'prohormone' processing enzyme(s) from other proteases found in lysozymes. Studies which have examined the conversion of proinsulin, proglucagon and prosomatostatin I to their corresponding mature hormones in the secretory granules of anglerfish pancreatic islets found that at least one of the enzymes involved in the conversion of these prohormones was different from other intracellular proteases. The enzyme was shown to be a thiol protease with a pH optimum near 5, cleave at pairs of basic amino acids and may possibly require the presence of segments of the prohormone for proper substrate recognition and/or binding[38-40]. It is associated with the membrane of the secretory granule. The enzyme is similar to cathepsin B, but has a more restricted substrate specificity and is not inhibited by N-p-tosyl-L-lysine chloromethylketone. This may be the enzyme with 'trypsin-like' substrate specificity originally proposed for the first step in insulin processing[41-46].

A 70 kDa glycoprotein has been purified from secretory vesicles of the intermediate lobe[47] and the neural lobe[48] of the bovine pituitary. The activity assay involves conversion of mouse proopiomelanocortin to 21 kDa ACTH, βLPH, and 23 kDa ACTH. The enzymes have acidic pH optima and the enzyme from the intermediate lobe cleaves on the carboxyl side of Arg in the Lys-Arg dipeptide when provasopressin is the substrate to produce arginine vasopressin extended at the carboxyl terminus by Gly-Lys-Arg. The Gly residue contributes the carboxyl amide during later steps of processing. The enzymes from both sources were not inhibited by diisopropyl-fluorophosphate, p-chloromercuribenzoate, or

EDTA, but were inhibited by pepstatin A, an inhibitor of aspartyl proteinases. Because the enzymes from both sources are very similar in physical properties and in substrate specificities, the possibility exists that the enzymes from both sources are identical or are closely related. Although the enzymes described by Loh and colleagues will carry out conversion of several prohormones and are located in secretory vesicles, experiments unambiguously demonstrating their requirement for prohormone conversion *in vivo* are unattainable until genetic manipulation of the activity becomes feasible.

A prohormone 'converting enzyme' activity described by Noe and his colleagues appears to be membrane-associated and is found in microsomes and secretory granules[49]. This membrane association of converting enzymes has been demonstrated in rat anterior pituitary[50], neurointermediate lobe[51], and rat hypothalamic synaptosomes[52]. It has been proposed that newly synthesized islet prohormones are membrane-associated in the microsome and secretory granule and that the RER/Golgi complex and secretory granule membranes serve as a matrix to bring the enzyme and substrate together[49].

Recently, partial purification of an endoprotease from bovine neurosecretory granules was reported which cleaved prooxytocin/neurophysin (OT/Np) peptides at a Lys-Arg doublet. This 58 kDa protease is strongly inhibited by divalent cation chelating agents and has a maximum activity around neutrality, suggesting the presence of an active site thiol. Replacement of either basic residue in synthetic substrates by a neutral or hydrophobic one abolished cleavage activity[53].

Biological consequences of altered proteolytic processing

Although the mammalian enzyme which selectively cleaves at two adjacent basic amino acid residues has not been characterized in detail, the importance of these dibasic sites for the proper processing of prohormones has been noted in several instances. Families with hyperinsulinemia have been described[54,55]. Normally, proinsulin is cleaved between Arg-65 and Gly-66, a peptide bond which connects the C-peptide and A chain of proinsulin (fig. 1). A point mutation in the Arg-65 codon changes this residue to a histidine which results in a blockage of the post-translational cleavage of proinsulin to insulin[55]. Alterations in the amino acid sequence of proalbumin can also result in variants which do not undergo conversion to albumin[56,57]. Normally proalbumin is processed at two adjacent arginine residues (Arg·Arg). In one of the variants, the alteration results in an Arg-Glu substitution while in another variant a His-Arg substitution is observed. These changes result in unprocessed proalbumin. Recently a variant of factor IX coagulant activity was described which resulted in hemophilia B[58,59]. Factor IX[Cambridge] has a

Factor IX – cDNA ACA GTT TTT CTT GAT CAT GAA AAC GCC AAC AAA ATT CTG AAT CGG CCA AAG AGG TAT AAT TCA GGT AAA TTG

Factor IX Cambridge Thr Val Phe Leu Asp His Glu Asn Ala Asn Lys Ile Leu Asn Arg Pro Lys Ser Tyr Asn Ser Gly Lys Leu

Factor IX Thr Val Phe Leu Asp His Glu Asn Ala Asn Lys Ile Leu Asn Arg Pro Lys Arg | Tyr Asn Ser Gly Lys Leu

-18   -15   -10   -5   -1 | +1   +5

Figure 4. Amino acid sequence of factor IX^{Cambridge}. Factor IX^{Cambridge} contains an 18-residue NH$_2$-terminal extension homologous to the predicted amino acid sequence of the propeptide of factor IX (stippled) extrapolated from the known cDNA sequence of factor IX (107,108). Factor IX has an NH$_2$-terminal sequence of Tyr-Asn-Ser. Residue 18 of factor IX^{Cambridge}, analogous to the arginine at residue -1 of the factor IX precursor, is mutated to a serine (solid). This point mutation (Arg^{-1}→Ser^{-1}) in the factor IX propeptide precludes normal proteolytic processing between residues -1 and $+1$. The codon for arginine-1 in the factor IX precursor, AGG, is altered to either AGT or AGC (solid), codons that code for serine. Reproduced from Diuguid et al.[59].

Lys-Ser substitution for the normally observed post-translational processing site Lys-Arg (fig. 4). This point mutation in a human protein precursor impairs proteolytic processing and results in a disease state (hemophilia B).

Another interesting example of mutation which has an important impact on proteolytic processing was noted in a child with a bleeding disorder. This child had circulating proalbumin with the normal Arg-Arg cleavage site intact. Thus, the proalbumin present in the serum appears normal. The reason for the lack of cleavage of proalbumin appears to reside with a mutation in a 'trypsin' protease inhibitor. This individual had a Met (358) →Arg change in the active site of an α-1 antitrypsin enzyme, which led to the hypothesis that the mutant α-1 antitrypsin was inhibiting the prohormone converting enzyme. It was found that the KEX2 protease recognized and cleaved the prosequence of proalbumin and was specifically inhibited by a mutant α-1 antitrypsin and not by other serine protease inhibitors[60].

Proteolytic cleavage at single basic residues and processing by exopeptidases

Flexibility in proteolytic processing is important for tissue-specific processing of prohormones. For example, the major physiologically active peptide produced as a result of proglucagon (fig. 1) processing in the pancreas appears to be glucagon. A larger peptide derived from the remainder of proglucagon is also produced[61], but has not yet been shown to possess biological activity. In contrast, four different physiologically active products[62] are produced in the gut (glicentin, oxyntomodulin, GLP-1, and GLP-2). One way of achieving this differential processing is through the use of multiple recognition sites for different converting enzymes.

Although prohormone conversion frequently occurs at basic dipeptides, a growing number of processing sites are being uncovered which involve cleavage at single arginine residues. One of these is a single Arg residue in proglucagon which is utilized during production of the GLP-1 mentioned above[63]. Other sites in proglucagon occur at adjacent basic residues. Another excellent example of the differential use of single versus double-basic conversion sites occurs in the processing of prosomatostatin (fig. 5) in mammals and in fish[64–66]. The processing of prosomatostatin in mammalian pancreas involves cleavage at an Arg-Lys dipeptide resulting in formation of a 14-residue form of somatostatin. In the gut, processing at an earlier, single Arg residue produces a 28-residue somatostatin[67–69] having different physiological activities[70–72]. Thus, two hormones or regulatory peptides may be obtained from a single precursor via use of different processing sites. The two forms of somatostatin are produced in

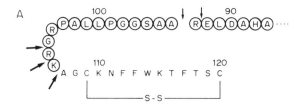

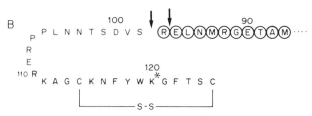

Figure 5. Processing sites for A): anglerfish preprosomatostatin I and B) anglerfish preprosomatostatin II. The residues belonging to the non-somatostatin portion of prosomatostatin are circled. The residues are numbered from the initiator methionine. The asterisk indicates the site of lysyl hydroxylation. The peptide bonds cleaved during conversion to the mature hormones are indicated by large arrows. Small arrows indicate minor processing sites.

separate cells in anglerfish pancreas[73]. The situation in the fish is complicated by the presence of two prosomatostatins, one of which is converted to the 14-residue somatostatin, and one which is converted predominantly to a 28-residue somatostatin (fig. 5). However, both forms contain putative monobasic and dibasic processing sites which are homologous. Enzyme activities which appear able to recognize the two different processing sites have been shown to be differentially localized in subcellular organelles of fish pancreas[49] and can be resolved chromatographically[74]. The activity which produces the 14-residue somatostatin from prosomatostatin-I (fig. 5) will also cleave prosomatostatin-II[74] to a 14-residue somatostatin. A separate protease will convert prosomatostatin-II to the 28-residue somatostatin, suggesting that in this case, specificity of cleavage may be due, in part, to differential expression of the proteases.

The protease(s) responsible for processing at single arginine residues has not yet been isolated. However, from the known structures of the substrates and products of processing some of the parameters important for substrate-specificity may be deduced. Although a given prohormone may contain several single arginine residues, only a limited number are utilized. Inspection of cleavage sites containing single arginine residues reveals no general similarities in primary sequences, although some of the cleavage sites may be grouped into those having nearby proline residues[75]. However, not all arginine residues in close proximity to proline are cleaved. Recently, Benoit et al.[76] pointed out that many known single basic processing sites are preceded by another basic residue 3, 5, and 7 residues before. Some single Arg processing sites overlap both classes of sites (i.e. have both a nearby Pro residue and a second basic residue at positions -3, -5, or -7 from the Arg). Unfortunately, inclusion of a site into one or both of these classes does not mean the site will be used by the converting enzymes. A quick perusal of prohormone sequences will identify a number of single Arg residues which fit either of these classes and yet there is not evidence for processing at these sites (for example Arg 87 of rat corticotropin[77]), Arg-36 and 41 of anglerfish preprosomatostatin-I[78,79] and Arg-48 and 50 of anglerfish preprosomatostatin-II[78,80], Arg-14 of canine motilin, Arg-41 and 43 of human growth hormone releasing factor (ref. 81 and many others). The existence of so many exceptions suggests that some additional factor is important such as differential expression of the processing enzymes or some aspect of the three-dimensional structure at the putative processing sites.

Three recent examples of prohormones containing apparent processing sites at single lysyl residues have recently appeared in the literature. The first example was noted in the precursor to the peptide, FMRF amide from *Aplysia*[82]. This proneuropeptide contains several copies of FMRF amide, some having typical Lys-Arg or single Arg processing sites at their carboxyl ends. Some of the FMRF amide sequences are followed by Lys-Ser. Although the implication is that the Lys-Ser represents a processing site, this has not yet been demonstrated. The precursor to the FMRF amide-like peptide from *D. melanogaster* also contains a potential processing site at a single lysyl residue in addition to typical processing sites[83]. Although the FRMF amide from *Drosophila* was isolated, the peptide arising from cleavage at the single lysyl residue has not yet been isolated. Finally, guinea pig pro-pancreatic polypeptide contains a single lysyl processing site[84]. Several precursors from species other than guinea pig have a Lys-Arg sequence at their cleavage site. The sequence in guinea pig is Lys-Ser. This single Lys residue is cleaved during prohormone conversion ultimately resulting in a mature 36-residue amidated peptide[85,86]. It is not known whether processing occurs before or after the Lys residue.

The limited number of examples where cleavage occurs at single lysyl residues all have the consensus sequence R(F/Y)GKSX, where X is a hydrophobic amino acid. As further examples of processing at single lysyl residues are identified, it will be interesting to note which of these residues remain in this consensus sequence. A possible cleavage at a single lysyl residue has recently been reported after isolation of the amino-terminal decapeptide of prosomatostatin from rat intestine[76]. This peptide has had two residues (Leu-Gln) removed from the carboxyl-terminus in addition to apparent cleavage at a Lys-Ser-Leu site. The removal of the two neutral residues is not typical of prohormone processing in general, thus, it is not clear that this peptide is a product of prohormone processing in the secretory granule.

It is interesting to note, regarding processing at single lysyl residues, that a protease capable of cleaving before single lysyl residues has been implicated in the post-secretion degradation of peptides from frog venom glands[87]. This protease appears to have a broader specificity, however, than would be expected for a prohormone-converting enzyme. Another recent publication described the isolation of a number of products derived from prosomatostatin processing in rat tissue[88]. This study thoroughly documented the importance of including the acid protease inhibitor pepstatin in acid extracts of tissue to prevent artifactual proteolysis by acid proteases.

Both classes of endopeptidase with specificity directed either towards single or to adjacent basic residues produce products having terminal basic residues. In many instances these residues must be removed to produce the active form of the peptide. Secretory vesicles have been reported to contain both an aminopeptidase[89] and a carboxypeptidase[90–92] with specificities directed towards basic residues.

The potent carboxypeptidase has been reported from a wide range of endocrine tissues. The enzyme is a metallopeptidase and has an acidic pH optimum in the range 5.0–6.0[91–95]. The carboxypeptidase from secretory granules is activated by Co^{2+} and can be distinguished from the lysosomal enzyme which is not. The secretory granule carboxypeptidase can be inhibited by chelating agents and by active-site directed inhibitors such as guanidinoethyl-mercaptosuccinic acid[91]. Two forms of the enzyme exist, a membrane-bound form and a soluble form having molecular weights of approximately 52,000 and 50,000 daltons respectively,[91,92,95]. The soluble form has been purified to homogeneity[96] and the cDNA has been cloned

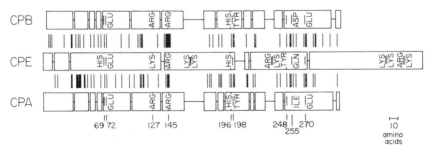

Figure 6. Schematic diagram of the carboxypeptidase isolated from secretory vesicles which is specific for basic residues. It is referred to as enkephalin converting enzyme (CPE) by Fricker and Snyder[96,97]. The structures of carboxypeptidases A and B (CPA and CPB, respectively) are also indicated for comparison. Deleted regions are indicated by a thin line and the residues important for substrate binding or activity are indicated by their three-letter codes. The His residues at position 69 in both CPA and CPB are not indicated but are present. Residues in CPA or CPB which are identical to those in CPE are indicated by lines in the spaces between the proteins. Figure adapted from Fricker and Snyder[91,92].

and sequenced[97]. The deduced protein sequence indicates some homology (fig. 6) with carboxypeptidases A and B (20% and 17%, respectively). The homology between carboxypeptidases A and B is 48%, suggesting that although the carboxypeptidases are related, A and B diverged from one another more recently than the secretory granule carboxypeptidase. Major differences between the converting enzyme and carboxypeptidase A and B include a carboxyl-terminal extension of 120 amino acids and changes in the substrate-binding regions (fig. 6).

An aminopeptidase has also been reported from secretory vesicles[89]. The enzyme is a membrane-bound metallopeptidase having a broad pH optimum in the range 5.5 to 7.5. Although the enzyme is specific for amino-terminal basic residues, its involvement in prohormone conversion has not yet been established.

Acknowledgment. This work was supported in part by grants from the National Institutes of Health. PCA was supported by a grant from the American Diabetes Association.

1 Douglass, J., Civelli, O., and Herbert, E., A. Rev. Biochem. *53* (1984) 665.
2 Gluschankof, P., and Cohen, P., Neurochem. Res. *12* (1987) 951.
3 Loh, Y. P., and Gainer, H., Biosynthesis and processing of neuropeptides, in: Brain Peptides, p. 79. Eds. D. T. Kreiger et al. John Wiley and Sons, New York 1983.
4 Funckes, C. L., Minth, C., Deschenes, R., Magazin, M., Tavianini, M., Sheets, M., Collier, K., Weith, H., Aron, D., Roos, B., and Dixon, J. E., J. biol. Chem. *258* (1983) 8781.
5 Potts, J. T., Jr., Kronenberg, H. M., Habener, J. F., and Rich, A., Ann. N.Y. Acad. Sci. *343* (1980) 38.
6 Wold, F., A. Rev. Biochem. *50* (1981) 783.
7 Steiner, D. F., Terris, S., Chan, S. J., and Rubenstein, A., in: Insulin: Islet Pathology — Islet Function and Insulin Treatment, p. 55. Ed. R. Luft. Lindergren and Soner, Sweden 1976.
8 Steiner, D. F., Quinn, P. S., Chan, S. J., Marsh, J., and Tager, H. S., Ann. N.Y. Acad. Sci. *341* (1980) 1.
9 Blobel, G., and Dobberstein, B., J. Cell Biol. *67* (1975) 835.
10 Blobel, G., and Dobberstein, B., J. Cell Biol. *67* (1975) 852.
11 Walter, P., and Blobel, G., Proc. natl Acad. Sci. USA *77* (1980) 7112.
12 Walter, P., and Blobel, G., J. Cell Biol. *91* (1981) 557.
13 Walter, P., and Blobel, G., Nature *299* (1982) 691.
14 Walter, P., Biegel, Lauffer, L., Garcia, P. N., Nilrich, A., and Harkins, R., in: Protein Transport and Secretion, p. 21. Ed. Mary-Jane Gething. Cold Spring Harbor Laboratory 1985.
15 Walter, P., Gilmore, R., and Blobel, G., Cell *38* (1984) 5.
16 Minth, C. D., Bloom, S., Polak, J., and Dixon, J. E., Proc. natl Acad. Sci. *81* (1984) 4577.
17 Dickerson, I., Dixon, J. E., and Mains, R. E., J. biol. Chem. *262* (1987) 13646.
18 Fraser, T. H., and Bruce, B. J., Proc. natl Acad. Sci. *75* (1978) 5936.
19 Talmadge, K., Stahl, S., and Gilbert, W., Proc. natl Acad. Sci. *77* (1980) 3369.
20 Wiedman, M., Huth, A., and Rapoport, T. A., Nature *309* (1984) 637.
21 Rice, M. C., and Wickner, W. T., Mechanisms of membrane assembly and protein secretion in *Escherichia coli*, in: Protein Transport and Secretion, p. 44. Ed. Mary-Jane Gething. Cold Spring Harbor Press 1985.
22 Wolfe, P., and Wickner, W., Cell *36* (1984) 1067.
23 Inouye, S., Soberon, X., Franceschini, T., Nakamura, K., Itakura, K., and Inouye, M., Proc. natl. Acad. Sci. USA *79* (1982) 3438.
24 Inouye, M., Inouye, S., Pollitt, S., Ghrayeb, J., and Lunn, C. A., Structure and functional analysis of the prolipoprotein signal peptide of *Escherichia coli*, in: Protein Transport and Secretion, p. 54. Ed. Mary-Jane Gething. Cold Spring Harbor Press 1985.

207

25 Vlasuk, G. P., Inouye, S., and Inouye, M., J. biol. Chem. *259* (1984) 6195.
26 Vlasuk, G. P., Inouye, S., Ito, H., Itakura, K., and Inouye, M., J. biol. Chem. *258* (1983) 7141.
27 Bedouelle, H., Bassford, P. J., Fowler, A. V., Zabin, I., Beckwith, J., and Hofnung, M., Nature *285* (1980) 78.
28 Emr, S. D., Hedgpeth, Clement, J., Silhavy, T. J., and Hofnung, M., Nature *285* (1980) 82.
29 Emr, S. D., Schwartz, M., and Silhavy, T. J., Proc. natl Acad. Sci. USA *75* (1983) 5802.
30 Michaelis, S. H., Inouye, S., Oliver, D., and Beckwith, J., J.- Bacteriol. *154* (1983) 366.
31 Evans, E. A., Gilmore, R., and Blobel, G., Proc. natl Acad. Sci. USA *83* (1986) 581.
32 Steiner, D. F., Kemmler, W., Tager, H. S., Rubenstein, A. H., Lernmark, A., and Zühlke, H., Proteolytic mechanism in the biosynthesis of polypeptide hormones, in: Proteases and Biological Control. Eds. E. Reich, D. Rifkin and E. Shaw. Cold Spring Harbor Laboratory 1975.
33 Julius, D., Brake, A., Blair, L., Kunisawa, R., and Thorner, J., Cell *37* (1984) 1075.
34 Fuller, R. S., Brake, A. J., Julius, D. J., and Thorner, J., in: Protein Transport and Secretion, p. 97. Ed. Mary-Jane Gething. Cold Spring Harbor Laboratory 1985.
35 DeMartino, G. N., and Croall, D. E., Biochemistry *22* (1983) 6287.
36 Murachi, T., in: Calcium and Cell Function, vol. 4, p. 377. Ed. W. Y. Cheung. Academic Press, New York 1983.
37 Sasaki, T., Kikuchi, T., Yumoto, N., Yoshimura, N., and Murachi, T., J. biol. Chem. *259* (1984) 12489.
38 Fletcher, D. J., Noe, B. D., Bauer, G. E., and Quigley, J. P., Diabetes *29* (1980) 593.
39 Fletcher, D. J., Quigley, J. P., Bauer, G. E., and Noe, B. D., J. Cell Biol. *90* (1981) 312.
40 Noe, B. D., J. biol. Chem. *256* (1981) 4940.
41 Kemmler, W., and Steiner, D. F., Biochem. biophys. Rev. Commun. *41* (1970) 1223.
42 Kemmler, W., Peterson, J. D., and Steiner, D. F., J. biol. Chem. *246* (1971) 6786.
43 Kemmler, W., Steiner, D. F., and Borg, J., J. biol. Chem. *248* (1973) 4544.
44 Sun, A. M., Lin, B. J., and Haist, R. F., Can. J. Physiol. Pharmac. *51* (1973) 175.
45 Zühlke, H., Steiner, D. F., Lernmark, A., and Lipsey, C., TBA Symp. *41* (1976) 183.
46 Zühlke, H., Kohnert, K.-D., Jahr, H., Schmidt, S., Kirschke, H., and Steiner, D. F., Acta biol. med. gen. *36* (1977) 1695–1703.
47 Loh, Y. P., Parish, D. C., and Tuteja, R., J. biol. Chem. *260* (1985) 7194.
48 Parish, D. C., Tuteja, R., Altstein, M., Gainer, H., and Loh, Y. P., J. biol. Chem. *261* (1986) 14392.
49 Noe, B. D., Debo, G., and Spiess, J., J. Cell Biol. *99* (1984) 578.
50 Chang, T.-L., and Loh, Y. P., Endocrinology *112* (1983) 1832.
51 Loh, Y. P., and Gainer, H., Proc. natl Acad. Sci. USA *79* (1982) 108.
52 Zingg, H. H., and Patel, Y. C., Life Sci. *33* (1983) 1241.
53 Clamagirand, C., Creminon, C., Fahy, C., Boussetta, H., and Cohen, P., Biochemistry *26* (1987) 6018.
54 Robbins, D. C., Blix, P. M., Rubenstein, A. H., Kanazawa, Y., Kosaka, K., and Tager, H. S., Nature (London) *291* (1981) 679.
55 Shibasaki, Y., Kawakami, T., Kanazawa, Y., Akanuma, Y., and Fumenaro, T., J. clin. Invest. *76* (1985) 378.
56 Abdo, Y., Rousseaux, J., and Dautrevaux, M., FEBS Lett. *131* (1981) 286.
57 Brennan, S. O., and Carrell, R. W., Nature (Lond.) *274* (1978) 908.
58 Bentley, A. K., Rees, D. J. G., Rizzo, C., and Brownlee, G. G., Cell *45* (1986) 343.
59 Diuguid, D. L., Rabiet, M. J., Furie, B. C., Liebman, H. A., and Furie, B., Proc. natl Acad. Sci. USA *83* (1986) 5803.
60 Bathurst, I. C., Brennan, S. O., Carrell, R. W., Cousens, L. S., Brake, A. J., and Barr, P. J., Science *235* (1987) 348.
61 Patzelt, C., and Schiltz, E., Proc. natl Acad. Sci. USA *81* (1984) 5007.
62 Ørskov, C., Holst, J. J., Knuhtsen, S., Baldissera, F. G. A., Poulsen, S. S., and Nielsen, O. V., Endocrinology *119* (1986) 1467.
63 Holst, J. J., Ørskov, C., Schwartz, T. W., Buhl, T., and Baldissera, F. G. A., 22nd Annual Meeting of Eur. Assoc. for the Study of Diabetes, Abstract #217. Diabetologia *29* (1986) 549A.

208

64 Andrews, P. C., Hawke, D., Shively, J. E., and Dixon, J. E., J. biol. Chem. *259* (1984) 15021.
65 Noe, B. D., Spiess, J., Rivier, J. E., and Valve, W., Endocrinology *105* (1979) 1410.
66 Spiess, J., and Noe, B. D., Proc. natl Acad. Sci. USA *82* (1985) 277.
67 Hobart, P., Crawford, R., Shen, L.-P., Pictet, R., and Rutter, W. J., Nature *288* (1980) 137.
68 Fischli, W., Goldstein, A., Hunkapiller, M. W., and Hood, L. E., Proc. natl Acad. Sci USA *79* (1982) 5435.
69 Pradayrol, L., Jornvall, H., Mutt, V., and Ribet, A., FEBS Lett. *109* (1980) 55.
70 Brown, M., Rivier, J., and Vale, W., Endocrinology *108* (1981) 2391.
71 Mandarino, L., Stenner, D., Blanchard, W., Nissen, S., Gerich, J., Ling, N., Brazeau, P., Bohlen, P., Esch. F., and Guillemin, R., Nature *291* (1981) 76.
72 Meyers, C. A., Murphy, W. A., Redding, T. W., Coy, D. H., and Schally, A., Proc. natl Acad. Sci. USA *77* (1980) 6171.
73 McDonald, J. K., Greiner, F., Gauer, G. E., Elde, R. P., and Noe, B. D., J. Histochem. Biochem. (1987) in press.
74 Mackin, R. B., Ph.D. Dissertation, Dept. of Anatomy and Cell Biology. Emory University School of Medicine 1987.
75 Schwartz, T., FEBS Lett. *200* (1986) 1.
76 Benoit, R., Ling, N., and Esch, F., Science *238* (1987) 1126.
77 Drouin, J., and Goodman, H. M., Nature *288* (1980) 610.
78 Hobart, P., Crawford, R., Shen, L.-P., Pictet, R., and Rutter, W. J., Nature *288* (1980) 137.
79 Andrews, P. C., and Dixon, J. E., Biochemistry *26* (1987) 4853.
80 Andrews, P. C., Nichols, R., and Dixon, J. E., J. biol. Chem. *262* (1987) 12692.
81 Guillemin, R., Brazeau, P., Böhlen, P., Esch, F., Ling, N., and Wehrenberg, W. B., Science *218* (1982) 585.
82 Schaefer, M., Picciotto, M. R., Kreiner, T., Kaldany, R.-R., Taussig, R., and Scheller, R. H., Cell *41* (1985) 457.
83 Nambu, J., Murphy-Erdosh, C., Andrews, P. C., Feistner, G. J., and Scheller, R. H., Neuron *1* (1988) in press.
84 Blackstone, C. D., Seino, S., Takeuchi, T., Yamada, T., and Steiner, D. F., J. biol. Chem. *263* (1988) in press.
85 Eng, J., Huang, J.-G. Pan, Y.-C. E., Hulmes, J. D., and Yalow, R. S., Peptides *8* (1987) 165.
86 Andrews, P. C. et al., unpublished results.
87 Giovannini, M. G., Poulter, L., Gibson, B. W., and Williams, O. H., Biochem. J. *243* (1987) 113.
88 Patel, Y. C., and O'Neil, W. O., J. biol. Chem. *263* (1988) 745.
89 Gainer, H., Russell, J. T., and Loh, Y. P., FEBS Lett. *175* (1984) 135.
90 Fricker, L. D., Plummer, T. H., and Snyder, S. H., Biochem. biophys. Res. Commun. *111* (1983) 994.
91 Fricker, L. D., and Snyder, S. H., Proc. natl Acad. Sci. USA *79* (1982) 3886.
92 Fricker, L. D., Suppattapone, S., and Snyder, S. H., Life Sci. *31* (1982) 1841.
93 Hook, V. Y. H., Eiden, L. E., and Brownstein, M. J., Nature *295* (1982) 341.
94 Hook, V. Y. H., and Loh, Y. P., Proc. natl Acad. Sci. *81* (1984) 2776.
95 Suppattapone, S., Fricker, L. D., and Snyder, S. H., J. Neurochem. *42* (1984) 1017.
96 Fricker, L. D., and Snyder, S. H., J. biol. Chem. *258* (1983) 10950.
97 Fricker, L. D., Evans, C. J., Esch, F. S., and Herbert, E., Nature *323* (1986) 461.
98 Bell, G. I., Sanchez-Pescador, R., Laybourn, P. J., and Najarian, R. C., Nature (1983) 368.
99 Bell, G. I., Swain, W. F., Pictet, R., Cordell, B., Goodman, H. M., and Rutter, W. J., Nature *282* (1979) 525.
100 Shibahara, S., Morimoto, Y., Furutani, Y., Notake, M., Takahashi, H., Shimizu, S., Horikawa, S., and Numa, S., EMBO J. *2* (1983) 775.
101 Seeburg, P. H., and Adelman, J. P., Nature *311* (1984) 666.
102 Lechan, R. M., Wu, P., Jackson, I. M. D., Wolf, H., Cooperman, S., Mandel, G. E., and Goodman, R. H., Science *231* (1986) 159.
103 Lewis, R. V., Stern, A. S., Kimura, S., Rossier, J., Stein, S., and Udenfriend, S., Science *208* (1980) 1459.
104 Mains, R. E., and Eipper, B. A., J. biol. Chem. *254* (1979) 7885.

105 Horikawa, S., Takai, T., Toyosato, M., Takahashi, H., Noda, M., Kakidani, H., Kubo, T., Hirose, T., Inayama, S., Hayashida, H., Miyata, T., and Numa, S., Nature *306* (1983) 611.
106 Nawa, H., Hirose, T., Takashima, H., Inayama, S., and Nakanishi, S., Nature *306* (1983) 32.
107 Choo, K. H., Gould, K. G., Rees, D. J. G., and Brownlee, G. G., Nature (Lond.) *299* (1982) 178.
108 Kurachi, K., and Davie, E. W., Proc. natl Acad. Sci. USA *79* (1982) 6461.

Transgenic mouse models and peptide producing endocrine tumours: morpho-functional aspects

G. Rindi, E. Solcia and J. M. Polak

Summary. Three transgenic mouse models which proved to develop endocrine tumours are reviewed and discussed.

The neoplasms were induced through the production of the transforming oncoprotein simian virus 40 (SV40) large T-antigen. The SV40/metallothionein-growth hormone (MGH), the insulin/SV40 (INS/SV40) and the vasopressin/SV40 (AVP/SV40) transgenic mice models all developed endocrine tumours of pancreas mainly composed of insulin-producing B cells, with a minor PP cell component. In the pancreata of INS/SV40 and AVP/SV40 transgenic mice, non-tumour lesions (hyperplasia and dysplasia) were also described. AVP/SV40 transgenic mice presented tumour genesis in anterior pituitary too.

The usefulness of transgenic mouse models in reproducing human pathology is outlined with special reference to genetically dependent tumours.

The transgenic models: general view

Transgenic mice are animals bearing exogenous DNA sequences in their genome which are obtained by the microinjection of foreign nucleic acid into fertilised mouse oocytes[1–5,8–12,15,18,19,21–23,27–31,33,36,38–40]. Successive implantation of such enriched oocytes into the oviduct of a pseudo-pregnant female mouse results in the birth of animals which present, to varying extents, the injected DNA in their genome. The integration into the host genome of the exogenous base pairs is randomised and the copy number of the transgene is usually variable within different animals.

The foreign DNA is called a 'transgene' and roughly resembles a natural gene, presenting sequences which are supposed to have regulatory properties followed by sequences encoding for a defined, easily detectable protein. The identification of the integrated transgene is performed with the usual techniques of blot hybridization on high molecular genomic DNA isolated from tissue of the offspring. The expression of the transgene is monitored by the identification of the transgene product, i.e. the protein encoded by the foreign DNA, in the tissues which are supposed to present expression of the transgene. This is most easily obtained by immunochemical or immunohistochemical methods on the target tissues[1,12,22,27,36].

The protein encoded by the transgene is preferably a transforming agent like the large T-antigen encoded by the early region of the DNA tumour virus simian virus 40 (SV40)[24,35,41]. This is a well-studied protein (mol. wt 94.000) which is required for the initiation and probably for the maintenance of the oncogenic process[35].

The sequences which are supposed to control the expression of the oncogenic protein are the real subject under study in the transgene, in order to elucidate their supposed regulatory functions. These sequences are derived from cloned genes and the rationale of the transgenic experiments resides in the attempt of targetting the regulatory specificity of these sequences[4,12,21,22,27,31,33,36]. It is supposed and expected that when the regulatory sequences exert their function, the production of the oncoprotein is switched on, inducing transformation in the tissues which express the hybrid oncogene.

Three transgenic models, all presenting endocrine tumour pathology as a result of the genetic manipulation[1,12,21,22,31,33,36], are discussed below.

The SV40/metallothionein-growth hormone (MGH) fusion gene model

This model has been described by Messing et al.[21] and Palmiter et al.[31]. In a number of experiments, transgenic mice bearing different constructs were produced, to assess their role in directing specifically the site of tumour production. Pathologic changes varied greatly, comprising choroid plexus papillomas, thymic hyperplasia, kidney lesions, peripheral neuropathy, liver carcinoma and pancreatic islet cells adenomas. It was demonstrated that the construct bearing the enhancer sequences of the SV40 genome was instrumental in directing tumorigenesis in choroid plexuses. In transgenic mice with this construct the frequency of pancreatic and/or liver growths was null, while it became significant in the absence of SV40 control elements, with constructs presenting only the SV40 large T-antigen coding sequences and the metallothionein-growth hormone (MGH) fusion gene. It was concluded tentatively that the MGH region had an important role in directing the site of oncogenesis. Given the high level of metallothionein synthesis in the liver[31], the liver pathology could have been expected once it was assumed that the metallothionein promoter exerted control over the SV40 large T-antigen coding sequences. On the contrary, the pancreatic tumours were unexpected and unexplained.

The pancreatic lesions were described as islet cell adenomas probably composed mainly of insulin producing B cells, although direct proof of their B cell nature was not given. The tumours were often multiple, varied greatly in size, and were made up of monomorphic elements with evident, eosinophilic cytolasm and organized in clusters frequently delimiting cystic spaces. Mitotic activity of these tumours was reported as variable.

Although lacking complete morpho-functional characterisation, the pancreatic growths described in the MGH/SV40 large T-antigen transgenic mice closely resemble some patterns well known to occur in human pancreatic adenomatosis, with or without associated extrapancreatic neoplasms (type 1 multiple endocrine neoplasia or MEN syndrome)[7,13,14,16,32,37,42].

The insulin/SV40 model

In this model described by D. Hanahan[12] the recombinant oncogene used consisted of the rat insulin II gene 5′ flanking region DNA linked to the SV40 large T-antigen coding sequences. Two different constructs, with the insulin regulatory sequences inverted or aligned to transcribe the large T-antigen, were used, resulting in similar phenotype of affected transgenic mice. Pathologic changes started between 10 and 20 weeks of age, were fully developed in the early generation and heritable. Multiple tumour formation in the endocrine pancreas, usually associated with enlarged/hyperplastic islets were reported. Moreover, by tissue immunohistochemistry, the author described a derangement of hormone expression in the remaining apparently normal islets, suggesting that the normal populations of A and D cells were sterically excluded or suppressed by altered B cells. These B cells of transgenic pancreata appeared to produce large T-antigen both in 'normal' and 'hyperplastic' islets besides in tumours. The hybrid oncogene proved to express only in B cells. It was suggested that all B cells produced detectable large T-antigen, thereby inducing their proliferation in a sequence from hyperplasia to neoplasia. By protein precipitation tests, no other site of expression of the hybrid oncogene was identified apart from the pancreas, thus proving the tissue specificity of the transgene[11].

It was concluded that the rat insulin II gene 5′ flanking region DNA is sufficient to direct the expression of the hybrid oncogene insulin/SV40 large T-antigen in a tissue and cell-specific manner.

A subsequent study on this insulin/SV40 model by Power et al.[33], focused on the careful characterisation of the pancreatic changes of these transgenic mice, drawing comparisons with the corresponding human pathology (insulinomas). The endocrine component of the transgenic pancreas revealed the usual profile of endocrine cells in the normal islets[16,26] with a progressive derangement of hormonal expression (decreased glucagon and somatostatin immunoreactants) in structurally modified 'hyperplastic' islets. A spectrum of morphologically altered islets was described from normal to neoplasia, thus confirming the original observations. However hyperplasia was defined as 'an increase in the size of the islets and number of B cells with accompanying cytological atypia', which may be called, more appropriately, 'atypical hyperplasia' or dysplasia.

The transgenic tumours showed the same morpho-functional characteristics as a group of human B cells neoplasms studied for comparison.

Significantly, PP-immunoreactivity was identified in about the same percentage of cells (approx. 30%) in both groups[13,14,32]. Ultrastructural patterns were also similar in the two kinds of tumours.

It is concluded that the insulin/SV40 transgenic mouse is a useful model for the corresponding human neoplastic pathology, because it shares common morpho-functional aspects.

The arginine-vasopressin/SV40 model

In this model, described by D. Murphy et al.[22], a hybrid oncogene made up of the bovine arginine-vasopressin (AVP) gene 5' derived upstream sequences promoting the SV40 large T-antigen coding sequences was constructed and used to produce transgenic mice. The vasopressin gene sequences were chosen to assess their regulatory properties. The bovine promoter proved to be functional in mouse by positive transfection experiments into a murine cell line using the same AVP/SV40 transgene[10].

50% of the animals obtained after microinjection were transgenic and one line (VT-C) showed an heritable pathology leading to the animals' death between 90 and 140 days. The lesions were found in the endocrine pancreas and the anterior pituitary. In the pancreas, a variable pattern of normal and hyperplastic islets, together with tumours were described (fig. 1). Immunohistochemistry showed that the tumours were mainly composed of insulin producing B cells with scattered A, D or PP elements. The neoplastic cells expressed large T-antigen, while no significant immunoreactivity was detected in B cells of normal or hyperplastic islets. Hyperplasia was defined as 'enlarged islets containing normal endocrine cells without abnormal *nuclei*, showing the same proportional relationship between the different types of peptide-immunoreactive cells as normal islets'[16,26]. In the anterior pituitary, the tumours showed no significant reactivity for any of the peptides found in the normal organ[25]. Large T-antigen immunoreactivity was demonstrated in neoplastic *nuclei* only. The electron microscopical study confirmed the B cells as main constituent of the pancreatic lesions, while poor granularity was demonstrated in the pituitary neoplasms.

The unexpected expression of the transgene in tissues which normally do not show vasopressin production, although a few vasopressin-immunoreactive cells have been reported in the rat adenohypophysis[20], lead to the conclusion that the vasopressin upstream sequences used within the transgene were insufficient to direct the expression in hypothalamic vasopressinergic cells.

Although in previous experiments including the INS/SV40 model the transgene expression could be driven by a few hundred base pairs[12,27,28], in the AVP/SV40 model further regulatory elements may be needed to

214

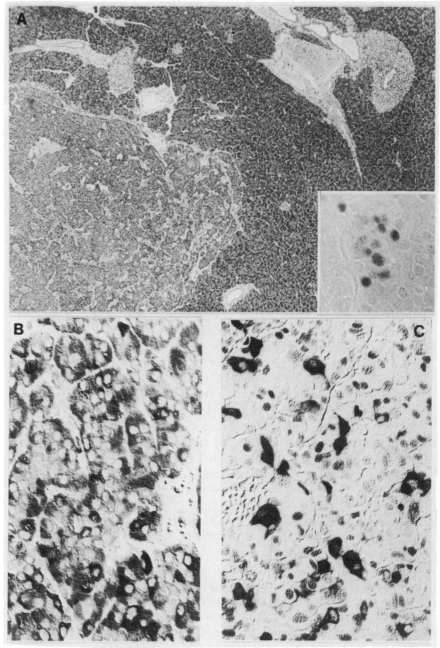

Figure 1. *A–C* Pancreas of 90–100 days old AVP/SV40 transgenic mouse. *A* General view of the organ with tumour, dysplastic and normal islets. Haematoxylin and eosin; × 170. Inset: large T-antigen immunoreactivity of cell *nuclei* in a dysplastic *microfocus* inside an otherwise normal islet. Immunoperoxidase, no counterstain; × 400. *B* Insulin-immunoreactive cells in a pancreatic tumour. Immunoperoxidase; no counterstain; Nomarski optics; × 500. *C* Large T-antigen nuclear immunoreactivity of variable intensity in tumour cells, some of which also react with anti-pancreatic polypeptide serum. Immunoperoxidase, double staining with peroxidase anti-peroxidase and avidin biotin procedures; no counterstain Nomarski optics; × 500.

direct site-appropriate expression, as suggested by the complicated site expression pattern demonstrated in other transgenic models[11,18,21,29-31]. Therefore, the AVP/SV40 expression described here may be appropriate to the regulatory elements contained within the hybrid oncogene. Moreover, possible effects of chromosomal position due to the integration site of the transgene cannot be ruled out[19]. It should also be outlined that vasopressinergic neurons, unlike most target cells of other transgenic models, are permanent, non-dividing elements, a condition which might interfere with the oncogene expression.

Nevertheless, the production of animals with heritable multiple endocrine tumorigenesis constituted the only available model for the human multiple endocrine neoplasia syndrome[7,14,16,37,42]. A subsequent study by Rindi et al.[35], focused on the characterisation of the AVP/SV40 mice lesions, in order to provide comparable morphological criteria with those of the corresponding human pathology. Moreover, animals of different age groups were also investigated to define the time course of the oncogenesis. In older animals, the pancreas revealed a peculiar pattern comprising overt neoplasms, abnormal dysplastic islets (fig. 2), normal islets and endocrine cell hyperplastic proliferation with insular neogenesis from ducts. Dysplasia was defined as foci of atypical cells organised in a solid structure inside an otherwise normal islet. Considering the size variability of the islets in normal mouse pancreas, enlargement was not considered a major morphological criterion. The transgene expression was limited to the dysplastic and neoplastic cells, both of which showed large T-antigen immunoreactivity and proved to be prevalently insulin-producing B cells. A consistent population of PP-immunoreactive cells, also showing large T-antigen immunoreactivity, was demonstrated in 25% of pancreatic tumours. The above observation may be indicative that the neoplastic transformation initiated in mature adult islet cells. Similar dysplastic lesions were found in older mouse pituitaries too, with similar cytological aspects and large T-antigen expression. Younger animals proved to have only dysplastic lesions in the pancreas, without neoplasms, and no lesions at all in the pituitary, thus suggesting late expression of the transgene in the latter organ. It was concluded that some tissue-specific factor drove preferential expression of the transgene in the pancreas.

AVP/SV40-induced tumours seem to represent an attractive model for the study of MEN I syndrome[7,14,16,37,42], perhaps suggesting mature islet cells as target of the transforming process, an hypothesis at variance with the currently accepted view that endocrine neoplasms grow from multipotential ductular stem cells[13,14]. This should stimulate a reconsideration of the morphology of MEN I, expecially in the pancreas, looking for similar lesions in human islets (fig. 2). Dysplastic changes in human islet cells have been already reported in patients genetically deficient for alpha-proteinase inhibitor[34]. The hyperplastic endocrine proliferation

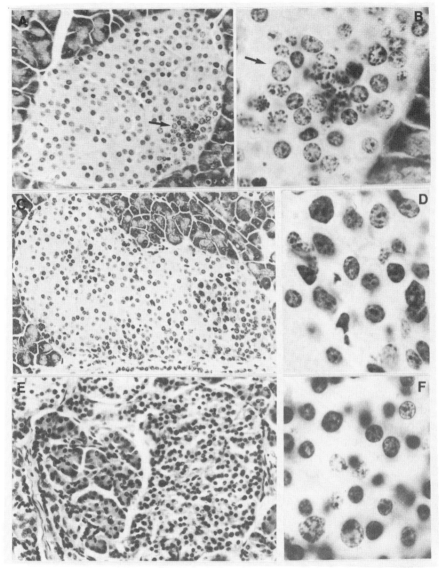

Figure 2. *A–D* Pancreas of 90–100 days old AVP/SV40 transgenic mouse. Single (*A, B*) and multiple (*C, D*) *foci* of dysplasia in normal sized (*A, B*) and enlarged (*C, D*) islets. The higher magnifications show the abnormal sizes of some nuclei with evident nucleoli and a characteristic spotted pattern of the chromatin (e.g. the same cell shown by arrows in *A* and *B*). Haematoxylin and eosin; ×400 (*A*); ×1200 (*B*); ×300 (*C*); ×1200 (*D*).

E–F: human pancreatic adenomatosis. Enlarged, hyperplastic islet (*E*) presenting crowding of nuclei with abnormal cytology (*F*). Note the nuclear polymetricism and spotted chromatin. Compare with lesions in *A–D*. Haematoxylin and eosin; ×300 (*E*); ×1200 (*F*).

from pancreatic ducts of transgenic mice, though lacking T-antigen expression as well as cytologic atypia, may be compared with the corresponding lesions in human MEN I, thus providing a further intriguing similarity and confirming the AVP/SV40 transgenic mice as a potentially useful model for human pathology.

Concluding remarks

Transgenic mice models represent one of the most recent and powerful tools for the evaluation and understanding of genes, providing unique *in vivo* opportunities for studying the functions of regulatory sequences[1–5,8–12,15,18,19,21,23,27–31,33,36,38–40]. On the other hand, hormones are optimal targets and subjects of this kind of experiment, generally providing highly segregated sites of expression[1,12,21,22,31,33,36]. However, only one of the three models discussed above (the insulin/SV40) fully satisfied the expectations, providing transgene expression only in the physiologically correct areas[1,12,33]. The discovery of rat insulin regulatory sequences capable of directing site specifically the expression of the transgene is an important step in our knowledge of the insulin gene.

The fact that two more models, quite different in their genetic construct, produce identical lesions of the endocrine pancreas, raises questions about the possibility that some regulatory function(s) may reside in the SV40 large T-antigen sequences, the only common part of the three different constructs. In this respect, the recent identification of episomal BK virus, which is related to SV40 [41], in human insulinomas[6], supports the hypothesis of a possible specific role for the Papova virus promoting endocrine tumours.

Nevertheless, as a partial side effect, these experimental models provided the unique opportunity of studying *in vivo* the oncological process of endocrine organs. Moreover, the AVP/SV40 transgenic mouse is the only available model for double localisation of endocrine neoplasms. The information obtained by the study of the morpho-functional aspects of the lesions, from early steps to open neoplasms, provided suggestions for a new evaluation of the corresponding human pathology[8,36]. More information will become available from the careful analysis of the subcellular damage to the hormonal synthetic pathways of these endocrine cells. The clarification of these pathological mechanisms may also help in understanding the subcellular functions of these elements.

Acknowledgments. This work was supported in part by grants from the Italian National Research Council (Special Project on Oncology and Biomedical Technologies; Gastroenterology Group) and the Italian Health and Education Ministry.

1 Adams, T. E., Alpert, S., and Hanahan, D., Non-tolerance and autoantibodies to a transgenic self-antigen expressed in pancreatic B-cells. Nature *324* (1987) 223–227.

218

2 Brinster, R. L., Chen, H. Y., and Trambauer, M., Somatic expression of herpes thymidine kinase in mice following injection of a fusion gene into eggs. Cell 27 (1981) 223–231.

3 Brinster, R. L., Chen, H. Y., Warren, R., Sarthy, A., and Palmiter, R. D., Regulation of metallothionein-thymidine kinase fusion plasmids injection into mouse eggs. Nature 296 (1982) 39–42.

4 Brinster, R. L., Chen, H. Y., Messing, A., van Dyke, T., Levine, A. J., and Palmiter, R. D., Transgenic mice harbouring SV40 T-antigen genes develop characteristic brain tumours. Cell 37 (1984) 367–379.

5 Bürki, K., and Ullrich, A., Transplantation of the human insulin gene into fertilized mouse eggs. EMBO J. 1 (1982) 127–131.

6 Caputo, A., Corallini, A., Grossi, M. P., Carra, L., Balboni, P. G., Negrini, M., Milanesi, G., Federspil, G., and Barbanti-Brodano, G., Episomal DNA of a BK Virus variant in human insulinoma. J. med. Virol. 12 (1983) 37–49.

7 De Lellis, R. A., Dayal, Y., Tischler, A. S., Lee, A. K., and Wolfe, J. H., Multiple endocrine neoplasia (MEN) syndromes: cellular origins and interrelationships. Int. Rev. exp. Path. 28 (1986) 163–215.

8 Evans, R. M., Swanson, L., and Rosenfeldt, M. G., Creation of transgenic animals to study development and as models for human disease. Rec. Progr. Horm. Res. 141 (1985) 317–332.

9 Gordon, J. W., Scango, G. A., Plotkin, D. J., Barbosa, J. A., and Ruddle, F. H., Genetic transformation of mouse embryos by microinjection of purified DNA. Proc. natl Acad. Sci. USA 77 (1980) 7380–7384.

10 Gorman, C., High efficiency gene transfer into mammalian cells, in: DNA cloning: a practical approach, pp. 143–190. Ed. D. Glover. IRL Press, London 1986.

11 Hammer, R. E., Krumlauf, R., Camper, S., Brinster, R. L., and Tilghman, J. M., Diversity of alpha-fetoprotein gene expression is generated by a combination of separate enhancer elements. Science 235 (1987) 53–58.

12 Hanahan, D., Heritable formation of pancreatic B-cell tumours in transgenic expression recombinant insulin/simian virus 40 oncogenes. Nature 315 (1985) 115–122.

13 Heitz, P. U., Kaspar, M., Polak, J. M., and Klöppel, G., Pancreatic endocrine tumours: immunocytochemical analysis of 125 tumours. Human Path. 13 (1982) 263–271.

14 Heitz, P. U., Pancreatic endocrine tumours, in: Pancreatic Pathology, pp. 206–232. Eds P. U. Heitz and G. Klöppel. Churchill Livingstone, Edinburgh 1984.

15 Hogan, B., Constantini, F., and Lacy, P. E., Manipulating the mouse embryo: a laboratory manual. Cold Spring Harbor, New York 1986.

16 Klöppel, G., Delling, G., Knipper, A., and Heitz, P. U., Immunocytochemical mapping of pancreatic apudoma in multiple endocrine adenomatosis with primary hyperparathyroidism. Acta endocr. 87 (1978) 57–58.

17 Klöppel, G., and Lewson, S., Anatomy and physiology of the endocrine pancreas, in: Pancreatic Pathology, pp. 133–153. Eds P. U. Heitz and G. Klöppel. Churchill Livingstone, Edinburgh 1984.

18 Krumlauf, R., Hammer, R. E., Brinster, R., Chapman, V. M., and Tilghman, S., Regulated expression of alpha-fetoprotein genes in transgenic mice. Cold Spr. Harb. Symp. Q. Biol. 50 (1985) 371–378.

19 Lacy, E., Roberts, S., Evans, E. P., Burtonshaw, M. D., and Constantini, F., A foreign B-globin gene in transgenic mice: integration at abnormal chromosomal positions and expression in inappropriate tissues. Cell 34 (1983) 343–358.

20 Lolait, S. J., Morwick, A. J., McNally, M., Abraham, J., Smith, A. I., and Funder, J. W., Anterior pituitary cells from Brattleboro (di/di), Long Evans and Sprague-Dawley rats contain immunoreactive arginine-vasopressine. Neuroendocrinology 43 (1986) 577–583.

21 Messing, A., Chen, H. Y., Palmiter, R. D., and Brinster, R. L., Peripheral neuropathies, hepatocellular carcinomas and islet cell adenomas in transgenic mice. Nature 316 (1985) 461–463.

22 Murphy, D., Bishop, A. E., Rindi, G., Murphy, M. N., Stamp, G. W. H., Hanson, J., Polak, J. M., and Hogan, B., Mice transgenic for a vasopressin-SV40 hybrid oncogene develop tumors of the endocrine pancreas and the anterior pituitary. A possible model for human multiple endocrine neoplasia type I. Am. J. Path. 129 (1987) 552–566.

23 Murphy, D., and Hanson, J., The production of transgenic mice by the microinjection of fertilised one-cell eggs. In: DNA Cloning: A Practical Approach, Ed. D. Glover. IRL Press, London (1988) in press.

24 Myers, R. M., Kligman, M., and Tjian, R., Does simian virus 40 T antigen unwind DNA? J. biol. Chem. *256* (1981) 10156–10160.
25 Nakane, P. K., Classification of anterior pituitary cell types with immunoenzyme histochemistry. J. Histochem. Cytochem. *18* (1970), 9–20.
26 Orci, L., Banting lecture 1981. Macro and micro domains in the endocrine pancreas. Diabetes *31* (1981) 538–565.
27 Ornitz, D. M., Palmiter, R. D., Messing, A., Hammer, R. E., Pinkert, C. A., and Brinster, R. L., Elastase I promoter directs expression of human growth hormone and SV40 T-antigen to pancreatic acinar cells in transgenic mice. Cold Spr. Harb. Symp. Q. Biol. *50* (1985) 399–409.
28 Overbeek, P. A., Chepelinsky, A. B., Khillan, J. S., Piatgorsky, J., and Westphal, H., Lens specific expression and developmental regulation of the bacterial chloramphenicol acetyltransferase gene driven by the murine aA-crystallin promoter in transgenic mice. Proc. natl Acad. Sci. USA *82* (1985) 7915–7918.
29 Palmiter, R. D., Brinster, R. L., Hammer, R. E., Traumbauer, M. E., Rosenfeld, M. G., Brinberg, N. C., and Evans, R. M., Dramatic growth of mice that develop from egg microinjected with metallothionein-growth hormone fusion genes. Nature *300* (1982) 611–615.
30 Palmiter, R. D., Chen, H. Y., and Brinster, R. L., Differential regulation of metallothionein-thymidine kinase fusion genes in transgenic mice and their offspring. Cell *29* (1982) 701–710.
31 Palmiter, R. D., Chen, H. Y., Messing, A., and Brinster, R. L., SV40 enhancer and large T-antigen are instrumental in development of choroid tumours in transgenic mice. Nature *316* (1985) 457–460.
32 Polak, J. M., Bloom, S. R., Adrian, T. E., Bryant, M. G., Heitz, P. U., and Pearse, A. G. E., Pancreatic polypeptide in insulinomas, gastrinomas, vipomas and glucagonomas. Lancet *i* (1976) 328–330.
33 Power, R. F., Holm, R., Bishop, A. E., Varndell, I. M., Alpert, S., Hanahan, D., and Polak J. M., Transgenic mouse model: a new approach for the investigation of endocrine pancreatic B-cell growth. Gut *28* (1987) 121–129.
34 Ray, M. B., and Zumwalt, R., Islet cells hyperplasia in genetic deficiency of alpha-1-proteinase inhibitor. Am. J. Clin. Path. (1986) 681–687.
35 Rigby, P. W. J., and Lane, D. E., Structure and function of Simian Virus 40 large T-antigen in: Advances in Viral Oncology, pp. 31–57. Ed. G. Klein. Raven Press, New York 1983.
36 Rindi, G., Bishop, A. E., Murphy, D., Solcia, E., Hogan, B., and Polak, J. M., A morphological analysis of endocrine tumour genesis in pancreas and anterior pituitary of AVP/SV40 transgenic mice. Vichow Arch. A *412* (1988) 255–266.
37 Schimke, R. N., Multiple endocrine neoplasia: search for the oncogenic trigger. New Engl. J. Med. *314* (1986) 1315–1316.
38 Small, J. A., Blair, D. G., Showalter, S. D., and Scangos, G. A., Analysis of a transgenic mouse containing simian virus 40 and v-*myc* sequences. Molec. Cell Biol. *5* (1985) 642–648.
39 Small, J. A., Khoury, G., Jay, J., Howley, P. M., and Scangos, G. A., Early regions of JC and BK virus induce distinct and tissue-specific tumors in transgenic mice. Proc. natl Acad. Sci. USA *83* (1986) 8288–8292.
40 Stewart, T. A., Patterngale, P. K., and Leder, P., Spontaneous mammary adenocarcinomas in transgenic mice that carry and express MTV/myc fusion genes. Cell *38* (1984) 627–637.
41 Tooze, J. (Ed.), Molecular Biology of Tumor Viruses, Part 2: DNA Tumor Viruses. Revised 2nd Edn. Cold Spring Harbor, New York 1982.
42 Wermer, P., Genetic aspects of adenomatosis of endocrine glands. Am. J. Med. *16* (1954) 363–371.

Endocrine cells producing regulatory peptides

E. Solcia, L. Usellini, R. Buffa, G. Rindi, L. Villani, A. Aguzzi and
E. Silini

Summary. Recent data on the immunolocalisation of regulatory peptides and related propeptide sequences in endocrine cells and tumours of the gastrointestinal tract, pancreas, lung, thyroid, pituitary (ACTH and opioids), adrenals and paraganglia have been revised and discussed. Gastrin, xenopsin, cholecystokinin (CCK), somatostatin, motilin, secretin, GIP (gastric inhibitory polypeptide), neurotensin, glicentin/glucagon-37 and PYY (peptide tyrosine tyrosine) are the main products of gastrointestinal endocrine cells; glucagon, CRF (corticotropin releasing factor), somatostatin, PP (pancreatic polypeptide) and GRF (growth hormone releasing factor), in addition to insulin, are produced in pancreatic islet cells; bombesin-related peptides are the main markers of pulmonary endocrine cells; calcitonin and CGRP (calcitonin gene-related peptide) occur in thyroid and extrathyroid C cells; ACTH and endorphins in anterior and intermediate lobe pituitary cells, α-MSH and CLIP (corticotropin-like intermediate lobe peptide) in intermediate lobe cells; met- and leu-enkephalins and related peptides in adrenal medullary and paraganglionic cells as well as in some gut EC (enterochromaffin) cells; NPY (neuropeptide Y)in adrenalin-type adrenal medullary cells, etc. Both tissue-appropriate and tissue-inappropriate regulatory peptides are produced by endocrine tumours, with inappropriate peptides mostly produced by malignant tumours.

Endocrine cells producing regulatory peptides are specialised epithelial cells characterised by their secretory granules of variable size, shape, density and inner structure enveloped by a unit membrane. The granules are formed at the trans side of the Golgi complex, from condensing vacuoles whose contents, simultaneously with the process of controlled proteolysis of prohormones, undergo progressive densification to form clathrin-coated 'immature' progranules and then 'mature' secretory granules storing the active hormones[107]. Besides hormonal peptides and related prohormone fragments, secretory granules store proteins like chromogranins/secretogranins and related peptides (pancreastatin[155], GAWK[7a]), monoamines, such as catecholamines and serotonin, polyamines and metal cations[122,178].

In addition to secretory granules storing peptides and chromogranins and resembling the large, dense-cored vesicles of nerves, a population of small clear vesicles, resembling the small synaptic vesicles which store classic neurotransmitters, has been described in some endocrine cells, such as paraganglionic, adrenal medullary and pulmonary endocrine cells[28,78,98]. Recently these small vesicles of nerves and endocrine cells (including adrenal medullary and pituitary cells) have been found to be selectively marked by a Ca^{2+}-binding membrane glycoprotein, the synaptophysin or P38 protein[102]. P38 protein immunoreactivity has been detected in adrenal medullary and paraganglionic cells, pancreatic islets, adenohypophysis and thyroid C cells as well as in pulmonary and gastric endocrine cells and related growths, while no reactivity has been observed in many intestinal endocrine cells, cardiac atrial cells producing natriuretic hormone and parathyroids[24,102,177]. Cholinergic[33,173], aminergic[110] and GABAergic[62] mechanisms have been found to operate in at least some of the P38-positive cells. Two other vesicle membrane proteins[21,97], neuron specific enolase[10], three distinct chromogranin proteins[122,125,143,178] and a number of regulatory peptides and amines are now known to be common markers of nerves and endocrine cells.

Thus, although the proposed neural crest origin of endocrine cells[110] has been confirmed only for adrenal medulla, carotid body and thyroid C cells[89], the ability of many (not all) endocrine cells, independently from their neural crest origin, to express morphologic and functional patterns characteristic of nerve cells is widely documented and may justify the designation of such cells as 'neuroendocrine' cells[111]. It seems interesting that during phylogenesis nerve cells first develop as peptidergic elements scattered in both the ectoderm and endoderm of coelenterates, partly as elongated 'sensory' cells contacting the epithelial surface, with processes at their basal part[56], a pattern resembling some paracrine cells of mammalian endodermal derivatives[137,141].

As a rule, in different endocrine cell types distinct genes are expressed coding for different propeptides. However, alternative processing of the same m-RNA may result in two distinct propeptides leading to different regulatory peptides, as in the case of calcitonin and calcitonin gene-related peptide (CGRP), coded by the same gene through different propeptides showing tissue specific, though partly overlapping, distributions[126,158]. More often, two or more active peptides, showing the same cellular distribution, may be the products of the same propeptide coded by a single gene, as in the case of ACTH and endorphins, both arising from proopiomelanocortin (POMC)[101]. Alternative posttranslational processing of the same prohormone in separate cells, due to different proteolytic cleavage, may also result in different regulatory peptides, as in the case of α-MSH and CLIP (corticotropin-like intermediate lobe peptide) produced in pituitary intermediate lobe cells, but not in ACTH/β-endorphin cells of the anterior pituitary, from further cleavage of ACTH[128]. Similarly, peptides of various molecular size enclosing the same biologically active sequence may result from different processing of the same prohormone in separate cell types (as for proglucagon in pancreatic A cells or intestinal L cells) or in the same cell type of different tissues (as for progastrin in duodenal or pyloric gastrin cells), or in many tumour cells in respect to normal parent cells.

Development of another enzyme activity, cleaving preferentially at Lys-Lys sites, in addition to the Lys-Arg specific enzyme[99] acting in cells of both the anterior and intermediate lobes of the pituitary might explain further cleavage of ACTH, β-endorphin (to β-endorphin 1–27) and δ-endorphin (to release β-MSH) in intermediate lobe cells[132]. Conversely, in pancreatic A cells, where only Lys-Arg sites seem to be cleaved consistently, unmasking of two such sites (blocked in intestinal L cells), possibly due to changes of interacting chromogranins[122], might promote further cleavage of glicentin (uncleaved in L cells) to glucagon.

Careful identification of hormonal and prohormonal peptides, including their cryptic peptides, is required as a basis for classification of the endocrine cells forming glands or scattered in various epithelia as components of the so-called diffuse endocrine system (DES, table). Structural

Table 1. Classification of peptide-characterised endocrine cells forming the DES and some endocrine glands

Tissue	Cell type	Main peptides	Amines
Carotid body	Type I	Enkephalins	NA, DA, 5HT
Sympathetic paraganglia	Main cell	Enkephalins	NA, DA
Sympathetic ganglia	SIF cell	Enkephalins	DA, 5HT
Adrenal medulla	A type	Enkephalins; NPY	A
	NA type	Dynorphins; bombesin	NA
(cat)	III type	Neurotensin	NA
Pituitary: anterior lobe	'ACTH' cell	ACTH, β-endorphin	
intermediate lobe	Main cell	α-MSH; CLIP; β-endorphin (1–27)	
Skin	Merkel cell	Enkephalin (rodents); VIP-like (other mammals)	
Thyroid	C cell	Calcitonin, PDN-21; CGRP; somatostatin	5HT
Lung	P cell	Bombesin; calcitonin; CGRP	5HT
Pancreas and gut	EC_1 cell	Substances P and K	5HT
	EC_2 cell	Enkephalin	5HT
	D cell	Somatostatin	
	B cell	Insulin; TRH	
	A cell	Glucagon; CRF	
	L cell	Glicentin; PYY	
	PP cell	PP; GRF	
	G cell	Gastrin	
	CCK cell	Cholecystokinin	
	M cell	Motilin	
	S cell	Secretin	
	GIP cell	GIP	
	N cell	Neurotensin	

characterisation at light and electron microscopical levels, topography, response to functional stimuli, detection of specific receptors and localisation of biogenic amines, individual enzymes or structural proteins are also important tools for precise cell characterisation at morphological and functional levels[137].

Regulatory peptides produced in the DES, adrenal medulla and paraganglionic cells as well as in POMC-producing pituitary cells will be dealt with in the following sections.

Bombesin, gastrin-releasing peptide (GRP) and related peptides

Endocrine cells reacting with antibodies against the amphibian peptide bombesin or its mammalian equivalent GRP have been detected in amphibian[88] and avian[157] stomach and in mammalian lung[174] and thyroid[132a]. Ultrastructurally, bombesin/GRP immunoreactivity has been localised to cells with small, round, thin-haloed granules of P type[28,157]. In both human lung[175] and chicken proventriculus[166], co-localisation of bombesin and 5HT immunoreactivities has been reported. In the lung, bombesin/GRP-immunoreactive cells may form small, intraepithelial bodies, the so-called neuroepithelial bodies, showing prominent innervation, including synapses between epithelial cells and nerves, possibly to be interpreted as an hypoxia-sensitive chemoreceptor apparatus[86,93].

No bombesin/GRP immunoreactivity has been demonstrated in P-type cells of the human gut and pancreas, the staining unequivocally reported in early studies[114] being possibly due to crossreactivity of bombesin antibodies with substance P and substance K, which are both stored in gut EC cells and share with bombesin the C-terminus sequence Leu-Met-NH$_2$, or with chromogranin B, suggested to react with N-terminally-directed bombesin antibodies[49].

Immunoreactivity against neuromedin B, the mammalian counterpart of the bombesin-like amphibian peptides ranatensin and litorin, has been shown in pituitary TSH (rat) and gonadotrophic (human) cells. Bombesin/GRP and neuromedin B (a bombesin-related peptide), immunoreactivities have been reported in noradrenalin-producing cells of the adrenal medulla[90]. Bombesin/GRP immunoreactivity has been reported in a number of 'neuroendocrine' tumours, including pulmonary tumourlets, carcinoids and small cell carcinomas[27,54,100,135,153] and thyroid medullary carcinoma[73]. In both human lung tissue and related tumours, C-terminus GRP fragments 14–27 and 18–27 seem to represent the major bombesin-related hormonal species[180]. However, antibodies directed against the C-flanking peptide of human pro-GRP[144] proved to be much more efficient than bombesin/GRP antibodies in staining small cell carcinomas arising in both pulmonary and extrapulmonary sites[58], suggesting either preferential preservation of this peptide during posttranslational processing of pro-GRP or defective pro-GRP cleavage by agranular and poorly granular tumour cells.

Due to the CRF-potentiating effect of bombesin-like peptides on ACTH release from pituitary cells, a Cushing syndrome has been reported in association with a thyroid medullary carcinoma producing both calcitonin and bombesin-like peptides(s) but neither ACTH nor CRF[68].

Substance P and related tachykinins

Substance P has been found to be co-localised with serotonin and chromogranin A in the argentaffin granules of a subpopulation of gut EC cells, called EC_1 cells[65,122]. Substance P producing EC_1 cells are a major component of midgut (ileum, jejunum, appendix, caecum), rectal, ovarian, testicular and pulmonary argentaffin carcinoids[57,96,116,135]. Other tachykinins, like substance K (or neurokinin-A) and neuropeptide K, known to be produced simultaneously with substance P during posttranslational processing of β-protachykinin[103], have been also detected in nontumour intestine and serotonin-producing EC cell carcinoids[35,104]. A major role for tachykinins in the genesis of the 'carcinoid syndrome', with special reference to flushing and diarrhoea, seems likely. Substance P has also been found in the adrenal medulla and some pheochromocytomas[52] as well as in some paraganglionic cells of the carotid body and cervical, coeliac and mesenteric ganglia[66].

ACTH-and/or opioid-related peptides (endorphins, enkephalins, dynorphins)

Two distinct types of pituitary cells have been shown to produce ACTH and opioid peptides from posttranslational processing of proopiomelanocortin (POMC). In the anterior lobe ACTH/β-endorphin cells, cleavage of the precursor at Lys-Arg sites produces ACTH (POMC 106–144), β-lipotropin (POMC 147–239), further cleaved to δ-lipotropin (POMC 147–206) and β-endorphin (POMC 209–239), as well as pro-δ-MSH (POMC 1–103). In intermediate lobe cells further cleavage at basic residues, with special reference to Lys-Lys sites, may produce, in addition, α-MSH (ACTH 1–13), CLIP (ACTH 17–38), β-MSH (lipotropin 43–60) and biologically inactive β-endorphin (1–27)[128,132]. Although in man intermediate lobe cells are poorly represented in foetal pituitary and either lacking or rudimentary in the adult gland, ACTH-secreting intermediate lobe tumours have been identified in some cases of Cushing's disease[80]. Both anterior lobe and intermediate lobe type of processing, including production of ACTH, β-endorphin, δ-MSH, β- and δ-lipotropins, α-MSH and CLIP, have been found to operate in extrapituitary tumours producing 'ectopic' hormones[3,120].

Cells reacting with β-endorphin, β-lipotropin and/or pro-δ-MSH anti-sera, though apparently lacking ACTH and α-MSH immunoreactivity, have been identified in the human intestine, especially in the small intestine[131]. ACTH-like and/or β-endorphin immunoreactivities have been detected in various endocrine cells of mammalian gut and pancreas[81,149,171], a finding supported by immunochemical studies on tissue extracts[46,108,154]. Part of this material might be co-localized with gastrin in the G cells, although a separate intestinal and pancreatic cell type seems also involved[149] resembling in its morphology and distribution the intestinal 'VL cell' identified ultrastructurally[138]. Whether authentic ACTH or some ACTH-related peptide is produced remains to be ascertained. Endocrine tumours producing ACTH and β-endorphin have been reported in the gut and pancreas, occasionally coupled with Cushing's disease.

ACTH and β-endorphin immunoreactivities have also been detected in occasional cells of normal human adrenal medulla as well as in some adrenal pheochromocytomas[30,91] (fig. 1). Apart from the pituitary, adrenals and gut, tumours producing ACTH, α-MSH, β-endorphin, pro-δ-MSH and other POMC-derived peptides have been observed in sites as the lung, thymus, thyroid and prostate, where related immunoreactivities are normally lacking, although they may appear in hyperplastic or dysplastic lesions[3,31]. POMC of extra-pituitary tumours may differ from pituitary POMC in the length of its mRNA and type of posttranslational processing[3,145].

Enkephalin immunoreactivity has been detected in a variety of cells, including adrenal medullary cells[94], cells of chromaffin and non-chromaffin paraganglia[176] (fig. 2), Merkel cells of rodent's skin[60], small intensely fluorescent (SIF) cells of sympathetic paraganglia[75], some endocrine cells of the human lung[37], a fraction of argentaffin EC cells of some mammals[4] and rat pituitary GH cells[172]. Human pheochromocytomas, paragangliomas and some pulmonary carcinoids have also been shown to produce enkephalins and related peptides[30,54,63,94]. In the adrenal medulla, enkephalin-related peptides have been shown to originate from a specific precursor molecule, proenkephalin, containing six met-enkephalin and one leu-enkephalin sequences, which are only partly released as free enkephalins, the rest being secreted as a mixture of larger enkephalin-containing peptides[160].

A second enkephalin precursor, called proenkephalin-B or prodynorphin, containing three leu-enkephalin and no met-enkephalin sequences as part of larger opioid peptides (dynorphins and neo-endorphins) has been identified in the pig hypothalamus and posterior pituitary[71]. Dynorphin immunoreactivity has been detected in noradrenalin cells of bovine adrenal medulla[42]. The contribution, if any, of this propeptide to the genesis of the 'leu-enkephalin' immunoreactivity co-localised with serotonin in some gut EC cells, lung endocrine cells and SIF cells remains to be investigated.

226

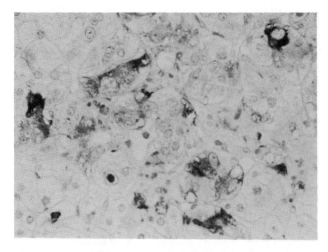

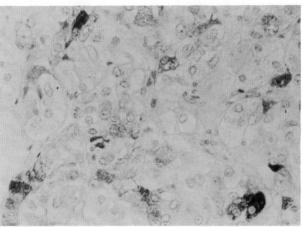

Figure 1. NPY (*a*, serum B48, Milab, Malmö, Sweden) and ACTH (*b*, serum 596010, Ortho Diagnostic Systems, Raritan, N.J.) immunoreactive cells scattered in the human adrenal medulla. Immunoperoxidase, × 240.

Corticotropin-releasing factor (CRF)

CRF-like immunoreactivity has been detected in glucagon-producing A cells of vertebrate pancreas[113], in endocrine cells of the cat, monkey and rat pyloric mucosa and small intestine distinct from gastrin- and glucagon-immunoreactive cells[113], and in the lung and adrenal tissue[148]. Only part of the CRF antisera staining hypothalamic CRF have been found to react with the peptide stored in extraneural tissues, suggesting that the latter peptide has some difference in structure or molecular species in respect to hypothalamic CRF.

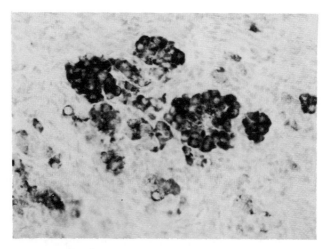

Figure 2. Type I cells forming 'Zell ballen' in the human carotid body, stained with anti-enkephalin serum B15 (Milab). Immunoperoxidase, × 230.

CRF-like immunoreactivity has been detected in endocrine tumours of the pancreas, bronchial carcinoids and pheochromocytomas as well as in small cell carcinomas of the lung[113,147]. A case of Cushing syndrome apparently due to inappropriate CRF secretion from metastatic carcinoma of the prostate, producing marked pituitary ACTH cell hyperplasia, has been reported[32].

Bioactive CRF-like peptides have been detected in a number of extrapituitary tumours producing 'ectopic' ACTH[9,161], a finding of interest in explaining the progressive growth and potential malignancy of such tumours. Besides CRF, vasopressin- and bombesin-like peptides might also contribute to CRF-like bioactivity[68] and tumour growth.

Thyrotropin-releasing hormone (TRH)

TRH-like immunoreactivity has been detected in pancreatic insulin-producing B cells and part of glucagon-producing A cells of the rat developing pancreas[76] as well as in insulin B cells and gastrin G cells of human developing pancreas and stomach (R. B., unpublished findings). Immunochemical studies suggest identity of the pancreatic peptide with the hypothalamic tripeptide pGlu-His-ProNH$_2$.

Calcitonin and calcitonin gene-related peptide (CGRP)

In thyroid C cells, calcitonin has been shown to be co-localised with its C-terminal flanking peptide (PDN-21 or katacalcin) and CGRP[2,126,158].

Moreover, co-localisation with somatostatin has been observed in majority (rabbit) and minority (rat, human) subsets of C cells[23,72,168] and with GRP in the majority of developing human C cells[132a].

Ultrastructurally, thyroid C cells are characterised by round, solid, medium-sized granules storing, besides neuroendocrine peptides, chromogranins A and C. Apart from the thyroid, C cells also occur in the ultimobranchial body, the 'upper' or 'inner' parathyroid (deriving from the fourth branchial pouch) and thymus IV. Calcitonin-[7] and CGRP-immunoreactive cells[34] have been detected in the human lung; however, their failure to react with some of the anti-calcitonin sera staining thyroid C cells, their frequent co-localisation on bombesin/GRP immunoreactivity[159] and the smaller size of their secretory granules[28] distinguish them from thyroid C cells.

Calcitonin, PDN-21 and CGRP have been detected regularly in thyroid medullary carcinoma[130] and, less frequently, in tumours from the lung, pancreas, adrenal medulla and other sites[29,41,63]. In many cases of thyroid medullary carcinomas, calcitonin, PDN-21 and CGRP have been found to coexist, in the same or distinct cells, with somatostatin, GRP, PP, ACTH and neurotensin[26,40,73,106,130].

Somatostatin

Somatostatin D cells have been observed in the pancreas and along the whole gastrointestinal tract, from cardia to rectum[5]; some cells have also been detected in chicken thymus[152]. Many of these cells show long cell processes contacting other endocrine and exocrine cells, a possible morphologic basis for a local paracrine function. Secretory granules of D cells are round, homogeneous, poorly dense and unreactive or poorly reactive with silver techniques (with the only exception of Davenport's alcoholic silver) and antisera to chromogranins, while reacting with lead-haematoxylin[133,137,141] and synaptophysin antibodies[24].

A case of extreme somatostatin cell hyperplasia of the gastro-duodenal mucosa causing dwarfism, obesity and goitre has been described[67]. Somatostatin-producing D cell tumours have been observed in the pancreas (with or without associated 'somatostatinoma' syndrome: diabetes, diarrhoea, steatorrhoea, hypo- or achlorhydria, anaemia and gallstones), duodenum, stomach, jejunum, ileum and rectum[39,79,140].

Somatostatin has been found to be co-localised with calcitonin in a large (rabbit) or minor (rat and human) proportion of thyroid C cells and in several thyroid medullary carcinomas[23,26], as well as with catecholamines in a fraction of human adrenal medullary cells and pheochromocytomas[94].

Peptides of glucagon and PP families

a) *Glucagon/glicentin, GLP₁/GLP₂; VIP/PHI; GRF.* By using C-terminally directed specific antibodies *glucagon* has been localised to A cells of the pancreas and gastric oxyntic and cardiac mucosa, whose secretory granules (α-granules) are characterised ultrastructurally by a solid, glucagon-storing core surrounded by an argyrophil halo containing chromogranin A and glicentin-related pancreatic peptide (GRPP) immunoreactivity[25,117,141]. Glicentin C-terminus hexapeptide, MPGF (major proglucagon fragment), containing glucagon-like peptides GLP_1 and GLP_2, and chromogranins C and B are also localised in α-granules[122,142,169]. Glucagon C-terminus immunoreactivity has also been reported in a few intestinal cells lacking the characteristic α granules of A cells and corresponding to a minor fraction (about 20%) of intestinal L cells[48,77,131]. The latter cells, which are scattered in both the large and small bowel, show solid, homogeneous granules with scarce or variable argyrophilia and chromogranin A or B immunoreactivity, consistent chromogranin C immunoreactivity and reactivity with antibodies directed against the proglucagon-related peptides *glicentin*, GRPP, *glucagon-37* (oxyntomodulin), GLP_1 and GLP_2[55,118,142,169].

Growth hormone releasing factor (GRF) immunoreactivity has been detected immunohistochemically in PP cells of the human and rat pancreas[16]. It has been extracted from the human pancreas and characterised immunochemically as an N-terminally extended molecule[129]. A possible relationship of GRF immunoreactivity to small-granulated P-type and mixed P/D_1 type cells occurring in human pancreas, especially during foetal life[28], remains to be investigated.

Although VIP immunoreactivities reported in endocrine cells of mammalian gut and pancreas are likely due to cross-reacting chemically-related peptides, especially of L cells (GLP_1 or GLP_2?) and D_1/PP cells (GRF?), VIP-storing cells seem to occur in the intestine of other vertebrates[83,121]. VIP has been found to be co-localised with enkephalins in adrenal chromaffin granules of the frog[87] while VIP- and met-enkephalin-immunoreactivities occur in Merkel cells of distinct species[61].

b) *PP, PYY and NPY*. Most L cells, besides storing glucagon-related peptides, also store *PYY* (peptide with N-terminal tyrosine and C-terminal tyrosine) in the same granules[17,22,135]. Moreover, a minority of L cells also store a PP-like peptide distinct (pro-PP?) from PYY and accounting for their reactivity with some PP antisera (i.e. Chance's hPP serum) lacking PYY crossreactivity, as well as with antisera directed against the icosapeptide fragment of proPP[48].

Cells producing both PP and PYY as well as chromogranin C, but lacking glucagon-related peptides and chromogranin A, occur in the pyloric mucosa of the dog and cat and, rarely, in other mammals[133,142].

As a rule pancreatic PP cells store *PP* as well as the icosapeptide (fig. 3); however, uncleaved pro-PP, lacking C-terminus icosapeptide immunoreactivity, seems to be produced in a minority of these cells[142,151]. PYY has been detected in a minority of pancreatic PP cells[142], while coexistence of PP-related and glucagon-related peptides occurs rarely in

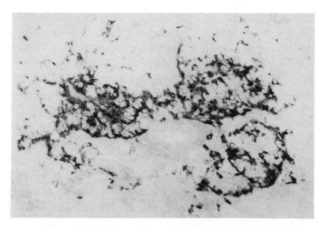

Figure 3. PP cells of PP-rich irregular islets in the posterior head of human pancreas, stained with rabbit anti-canine pro-PP icosapeptide serum 3204 (from T. W. Schwartz, Dept. Clinical Chemistry, Rigshospitalet, Copenhagen, Denmark). Immunoperoxidase, × 70.

the pancreas, apart from during foetal life[1]. So far, no difference of peptide immunoreactivity has been identified in the two ultrastructural subtypes of human PP cells, the F and D_1 subtype[139].

The various associations of glucagon-related and PP-related peptides or propeptides in the gut and pancreas suggest that intestinal L cells may represent the phylogenetic and ontogenetic ancestor cells of both glucagon A cells and PP cells. ProPP immunoreactive or C-terminus glucagon reactive subsets of intestinal L cells, pyloric and pancreatic cells producing both PP and PYY, immature A cells of human early foetal pancreas showing glicentin immunoreactivity while lacking glucagon C-terminus immunoreactivity[146] and/or showing co-localized PYY immunoreactivity[1], pancreatic cells storing both PP and GRF[16], are all findings suggesting a continuous spectrum of cells (and peptides) evolving from typical intestinal L cells (storing PYY and glicentin/glucagon-37/GLP_1/GLP_2) to classical pancreatic glucagon A cells (producing glucagon-29, GRPP, glicentin C-terminus hexapeptide and MPGF) or to PP cells (producing PP and pro-PP icosapeptide together with GRF).

Neuropeptide Y (NPY) immunoreactivity has been detected in adrenal chromaffin cells of adrenalin-producing type (fig. 1) and pheochromocytomas[30,95].

c) *Endocrine tumours* producing glucagon-related and/or PP-related hormones have been found in the pancreas, intestine (fig. 4), ovary, kidney and thyroid[14,15,47,64,135]. PYY, PP, pro-PP-icosapeptide, glicentin/glucagon-37, GLP_1, GLP_2 and occasionally even glucagon-29 immunoeactivities have been detected in L cell tumours of the rectum, colon, appendix, ileum, ovary and kidney, in the same or separate cellular subsets and with or without associated EC cells producing serotonin and

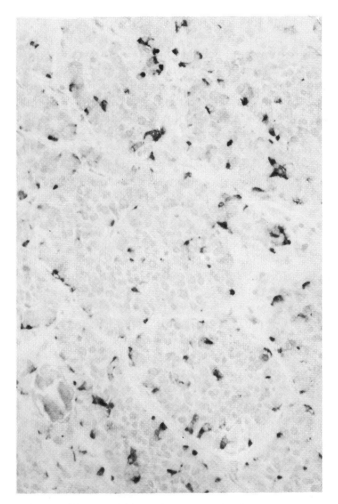

Figure 4. Rectal carcinoid stained with rabbit serum PP221 (from M. M. T. O'Hare, Dept. Medicine, Queen's University of Belfast, U.K.) directed against C-terminus PP hexapeptide, reacting with both PP and PYY. Note scattered immunoreactive tumour cells. Immunoperoxidase, × 235.

substance P[48]. The only L cell tumour apparently producing an hyperfunctional syndrome was a kidney tumour associated with decreased intestinal mucosa, all of which disappeared after tumour resection[12,53].

Both mature A cells with typical α-granules and immature, foetal-type A cells with solid, L-like granules lacking glucagon C-terminal immunoreactivity, have been detected in pancreatic glucagonomas, some of which seem to recapitulate A cell ontogenesis as observed in early foetal pancreas[14,142]. PP cells are often found to coexist in such tumours while pure PP cell tumours are observed rather rarely in the pancreas[15,64]. In most cases tumour PP cells resemble ultrastructurally the D_1 subtype, although the F subtype, normally prevailing in the PP-rich tissue originating from the ventral pouch, has also been observed sometimes[139]. PP immunoreactive cells have also been detected in some duodenal tumours, including so-called 'gangliocytic paragangliomas' or 'neurocarcinoids'[140] and in some thyroid medullary carcinomas, especially of familiar type[106]. The usefulness of PP as a marker of multiple endocrine neoplasia families has been stressed[50].

VIP-producing tumours (vipomas) of the pancreas and intestine are epithelial endocrine tumours, mainly of low grade malignancy, associated with watery diarrhoea, hypokalaemia and achlorhydria (WDHA) syndrome[13,29]. As expected, besides VIP, tumour cells produce and secrete PHI (peptide histidine isoleucine), a VIP-like peptide encoded by the same gene as VIP itself, as an integral part of pro-VIP[11]. PP cells have been observed in 11 out of 27 pancreatic vipomas investigated[29]; in some cases, PP-and VIP-immunoreactivity coexisted in the same tumour cell[74]. These findings further support a possible relationship of vipomas with cells of PP, A and L lines[142]. Glucagon, somatostatin, neurotensin, and calcitonin have been also detected in several of these tumours[29,43]. VIP and PHI production has been also observed in adrenal pheochromocytoma[63] and ganglioneuroblastoma, sometimes coupled with the 'vipoma' syndrome[11,92] and, occasionally, in thyroid medullary carcinoma and lung small cell carcinoma[127].

VIP has been localised immunocytochemically to very small, round, thin-haloed (P-type) granules stored in pancreatic and intestinal tumour cells[29]. Small, round, thin-haloed granules resembling those of vipoma cells have been also detected in endocrine tumours of the lung, pancreas and gut (with or without associated acromegaly) producing GRF[6,38,135], a peptide known to display consistent homolgy with PHI and VIP[69,123].

Secretin and GIP

To obtain specific detection of secretin or GIP in immunohistochemical and immunocytochemical tests special care must be taken to avoid crossreactivity of pertinent antisera with chemically related hormone and prohormone sequences (such as glucagon, glicentin, GLP_1, GLP_2, VIP, PHI and GRF). In all mammals so far investigated both secretin and GIP cells proved to be exclusive to the small intestine, usually with preference for the duodenum and upper jejunum. Only in the rat and mouse are these cells about as numerous in the ileum as in the upper small intestine. As a rule, secretin cells occur preferentially in the villi and upper crypts while GIP cells are more deeply situated in the crypts[13,131,165]. Most secretin cells show cytoplasmic granules with intense argyrophilia and

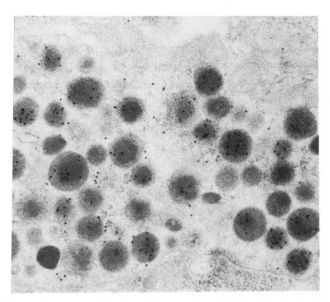

Figure 5. Electron immunocytochemistry of secretin in an S cell of the dog duodenum using rabbit anti-porcine secretin serum R.7875-02-2 (Milab) free of GIP, glucagon, glicentin, GLP_1 and GLP_2 crossreactivity. Note selective deposition of gold particles over target-like secretory granules, with preference for their dense, osmiophilic core. Protein A-immunogold technique, uranyl counterstaining. × 28,000.

chromogranin A and B immunoreactivity while GIP cell granules react poorly to both these tests[133,163].

Ultrastructurally, secretin cells show considerable pleomorphism of secretory granules, ranging from solid, round to ovoid and thin-haloed patterns to target-like granules with a round, osmiophilic core storing secretin surrounded by a more or less dense argyrophil matrix, likely to contain chromogranins[163]. Target-like granules are prominent in the dog and human secretin cells (fig. 5).

GIP cells, whose specific detection, free of crossreacting A and L cells, has been obtained with monoclonal antibodies devoid of any peptide crossreactivity[18], are characterized ultrastructurally by solid, relatively small granules showing uniform reactivity with GIP antibodies and differing clearly from secretin cell granules[165].

So far, neither secretin nor GIP cells have been convincingly detected in endocrine tumours. Nutrient-mediated regulation of exocrine (secretin) and endocrine (GIP) secretions of the pancreas seems to be the main function of secretin and GIP cells.

Gastrin, CCK and C-terminus gastrin/CCK

Gastrin G cells are medium-sized, ovoid to bottle- or pear-shaped cells concentrated in the deep neck and upper body of pyloric glands.

234

Mammalian G cells react with antibodies directed against all sequences of gastrin-17, gastrin-34, C-terminally extended gastrin and progastrin, including its N-terminal and C-terminal cryptic peptides[59,70]. Many of their moderately argyrophil granules storing chromogranins A and B together with gastrin-17 and progastrin fragments, are characterised ultrastructurally by a typical vesicular pattern with floccular content[141]. More solid granules, apparently storing an increased proportion of large gastrin molecules, are also present in G cells, especially in the Golgi area[141,170]. Xenopsin, a neurotensin-like octapeptide, has also been detected in mammalian G cells[124], while neurotensin itself seems present in chicken pyloric G cells[119].

In the human duodenum, but not in dog or cat duodenum, few gastrin cells resembling G cells are scattered in Brunners glands and deep crypts. A few cells with small, round, solid granules reacting with both C-terminal and N-terminal gastrin-34 antibodies (so-called intestinal gastrin or IG cells) have also been observed in the human duodenum[19,167]. In the pancreas of some species (rat, cat), but not of man, gastrin G cells have been detected during foetal and neonatal life[85].

An abundant population of cells reacting intensely with antibodies (TG cells) has been detected in the duodenum, jejunum and ileum of man and other mammals. Ultrastructurally, these cells show large, round to irregular secretory granules, sometimes with inner dense bodies, occasionally giving target-like patterns[82,167].

Cholecystokinin (CCK)-producing cells have been characterised with non-C terminal reactive antibodies, lacking histochemical crossreactivity with gastrin (fig. 6). They are scattered in the crypts and villi of the duodenum and jejunum. Their secretory granules are small, round and

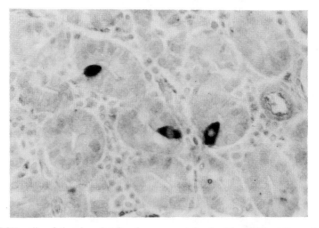

Figure 6. CCK cells of the dog duodenal mucosa stained with rabbit anti-porcine CCK-33 serum AB01 (CRB, Cambridge, U.K.), with middle to N-terminus specificity and no gastrin crossreactivity. Immunoperoxidase, × 350.

thin-haloed[164,167]. In the human duodenum (but not in the duodenum of the dog and other mammals or in the human jejunum) a large fraction of these cells, besides reacting with non-C terminal CCK antibodies and C-terminal gastrin/CCK antibodies, also show immunoreactivity with C-terminus gastrin-34 antibodies, co-localised with CCK in a variable proportion of secretory granules[167]. CCK/gastrin cells have been also reported in human foetal duodenum and newborn rat duodenum[82]. These 'mixed CCK/gastrin cells', resemble more CCK cells than gastrin G cells both histologically and ultrastructurally, and seem to be considered as CCK cells developing (or retaining from foetal life) partial gastrin coexpression, a mode of behaviour possibly reminiscent of a common phylogenetic origin of CCK and gastrin cells[84].

Hyperplasia and hyperfunction of pyloric gastrin cells has been reported in patients with peptic ulcer disease, hyperchlorhydria and food-stimulated hypergastrinaemia or in achlorhydric patients due to type A chronic atrophic gastritis and secondary hypergastrinaemia[115,141]. Gastrin cell tumours, with and without associated hypergastrinaemia, hyperchlorhydria and peptic ulcer disease (Zollinger-Ellison syndrome) have seldom been found in the stomach, jejunum and biliary tree or liver, more frequently in the duodenum and, especially, in the pancreas[134]. In tumour cells secretory granules were often fewer than in normal cells, more solid and smaller in sise, thus resembling progranules of normal cells, in keeping with the higher proportions of gastrin-34 and progastrin they produce[36,109]. Cells with large, often irregularly shaped granules resembling those of C-terminal gastrin immunoreactive cells (TG/VL cells) have also been observed in pancreatic and intestinal gastrinomas[8,134]. So far, CCK-immunoreactive tumour cells have only been observed as a minority population of a single duodenal gastrinoma[140].

Motilin

Motilin cells, like secretin and GIP cells, are exclusive to the small intestine, particularly to its upper part[112]. Ultrastructurally, motilin-immunoreactive cells are characterised by small, round, solid and fairly osmiophilic granules and abundant microfilaments[162]. They seem to play an important part in the modulation of gut motility, especially in the inter-digestive phase. Motilin cells have been found only occasionally in intestinal endocrine tumours[179].

Neurotensin and xenopsin

In mammals, neurotensin N cells are mostly confined to the small intestine, especially the ileum[51] and lower jejunum; only occasional cells have been detected in the large bowel[177]. Ultrastructurally, they correspond to a cell with large, solid, dense granules[150] (fig. 7) of consistent

236

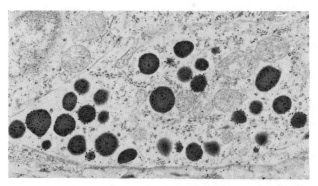

Figure 7. Electron immunocytochemistry of neurotensin in large, solid secretory granules of an N cell in the dog ileum, using anti-neurotensin serum 122/3 (from Dr G. E. Feurle, Medisinische Poliklinik, University of Heidelberg, FGR). Protein A-immunogold technique, uranyl counter-staining; × 14,500.

argyrophilia and chromogranin A immunoreactivity[133]. In the chicken antrum, neurotensin-immunoreactive cells have been observed[150] which may correspond to a fraction of G cells storing both gastrin- and neurotensin-like peptides[119]. While neurotensin immunoreactivity has never been detected in mammalian stomach, the presence of xenopsin (a neurotensin-like octapeptide first identified in the skin of *Xenopus laevis*) in mammalian pyloric G cells has been ascertained[124]. Neurotensin cells have also been detected in chicken thymus[152] and in a subpopulation of noradrenalin-containing cells (the so-called II type cells) of the cat adrenal medulla[156].

Neurotensin cells have been repeatedly reported in pancreatic tumours, with special reference to those associated with the watery diarrhoea syndrome, sometimes as an overwhelming cellular component, usually with concomitant VIP secretion[29,44]. Concomitant production of xenopsin and gastrin by the same tumour cells[45] and of neurotensin and gastrin by the same pancreatic tumour[43] has been reported, a finding in keeping with the coexistence of gastrin and neurotensin-like peptides in normal pyloric G cells.

Neurotensin cells have occasionally been found also in endocrine tumours of the appendix and rectum[105,179]. Generalised pruritus and dermatographism have been observed in a patient showing an inoperable rectal tumour associated with very high plasma levels of neurotensin. The powerful histamine-releasing action of neurotensin on mast cells might be involved in causing these symptoms[20].

Concluding remarks

Despite the impressive progress made during the past few years, functional characterisation of the manifold population of peptide-producing endocrine cells is far from being fully achieved. The products of some ultrastructurally-characterised cells in the gastrointestinal mucosa, skin and urethra remain to be ascertained[135,137], the exact precursor molecules of several regulatory peptides localised to endocrine cells are still to be identified, the intragranular enzymes involved in their posttranslational processing, as well as pertinent regulatory mechanisms, are largely unknown, while specific receptors and intracellular mediators involved in endocrine cell activation remain mostly to be characterised. Clarification of these points is essential not only for the understanding of many endocrine functions but also for appropriate interpretation of pertinent pathological processes, with special reference to tumour pathology. At present, it seems clear that endocrine tumours produce more frequently those peptides (or related propeptides) that are normally expressed by their tissue of origin. Examples include bombesin/GRP in lung, calcitonin in thyroid, substance P in midgut, glicentin in rectal and enkephalins in adrenal medulla and paraganglionic tumours. However, tumour cells show an increased plasticity of peptide expression, especially in malignant tumours, leading to production of several inappropriate peptides (among which ACTH and calcitonin are those more frequently reported), sometimes with clear-cut site-dependent preference, as for gastrin and VIP expression in pancreatic tumours[136]. A tendency to release an increased proportion of larger molecular forms or even the entire, uncleaved propeptide, has also been noted in many tumours. This is likely to be due to defective intragranular prohormone storage and/or processing because of selective loss of intragranular cleaving enzymes or lack of granule formation and storage resulting in agranular tumour cells. Clarification of molecular and cellular mechanisms underlying these phenomena may help in understanding the biology and clinical behaviour of such growths.

Acknowledgments. This work was supported in part by grants from the Italian National Research Council (Special Projects on Oncology and Biomedical Technologies; Gastroenterology Group) and the Health and Education Ministry.

1 Ali-Rachedi, A., Varndell, I. M., Adrian, T. E., Gapp, D. A., Van Noorden, S., Bloom, S. R., and Polak, J. M., Peptide YY (PYY) immunoreactivity is co-stored with glucagon-related immunoreactants in endocrine cells of the gut and pancreas. Histochemistry 80 (1984) 487–491.

2 Ali-Rachedi, A., Varndell, I. M., Facer, P., Hillyard, C. J., Craig, R. K., MacIntyre, I., and Polak, J. M., Immunohistochemical localisation of katacalcin, a calcium-lowering hormone cleaved from the human calcitonin precursors. J. clin. Endocr. Metab. 57 (1983) 280–282.

3 Alumets, J., Ekman, R., Håkanson, R., and Sundler, R., Evidence for the presence of pro-σ-melanotropin, the NH2-terminal fragment of the corticotropin-β-lipotropin precursor, in corticotropin-producing tumours. Virch. Arch. Path. Anat. 394 (1981) 143–150.

4 Alumets, J., Håkanson, R., Sundler, F., and Chang, K.-J., Leu-enkephalin-like material in nerves and enterochromaffin cells in the gut. An immunohistochemical study. Histochemistry 56 (1978) 187–196.

238

5 Alumets, J., Sundler, F., and Håkanson, R., Distribution, ontogeny and ultrastructure of somatostatin immunoreactive cells in the pancreas and gut. Cell Tiss. Res. *185* (1977) 465–479.

6 Asa, S. L., Kovacs, K., Thorner, M. O., Leong, D. A., Rivier, J., and Vale, W., Immuno-histological localization of growth hormone-releasing hormone in human tumours. J. clin. Endocr. Metab. *60* (1985) 423–427.

7 Becker, K. L., Monaghan, K. G., and Silva, O. L., Immunocytochemical localization of calcitonin in Kulchitsky cells of human lung. Archs Path. Lab. Med. *104* (1980) 196–198.

7a Benjannet, S., Leduc, R., Adrouche, N., Falgueyret, J. P., Marcinkiewicz, M., Seidah, N. J., Mbikay, M., Lazure, C., and Chretien, M., Chromogranin B (secretogranin I), a puta-tive precursor of two novel pituitary peptides through processing at paired basic residues FEBS Lett. *224* (1987) 142–148.

8 Berger, G., Berger, F., Boman, F., Chayvialle, J. A., and Feroldi, J., Localisation of C-terminal gastrin immunoreactivity in gastrinoma cells. Virch. Arch. Path. Anat. *406* (1985) 223–236.

9 Birkenhäger, J. C., Upton, G. V., Seldenrath, H. J., Krieger, D. T., and Tashjian, A. H., Medullary thyroid carcinoma-ectopic production of peptides with ACTH-like corti-cotrophin releasing factor-like and prolactin production stimulating activities. Acta en-docr. *83* (1976) 280–292.

10 Bishop, A. E., Polak, J. M., Facer, P., Ferri, G. L., Marangos, P. J., and Pearse, A. G. E., Neuron specific enolase: a common marker for the endocrine cells and innervation of the gut and pancreas. Gastroenterology *83* (1982) 902–915.

11 Bloom, S. R., Christofides, N. D., Delamarter, J., Buell, G., Kawashima, E., and Polak, J. M., Diarrhoea in vipoma patients associated with cosecretion of a second active peptide (peptide histidine isoleucine) explained by single coding gene. Lancet *2* (1983) 1163–1165.

12 Bloom, S. R., An enteroglucagon tumour. Gut *13* (1972) 520–523.

13 Bloom, S. R., Polak, J. M., and Pearse, A. G. E., Vasoactive intestinal peptide and watery-diarrhea syndrome. Lancet *2* (1973) 14–16.

14 Bordi, C., Togni, R., Baetens, D., Baetens, D., Gorden, P., Unger, R. H., and Orci, L., A study of glucagonomas by light and electron microscopy and immunofluorescence. Dia-betes *28* (1979) 925–936.

15 Bordi, C., Togni, R., Baetens, D., Ravazzola, M., Malaisse-Lagae, F., and Orci, L., Human islet cell tumour storing pancreatic polypeptide (PP): a light and electron microscopy study. J. clin. Endocr. Metab. *46* (1978) 215–219.

16 Bosman, F. T., Van Assche, C., Kruseman, A. C. N., Jackson, S., and Lowry, P. J., Growth hormone releasing factor (GRF) immunoreactivity in human and rat gastrointestinal tract and pancreas. J. Histochem. Cytochem. *32* (1984) 1139–1144.

17 Böttcher, G., Sjölung, K., Ekblad, E., Håkanson, R., Schwartz, T. W., and Sundler, F., Coexistence of peptide YY and glicentin immunoreactivity in endocrine cells of the gut. Reg. Peptides *8* (1984) 261–266.

18 Buchan, A. M. J., Ingman-Baker, J., Levy, J., and Brown, J. C., A comparison of the ability of serum and monoclonal antibodies to gastric inhibitory polypeptide to detect immunoreactive cells in the gastroenteropancreatic system of mammals and reptiles. Histo-chemistry *76* (1982) 341–349.

19 Buchan, A. M. J., Polak, J. M., Solcia, E., and Pearse, A. G. E., Localization of intestinal gastrin in a distinct endocrine cell type. Nature *277* (1979) 138–140.

20 Buchanan, K. C., and Shaw, Ch., Neuroendocrine tumour associated with elevated circu-lating levels of neurotensin and generalized pruritis, in: Sixth International Symposium on Gastrointestinal Hormones, abstr, p. 409. Cand. J. Physiol. Pharmac., 1986.

21 Buckley, K., and Kelly, R. B., Identification of a transmembrane glycoprotein specific for secretory vesicles of neural and endocrine cells. J. Cell Biol. *100* (1985) 1284–1294.

22 Buffa, R., Capella, C., Fontana, P., Usellini, L., and Solcia, E., Types of endocrine cells in the human colon and rectum. Cell Tiss. Res. *192* (1978) 227–240.

23 Buffa, R., Chayvialle, J. A., Fontana, P., Usellini, L., Capella, C., and Solcia, E., Parafol-licular cells of rabbit thyroid store both calcitonin and somatostatin and resemble gut D cells ultrastructurally. Histochemistry *62* (1979) 281–288.

24 Buffa, R., Rindi, G., Sessa, F., Gini, A., Capella, C., Jahn, R., Navone, F., De Camilli, P., and Solcia, E., Synaptophysin immunoreactivity and small clear vesicles in neuroendocrine cells and related tumours. Molec. Cell. Probes *1* (1987) 367–381.

25 Bussolati, G., Capella, C., Vassallo, G., and Solcia, E., Histochemical and ultrastructural studies on pancreatic A cells. Evidence for glucagon and non-glucagon components of the α-granule. Diabetologia 7 (1971) 181–188.

26 Capella, C., Bordi, C., Monga, G., Buffa, R., Fontana, P., Bonfanti, S., Bussolati, G., and Solcia, E., Multiple endocrine cell types in thyroid medullary carcinoma. Evidence for calcitonin, somatostatin, ACTH, 5HT and small granule cells. Virch. Arch. Path. Anat. A 377 (1978) 111–128.

27 Capella, C., Frigerio, B., Usellini, L., Jehenson, P., and Solcia, E., Tumori endocrini del polmone. Atti Accad. Peloritana. Classe Scienze Med.-Biol. 69/Suppl. 2 (1981) 405–418.

28 Capella, C., Hage, E., Solcia, E., and Usellini, L., Ultrastructural similarity of endocrine-like cells of the human lung and some related cells of the gut. Cell Tiss. Res. 186 (1978) 25–37.

29 Capella, C., Polak, J. M., Buffa, R., Tapia, F. J., Heitz, Ph., Bloom, S. R., and Solcia, E., Morphological patterns and diagnostic criteria of VIP-producing endocrine tumours. A histological, histochemical, ultrastructural and biochemical study of 32 cases. Cancer 52 (1983) 1860–1874.

30 Capella, C., Riva, C., Cornaggia, M., Chiaravalli, A. M., Frigerio, B., and Solcia, E., Histopathology, cytology and cytochemistry of pheochromocytomas and paragangliomas, including chemodectomas. Path. Res. Pract. 183 (1988) 176–187.

31 Capella, C., Usellini, L., Buffa, R., Frigerio, B., and Solcia, E., The endocrine component of prostatic carcinomas, mixed adenocarcinoma-carcinoid tumours and non-tumour prostate. Histochemical and ultrastructural identification of the endocrine cells. Histopathology 5 (1981) 175–192.

32 Carey, R. M., Varma, S. K., Drakl, C. R., Thorner, M. O., Kovacs, K., Rivier, J., and Vale, W., Ectopic secretion of corticotropin-releasing factor as a cause of Cushing's syndrome. New Engl. J. Med. 311 (1984) 13–20.

33 Carvalheira, A. F., Welsch, U., and Pearse, A. G. E., Cytochemical and ultrastructural observations on the argentaffin and argyrophil cells of the gastrointestinal tract in mammals, and their place in the APUD series of polypeptide-secreting cells. Histochemie 14 (1968) 33–46.

34 Collina, G., Springall, D. R., Barer, G., Bee, D., and Polak, J. M., Increased numbers of CGRP-immunoreactive endocrine cells of the rat respiratory tract in hypoxia. Reg. Peptides 15 (1986) 171.

35 Conlon, J. F., Deacon, C. F., Richter, G., Schmidt, W. E., Stockmann, F., and Creutzfeldt, W., Measurement and partial characterization of the multiple forms of neurokinin A-like immunoreactivity in carcinoid tumours. Reg. Peptides 13 (1986) 183–196.

36 Creutzfeldt, W., Arnold, R., Creutzfeldt, C., and Track, N. S., Pathomorphologic, biochemical and diagnostic aspects of gastrinomas (Zollinger-Ellison syndrome). Hum. Path. 6 (1975) 47–76.

37 Cutz, E., Chan, W., and Track, N. S., Bombesin, Calcitonin and leu-enkephalin immunoreactivity in endocrine cells of the human lung. Experientia 37 (1981) 765–767.

38 Dayal, Y., Lin, H. D., Tallberg, K., Reichlin, S., De Lellis, R. A., and Wolfe, H. J., Immunocytochemical demonstration of growth hormone-releasing factor in gastrointestinal and pancreatic endocrine tumours. Am. J. clin. Path. 85 (1986) 13–20.

39 Dayal, Y., Nunnemacher, G., Doos, W. G., De Lellis, R. A., O'Brien, M. J., and Wolfe, H. J., Psammomatous somatostatinomas of the duodenum. Am. J. Surg. Path. 7 (1983) 653–665.

40 Deftos, L. J., Bone, H. G., Parthemore, J. G., and Burton, D. W., Immunohistological studies of medullary thyroid carcinoma and C cell hyperplasia. J. clin. Endocr. Metab. 51 (1980) 857–862.

41 Deftos, L. J., and Burton, D. W., Immunohistological studies of nonthyroidal calcitonin-producing tumours. J. clin. Endocr. Metab. 50 (1980) 1042–1045.

42 Dumont, M., Day, R., and Lemaire, S., Distinct distribution of immunoreactive dynorphin and leucine enkephalin in various populations of isolated adrenal chromaffin cells. Life Sci. 32 (1983) 287–294.

43 Feurle, G. E., Physiological and pathological aspects of a neuroendocrinological principle: neurotensin. Front. Horm. Res. 12 (1984) 157–167.

44 Feurle, G. E., Helmstaedter, V., Tischbirek, K., Carraway, R., Forssman, W. F., Grube,

D., and Roher, H., A multihormonal tumor of the pancreas producing neurotensin. Dig. Dis. Sci. 26 (1981) 1125–1133.

45 Feurle, G. E., and Rix, E., Localization of xenopsin immunoreactivity to gastric antral G-cells and gastrinoma G cells, in: Sixth International Symposium on Gastrointestinal Hormones, abstr., p. 155. J. Physiol. Pharmac. 1986.

46 Feurle, G. E., Weber, U., and Helmstaedter, V., Corticotropinlike substances in human gastric antrum and pancreas. Biochem. biophys. Res. Commun. 95 (1980) 1656.

47 Fiocca, R., Capella, C., Buffa, R., Fontana, P., Solcia, E., Hage, E., Chance, R. E., and Moody, A. J., Glucagon-, glicentin- and pancreatic polypeptide-like immunoreactivities in rectal carcinoids and related colorectal cells. Am. J. Path. 100 (1980) 81–92.

48 Fiocca, R., Rindi, G., Capella, C., Grimelius, L., Polak, J. M., Schwartz, T. W., Yanaihara, N., and Solcia, E., Glucagon, glicentin, proglucagon, PYY, PP and proPP-icosapeptide immunoreactivities of rectal carcinoid tumors and related non-tumor cells. Reg. Peptides 17 (1986) 9–29.

49 Fischer-Colbrie, R., Diez-Guerra, J., Emson, P. C., and Winkler, H., Bovine chromaffin granules: immunological studies with antisera against neuropeptide Y, (met)enkephalin and bombesin. Neuroscience 18 (1986) 167–174.

50 Friesen, S. R., Tomita, T., and Kimmel, J. R., Pancreatic polypeptide update: its roles in detection of the trait for multiple endocrine adenopathy syndrome, type I and pancreatic-polypeptide-secreting tumors. Surgery 94 (1983) 1028–1037.

51 Frigerio, B., Ravazzola, M., Ito, S., Buffa, R., Capella, C., Solcia, E., and Orci, L., Histochemical and ultrastructural identification of neurotensin cells in the dog ileum. Histochemistry 54 ((1977) 123–131.

52 Gamse, R., Saria, A., Bucsics, A., and Lambeck, F., Substance P in tumors: pheochromocytoma and carcinoid. Peptides 2, Suppl. 2 (1981) 275–280.

53 Gleeson, M. H., Bloom, S. R., Polak, J. M., Henry, K., and Dowling, R. H., Endocrine tumour in kidney affecting small bowel structure, mobility, and absorptive function. Gut 12 (1971) 773–782.

54 Gould, V. E., Linnoila, I., Memoli, V. A., and Warren, W. H., Neuroendocrine components of the bronchopulmonary tract: hyperplasias, dysplasias and neoplasms. Lab. Invest. 49 (1983) 519–537.

55 Grimelius, L., Capella, C., Buffa, R., Polak, J. M., Pearse, A. G. E., and Solcia, E., Cytochemical and ultrastructural differentiation of enteroglucagon and pancreatic-type glucagon cells of the gastro-intestinal tract. Virch. Arch. Cell Path. B 20 (1976) 217–228.

56 Grimmelikhuijzen, C. J. P., Peptides in the nervous system of coelenterates, in: Evolution and Tumour Pathology of the Neuroendocrine System, pp. 39–58. Eds S. Falkmer, R. Håkanson and F. Sundler. Elsevier, Amsterdam 1984.

57 Håkanson, R., Bengmark, S., Brodin, E., Ingemansson, S., Larsson, L.-I., Nilsson, G., and Sundler, F., Substance P-like immunoreactivity in intestinal carcinoid tumors, in: Substance P, pp. 55–58. Eds U.S. von Euler and B. Pernow. Raven Press, New York 1977.

58 Hamid, Q., Springall, D. R., Ghatei, M. A., Fountain, B. A., Addis, B., Ibrahim, B. N., Bloom, S. R., and Polak, J. M., Expression of C-flanking peptide of human pro-bombesin in pulmonary and extra-pulmonary small cell carcinoma. Reg. Peptides 15 (1986) 180.

59 Hara, M., Varndell, I. M., Bishop, A. E., Aitchison, M., Rode, J., Yamada, T., Green, D. M., Bloom, S. R., and Polak J. M., Expression of C-terminal flanking peptide of human progastrin in human gastroduodenal mucosa, G-cell hyperplasia and islet cell tumours producing gastrin. Molec. cell. Probes 1 (1987) 95–108.

60 Hartschuh, W., Weihe, E., Buchler, M., Helmstaedter, V., Feurle, G. E., and Forssmann, W. G., Met-enkephalin-like immunoreactivity in Merkel cells. Cell Tissue Res. 201 (1979) 343–348.

61 Hartschuh, W., Weihe, E., Yanaihara, N., and Reinecke, M., Immunohistochemical localization of vasoactive intestinal polypeptide (VIP) in Merkel cells of various mammals: evidence for a neuromodulator function of the Merkel cell. J. Invest. Dermat. 81 (1983) 361–364.

62 Harty, R. F., and Franklin, P. A., GABA affects antral gastrin and somatostatin release. Nature. 303 (1983) 623–624.

63 Hassoun, J., Monges, G., Giroud, P., Henry, J. F., Charpin, C., Payan, H., and Toga, M., Immunohistochemical study of pheochromocytomas. An investigation of methio-

ineenkephalin, vasoactive intestinal peptide, somatostatin, corticotropin, β-endorphin, and calcitonin in 16 tumours. Am. J. Path. *14* (1984) 56–63.

64 Heitz, Ph. U., Kasper, M., Polak, J. M., and Kloppel, G., Pancreatic endocrine tumors: immunocytochemical analysis of 125 tumors. Hum. Path. *13* (1982) 263–271.

65 Heitz, Ph. U., Polak, J. M., Timson, C. M., and Pearse, A. G. E., Enterochromaffin cells as the endocrine source of gastrointestinal substance P. Histochemistry *49* (1976) 343–347.

66 Heym, C., and Reinecke, M., Immunohistochemistry of neuropeptides in cat paraganglia. Front. Horm. Res. *12* (1984) 91–94.

67 Holle, G. E., Spann, W., Eisenmenger, W., Riedel, J., and Pradayrol, L., Diffuse somatostatin-immunoreactive D-cell hyperplasia in the stomach and duodenum. Gastroenterology *91* (1986) 733–739.

68 Howlett, T. A., Price, J., Hale, A. C., Doniach, I., Rees, L. H., Wass, J. A. H., and Besser, G. M., Pituitary ACTH dependent Cushing's syndrome due to ectopic production of a bombesin-like peptide by a medullary carcinoma of the thyroid. Clin. Endocr. *22* (1985) 91–101.

69 Itoh, N., Obata, K.-I., Yanaihara, N., and Okamoto, H., Human preprovasoactive intestinal polypeptide contains a novel PHI-27-like peptide, PHM-27. Nature *304* (1983) 547–549.

70 Jonsson, A. C., and Dockray, G. J., Immunohistochemical localization to pyloric antral G cells of peptides derived from porcine preprogastrin. Reg. Peptides *8* (1984) 283–290.

71 Kakidani, H., Furutani, Y., Takahashi, H., Noda, M., Morimoto, Y., Hirose, T., Asai, M., Inayama, S., Nakanishi, S., and Numa, S., Cloning and sequence analysis of cDNA for porcine β-neo-endorphin/dynorphin precursor. Nature *298* (1982) 245–249.

72 Kameda, Y., Oyama, H., Endoh, M., and Horino, M., Somatostatin immunoreactive C cells in thyroid glands from various mammalian species. Anat. Rec. *204* (1982) 161–170.

73 Kameya, T., Bessho, T., Tsumuraya, M., Yamaguchi, K., Abe, K., Shimosato, Y., and Yanaihara, N., Production of gastrin releasing peptide by medullary carcinoma of the thyroid. An immunohistochemical study. Virch. Arch. Path. Anat. A *401* (1983) 99–108.

74 Kameya, T., Tsmuraya, M., Shimosato, Y., Abe, K., and Yanaihara, N., Demonstration of multiple hormone production by single cells in neoplasia. J. Histochem. Cytochem. *30* (1982) 554.

75 Kanagawa, Y., Matsuyama, T., Wanka, A., Yoneda, S., Kimura, K., Kamada, T., Steinbusch, H. W. M., and Tohyama, M., Coexistence of enkephalin- and serotonin-like substances in single small intensely fluorescent cells of the guinea pig superior cervical ganglion. Brain Res. *379* (1986) 377–379.

76 Kawano, H., Daikoku, S., and Saito, S., Location of thyrotropin-releasing hormone-like immunoreactivity in rat pancreas. Endocrinology *112* (1983) 951–955.

77 Knudsen, J. B., Holst, J. J., Asnaes, S., and Johansen, A., Identification of cells with pancreatic-type and gut-type glucagon immunoreactivity in the human colon. Acta path. microbiol. scand. *83* (1975) 741–743.

78 Kobayashi, S., and Coupland, R. E., Two populations of microvesicles in the SGC (small granule chromaffin) cells of the mouse adrenal medulla. Arch. histol. jap. *40* (1977) 251–259.

79 Krejs, G. J., Orci, L., Conlon, J. M., Ravazzola, M., Davis, G. R., Raskin, P., Collins, S. M., McCarthy, D. M., Baetens, D., Rubenstein, A., Aldor, T. A. M., and Unger, R. H., Somatostatinoma syndrome. Biochemical, morphologic and clinical features. N. Engl. J. Med. *301* (1979) 285–292.

80 Lamberts, S. W. J., Lange, S. A., and Stefanko, S. Z., Adrenocorticotropin-secreting pituitary adenomas originate from the anterior or the intermediate lobe in Cushing's disease: differences in the regulation of hormone secretion. J. clin. Endocr. Metab. *54* (1982) 286–291.

81 Larsson, L.-I., Corticotropin-like peptide in central nerves and in endocrine cells of gut and pancreas. Lancet *2* (1977) 1321.

82 Larsson, L.-I., and Jorgensen, L. M., Ultrastructural and cytochemical studies on the cytodifferentiation of duodenal endocrine cells. Cell Tiss. Res. *194* (1978) 79–102.

83 Larsson, L.-I., Polak, J. M., Buffa, R., Sundler, F., and Solcia, E., On the immunocytochemical localization of the vasoactive intestinal polypeptide. J. Histochem. Cytochem. *27* (1979) 936–938.

242

84 Larsson, L.-I., and Rehfeld, J. F., Evidence for a common evolutionary origin of gastrin and cholecystokinin. Nature *269* (1977) 335–338.

85 Larsson, L.-I., Rehfeld, J. F., Sundler, F., and Håkanson, R., Pancreatic gastrin in foetal and neonatal rats. Nature *262* (1976) 609–610.

86 Lauweryns, J. M., Cokelaere, M., and Theunynck, P., Neuroepithelial bodies in the respiratory mucosa of various mammals: a light optical, histochemical and ultrastructural investigation. Z. Zellforsch. *135* (1972) 569–592.

87 Leboulenger, F., Leroux, P., Tonon, M.-C., Coy, D. H., Vaudry, H., and Pelletier, G., Coexistence of vasoactive intestinal peptide and enkephalins in the adrenal chromaffin granules of the frog. Neurosci. Lett. *37* (1983) 221–225.

88 Lechago, J., Holmquist, A. L., Rosenquist, G. L., and Walsh, J. H., Localization of bombesin like peptides in frog gastric mucosa. Gen. comp. Endocr. *36* (1978) 553–558.

89 Le Douarin, N. M., The embryological origin of the endocrine cells associated with the digestive tract: experimental analysis based on the use of a stable cell marking technique, in: Gut Hormones, pp. 49–56. Ed. S. R. Bloom. Churchill Livingstone, Edinburgh 1978.

90 Lemaire, S., Chouinard, L., Mercier, P., and Day, R., Bombesin-like immunoreactivity in bovine adrenal medulla. Reg. Peptides *13* (1986) 133–146.

91 Lloyd, R. V., Shapiro, B., Sisson, J. C., Kalff, V., Thompson, N. W., and Beierwaltes, W. A., An immunohistochemical study of pheochromocytomas. Archs Path. Lab. Med. *108* (1984) 541–544.

92 Long, R. G., Bryant, M. G., Mitchell, S. J., Adrian, T. E., Polak, J. M., and Bloom, S. R., Clinicopathological study of pancreatic and ganglioneuroblastoma tumours secreting vasoactive intestinal polypeptide (vipomas). Br. med. J. *282* (1981) 1767–1771.

93 Luciano, L., Solcia, E., and Reale, E., The fine structure of the neuroepithelial bodies in the adult rat. Verh. anat. Ges. *75* (1981) 641–642.

94 Lundberg, J. M., Hamberger, B., Schultzberg, M., Hökfelt, T., Granberg, P. D., Efendic, S., Terenius, L., Goldstein, M., and Luft, R., Enkephalin- and somatostatin-like immunoreactivities in human adrenal medulla and pheochromocytoma. Proc. natn. Acad. Sci. USA *76* (1979) 4079–4083.

95 Lundberg, J. M., Hökfelt, T., Hemsen, A., Theodorsson-Norheim, E., Pernow, J., Hamberger, B., and Goldstein, M., Neuropeptide Y-like immunoreactivity in adrenaline cells of adrenal medulla and in tumors and plasma of pheochromocytoma patients. Reg. Peptides *13* (1986) 169–182.

96 Märtensson, H., Nobin, A., Sundler, F., and Falkmer, S., Endocrine tumors of the ileum. Cytochemical and clinical aspects. Path. Res. Pract. *180* (1985) 356–363.

97 Matthew, W. D., Tsavaler, L., and Reichardt, L. F., Identification of a synaptic vesicle-specific membrane protein with a wide distribution in neuronal and neurosecretory tissue. J. Cell Biol. *91* (1981) 257–269.

98 McDonald, M. D., and Mitchell, R. A., The innervation of glomus cells, ganglion cells and blood vessels in the rat carotid body: a quantitative ultrastructural analysis. J. Neurocytol. *4* (1975) 177–230.

99 Mizuno, K., Kojima, M., and Matsuo, H., A putative prohormone processing protease in bovine adrenal medulla specifically cleaving in between Lys-Arg sequences. Biochem. biophys. Res. Commun. *128* (1985) 884–891.

100 Moody, T. W., Pert, C. B., Gazdar, A. F., Carney, D. N., and Minna, J. D., High levels of intracellular bombesin characterized human small-cell lung carcinoma. Science *214* (1981) 1246–1248.

101 Nakanishi, S., Inoue, A., Kita, T., Nakamura, M., Chang, A. C. Y., Cohen, S. N., and Numa, S., Nucleotide sequence of cloned cDNA for bovine corticotropin-β-lipotropin precursor. Nature *278* (1979) 423–427.

102 Navone, E., Jahn, R., Di Gioia, G., Stukenbrok, H., Greengard, P., and De Camilli, P., Protein P38: an integral membrane protein specific for small clear vesicles of neurons and neuroendocrine cells. J. Cell Biol. *103* (1986) 2511–2527.

103 Nawa, H., Kotani, H., and Nakanishi, S., Tissue-specific generation of two preprotachykinin mRNAs from gene by alternative RNA splicing. Nature *312* (1984) 729–735.

104 Norheim, I., Theodorsson-Norheim, E., Brodin, E., Öberg, K., Lundqvist, G., and Rosell, S., Antisera raised against eledoisin and kassinin detect elevated levels of immunoreactive material in plasma and tumor tissues from patients with carcinoid tumors. Reg. Peptides *9* (1984) 245–257.

105 O'Brian, D. S., Dayal, Y., De Lellis, R. A., Tischler, A. S., Bendon, R., and Wolfe, H. J., Rectal carcinoids as tumors of the hindgut endocrine cells. A morphological and immuno-histochemical analysis. Am. J. Surg. Path. 6 (1982) 131–142.

106 O'Hare, M. M. T., Shaw, C., Johnston, C. F., Russell, C. F. J., Sloan, J. M., and Buchanan, K. D., Pancreatic polypeptide immunoreactivity in medullary carcinoma of the thyroid: identification and characterisation by radioimmunoassay, immunocytochemistry and high performance liquid chromatography. Reg. Peptides 14 (1986) 169–180.

107 Orci, L., Ravazzola, M., Amherdt, M., Madsen, O., Vassalli, J-D., and Perrelet, A., Direct identification of prohormone conversion site in insulin-secreting cells. Cell 42 (1985) 671–681.

108 Orwoll, E. S., and Kendall, J. W., β-Endorphin an adrenocorticotropin in extrapituitary sites: gastrointestinal tract. Endocrinology 107 (1980) 438–442.

109 Pauwels, S., Desmond, H., Dimaline, R., and Dockray, G. J., Identification of progastrin in gastrinoma, antrum and duodenum by a novel radioimmunoassay. J. clin. Invest. 77 (1986) 376–381.

110 Pearse, A. G. E., The cytochemistry and ultrastructure of polypeptide hormone-producing cells of the APUD series, and the embryologic, physiologic and pathologic implications of the concept. J. Histochem. Cytochem. 17 (1969) 303–313.

111 Pearse, A. G. E., Peptides in brain and intestine. Nature 262 (1976) 92–94.

112 Pearse, A. G. E., Polak, J. M., and Bloom, S. R., The newer gut hormones. Cellular sources, physiology, pathology and clinical aspects. Gastroenterology 72 (1977) 746–761.

113 Petrusz, P., Merchenthaler, I., Maderdrut, J. L., and Heitz, Ph. U., Central and peripheral distribution of corticotropin-releasing factor. Fedn. Proc. 44 (1985) 229–235.

114 Polak, J. M., Bloom, S. R., Hobbes, S., Solcia, E., and Pearse, A. G. E., Distribution of a bombesin-like peptide in human gastrointestinal tract. Lancet 1 (1976) 1109–1110.

115 Polak, J. M., Stagg, B., and Pearse, A. G. E., Two types of Zollinger-Ellison syndrome. Immunofluorescent, cytochemical and ultrastructural studies of the antral and pancreatic gastrin cells in different clinical states. Gut 13 (1972) 501–512.

116 Ratzenhofer, M., Gamse, R., Höfler, H., Auböck, L., Popper, H., Pohl, H., and Lembeck, F., Substance P in an argentaffin carcinoid of the caecum: biochemical and biological characterization. Virchows Arch. Path. Anat. A 392 (1981) 21–31.

117 Ravazzola, M., and Orci, L., Glucagon and glicentin immunoreactivity are topologically segregated in the α-granule of the human pancreatic A cell. Nature 284 (1980) 66–67.

118 Ravazzola, M., Siperstein, A., Moody, A. J., Sundby, F., Jacobsen, H., and Orci, L., Glicentin immunoreactive cells: their relationship to glucagon-producing cells. Endocrinology 105 (1979) 499–508.

119 Rawdon, B. B., and Andrew, A., An immunocytochemical survey of endocrine cells in the gastrointestinal tract of chicks at hatching. Cell Tiss. Res. 220 (1981) 279–292.

120 Rees, L. H., and Ratcliffe, J. G., Ectopic hormone production by non-endocrine tumors. Clin. Endocr. 3 (1974) 263–299.

121 Reinecke, M., Schlüter, P., Yanaihara, N., and Forssmann, W. G., VIP immunoreactivity in enteric nerves and endocrine cells of the vertebrate gut. Peptides 2, Suppl. 2 (1981) 149–156.

122 Rindi, G., Buffa, R., Sessa, F., Tortora, O., and Solcia, E., Chromogranin A, B and C immunoreactivities of mammalian endocrine cells. Distribution distinction from costored hormones/prohormones and relationship with the argyrophil component of secretory granules. Histochemistry 85 (1986) 19–28.

123 Rivier, J., Spiess, J., Thorner, M., and Vale, W., Characterization of a growth hormone-releasing factor from a human pancreatic islet tumour. Nature 300 (1982) 276–278.

124 Rix, E. W., Feurle, G. E., and Carraway, R. E., Co-localization of xenopsin and gastrin immunoreactivity in gastric antral G-cells. Histochemistry 85 (1986) 135–138.

125 Rosa, P., Hille, A., Lee, R. W. H., Zanini, A., and De Camilli, P., Secretogranins I and II. Two tyrosine-sulfated secretory proteins common to a variety of cells secreting peptides by the regulated pathway. J. Cell Biol. 101 (1985) 1999–2011.

126 Sabate, M. I., Carpani, M., Varndell, I. M., Ghatei, M. A., Rosenfeld, M. G., Bloom, S. R., and Polak, J. M., Calcitonin gene-related peptide in normal thyroid and medullary carcinoma of thyroid. J. Path. 142 (1984) A29.

244

127 Said, S. I., and Faloona, G. R., Elevated plasma and tissue levels of vasoactive intestinal polypeptide in the watery-diarrhea syndrome due to pancreatic, bronchogenic and other tumors. N. Engl. J. Med. *293* (1975) 155–160.

128 Scott, A. P., Ratcliffe, J. G., Rees, L. H., Landon, J., Bennett, H. P. J., Lowry, P. J., and McMartin, C., Pituitary peptide. Nature, New Biol. *244* (1973) 65–67.

129 Shibasaki, T., Kiyosawa, Y., Masuda, A., Nakahara, M., Imaki, T., Wakabayashi, I., Demura, H., Shizume, K., and Ling, N., Distribution of growth hormone-releasing hormone-like immunoreactivity in human tissue extracts. J. clin. Endocr. Metab. *59* (1984) 263–268.

130 Sikri, K. L., Varndell, I. M., Hamid, Q. A., Wilson, B. S., Kameya, T., Ponder, B. A. J., Lloyd, R. V., Bloom, S. R., and Polak, J. M., Medullary carcinoma of the thyroid. An immunocytochemical and histochemical study of 25 cases using eight separate markers. Cancer *56* (1985) 2481–2491.

131 Sjölund, K., Sanden, G., Håkanson, R., and Sundler, F., Endocrine cells in human intestine: an immunocytochemical study. Gastroenterology *85* (1983) 1120–1130.

132 Smyth, D. G., and Zakarian, S., Selective processing of β-endorphin in regions of porcine pituitary. Nature *288* (1980) 613–615.

132a Solcia, E., Buffa, R., Gini, A., Capella, C., Rindi, G., and Polak, J. M., Bombesin peptides in the diffuse neuroendocrine system. Annals N.Y. Acad. Sci. (1988), in press.

133 Solcia, E., Buffa, R., Sessa, F., and Rindi, G., Distinct patterns of chromogranin A, B and C immunoreactivities in different types of gastroenteropancreatic endocrine cells, in: Sixth International Symposium on Gastrointestinal Hormones, abstr., p. 90. Can. J. Physiol. Pharmac., 1986.

134 Solcia, E., Capella, C., Buffa, R., Frigerio, B., and Fiocca, R., Pathology of the Zollinger-Ellison syndrome, in: Progress in Surgical Pathology, pp. 119–133. Ed. C. M. Fenoglio. Masson Publishing USA, New York 1980.

135 Solcia, E., Capella, C., Buffa, R., Frigerio, B., Usellini, L., Fiocca, R., Tenti, P., Sessa, F., and Rindi, G., Cytology of tumours in the gastroenteropancreatic and diffuse (neuro)endocrine system, in: Evolution and Tumour Pathology of the Neuroendocrine System, pp. 453–480. Eds R. Håkanson and S. Falkmer. Elsevier, Amsterdam 1984.

136 Solcia, E., Capella, C., Buffa, R., Tenti, R., Rindi, G., and Cornaggia, M., Antigenic markers of neuroendocrine tumors: their diagnostic and prognostic value, in: New Concepts in Neoplasia as Applied to Diagnostic Pathology, pp. 242–261. Eds C. M. Fenoglio, R. S. Weinstein and N. Kaufman. Williams and Wilkins, Baltimore 1986.

137 Solcia, E., Capella, C., Buffa, R., Usellini, L., Fiocca, R., and Sessa, F., Endocrine cells of the digestive system, in: Physiology of the Gastrointestinal Tract, 2nd edn, pp. 401–420. Ed. L. R. Johnson. Raven Press, New York 1986.

138 Solcia, E., Capella, C., Buffa, R., Usellini, L., Frigerio, B., and Fontana, P., Endocrine cells of the gastrointestinal tract and related tumors. Pathobiol. Ann. *9* (1979) 163–203.

139 Solcia, E., Capella, C., Fiocca, R., Sessa, F., Tenti, P., Rindi, G., and Tortora, O., Ultrastructural and immunohistochemical characterization of F-type and D-type PP cells and their distribution in normal, annular, chronically inflamed, heterotopic or tumor pancreas, in: Frontiers of Hormone Research, vol. 12, pp. 31–40. Ed. M. Ratzenhofer. Karger, Basel 1984.

140 Solcia, E., Capella, C., Fiocca, R., Tenti, P., Sessa, F., and Riva, C., Disorders of endocrine system, in: Pathology of the Gastrointestinal Tract, Chapter 13. Eds S.-I. Ming. and H. Harvey. W. B. Saunders, Philadelphia 1987.

141 Solcia, E., Capella, C., Vassallo, G., and Buffa, R., Endocrine cells of the gastric mucosa. Int. Rev. Cytol. *42* (1975) 223–286.

142 Solcia, E., Fiocca, R., Capella, C., Usellini, L., Sessa, F., Rindi, G., Schwartz, T. W., and Yanaihara, N., Glucagon- and PP-related peptides of intestinal L cells and pancreatic/gastric A or PP cells. Possible interrelationship of peptides and cells during evolution, fetal development and tumor growth. Peptides *6*, Suppl. (1985) 223–229.

143 Somogyi, P., Hodgson, A. J., De Potter, R. W., Fischer-Colbrie, R., Schober, M., Winkler, H., and Chubb, W., Chromogranin immunoreactivity in the central nervous system. Immunochemical characterization, distribution and relationship to catecholamine and enkephaline pathways. Brain Res. Rev. *8* (1984) 193–230.

144 Spindel, E. R., Chin, W. W., Price, J., Rees, L. H., Besser, G. M., and Habener, J., Cloning

and characterization of cDNAs encoding human gastrin-releasing peptide. Proc. natn. Acad. Sci. USA *81* (1984) 5699–5703.

145 Steenbergh, P. H., Höppener, J. W. M., Zandberg, J., Roos, B. A., Jansz, H. S., and Lips, C. J. M., Expression of the propiomelanocortin gene in human medullary thyroid carcinoma. J. clin. Endocr. Metab. *58* (1984) 904–908.

146 Stefan, Y., Ravazzola, M., Grasso, S., Perrelet, A., and Orci, L., Glicentin precedes glucagon in the developing human pancreas. Endocrinology *110* (1982) 2189–2191.

147 Suda, T., Tomori, N., Tozawa, F., Demura, H., Shizume, K., Mouri, T., Miura, Y., and Sasano, N., Immunoreactive corticotropin and corticotropin-releasing factor in human hypothalamus, adrenal, lung cancer, and pheochromocytoma. J. clin. Endocr. Metab. *58* (1984) 919–924.

148 Suda, T., Tomori, N., Tozawa, F., Mouri, T., Demura, H., and Shizume, K., Distribution and characterization of immunoreactive corticotropin-releasing factor in human tissues. J. clin. Endocr. Metab. *59* (1984) 861–866.

149 Sundler, F., Alumets, J., Ekman, R., Håkanson, R., and Van Wimersma Greidanus, T. J. B., Immunoreactive adrenocorticotropic hormone (ACTH) in porcine gut and pancreas: fact or artifact? J. Histochem. Cytochem. *29* (1981) 1328–1335.

150 Sundler, F., Alumets, J., Håkanson, R., Carraway, R., and Leeman, S. E., Ultrastructure of the gut neurotensin cell. Histochemistry *53* (1977) 25–34.

151 Sundler, F., Böttcher, G., Håkanson, R., and Schwartz, T. W., Immunocytochemical localization of the icosapeptide fragment of the PP precursor: a marker for "true" PP cells? Reg. Peptides *8* (1984) 217–224.

152 Sundler, F., Carraway, R. E., Håkanson, R., Alumets, J., and Dubois, M. P., Immunoreactive neurotensin and somatostatin in the chicken thymus. A chemical and histochemical study. Cell Tiss. Res. *194* (1978) 367–376.

153 Tamai, S., Kameya, T., Yamaguchi, K., Yanai, N., Abe, K., Yanaihara, N., Yamazaka, H., and Kageyama, K., Peripheral lung carcinoid tumor producing predominantly gastrin releasing peptide (GRP). Morphological and hormonal studies. Cancer *52* (1983) 273–281.

154 Tanaka, I., Nakai, Y., Nakao, K., Oki, S., Masaki, N., Ohtsuki, H., and Imura, H., Presence of immunoreactive y-melanocyte-stimulating hormone, adrenocorticotropin, and β-endorphin in human gastric antral mucosa. J. clin. Endocr. Metab. *54* (1982) 392–396.

155 Tatemoto, K., Efendić, S., Mutt, V., Makk, G., Feistner, G. J., and Barchas, J. D., Pancreastatin, a novel pancreatic peptide that inhibits insulin secretion. Nature *324* (1986) 476–478.

156 Terenghi, G., Polak, J. M., Varndell, I., Lee, Y. C., Wharton, J., and Bloom, S. R., Neurotensin-like immunoreactivity in a subpopulation of noradrenaline-containing cells of the cat adrenal gland. Endocrinology *112* (1983) 226–233.

157 Timson, C. M., Polak, J. M., Wharton, J., Ghatei, M. A., Bloom, S. R., Usellini, L., Capella, C., Solcia, E., Brown, M. R., and Pearse, A. G. E., Bombesin-like immunoreactivity in the avian gut and its localisation to a distinct cell type. Histochemistry *61* (1979) 213–221.

158 Tschopp, F. A., Tobler, P. H., and Fischer, J. A., Calcitonin gene-related peptide in the human thyroid, pituitary and brain. Molec. cell. Endocr. *36* (1984) 53–57.

159 Tsutsumi, Y., Osamura, R. Y., Watanabe, K., and Yanaihara, N., Simultaneous immunohistochemical localization of gastrin releasing peptide (GRP) and calcitonin (CT) in human bronchial endocrine-type cells. Virch. Arch. Path. Anat. *400* (1983) 163–171.

160 Undenfriend, S., and Kilpatrick, D. L., Biochemistry of the enkephalins and enkephalin-containing peptides. Archs Biochem. Biophys. *221* (1983) 309–323.

161 Upton, G. V., and Amatruda, T. T., Evidence for the presence of tumor peptides with corticotrophin-releasing-factor-like activity in the ectopic ACTH syndrome. N. Engl. J. Med. *285* (1971) 419–424.

162 Usellini, L., Buchan, A. M. J., Polak, J. M., Capella, C., Cornaggia, M., and Solcia, E., Ultrastructural localization of motilin in endocrine cells of human and dog intestine by the immunogold technique. Histochemistry *81* (1984) 363–368.

163 Usellini, L., Capella, C., Frigerio, B., Rindi, G., and Solcia, E., Ultrastructural localization of secretin in endocrine cells of the dog duodenum by the immunogold technique. Comparison with ultrastructurally characterized S cells of various mammals. Histochemistry *80* (1984) 435–441.

246

164 Usellini, L., Capella, C., Malesci, A., Rindi, G., and Solcia, E., Ultrastructural localization of cholecystokinin in endocrine cells of the dog duodenum by the immunogold technique. Histochemistry *83* (1985) 331–336.

165 Usellini, L., Capella, C., Solcia, E., Buchan, A. M. J., and Brown, J. C., Ultrastructural localization of gastric inhibitory polypeptide (GIP) in a well-characterized endocrine cell of canine duodenal mucosa. Histochemistry *80* (1984) 85–89.

166 Usellini, L., Tenti, P., Fiocca, R., Capella, C., Buffa, R., Terenghi, C., Polak, J. M., and Solcia, E., The endocrine cells of the chicken proventriculus. Bas. appl. Histochem. *27* (1983) 87–102.

167 Usellini, L., Riva, C., Rindi, G., Capella, C., and Solcia, E., Gastrin, cholecystokinin (CCK), mixed gastrin/CCK and C-terminus gastrin/CCK immunoreactive cells of the human small intestine. A light and electron immunocytochemistry study. (1988) in preparation.

168 Van Noorden, S., Polak, J. M., and Pearse, A. G. E., Single cellular origin of somatostatin and calcitonin in the rat thyroid gland. Histochemistry *53* (1977) 243–247.

169 Varndell, I. M., Bishop, A. E., Sikri, K. L., Uttenthal, L. O., Bloom, S. R., and Polak, J. M., Localization of glucagon-like peptide (GLP) immunoreactants in human gut and pancreas using light and electron microscopic immunocytochemistry. J. Histochem. Cytochem. *33* (1985) 1080–1086.

170 Varndell, I. M., Harris, A., Tapia, F. J., Yanaihara, N., De Mey, J., Bloom, S. R., and Polak, J. M., Intracellular topography of immunoreactive gastrin demonstrated using electron immunocytochemistry. Experientia *39* (1983) 713–717.

171 Watkins, W. B., Bruni, J. F., and Yen, S. S. C., β-Endorphin and somatostatin in the pancreatic D-cell. Colocalization by immunocytochemistry. J. Histochem. Cytochem. *28* (1980) 1170–1174.

172 Weber, E., Voigt, K. H., and Martin, R., Pituitary somatotrophs contain (met)enkephalin-like immunoreactivity. Proc. natn. Acad. Sci. USA *75* (1978) 6134–6138.

173 Welsch, U., and Pearse, A. G. E., Electron cytochemistry of BuChE and AChE in thyroid and parathyroid C cells, under normal and experimental conditions. Histochemie *17* (1969) 1–10.

174 Wharton, J., Polak, J. M., Bloom, S. R., Ghatei, M. A., Solcia, E., Brown, M. R., and Pearse, A. G. E., Bombesin-like immunoreactivity in the lung. Nature *273* (1978) 769–770.

175 Wharton, J., Polak, J. M., Cole, G. A., Marangos, P. J., and Pearse, A. G. E., Neuron-specific enolase as an immunocytochemical marker for the diffuse neuroendocrine system in human fetal lung. J. Histochem. Cytochem. *29* (1981) 1359–1364.

176 Wharton, J., Polak, J. M., Pearse, A. G. E., McGregor, G. P., Bryant, M. G., Bloom, S. R., Emson, P. C., Bisgard, G. E., and Will, J. A., Enkephalin-, VIP- and substance P-like immunoreactivity in the carotid body. Nature *284* (1980) 269–271.

177 Wiedenmann, B., Franke, W. W., Kuhn, C., Moll, R., and Gould, V. E., Synaptophysin. A marker protein for neuroendocrine cells and neoplasms. Proc. natn. Acad. Sci. USA *83* (1986) 3500–3504.

178 Winkler, H., Apps, D. K., and Fisher-Colbrie, R., The molecular function of adrenal chromaffin granules: established facts and unresolved topics. Neuroscience *18* (1986) 261–290.

179 Yang, K., Ulrich, T., Chen, G. L., and Lewin, K. J., The neuroendocrine products of intestinal carcinoids. Cancer *51* (1983) 1918–1926.

180 Yoshizaki, K., de Bock, V., Takai, I., Wang, N. S., and Solomon, S., Bombesin-like peptides in human small cell carcinoma of the lung. Reg. Peptides *14* (1986) 11–20.

Neuropeptides and the microcircuitry of the enteric nervous system

I. J. Llewellyn-Smith

Summary. The discovery of neuropeptides in enteric neurons has revolutionized the study of the microcircuitry of the enteric nervous system. From immunohistochemistry, it is now clear that some individual enteric neurons contain several different neuropeptides with or without other transmitter-specific markers and that these markers occur in different combinations. There is evidence from experiments in which nerve pathways are interrupted that populations of enteric neurons with various combinations of markers have different projection patterns, sending their processes to distinct targets using different routes. Correlations between the neurochemistry of enteric neurons and the types of synaptic inputs they receive are also beginning to emerge from electrophysiological studies. These findings imply that enteric neurons are chemically coded by the combinations of peptides and other transmitter-related substances they contain and that the coding of each population correlates with its role in the neuronal pathways that control gastrointestinal function.

The mammalian gastrointestinal tract contains about as many neurons as the spinal cord[29]. These intrinsic (enteric) neurons, along with the processes of sympathetic, parasympathetic and sensory neurons supplying the gut, and enteric glial cells, make up the enteric nervous system (ENS), which is now generally classified as a third division of the autonomic nervous system. The cell bodies of the enteric neurons are grouped into ganglia in two main plexuses: the submucous plexus, discovered by Meissner[56] in 1857, in the loose connective tissue of the submucosa, and the myenteric plexus, described by Auerbach[1] in 1864, which lies between the longitudinal and circular layers of the muscularis externa. Non-ganglionated plexuses, which are continuous with the main ganglionated plexuses, supply the muscle layers, the mucosa and blood vessels.

The ENS influences or controls a variety of functions, including movement of digesta along the gastrointestinal tract, gastric acid secretion, transport of water and electrolytes, release of gastrointestinal hormones, and blood flow[30,31]. Enteric reflexes persist even when the gut is disconnected from the central nervous system[2,3]. Hence, the ENS must contain several different functional types of neurons, namely motor neurons to the muscle, secretomotor and vasomotor neurons, interneurons and sensory neurons. Furthermore, these neurons must be arranged in an orderly fashion to form circuits that govern the different enteric reflexes. The need to unravel the internal circuitry of the ENS was recognized early, but this analysis was frustrated until the past decade by the inadequacy of neurohistological and pathway tracing techniques. Since 1975, however, our knowledge of neuronal pathways in the ENS has increased significantly because of the discovery of neuropeptides in enteric neurons and because of technical advances, including light and electron microscopical methods for immunohistochemistry on whole mount preparations of separated gut layers, techniques for lesioning enteric nerve pathways, and correlated physiological, pharmacological and electrophysiological studies. The ways in which these developments have improved our understanding of the microcircuitry of the ENS are outlined below. It should be noted that most of the information provided here comes from experiments on the guinea pig small intestine, which has been the main model for studying enteric nerve pathways.

Histological and histochemical studies before 1975

Early investigators, using methylene blue or silver or osmium impregnation, stained either all enteric neurons and their processes or variable subsets that did not appear to correspond to any single anatomical or functional class[14]. These studies were useful for establishing the arrangements of the enteric plexuses, their interconnections and the shapes of enteric neurons in various regions of the gut and various species[30,39] but were not helpful for defining neuronal circuitry within the ENS. A notable series of observations on methylene blue-stained enteric ganglia from several species was made by Dogiel at the end of the last century[17]. Dogiel was able to define three types of enteric neurons on the basis of their shapes: Type I neurons had many short irregular processes and one long process; Type II neurons had many long processes, whereas Type III cells had intermediate length branching processes[30]. Dogiel believed that his different neuronal types corresponded to functional classes, with Type I cells being motor and Type II cells being sensory. Debate continued for many years about the validity of this proposal and even about the validity of Dogiel's classification system. However, recent experiments have shown that most enteric neurons have shapes that allow them to be fitted into a classification scheme similar to Dogiel's and that there are correlations between cell shape and function (see below).

The development and application of histochemical techniques in the 1950's and 1960's[14] brought only a few advances in understanding the neuronal circuitry of the ENS. Reactions to localize the neurotransmitter-degrading enzymes, acetylcholinesterase and monoamine oxidase, began to be used to examine the ENS in the 1950's, but unfortunately these enzymes were not confined to a single type of enteric neuron. In the 1960's it became possible to demonstrate catecholamine-containing neurons with fluorescence histochemistry. Noradrenergic fibers arising from postganglionic sympathetic neurons of prevertebral ganglia were found to supply primarily the enteric ganglia in non-sphincter regions of the gastrointestinal tract[27], but catecholamine nerve cell bodies were absent from the ENS in most regions of the gut in most species. These histochemical findings and information from physiological and pharmacological studies provided an understanding of how sympathetic nerves controlled motility and blood flow, which has seen little refinement since[30]. However, no information was gained from catecholamine fluorescence histochemistry about intrinsic reflex pathways involving the enteric neurons.

The presence of neuropeptides in enteric nerves

The finding that enteric neurons contain neuropeptides was a major breakthrough for the study of the ENS. Using immunohistochemical techniques, Pearse and Polak[60] and Nilsson and colleagues[57] showed in 1975 that some enteric neurons were immunoreactive for substance P, and in the same year Hökfelt and co-workers[44] found that some enteric neurons contained somatostatin-like immunoreactivity. Since these discoveries little more than a decade ago, many neuropeptides have been localized immunohistochemically within nerve cell bodies and varicose nerve fibers in the ENS[38]. The substances detected immunohistochemically include calcitonin gene-related peptide (CGRP), cholecystokinin (CCK), dynorphin (DYN), enkephalin (ENK), galanin (GAL), gastrin-releasing peptide (GRP; also known as mammalian bombesin), neuropeptide Y (NPY), neurotensin, peptide HI, somatostatin (SOM), substance P (SP), neurokinin A (substance K) and vasoactive intestinal peptide (VIP). For many of these peptides, the molecular form of the antigen that is localized immunohistochemically has been determined[16]. Thus, immunohistochemistry has allowed many different populations of enteric neurons to be defined on the basis of their neuropeptide content. As well as being present in enteric neurons, neuropeptide immunoreactivity is also found in the processes of extrinsic neurons supplying the gut. In guinea pigs, noradrenergic nerves to gastrointestinal blood vessels contain NPY[32], and those to submucous ganglia of the small intestine contain SOM[11]; capsaicin-sensitive, presumably sensory, fibers of extrinsic origin contain SP and CGRP[40].

From whole mount preparations of myenteric and submucous plexuses of guinea pig ileum immunohistochemically labeled for single neuropeptides[9]; it became clear that there was a correlation between cell shape and neuropeptide content, at least for some neuropeptides. In control preparations that had not been treated with colchicine, neuropeptide immunoreactivity was intense in some enteric neurons so that their shapes could be described easily. Thus, intensely ENK-positive myenteric neurons had a typical Dogiel Type I morphology with a single long axon and many short dendrites[36], whereas NPY-immunoreactive myenteric neurons were Dogiel Type III with a long process and many intermediate length dedrites[32]. For other peptides, such as substance P, the shapes of the nerve cell bodies could not be ascertained because processes did not contain enough immunoreactivity to be visualized clearly. The whole mount technique for immunohistochemistry also made it possible to count the number of immunoreactive neurons in the ganglionated plexuses so that the proportion of neurons immunoreactive for a particular neuropeptide in either myenteric or submucous ganglia could be calculated[14].

The various types of neuropeptide-immunoreactive nerves in the ENS have different distributions[38]. Conversely, many different types of peptide-containing nerves are present in each of the enteric plexuses. For example, in the guinea pig deep muscular plexus there are dense networks of varicose fibers immunoreactive for ENK, GAL, GRP, DYN, NPY, SP, or VIP. In the small intestine, the mucosa is densely supplied with nerve fibers immunoreactive for SP, VIP and NPY. This suggested that many different populations of neurons, as defined by their content of a single peptide, may participate in the control of an effector such as the muscle or the mucosal epithelium. However, some types of peptide nerves are absent from some of the enteric plexuses. For instance, in most species, ENK neurons are absent from submucous ganglia and ENK fibers do not occur in the mucosal plexuses. SOM fibers, on the other hand, do not usually occur in the muscle plexuses. These differences in distribution suggested differences in function. Hence, ENK neurons are unlikely to be secretomotor since they are not associated with the mucosal epithelium. On the other hand, some are probably motor neurons to the muscle since their processes are found in the circular and deep muscular plexuses. SOM myenteric neurons do not appear to be motor neurons which affect the muscle directly but are probably interneurons for pathways travelling through the myenteric plexus since they have processes that supply other myenteric ganglia and form baskets around other myenteric neurons.

Neuropeptide immunohistochemistry suggested that there were multiple types of enteric neurons and that many of them supplied the same effectors. The complex array of enteric reflexes appeared to be mirrored by an equally complex array of different neurochemical types of enteric neurons. However, the finding of multiple neurochemical messengers in the same neuron (see the article by Hökfelt in this volume) suggested that the complexity of the ENS might be overestimated by examining material that was immunohistochemically treated to reveal only a single peptide.

Co-localization of neuropeptides in enteric neurons

The first evidence that enteric neurons contained more than one neuropeptide was presented in 1980 by Schultzberg et al.[62], who showed that, in proximal colon of guinea pig, SOM-immunoreactive submucous neurons also contained CCK. However, the extent of neuropeptide coexistence in submucous neurons became clear only when double labeling techniques with a variety of antisera were applied to guinea pig small intestine. Four populations of submucous neurons were identified on the basis of their immunoreactivity for VIP, NPY, CCK, SOM, SP and choline acetyltransferase (ChAT), the acetylcholine synthesizing enzyme[35]. About 45% of submucous neurons were immunoreactive for VIP and the remainder (about 55%) for ChAT, a finding that correlated well with the physiological observations that there were cholinergic and non-cholinergic secretomotor neurons[46]. The ChAT neurons were subdivided into 3 classes: those that contained NPY, CCK and SOM (29% of total neurons), those that contained SP (11% of neurons) and those that were not immunoreactive for any of the peptides studied (14% of neurons). Subsequently, the ChAT/NPY/SOM/CCK submucous neurons have been shown to contain CGRP and sometimes GAL[34,37] and the VIP neurons to contain DYN and GAL[14,37]. The different types of submucous neurons were topographically organized: in smaller submucous ganglia the VIP/DYN/GAL neurons occurred in groups and in larger ganglia the groups were usually located centrally with ChAT neurons surrounding them.

Individual myenteric neurons also contain several neurochemical markers. Colchicine has been important for demonstrating this in guinea pig myenteric neurons because colchicine treatment *in vitro* raises neuropeptide immunoreactivity to a detectable level in many myenteric nerve cell bodies[12]. In guinea pig small intestine, the percentages of nerve cell bodies that are immunoreactive for individual neuropeptides after colchicine implies extensive coexistence[14]. The patterns of neuropeptide coexistence, as shown by double immunofluorescent staining[13] or by localization of individual neuropeptides in serial semithin sections (fig. 1), are more complex in myenteric than in submucous neurons[30]. Consequently, myenteric neurons can be divided into many more than four types on the basis of their peptide content. Much of this complexity arises because several neurochemically-distinct populations of myenteric neurons contain the same neuropeptide. For example, in guinea pig small intestine there are four populations of myenteric neurons that contain VIP and DYN with other neuropeptides, VIP/DYN/GAL, VIP/DYN/GRP, VIP/DYN/ENK/NPY and VIP/DYN/ENK/GRP/CCK neurons, and at least two populations of SP neurons, SP and SP/ENK neurons. The different types of neurons are probably also topographically organized within myenteric ganglia. VIP/ENK neurons with Dogiel Type I morphology, for example, usually lie on the surfaces of the ganglia, mainly facing the circular muscle. Small Type I SP/ENK neurons often have their cells bodies where internodal strands join the ganglia.

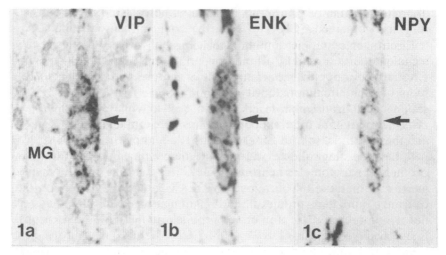

Figure 1. Serial semithin (1.5 μm) sections through a myenteric ganglion (MG) from guinea pig small intestine that was treated with colchicine *in vitro*. The sections were stained with antiserum to vasoactive intestinal peptide (VIP, fig. 1A), enkephalin (ENK, fig. 1B) or neuropeptide Y (NPY, fig. 1C). The arrowed myenteric neuron, which lies at the edge of the ganglion, is immunoreactive for all three neuropeptides. × 920.

The link between the neurochemistry of guinea pig myenteric neurons and the shapes of their somas, which was apparent for some neuropeptide-containing neurons in untreated tissue, has been strengthened by observations on colchicine-treated tissue double labeled to reveal more than one neuropeptide. The majority of myenteric neurons stained for neuropeptides after colchicine fall into one of Dogiel's morphological categories[30]. For instance, the VIP/DYN/ENK/NPY myenteric neurons have a typical Dogiel Type I morphology with many short stubby dendrites and one long axon. The VIP/DYN/ENK/GRP/CCK neurons and the VIP/DYN/GRP neurons are also Dogiel Type I. There are ChAT/NPY/CCK/SOM/CGRP neurons in the myenteric plexus as well as in the submucous plexus[34] and these have a Dogiel Type III morphology with intermediate length processes. VIP/DYN/GAL myenteric neurons are also Dogiel Type III.

Although most of the evidence for co-existence of neuropeptides in enteric neurons comes from studies on guinea pig ileum, there is also evidence for coexistence of neuropeptides in enteric neurons from other species[14]. However, the combinations of peptides that occur together are not always the same as in guinea pig small intestine. In rats and pigs, for example, NPY is present in submucous neurons that contain VIP[18] whereas VIP-containing and NPY-containing submucous neurons form separate populations in the guinea pig small intestine.

The ultrastructure of neuropeptide-immunoreactive enteric nerves

Electron microscopic immunocytochemistry is a very useful tool for studying the connectivity of neuropeptide-containing neurons in the ENS. It provides two important kinds of information that light microscopic methods cannot. Since non-immunoreactive structures can be seen as well as immunoreactive ones by electron microscopy, it is possible to account for total populations of nerve fibers. Since synapses or close contacts can be observed with the electron microscope, it is possible to distinguish pre and post-synaptic connections. Information on the ultrastructure of neuropeptide nerves, particularly in myenteric ganglia, is increasing[14] but most workers have examined enteric nerve cell bodies and nerve profiles that were immunoreactive for only a single neuropeptide. Since it is clear from light microscopic immunohistochemistry that enteric neurons contain multiple neuropeptides and that the same peptide can be present in several neurochemically-distinct populations of neurons, any ultrastructural study that deals with tissue labeled for a single neuropeptide may provide information on several different populations of enteric neurons.

The first electron microscopic immunocytochemical observations on enteric nerves were published in 1977 by Larsson[48], who demonstrated VIP-immunoreactive vesicles in nerve profiles in cat colon with a post-embedding staining technique. Since that time, there have been a number of ultrastructural investigations with pre- and post-embedding staining methods on VIP-, SP-, and ENK-positive nerve fiber profiles and nerve cell bodies in the enteric plexuses of various gut regions in several species[14]. SOM-[22,61], CGRP-[25] and NPY-containing[24] enteric neurons have also been examined at the electron microscope level. These studies describe the occurrence, location and appearance of immunoreactive nerve cell bodies and nerve fiber profiles, the distribution of immunoreactivity within the fibers and cell bodies and the types of vesicles the immunoreactive fibers contain.

Neuropeptide-immunoreactive nerve profiles have been shown sometimes to form synaptic contacts on the cell bodies or processes of enteric neurons (for example, fig. 2). In most cases, synapses were found in random ultrathin sections through immunocytochemically-stained tissue. With this technology, NPY-immunoreactive[24] and SOM-immunoreactive[22] synapses have been detected in guinea pig myenteric ganglia as have GRP- and DYN-immunoreactive synapses (own unpublished observations); VIP-immunoreactive synapses have been found in rat myenteric ganglia[26], and SP-immunoreactive synapses have been observed in human myenteric ganglia[53]. A more informative approach in terms of understanding ENS circuitry is to combine light and electron microscopy on the same tissue so that the location of synapses can be correlated with the distribution of immunoreactive fibers within a

254

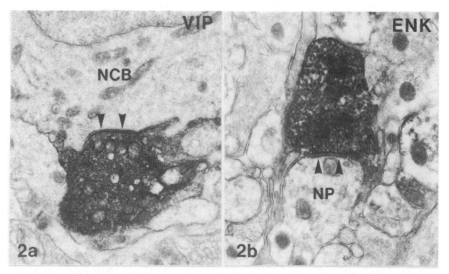

Figure 2. Electron micrographs showing synapses by neuropeptide-immunoreactive nerve fiber profiles in guinea pig myenteric ganglia. Figure 2a shows a VIP-immunoreactive nerve fiber profile forming a synapse (arrowheads) on a nerve cell body (NCB). × 24,000. Figure 2b shows an ENK-immunoreactive synapse (arrowheads) on a nerve process (NP). × 36,000.

ganglion. In two studies, baskets of immunoreactive fibers that were identified by light microscopy have been studied ultrastructurally. Synapses were found to occur where VIP-immunoreactive nerve fibers formed baskets around immunoreactive and non-immunoreactive nerve cell bodies in whole mount preparations of guinea pig myenteric ganglia[50] and when VIP-immunoreactive fibers formed baskets around submucous neurons in sections through rat submucous ganglia[55]. In guinea pig myenteric ganglia, ENK-immunoreactive nerve fibers in baskets also make synapses (own unpublished observations).

Only one study to date has provided a systematic investigation of enteric nerves immunoreactive for a single peptide[51]. Whole mount preparations were stained to reveal all nerve fibers immunoreactive for SP[50] and studied ultrastructurally. Since the distribution of SP-positive fibers in the ganglia is homogeneously dense by light microscopy, quantitative measurements were made on random ultrathin sections. At electron microscope level, two thirds of the vesicle-containing nerve fiber profiles in myenteric ganglia were found to be immunoreactive for SP. Only about 0.5% of the positive vesiculated profiles formed morphologically identifiable synaptic contacts, with clustering of vesicles pre-synaptically and membrane densities post-synaptically. However, almost half of the total number of synapses in the ganglia involved SP-immunoreactive profiles. SP synapses occurred near the surfaces of the ganglia on non-vesiculated nerve profiles and on small SP-positive and SP-negative nerve cell bodies. The cell bodies with synapses often had processes that reached the surfaces of the ganglia or lay near the junctions of internodal strands and ganglia. The cell bodies of motor neurons to the muscle have similar characteristics so SP synapses may be involved in the control of gastrointestinal motility.

Electron microscopic techniques have been used to demonstrate the co-existence of neuropeptides in the same enteric nerve fiber profile. Larsson and Stengaard-Pedersen[49] showed that met- and leu-ENK were co-localized in nerve fibers in the gastrointestinal tract of cats by correlating fluorescent antibody labeling and ultrastructure on adjacent ultrathin sections. Uchida et al.[66] found met ENK-Arg-Gly-Leu to be present in some SP- and some PHI-nerve fibers in the enteric plexuses of guinea pig duodenum using ultrathin sections labeled with protein A-gold. Llewellyn-Smith et al.[52] examined the extent of co-existence of SP, VIP and ENK in nerve fibers in the circular muscle of guinea pig ileum. Whole mount preparations were labeled with anti-ENK, anti-SP or anti-VIP sera, with pairs of antisera or with all three antisera together and then the number of immunoreactive nerve fiber profiles were counted for each treatment. Simultaneous application of antisera to SP and VIP labeled 95% of vesicle-containing nerve fiber profiles as did staining with all three antisera together. Statistical analysis indicated that SP- and VIP-immunoreactive profiles occurred in equal numbers and that the two peptides were not co-localized. ENK was always co-localized with either SP or VIP, being present in about two-thirds of the SP profiles and about two-thirds of the VIP profiles. Since there is good evidence that SP is an excitatory transmitter and VIP an inhibitory transmitter to circular muscle, these results imply that SP marks the excitatory fibers and VIP the inhibitory fibers to this smooth muscle layer.

Projections of neuropeptide-immunoreactive enteric nerves

The development of lesioning techniques which allow intrinsic nerve pathways to be analyzed was another significant step towards understanding the microcircuitry of the ENS[28,29]. To define the projections of myenteric neurons, cuts are made around the circumference of the intestine through the longitudinal muscle and myenteric plexus (myotomy). If there are two cuts, the myenteric plexus can be left intact between them (double myotomy) or removed (myectomy). After these operations, immunoreactive material accumulates in the cut stumps of nerve fibers that are still attached to their cell bodies whereas nerve fibers severed from their cell bodies degenerate. In whole mount preparations of separated gut layers from operated animals processed for immunohistochemistry, the accumulations of immunoreactivity and the disappearance of fibers allow the polarity, lengths and minimum and maximum areas of innervation of myenteric neurons to be established. To aid in the study of the

Table 1. Projections of enteric neurons immunoreactive for single neuropeptides in guinea pig small intestine

Neuropeptide	Cell body location	Projection
VIP	Myenteric ganglia	Anally to other myenteric ganglia Anally through the myenteric plexus and then to submucous ganglia To prevertebral ganglia Anally through the myenteric plexus and then to circular muscle Directly to circular muscle
	Submucous ganglia	To other submucous ganglia To the mucosa To submucous blood vessels
DYN	Myenteric ganglia	Anally to other myenteric ganglia Anally through the myenteric plexus and then to submucous ganglia To prevertebral ganglia Anally through the myenteric plexus and then to circular muscle Directly to circular muscle
	Submucous ganglia	To the mucosa
CCK	Myenteric ganglia	Anally to other myenteric ganglia Directly to submucous ganglia To prevertebral ganglia Directly to the mucosa
	Submucous ganglia	Directly to the mucosa
ENK	Myenteric ganglia	Orally to other myenteric ganglia Directly to submucous ganglia To prevertebral ganglia Orally to the circular muscle Anally to the circular muscle
GAL	Myenteric ganglia	Anally to other myenteric ganglia Directly to circular muscle
	Submucous ganglia	To other submucous ganglia To the mucosa To submucous blood vessels
SOM	Myenteric ganglia	To local myenteric ganglia Anally to other myenteric ganglia Directly to submucous ganglia Directly to the mucosa
	Submucous ganglia	To the mucosa
SP	Myenteric ganglia	To local myenteric ganglia Directly to submucous ganglia Directly to circular muscle Directly to the mucosa
	Submucous ganglia	To the mucosa

Table 1. (*continued*)

Neuropeptide	Cell body location	Projection
GRP	Myenteric ganglia	Anally to other myenteric ganglia Anally through the myenteric plexus and then to submucous ganglia To prevertebral ganglia Anally through the myenteric plexus and then to circular muscle
NPY	Myenteric ganglia	To ganglion of origin and more anal myenteric ganglia Directly to circular muscle Directly to the mucosa
	Submucous ganglia	To the mucosa
CCRP	Myenteric ganglia	Directly to the mucosa
	Submucous ganglia	To the mucosa

projections of submucous neurons, a short segment of gut is completely severed from and then rejoined to the rest of the intestine (homotopic autotransplant) and, after a few days, processed for immunohistochemistry[47]. The projections of enteric neurons outside the gastrointestinal tract are studied after disruption of pathways traveling to and from the small intestine through the mesenteric nerves (extrinsic denervation) or by retrograde transport.

These lesioning methods have been used most extensively for investigating the projections of chemically-identified neurons in the guinea pig ileum. Similar studies are now beginning to appear on neuropeptide-immunoreactive enteric neurons in rats[19–21] and dogs[15]. In guinea pig small intestine, immunoreactivity for a single neuropeptide is observed in several neuronal projections[14,30,37] (Table 1). For example, VIP immunoreactivity seems to occur in 8 different types of enteric neuron[10,14,28,30,37]. Myenteric VIP neurons have three projections to ganglia: to other myenteric ganglia located more anally, to submucous ganglia after traveling anally through the myenteric plexus for several millimeters and to prevertebral ganglia. There are two projections of myenteric VIP neurons to the circular muscle: directly to circular muscle near the ganglion of origin or to circular muscle after traveling anally for several millimeters through the myenteric plexus. VIP-immunoreactive submucous neurons send processes to the mucosa, to other submucous ganglia and to submucosal arterioles. Of the other neuropeptide-containing neurons in the ENS, DYN-immunoreactive enteric neurons have been shown to have 6 different projections within the guinea pig small intestine; CCK-, ENK-, GAL-, SOM- and SP-immunoreactive neurons each have 5 different projections; GRP and NPY neurons each have 4 different projections; and

CGRP neurons have 2 projections. Since in lesion experiments the connections between cell bodies and terminals are deduced from patterns of degeneration, it cannot always be ascertained whether a particular nerve cell supplies a single target or whether some types of neuron could send collaterals to several different targets. This question is beginning to be answered through lesion experiments in which results from double immunofluorescent labeling are cross-correlated. For all the cases studied so far, it has been found that enteric neurons containing a specific combination of neuropeptides have a well-defined projection, sending their axons in a specific direction for a specific distance to a specific target[13]. For example, the VIP/DYN/ENK/NPY myenteric neurons of the guinea pig small intestine are the VIP neurons that project directly to the circular muscle supplying varicose axons to the circular muscle and deep muscular plexuses. The VIP/DYN/ENK/GRP/CCK myenteric neurons are the population that send their axons to the prevertebral ganglia where baskets are formed around some sympathetic neurons[54]. The VIP/DYN/GRP neurons have processes that travel anally through the myenteric plexus before sypplying the circular muscle. To date, the branching pattern of only one type of neuropeptide-immunoreactive enteric neuron, the ChAT/NPY/CCK/SOM/CGRP neurons with cell bodies in both myenteric and submucous ganglia, has been visualized in its entirety[34]. This neuron type supplies terminals to only a single target, the mucosal epithelium.

Lesion experiments on guinea pig small intestine have also been valuable for defining the locations of the cell bodies of motor neurons. The motor neurons to the circular muscle must have their cell bodies in myenteric ganglia since no nerve fibers are present in the circular muscle of myectomized, extrinsically-denervated animals on ultrastructural examination[67]. The cell bodies of the secretomotor neurons lie primarily in the submucous ganglia since the density of terminals in the mucosal plexus is not altered significantly after homotopic autotransplant, myectomy or extrinsic denervation[47].

Neuropeptide content and electrophysiology

A necessary step in the analysis of enteric circuitry is to define the functions of neurochemically-distinct populations of enteric neurons.

One method for doing this is to combine electrophysiology with immunohistochemistry (for example, fig. 3). Intracellular microelectrodes are used to record the properties of enteric neurons, which are then filled with dye[43] and processed to reveal their neuropeptide content[6,45]. This approach has shown that in both myenteric and submucous plexuses the peptide content of neurons correlates with the synaptic inputs they receive.

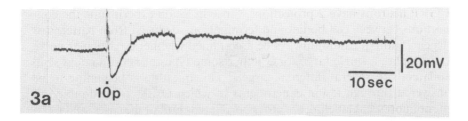

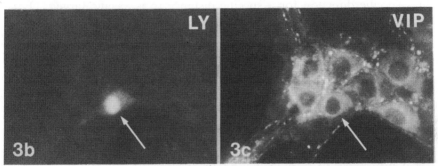

Figure 3. Correlation of electrophysiology and immunohistochemistry in a submucous neuron from guinea pig small intestine. Figure 3a shows an electrophysiological recording of an inhibitory synaptic potential and a slow excitatory synaptic potential in a submucous neuron after a stimulus of 10 pulses. The recording electrode contained Lucifer Yellow (LY) and 0.5 M KCl. The neuron was filled with LY after electrophysiological characterization and processed for immunohistochemistry to reveal VIP. The fluorescence micrographs show that the LY-filled neuron (arrow in fig. 3b) is immunoreactive for VIP (arrow in fig. 3c). The trace and micrographs were kindly supplied by Dr Joel Bornstein.

Myenteric neurons from guinea pig small intestine can be classified into two types on the basis of their responses to stimulation. S (for synaptic) neurons receive inputs through which fast excitatory synaptic potentials (esps) are mediated; AH (for after-hyperpolarizing) neurons rarely have fast esps but show hyperpolarizations that last for many seconds following action potentials in their somas[41,58]. After electrophysiological characterization and injection of fluorescent dye[6] or horseradish peroxidase[23], most S cells have been found to have Dogiel Type I morphologies whereas AH neurons were found to be Dogiel Type II. Furthermore, about half of the S cells in an electrophysiologically-identified sample of myenteric neurons were found to be immunoreactive for ENK whereas AH cells were always ENK-negative[6]. VIP-immunoreactivity is also absent from AH cells but some S cells contain this peptide[45]. Since VIP and ENK co-exist in many Dogiel Type I myenteric neurons, it seems likely that these two studies were examining overlapping populations of S cells.

Almost all submucous neurons have fast esps and a proportion also show inhibitory synaptic potentials (isps) and slow esps[42,65]. Isps were

260

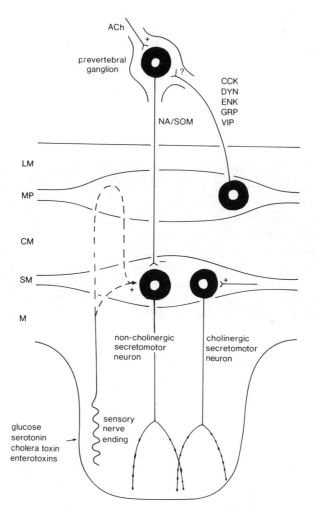

Figure 4. Secretomotor pathways in the intestine deduced from physiological, pharmacological, electrophysiological and immunohistochemical evidence. The chemical coding of the noradrenergic and intestinofugal neurons applies to the guinea pig small intestine.

Exposure of the mucosa (M) to glucose, serotonin, cholera toxin or enterotoxins stimulates sensory nerve endings. A reflex is initiated that activates primarily the non-cholinergic secretomotor neurons, whose cell bodies, along with those of the cholinergic secretomotor neurons, are concentrated in the submucous plexus (SM). The intrinsic neurons that are involved in the transmission of the signal from the mucosa to the secretomotor neurons have not been identified. Noradrenergic neurons in prevertebral ganglia tonically suppress the non-cholinergic secretomotor neurons. Neurons in the myenteric plexus (MP) project to the prevertebral ganglia where they are presumed to excite the noradrenergic neurons. LM, longitudinal muscle; CM, circular muscle, ACh, acetylcholine; CCK, cholecystokinin; DYN, dynorphin; ENK, enkephalin; GRP, gastrin-releasing peptide; NA, noradrenaline; SOM, somatostatin; VIP, vasoactive intestinal peptide. Reproduced with permission from Furness and Costa[30].

found almost exclusively in VIP-immunoreactive neurons (which are non-cholinergic, see above) when dye-filled, characterized submucous neurons were processed to reveal neuropeptide immunoreactivity[4]. VIP neurons also had fast esps and usually slow esps as well. The source of the isp-producing fibers has been unclear until recently. The fibers producing the isps were originally thought to arise from cell bodies in the enteric ganglia because some submucous neurons showed isps after extrinsic denervation[42]. However, isps in submucous neurons could be blocked pharmacologically by guanethidine or alpha adrenoreceptor antagonist[42,59], suggesting that the isps were initiated by the extrinsic sympathetic nerves. It is now known from experiments after myectomy and extrinsic denervation that both intrinsic fibers and extrinsic noradrenergic fibers can cause isps in VIP-immunoreactive submucous neurons[5]. Unlike VIP neurons, NPY submucous neurons (which are cholinergic, see above) do not have inhibitory inputs but do show fast esps[4]. After removal of the myenteric plexus, some of the fast inputs to submucous neurons persist[7], indicating that some of the fast esps on submucous neurons arise from myenteric neurons and some come from other sub-mucous neurons.

There is now good physiological evidence that intrinsic secretomotor reflexes control water and electrolyte transport across the mucosal epithelium[8,63,64]. VIP neurons, which enhance secretion, are involved and there is an inhibition of secretion by sympathetic nerves. These findings, in conjunction with the anatomical and electrophysiological results summarized above, allow a partial circuit diagram to be drawn for the neuronal control of secretomotor reflexes[30,46] (fig. 4).

Concluding remarks

Analysis of the microcircuitry of the enteric nervous system is at an exciting stage. Light microscopical immunohistochemistry for neuro-transmitter-specific markers, including neuropeptides, has revealed the existence of many chemically-distinct populations of enteric neurons. Neuropeptide-immunoreactive nerves that have been selected by light microscopy in whole mount preparations can be studied electron microscopically and quantitative methods are beginning to be used to define their synaptology and neuroeffector relationships. Lesioning studies have indicated that each of the neurochemically-distinct populations of enteric neurons has a precise projection to a specific target following a specific path. Immunohistochemistry is beginning to be used in conjunction with electrophysiology to show the relationship between the neurochemistry of an enteric neuron and its synaptic input. The data gained from all these techniques make it clear that enteric neurons are chemically-coded and indicate that the neurochemical coding of an enteric neuron may correlate with its position in the intrinsic circuitry of the ENS. By combining

262

the information gained from the kinds of anatomical and electrophysiological studies discussed in this article with physiological data, it is becoming possible to establish the positions of the different populations of enteric neurons within neuronal circuits in the ENS and thereby define their roles in the pathways that control gastrointestinal function.

Acknowledgments. The financial support of the National Health and Medical Research Council of Australia is gratefully acknowledged. Profs John Furness and Marcello Costa and Drs Joel Bornstein, Judy Morris and Roger Murphy provided helpful ciriticisms of the manuscript.

1 Auerbach, L., Fernere vorlaufige Mittheilung uber den Nervenapparat des Darmes. Arch. Pathol. Anat. Physiol. *30* (1864) 457–460.
2 Bayliss, W. M., and Starling, E. H., The movements and innervation of the small intestine. J. Physiol. *24* (1899) 100–143.
3 Bayliss, W. M., and Starling, E. H., The movements and innervation of the small intestine. J. Physiol. *26* (1900) 125–138.
4 Bornstein, J. C., Costa, M., and Furness, J. B., Synaptic inputs to immunohistochemically identified neurones in the submucous plexus of the guinea-pig small intestine. J. Physiol. *381* (1986) 465–482.
5 Bornstein, J. C., M., and Furness, J. B., Intrinsic and extrinsic inhibitory synaptic inputs to submucous neurones of the guinea-pig small intestine. J. Physiol. *398* (1988) 371–390.
6 Bornstein, J. C., Costa, M., Furness, J. B., and Lees, G. M., Electrophysiology and enkephalin immunoreactivity of identified myenteric plexus neurones of guinea-pig small intestine. J. Physiol. *351* (1984) 313–325.
7 Bornstein, J. C., Furness, J. B., and Costa, M.. Sources of excitatory synaptic inputs to neurochemically identified submucous neurons of guinea-pig small intestine. J. autonom. nerv. Sys. *18* (1986) 83–91.
8 Cassuto, J., Siewert, A., Jodal, M., and Lundgren, O., The involvement of intramural nerves in cholera toxin induced intestinal secretion. Acta physiol. scand. *117* (1983) 195–202.
9 Costa, M., Buffa, R., Furness, J. B., and Solcia, E., Immunohistochemical localization of polypeptides in peripheral autonomic nerves using whole mount preparations. Histochemistry *65* (1980) 157–165.
10 Costa, M., and Furness, J. B., The origins, pathways and terminations of neurons with VIP-like immunoreactivity in the guinea-pig small intestine. Neuroscience *8* (1983) 665–676.
11 Costa, M., and Furness, J. B., Somatostatin is present in a subpopulation of noradrenergic nerve fibers supplying the intestine. Neuroscience *13* (1984) 911–920.
12 Costa, M., Furness, J. B., and Cuello, A. C., Separate populations of opioid containing neurons in the guinea-pig intestine. Neuropeptides *5* (1985) 445–448.
13 Costa, M., Furness, J. B., and Gibbins, I. L., Chemical coding of enteric neurons. Prog. Brain Res. *68* (1986) 217–239.
14 Costa, M., Furness, J. B., and Llewellyn-Smith, I. J., Histochemistry of the enteric nervous system in: Physiology of the Gastrointestinal Tract, 2nd Edn, pp. 1–41. Ed. L. R. Johnson. Raven Press, New York 1986.
15 Daniel, E. E., Furness, J. B., Costa, M., and Belbeck, L., The projections of chemically identified nerve fibers in canine ileum, Cell Tiss. Res. *247* (1987) 377–384.
16 Dockray, G. J., Physiology of enteric neuropeptides, in: Physiology of the Gastrointestinal Tract, 2nd Edn, pp. 42–63. Ed. L. R. Johnson. Raven Press, New York 1986.
17 Dogiel, A. S., Ueber den Bau der Ganglien in den Geflechten des Darmes und der Gallenblase des Menschen und der Säugetiere. Arch. Anat. Physiol. Leipzig, Anat. Abt. (Jg 1899) 130–158.
18 Ekblad, E., Håkanson, R., and Sundler, F., VIP and PHI coexist with an NPY-like peptide in intramural neurones of the small intestine. Reg. Pep. *10* (1984) 47–55.
19 Ekblad, E., Ekman, R., Håkanson, R., and Sundler, F., GRP neurones in the rat small intestine issue long anal projections. Reg. Pep. *9* (1984) 279–287.

20 Ekblad, E., Rökaeus, A., Håkanson, R., and Sundler, F., Galanin nerve fibers in the rat gut: Distribution, origin and projections. Neuroscience 16 (1985) 355–363.

21 Ekblad, E., Winther, C., Ekman, R., Håkanson, R., and Sundler, F., Projections of peptide-containing neurons in rat small intestine. Neuroscience 20 (1987) 169–188.

22 Endo, Y., Uchida, T., and Kobayashi, S., Somatostatin neurons in the small intestine of the guinea pig: A light and electron microscopic immunocytochemical study combined with nerve lesion experiments by laser irradiation. J. Neurocytol. 15 (1986) 725–731.

23 Erde, S. M., Sherman, D., and Gershon, M. D., Morphology and serotonergic innervation of physiologically identified cells of the guinea pig's myenteric plexus. J. Neurosci. 5 (1985) 617–633.

24 Fehér, E., and Burnstock, G., Electron microscopic study of neuropeptide Y-containing nerve elements of the guinea-pig small intestine. Gastroenterology 91 (1986) 956–965.

25 Fehér, E., Burnstock, G., Varndell, I. M., and Polak, J. M., Calcitonin gene-related peptide-immunoreactive nerve fibers in the small intestine of the guinea-pig: electron-microscopic immunocytochemistry. Cell Tiss. Res. 245 (1986) 353–358.

26 Fehér, E., and Léránth, C., Light and electron microscopic immunocytochemical localization of vasoactive intestinal polypeptide VIP-like activity in the rat small intestine. Neuroscience 10 (1983) 97–106.

27 Furness, J. B., and Costa, M., The adrenergic innervation of the gastrointestinal tract. Ergebn. Physiol. 69 (1974) 1–51.

28 Furness, J. B., and Costa, M., Projections of intestinal neurons showing immunoreactivity for vasoactive intestinal polypeptide are consistent with these neurons being the enteric inhibitory neurons. Neurosci. Lett. 15 (1979) 199–204.

29 Furness, J. B., and Costa, M., Types of nerves in the enteric nervous system. Neuroscience 5 (1980) 1–20.

30 Furness, J. B., and Costa, M., The Enteric Nervous System. Churchill Livingstone, Edinburgh 1987.

31 Furness, J. B., and Costa, M., Identification of transmitters of functionally defined enteric neurons. Hb. Physiol. (in press) 1988.

32 Furness, J. B., Costa, M., Emson, P. C., Håkanson, R., Moghimzadeh, E., Sundler, F., Taylor, I. L., and Chance, R. E., Distribution, pathways and reactions to drug treatment of nerves with neuropeptide Y- and pancreatic polypeptide-like immunoreactivity in the guinea-pig digestive tract. Cell Tiss. Res. 234 (1983) 71–92.

33 Furness, J. B., Costa, M., Franco, R., and Llewellyn-Smith, I. J., Neuronal peptides in the intestine: Distribution and possible functions, in: Neural Peptides and Neuronal Communication (Adv. Biochem. Pharmac., vol. 22), pp. 601–617. Eds E. Costa and M. Trabucchi. Raven Press, New York 1980.

34 Furness, J. B., Costa, M., Gibbins, I. L., Llewellyn-Smith, I. J., and Oliver, J. R., Neurochemically similar myenteric and submucous neurons directly traced to the mucosa of the small intestine. Cell Tiss. Res. 241 (1985) 155–163.

35 Furness, J. B., Costa, M., and Keast, J. R., Choline acetyltransferase- and peptide immunoreactivity of submucous neurons in the small intestine of the guinea-pig. Cell Tiss. Res. 237 (1984) 329–336.

36 Furness, J. B., Costa, M., and Miller, R. J., Distribution and projections of nerves with enkephalin-like immunoreactivity in the guinea-pig small intestine. Neuroscience 8 (1983) 653–664.

37 Furness, J. B., Costa, M., Rökaeus, A., McDonald, T. J., and Brooks, B., Galanin-immunoreactive neurons in the guinea-pig small intestine: Their projections and relationships to other enteric neurons. Cell Tiss. Res. 250 (1987) 607–615.

38 Furness, J. B., Llewellyn-Smith, I. J., Bornstein, J. C., and Costa, M., Chemical neuroanatomy and analysis of neuronal circuitry in the enteric nervous system, in: Handbook of Chemical Neuroanatomy, in press. Eds C. Owman, A. Bjorklund and T. Hökfelt. Elsevier, Amsterdam 1988.

39 Gabella, G., Innervation of the gastrointestinal tract. Int. Rev. Cytol. 59 (1979) 130–193.

40 Gibbins, I. L., Furness, J. B., Costa, M., MacIntyre, I., Hillyard, C. J., and Girgis, S., Co-localization of calcitonin gene-related peptide-like immunoreactivity with substance P in cutaneous, vascular and visceral sensory neurons of guinea-pigs. Neurosci. Lett. 57 (1985) 125–130.

264

41 Hirst, G. D. S., Holman, M. E., and Spence, I., Two types of neurones in the myenteric plexus of duodenum in the guinea-pig. J. Physiol. *236* (1974) 303–326.

42 Hirst, G. D. S., and McKirdy, H. C., Synaptic potentials recorded from neurones of the submucous plexus of guinea-pig small intestine. J. Physiol. *249* (1975) 369–385.

43 Hodgkiss, J. P., and Lees, G. M., Morphological studies of electrophysiologically-identified myenteric plexus neurons of the guinea-pig ileum. Neuroscience *8* (1983) 593–608.

44 Hökfelt, T., Johansson, O., Efendic, S., Luft, R., and Arimura, A., Are there somatostatin-containing nerves in the rat gut? Immunohistochemical evidence for a new type of peripheral nerves. Experientia *31* (1975) 852–854.

45 Katayama, Y., Lees, G. M., and Pearson, G. T., Electrophysiology and morphology of vasoactive-intestinal-peptide-immunoreactive neurones of the guinea-pig ileum. J. Physiol. *378* (1986) 1–11.

46 Keast, J. R., Mucosal innervation and control of water and ion transport in the intestine. Rev. Physiol. Biochem. Pharmac. *109* (1987) 1–59.

47 Keast, J. R., Furness, J. B., and Costa, M., Orgins of peptide and norepinephrine nerves in the mucosa of the guinea pig small intestine. Gastroenterology *86* (1984) 637–644.

48 Larsson, L.-I., Ultrastructural localization of a new neuronal peptide (VIP). Histochemistry *54* (1977) 173–176.

49 Larsson, L.-I., and Stengaard-Pedersen, K., Immunocytochemical and ultrastructural differentiation between met-enkephalin-, leu-enkephalin-, and met/leu-enkephalin-immunoreactive neurons of feline gut. J. Neurosci. *2* (1982) 861–878.

50 Llewellyn-Smith, I. J., Costa, M., and Furness, J. B., Light and electron microscopic immunocytochemistry of the same nerves from whole mount preparations. J. Histochem. Cytochem. *33* (1985) 857–866.

51 Llewellyn-Smith, I. J., Furness, J. B., and Costa, M., Ultrastructural analysis of substance P-immunoreactive nerve fibers in myenteric ganglia of guinea-pig small intestine. J. Neurosci. (in press) 1988.

52 Llewellyn-Smith, I. J., Furness, J. B., Gibbins, I. L., and Costa, M., Quantitative ultrastructural analysis of enkephalin-, substance P, and VIP-immunoreactive nerve fibers in the circular muscle of the guinea-pig small intestine. J. comp. Neurol. *272* (1988) 139–148.

53 Llewellyn-Smith, I. J., Furness, J. B., Murphy, R., O'Brien, P. E., and Costa, M., Substance P-containing nerves in the human small intestine. Distribution, ultrastructure and characterization of the immunoreactive peptide. Gastroenterology *86* (1984) 421–435.

54 Macrae, I. M., Furness, J. B., and Costa, M., Distribution of subgroups of noradrenaline neurons in the coeliac ganglion of the guinea-pig. Cell Tiss. Res. *244* (1986) 173–180.

55 Maeda, M., Takagi, H., Kubota, Y., Morishima, Y., Akai, F., Hashimoto, S., and Mori, S., The synaptic relationship between vasoactive intestinal polypeptide (VIP)-like immunoreactive neurons and their axon terminals in the rat small intestine: Light and electron microscopic study. Brain Res. *329* (1985) 356–359.

56 Meissner, G., Ueber die Nerven der Darmwand. Z. Ration. Med. N.F. *8* (1857) 364–366.

57 Nilsson, G., Larsson, L.-I., Håkanson, R., Brodin, E., Pernow, B., and Sundler, F., Localization of substance P-like immunoreactivity in mouse gut. Histochemistry *11* (1975) 97–99.

58 Nishi, S., and North, R. A., Intracellular recording from the myenteric plexus of the guinea-pig ileum. J. Physiol. *231* (1973) 471–491.

59 North, R. A., and Surprenant, A., Inhibitory synaptic potentials resulting from α_2-adrenoreceptor activation in guinea-pig submucous plexus neurones. J. Physiol. *358* (1985) 17–33.

60 Pearse, A. G. E., and Polak, J. M., Immunocytochemical localization of substance P in mammalian intestine. Histochemistry *41* (1975) 373–375.

61 Probert, L., De Mey, J., and Polak, J. M., Ultrastructural localization of four different neuropeptides within separate populations of p-type nerves in the guinea-pig colon. Gastroenterology *85* (1983) 1094–1104.

62 Schultzberg, M., Hökfelt, T., Nilsson, G., Terenius, L., Rehfeld, J. F., Brown, M., Elde, R., Goldstein, M., and Said, S., Distribution of peptide- and catecholamine-containing neurons in the gastro-intestinal tract of rat and guinea-pig: Immunohistochemical studies with antisera to substance P, vasoactive intestinal polypeptide, enkephalins, somatostatin, gastrin/cholecystokinin, neurotensin and dopamine β-hydroxylase. Neuroscience *5* (1980) 689–744.

63 Sjövall, H., Jodal, M., and Lundgren, O., Further evidence for a glucose-activated secretory mechanism in the jejunum of the cat. Acta physiol. scand. *120* (1984) 437–443.

64 Sjövall, H., Redfors, S., Jodal, M., and Lundgren, O., On the mode of action of the sympathetic fibers on intestinal fluid transport: Evidence for the existence of a glucose-stimulated secretory nervous pathway in the intestinal wall. Acta physiol. scand. *119* (1983) 39–48.

65 Surprenant, A., Slow excitatory synaptic potentials recorded from neurones of guinea-pig submucous plexus. J. Physiol. *351* (1984) 343–361.

66 Uchida, T., Kobayashi, S., and Yanaihara, N., Occurrence and projections of three subclasses of met-enkephalin-Arg^6-Gly^7-Leu^8 neurons in the guinea-pig duodenum: Immunoelectron microscopic study on the co-storage of met-enkephalin-Arg^6-Gly^7-Leu^8 with substance P or PHI (1–15). Biomed. Res. *6* (1985) 415–422.

67 Wilson, A. J., Llewellyn-Smith, I. J., Furness, J. B., and Costa, M., The source of the nerve fibers forming the deep muscular and circular muscle plexuses in the small intestine of the guinea-pig. Cell Tiss. Res. *247* (1987) 497–504.

Neuropeptides and the ocular innervation

R. A. Stone, Y. Kuwayama and A. M. Laties

Summary. The eye is designed to provide the retina with protection and nutrition as well as a focussed image. The present review concentrates on regulatory peptides in nerves to the supportive structures of the eye. The retina itself will be discussed but briefly. To understand the innervation of the eye requires an overview of ocular physiology and anatomy.

Structure of the eye

The eye is made up of three coats: a tough, outer collagenous coat of cornea and sclera; an intermediate vascular coat, the uvea; and an inner epithelial coat, highly differentiated by region. The transparent and convex cornea not only contributes to the form of the eye but also serves as a powerful lens. It requires protection and nutrition. A multicomponent tear film secreted by the lacrimal gland and by small glands in the eyelids wets and lubricates the corneal surface. A rich vascular system supplies the junction of cornea and sclera, the limbus, and provides nutrition to the cornea. Just deep to the corneo-scleral limbus are the filtration tissues responsible for the drainage of aqueous humor from the anterior chamber of the eye into the limbal blood vessels. The secretion of aqueous humor generates intraocular pressure and maintains the shape of the eye.

The densely pigmented middle or uveal coat contains many specialized structures. Within the iris are the sphincter and dilator muscles to regulate pupillary diameter. Just behind the iris, the ciliary body contains the ciliary muscles and the ciliary processes. Acting through zonules connected to the equatorial region of the lens, the ciliary muscle controls accommodation. The innermost part of the ciliary body is thrown up into heavily vascularized, epithelial-lined radial projections termed ciliary processes; these secrete the aqueous humor. In combination, the secretion of aqueous humor by the ciliary processes and the resistance to its drainage in the chamber angle regulate the internal pressure of the eye. The aqueous humor also provides nutrition to the lens and cornea. Posteriorly, the uvea is called the choroid. A heavily vascularized tissue, it provides blood flow so vastly in excess of the nutritional requirement of the retina that it not only guarantees a wide margin of safety but also maintains the retinal temperature within strict limits.

The third or epithelial layer of the eye has remarkable regional differences. At first, it is two cells deep and results from the invagination of the optic vesicle during embryogenesis. At the front of the eye, one layer differentiates into the iris smooth muscle cells; the other is heavily pigmented and lines the back surface of the iris. The two cell layers continue back onto the ciliary body where they are responsible for secretion of aqueous humor. Posteriorly, the outer cell layer is called the retinal pigmented epithelium while the inner differentiates into the neurosensory retina. The retinal pigmented epithelium separates the choroidal circulation and the neurosensory retina. It provides nutrition to the outer retinal layers, engulfs shed outer segment discs and constitutes an essential part of the blood-retinal barrier. The mature neurosensory retina contains three layers of neurons. The outermost produce photosensitive outer segments. Neural signalling is influenced greatly by laterally integrated neurons, termed horizontal and amacrine cells, in the middle layer. The ganglion cells at the inner layer issue the axons responsible for transmission to brain. In keeping with the optical needs of the eye, the retina is transparent while beyond it lie two types of pigmented cells, the pigmented epithelium as well as the melanocytes in the choroid. Both absorb stray light to reduce random reflection and thus prevent glare.

Sources of peripheral innervation to the eye

The peripheral nerves to the eye derive from several sources. The parasympathetic innervation is the most complex. It includes postganglionic nerves of the ciliary ganglion which provide the eye with important parasympathetic supply to the ciliary and iris muscles. It also includes the pterygopalatine ganglion, from which postganglionic nerve fibers enter the orbit as *rami orbitales*. Intraocular nerve fibers from the pterygopalatine ganglion innervate chiefly the choroid[106]. In addition, microganglia and ganglion cells are scattered within the orbit and are even found occasionally within the eye itself[17,84,129,139]. The number and location of these ill-defined neural clusters vary among species and even among individuals within a species. They are mainly located in such places as the third nerve near the ciliary ganglion, the optic nerve sheath, the ciliary nerves near their penetration into the sclera, the choroid and the iris[63,133]. Sympathetic nerves to the eye are supplied almost exclusively from the ipsilateral superior cervical ganglion. Sensory nerves to the eye derive from the trigeminal ganglion. Those from the first division enter the orbit within the ophthalmic nerve. The maxillary nerve also makes a contribution to the sensory component of the ciliary nerves in monkeys[103]. Most likely, it does the same in man as well[4].

Cholinergic innervation to the eye

The thiocholine technique for acetylcholinesterase[59], though indirect, remains the only histochemical method used to date for the demonstration of cholinergic nerves in the eye[26,70]. Because this technique demonstrates the enzyme that metabolizes acetylcholine instead of the neurotransmitter itself, the method is indirect. Immunohistochemical techniques using antisera against both choline acetyltransferase or acetylcholine itself have been developed and promise greater specificity, but these newer methods have not yet been applied successfully to peripheral ocular nerves. The issue of the specificity of the thiocholine technique is complex. Perhaps it is best simply to assert that it has proved to have practical value in ocular studies. Specifically, findings with it correlate well both with physiological responses and biochemical measurements of choline acetyltransferase[85]. On this basis, it is believed to provide an accurate description of cholinergic innervation within the eye. Interestingly, the thiocholine technique works best in albino animals, particularly the albino rabbit. The presence of melanin in tissue sections tends to dampen the histochemical reaction for cholinesterase and thus limits the effectiveness of this method for pigmented eyes.

In the cornea, nerve bundles show acetylcholinesterase activity, chiefly in the middle and anterior stroma and in the basal layers of the epithelium where, in favorable tissue sections, multiple fine beaded fibers are seen. With regard to the drainage angle for aqueous humor, the friability of cryostat-cut sections in this region makes it difficult to preserve tissue relationships. To add to the difficulty, considerable species variation has been observed in density of cholinergic nerves to this area[70]. In spite of these problems, in the chamber angle of rabbit for instance, a moderate number of cholinesterase reactive nerves have been seen. For monkey and cat, however, only occasional reactive fibers are visualized; no studies have been reported in man.

For ciliary body, the issue of innervation of the ciliary muscle must be separated from that to the ciliary processes. Innervation of the former relates to accommodation; of the latter, to aqueous humor formation. For the ciliary muscle, the density of acetylcholinesterase enzyme is so great that individual nerve fibers can be distinguished only at short incubation times. For the ciliary processes, pigmentation is important. Cholinesterase positive nerve fibers are found in each and every ciliary process of the albino rabbit; but in pigmented rabbits, rats and monkeys, a careful search must be undertaken to visualize just a few nerve fibers in the same region. This apparent discrepancy most likely is explained by melanin dampening of the histochemical reaction rather than a true difference in innervation density.

For the iris, the thiocholine technique reveals a rich network of cholinesterase positive nerve fibers in the sphincter muscle, a parallel to the ciliary muscle. In the dilator muscle, the reaction indicates far fewer nerve fibers than in the sphincter, but it importantly still indicates a dual innervation to a muscle once presumed to be served exclusively by sympathetic nerves.

In the choroid[26,51], histochemical studies have revealed cholinesterase positive nerve fibers surrounding large and medium-sized arterioles. Nerve fibers also run within the choroidal stroma and, on occasion, are seen at the choriocapillaris.

Adrenergic innervation to the eye

Of all the components of the ocular innervation, the adrenergic is the most extensively studied. Since the introduction of the histofluorometric method for tissue localization of catecholamines[38], numerous reports have appeared on diverse species[27,28,70]. The adrenergic nerves within the eye derive from sympathetic nerve fibers originating in the ipsilateral superior cervical ganglion[33]. Within species, observed distributions of nerve fibers are relatively constant, implying a uniform innervational complement. While considerable variations among species have been recorded, some general observations can be made.

As a rule, innervation is plentiful to blood vessels within the eye, but the retina presents an exception. The dense adrenergic innervation to the central retinal artery drops markedly as the vessel enters the globe[68]. A few adrenergic nerve fibers are seen surrounding blood vessels on and near the optic nerve head within the eye, but the rest of the retinal vasculature is devoid of adrenergic innervation[68]. Some have taken exception and dispute this observation[40,41]; the disagreement may reflect species differences.

In the ciliary body, the ciliary muscle has a modest adrenergic innervation that takes part in the relaxation of accommodation. Each ciliary process ordinarily contains a visible array of adrenergic nerves, providing an anatomical counterpart to the adrenergic sensitivity of its secretory function. From the results of tissue assays, it is generally held that the adrenergic innervation of the anterior segment of the eye is noradrenergic. However, even though assays show a 20 : 1 ratio of noradrenalin to dopamine, there are indications that the matter may be more complex. For instance, the recognition of a D-1 receptor-related phosphoprotein, DARPP-32, in the non-pigmented epithelium of the ciliary process might well hint at a role for dopamine at this particular location[126]. A sparse but definite adrenergic innervation serves the junction of choroid and sclera at the lamina fusca.

In the iris, the radially directed dilator muscle is heavily innervated by adrenergic nerve fibers and is very responsive to catecholamines. As already noted, the thiocholine technique gives presumptive evidence for dual, albeit unbalanced, innervation to the dilator. The scant but definite adrenergic innervation to the opposing pupillary sphincter muscle, taken with the results of cholinesterase staining, thus indicates a dual and reciprocal adrenergic/cholinergic innervation to both iris muscles.

Surprisingly, melanocytes within the eye are innervated[31,53] even though outside the eye this is a rarity. Especially on the anterior surface of the iris, melanocytes have a plentiful adrenergic innervation. In this location, they are dependent on it for the maintenance of their pigment content; iris tyrosinase activity drops within days of superior cervical denervation[69]. Additionally, topical application of the α-adrenergic blocker thymoxamine but not the β-adrenergic blocker timolol leads to a pale iris in juvenile pigmented rabbits, an observation that confirms adrenergic dependency while defining an α-adrenergic receptor mechanism[94].

The density of adrenergic innervation to the aqueous humor outflow apparatus varies widely among species. It is generally plentiful in animals such as guinea pig and rabbit. In humans it is sparse but usually present in the young[27,28]. In most instances it cannot be seen in the old, but this observation must be qualified because an age-related increase in autofluorescence confounds visibility. If meshwork innervation truly is diminished or lacking in aged humans, the observation has potential importance in understanding the well-known age-related increase in ocular hypertension and glaucoma.

Variability among species is especially pronounced for the adrenergic innervation of the cornea. In rabbit and guinea pig, the adult cornea harbors a moderate but predictable adrenergic innervation, most of it anteriorly disposed. In monkey and man, histochemically demonstrable adrenergic corneal innervation is for practical purposes absent; only rarely are nerve fibers seen and then mostly at the junction of cornea and sclera. Interestingly, a remarkable adrenergic innervation is found during embryonic development of the cornea in primates[34], suggesting that the rare nerve fiber visible in adult is vestigial.

The importance of the tear film to the preservation of the corneal surface is widely acknowledged. Tears originate chiefly from the main lacrimal gland although various accessories are of significance. Dense networks of acetylcholinesterase-reacting nerve fibers are described by light microscopy to the lacrimal gland in humans[88] as well as to a broad variety of mammals[26]. Such uniformity of cholinergic innervation stands in contrast to the species variability of the adrenergic as revealed by histofluorometric methods. Adrenergic innervation to the lacrimal acini is noted in monkey, cat and dog; except to blood vessels, it is absent in rabbits and rodents[26]. For the human and monkey, dual innervation with the cholinergic more pronounced has been observed in careful electron microscopic studies of terminal nerve fields by Ruskell[104,105].

Neuropeptides localized to peripheral ocular nerves

Many neuropeptides already have been identified in ocular nerves by the use of immunohistochemical techniques. The potential for unrecognized crossreactivity justifies the suffix '-like immunoreactivity' (-LI) in describing peptidergic nerves; in a few instances, confirming biochemical isolation of the antigen has been performed in ocular nerves.

Because of the diversity of tissues and the regional differences in function within individual coats of the eye, highly detailed descriptions of the distribution of peptidergic nerves are required both for accuracy and for understanding (fig. 1). Among eyes of different mammalian species, the distribution of nerve fibers containing individual neurotransmitters and neuropeptides tends to be reasonably similar. The chief differences concern nerve fiber density. The table contains a general summary of the distribution of peptidergic nerves most extensively studied to date in the mammalian eye. The precise details for particular species are provided in the cited immunohistochemical references.

Vasoactive intestinal polypeptide

Vasoactive intestinal polypeptide (VIP) was the first neuropeptide localized to ocular autonomic nerves[144], and ocular nerves containing this peptide are an important part of the parasympathetic system to the eye[130] (figs 2 and 3). A rich VIP-LI innervation supplies ocular blood vessels, most prominently in the choroid[139,144,149]. More recently a non-vascular VIP-LI innervation has also been recognized as a forward extension with supply that includes the ciliary muscle, both iris muscles, and the chamber angle[84,109,115,129,150]. A pterygopalatine origin for choroidal vascular nerves has been established in rabbit, cat and monkey[106] and has been shown to be the source of the VIP-LI nerve fibers in the cat choroid[144]. This component of the vascular innervation to the eye is a part of the pterygopalatine-based innervation to blood vessels of the face and anterior cranium[42]. In addition, VIP-LI innervation extends to accessory structures, prominent among which are the lacrimal and salivary glands. In the lacrimal gland, VIP-LI nerves similarly derive from the pterygopalatine ganglion;[144] the VIP immunoreactive network of nerves to the acini is coextensive with the cholinergic as defined by acetylcholinesterase staining[110].

VIP immunoreactivity also occurs in a few ciliary ganglion and accessory ciliary neurons of rat[63], and in uveal ganglion cells of guinea pig[139] and man[84,129]. Whether nerves supplying specific structures within the eye originate from one or more particular parasympathetic sources is not fully known. Nor is the matter readily clarified because any orbital parasympathetic denervation of the eye is complicated by the

Distribution of peptide-containing peripheral nerves in mammalian eye

| Neuropeptide | Corneal innervation | Intraocular muscle innervation | | | Vascular and tissue innervation | | | | | | |
		Iris sphincter	Iris dilator	Ciliary muscle	Limbal blood vessels	Drainage angle	Iris blood vessels	Ciliary body blood vessels	Ciliary process	Cho-roid	Melano-cytes
VIP	0	+	+	+	++	+	+	++	+	+++	+
Neuropeptide Y	0	++	+++	++	++	+	+	++	+++	+++	+
Substance P	++	++	++	++	+++	++	+	++	++	++	+
CGRP	++	+	±	++	++	+	++	++	++	++	+

The distribution of other neuropeptides found in peripheral ocular nerves, peptide histidine isoleucine, cholecystokinin and enkephalin, have not been described in sufficient detail to include in this summary. See text. Arbitrary scale of 0 to 3+, generalized from published reports.

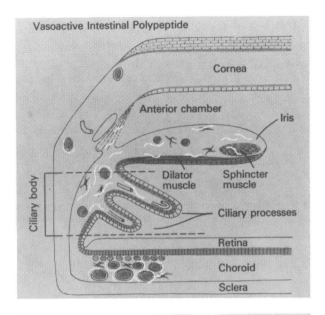

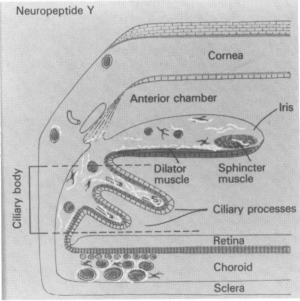

Figure 1. Schematic illustrations of the comparative distribution of two types of peptidergic nerve fibers in the human eye. For precise descriptions, see text and references. Top: Nerve fibers immunoreactive to vasoactive intestinal polypeptide represent a sub-population of parasympathetic nerve fibers. They occur with greatest density in the choroid but are also found more anteriorly in the eye. Bottom: Nerve fibers immunoreactive to neuropeptide Y, likely representing a substantial portion of sympathetic nerve fibers, are distributed differently within the eye. While also occurring in high density in the choroid, a richer anterior segment distribution is evident for nerve fibers containing this neuropeptide. Particularly prominent is the innervation to the ciliary processes and to the iris dilator muscle.

274

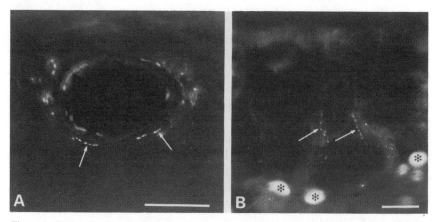

Figure 2. Vasoactive intestinal peptide-like immunoreactive nerve fibers in the corneo-scleral limbus. (Fluorescence micrographs; bars, 50 μm.) *A* Immunoreactive nerve fibers (arrows) surround a superficial limbal blood vessel in the rhesus monkey eye. *B* Immunoreactive nerve fibers (arrows) lie in the cat ciliary cleft through which aqueous humor drains. Chromatophores (*) in the cat uvea show intense autofluorescence.

mixed nature and overlapping paths of the many nerves in this region. For instance, sympathetic and sensory fibers pass through the ciliary ganglion in many species. While the rat is an exception to this generalization, the ciliary ganglion of this species contains only a limited proportion of the parasympathetic ganglion cells within the orbit[63]. Studies undertaken so far to ascertain the relative contribution and

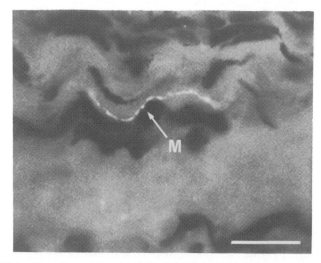

Figure 3. Nerve fibers immunoreactive to individual neuropeptides have been observed to drape around melanocytes in the uvea. Illustrated is a vasoactive intestinal polypeptide-like immunoreactive nerve fiber in apposition to a melanocyte (M) within the posterior ciliary body of the bovine eye. (Fluorescence micrograph; bar, 25 μm.)

intraocular contribution from different parasympathetic sources by anatomical methods have just begun to yield definitive results[45,128].

Peptide histidine isoleucine

Peptide histidine isoleucine (PHI) co-localizes with VIP in many peripheral autonomic neurons[7,37,101]. This neuropeptide has marked structural similarities to VIP and in fact derives from the same precursor molecule[136]. Immunoreactivity to PHI has been observed both in the guinea pig choroid[101] and in the rat iris[9]. In the rat iris, the distribution of PHI-LI nerve fibers closely parallels that of VIP-LI nerves, suggesting co-localization. No more detailed studies of the distribution of PHI in the eye have appeared to date.

Neuropeptide Y

A peptide of the pancreatic polypeptide family occurs in ocular nerves[15,54,116,123,137] and now has been identified biochemically as neuropeptide Y (NPY) in rat and guinea pig[125]. The distribution of nerves immunoreactive for NPY closely mirrors the distribution of sympathetic nerves containing catecholamines as revealed by histofluorometric methods (fig. 4). Notable differences include the absence of NPY from corneal nerves and a lower density of NPY-LI nerves in the ciliary body and around anterior uveal vessels. Following superior cervical ganglionectomy, most NPY-LI ocular nerves disappear from the ipsilateral eye just as do the adrenergic[15,137] and radioimmunoassay levels of the peptide fall[1]. Unlike the observations with histofluorometric techniques, however, some choroidal nerve fibers persist[15,137]. Recently, NPY has been isolated biochemically from the rat pterygopalatine[62] and otic[72] ganglia, and NPY-LI cells have been observed in the ciliary ganglion by immunohistochemistry[72,128]. While most NPY-LI ocular nerves likely derive from the superior cervical ganglion, some are of parasympathetic origin as well.

Neuropeptides in ocular sensory nerves

Four neuropeptides have been identified to date in ocular nerves likely originating in the trigeminal ganglion (figs 5 and 6): substance P (SP)[83,118,124,139–143], calcitonin gene-related peptide (CGRP)[74,80,122,138], a peptide of the cholecystokinin-gastrin (CCK) family[9,64,98,120] and galanin[121,131]. Only for substance P has biochemical identification been performed[114]. Whether the newly discovered tachykinin, substance K[77],

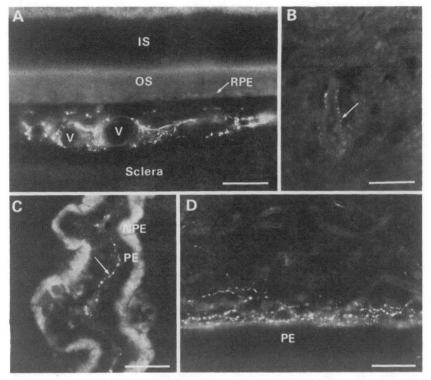

Figure 4. Neuropeptide Y-like immunoreactive nerve fibers supply both vascular and non-vascular structures in the eye. (Fluorescence micrographs; bars, 50 μm.) *A* In the albino rat choroid, a rich innervation is present and is clearly related to the vascular supply. IS, photoreceptor inner segment; OS, photoreceptor outer segment; RPE, retinal pigmented epithelium; V, blood vessel in choroid, *B* An immunoreactive nerve fiber (arrow) surrounds an arteriole in the ciliary body of the rhesus monkey. *C* An immunoreactive nerve fiber (arrow) within the cilary process of the cat. At the light microscopic level, it is not possible to resolve whether immunoreactive nerve fibers in this structure relate primarily to the epithelial layer or to the blood vessels within the core of the process. NPE, non-pigmented epithelial layer: PE, pigmented epithelial layer. *D* A rich innervation lies along the dilator muscle in the rhesus monkey iris. Nerve fibers also are seen more anteriorly in the stroma. PE, pigmented epithelial layer.

also occurs in ocular sensory nerves is not known. The evidence for the occurrence in the eye of CGRP, CCK, and galanin rests solely on immunohistochemical identification. The validity of identification of these peptides by immunohistochemistry is discussed at length in the cited manuscripts. Based on analysis of trigeminal ganglia, SP, CCK, and galanin-like immunoreactivities have been observed primarily in small trigeminal ganglion cells; CGRP immunoreactivity is present in small but also in large trigeminal neurons.

About 20% of trigeminal ganglion cells are immunoreactive to SP, about 40% are immunoreactive to CGRP[71,73,111,138] and about 2% immunoreactive to CCK[66]. Their pattern of co-localization is complex.

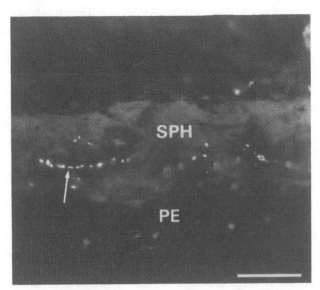

Figure 5. A substance P-like immunoreactive nerve fiber (arrow) in the sphincter muscle (SPH) of the human iris. Despite the widespread distribution throughout the eye of substance P-like immunoreactive nerve fibers, the only established ocular effect of this peptide is miosis in selective species. Despite the innervation present, substance P seems inactive to the human iris sphincter muscle.[121] PE, pigmented epithelial layer. (Fluorescence micrograph; bar, 50 μm.)

Based on co-localization studies in the trigeminal ganglion of rat[73,111,138] and guinea pig[66], essentially all SP-LI cells also are immunoreactive to CGRP. Consistent with their greater number, many CGRP-LI cells are negative for SP. For SP and CCK in the guinea pig trigeminal ganglion, some immunoreactive cells are positive for only one of these peptides

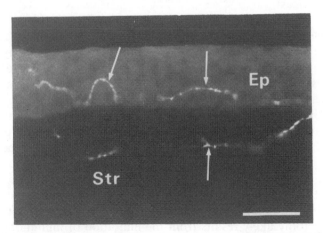

Figure 6. Calcitonin gene-related peptide-like immunoreactive nerve fibers (arrow) are present in the epithelium (Ep) and anterior stroma (Str) of the rat cornea. (Fluorescence micrograph; bar, 50 μm.)

and some are positive for both[64]. While not directly shown, it seems reasonable to suspect that some trigeminal ganglion cells are immuno-reactive to all three peptides. A few small trigeminal neurons are immunoreactive to galanin[112,131], only some of which are simultaneously reactive for CGRP[131].

The pattern of co-localization observed in ocular nerve fibers parallels in a general sense the observation in trigeminal ganglion. By the use of simultaneous immunohistochemical labelling techniques, nerve fibers immunoreactive to both SP and CCK[64] or to both SP and CGRP[44,80] have been observed in the eye. Although not studied in the eye, it is now known that the latter two can coexist in the same secretory vesicle[46]. Consistent with the co-localization pattern of SP and CCK in the trigeminal ganglion, nerve fibers have been observed in the eye that are immunoreactive to both or to just one of these peptides[64]. In co-localization studies with CGRP and SP, ocular nerve fibers immunoreactive to CGRP but negative for SP are observed, also consistent with the pattern in the trigeminal ganglion[80]. In the porcine eye, co-localization of galanin and SP is seen in ocular nerves[121], but further information on galanin co-localization in the eye is not yet available.

The correspondence of patterns of peptide co-localization in the trigeminal ganglion and in the eye, however, is not precise. The most notable present discrepancy is in the guinea pig iris. Here, in contrast to the ganglion, many SP-LI nerve fibers are negative for CGRP[65]. Selective denervations have indicated that the SP positive, CGRP negative fibers do not derive from the trigeminal ganglion, but their anatomical source is not yet established.

Opiate peptides

Enkephalin. Fine nerve fibers immunoreactive to leu-enkephalin have been observed in the eye of the rat[10] and the guinea pig (Kuwayama and Stone, unpublished observations) by immunohistochemical techniques. They present a novel problem. Standard selective denervations do not affect their integrity[10], and their origin has been speculative. Recently, enkephalin immunoreactivity has been identified in some of the neurons of the rat ciliary ganglion and its accessory cells in the optic nerve sheath[120], indicating a parasympathetic origin for at least some of these neurons. Because of the large number of accessory ganglion cells in the rat[63], ciliary ganglionectomy as commonly performed leaves intact much of this animal's parasympathetic ocular innervation; the apparent persistence of enkephalin immunoreactivity in the eye after ciliary ganglionectomy likely results from the limitation of immunohistochemistry to detect partial denervation in sparsely innervated peripheral tissues.

Dynorphin. A recent immunohistochemical survey of guinea pig tissues provides brief mention of dynorphin-LI nerve fibers in the eye, both in iris and cornea[43]. While dynorphin-LI neurons are plentiful in the trigeminal and sparse in the superior cervical ganglion, the origin of these ocular nerves is not yet apparent because some dynorphin-LI iris nerve co-localize with SP, others with NPY. Perhaps multiple ganglia contribute dynorphin-LI nerves to the eye, as already known for NPY and SP.

Intraganglionic organization

By intraocular injection of probes of retrograde axoplasmic transport, it is possible to identify neurons both in the superior cervical ganglion and the trigeminal ganglion that project to the eye[2]. In the case of the superior cervical ganglion, neurons projecting to the eye are distributed in the caudal portion of the ganglion. No specific transmitter-related histochemistry has yet been reported on these cells. For the trigeminal ganglion in guinea pig, the specific issue of intraganglionic origin of cells projecting to the eye has been addressed[66]. They are located within the anteromedial region of the ophthalmic division of the ganglion[2]. When studied by immunohistochemical techniques, there is no recognizable somatotopical organization to peptidergic neurons containing SP, CCK, or CGRP either for the ganglion as a whole or for cells specifically projecting to the eye. When the proportion of neurons immunoreactive for each peptide and projecting to the eye is compared to the immunoreactive neurons of the ganglion as a whole, proportions are similar[66]. Similar studies are yet to be done for other peptides. Because of the diversity of innervated intraocular structures and because retrograde tracers label neurons non-selectively, such techniques applied to intraocular tissues may not adequately identify neurons of specific functional subtype. Recently developed methods to label selectively neurons from the cornea, however, show promise for evaluating the neurons innervating this particular tissue[79].

Interactions of parasympathetic, sympathetic and sensory nerves

The eye has proved a usable model for demonstrating the plasticity of the peripheral nervous system, particularly with respect to neuropeptides. For instance, two changes occur to the ocular innervation following sympathectomy besides the apparent loss of adrenergic nerves. First, unilateral sympathectomy results in a gradual enhancement of sensory neuropeptides in the ipsilateral eye judged both by immunohistochemistry and by radioimmunoassay[18,58,64,107,131,147,155]. Second, nonsympathetic nerve fibers in the eye gradually develop tyrosine hydroxylase and

NPY immunoreactivity, each of which is considered a 'sympathetic' marker[11]. Yet despite the presence of tyrosine hydroxylase immunoreactivity in sympathectomized eyes, there is no evidence of catecholamine synthesis. Some nerve fibers induced to express tyrosine hydroxylase or neuropeptide Y derive from the ciliary ganglion; it has not been possible to identify the source of all such nerve fibers[11]. Nor is it clear whether the enhanced NPY derives from elevation of peptide levels in parasympathetic cells already containing the peptide or from the induction of *de novo* synthesis in other neurons. Also this phenomenon is not necessarily restricted to sympathectomy alone: ocular galanin immunoreactivity[131] increases after extirpation of pterygopalatine or ciliary ganglia as well as after removal of superior cervical ganglion.

In total, these observations indicate complex interactions of parasympathetic, sympathetic and sensory nerves and substantiate observations on the inductive plasticity of neurotransmitters/neuromodulators in peripheral nerves as studied in other systems. The functional implications of these interactions are not currently known; certainly the eye continues to be a useful end organ for their elucidation.

Physiological functions

Physiology trails behind immunohistochemistry in the eye. Despite the large number of neuropeptides described by immunohistochemical and biochemical techniques, relatively little is known about their physiological function.

For some there is a good start. One of the most extensively studied neuropeptides is VIP. It acts as a vasodilator in the eye[93], consistent with the same function elsewhere in the body. VIP enhances prejunctional cholinergic transmission in the ciliary muscle[134], and both iris muscles relax under its influence[49,50,135]. VIP stimulates adenylate cyclase in the rabbit iris ciliary body[87], suggesting a role in the regulation of aqueous humor formation. VIP also stimulates the production of cyclic AMP by cultured retinal pigmented epithelial cells[60]; therefore, it could help regulate the retinal environment. Since VIP localizes to retinal amacrine cells[13] as well as autonomic nerves of the choroid, either or both sources of VIP might be involved in the regulation of retinal pigmented epithelial function as it lies between sensory retina and choroid.

Despite the rich distribution of NPY-LI immunoreactive ocular nerves and the well-established and varied physiologic functions of this peptide, little physiologic work has been performed with it in the eye. NPY augments the contractile effects of 1-phenylephrine on the iris dilator but has no direct effect on the muscle[100], an observation consistent with the rich innervation of the iris dilator by nerves containing this neuropeptide as well as the concomitant adrenergic innervation.

Consistent with its known vasoconstrictor action, intravenous administration of NPY reduces uveal blood flow[92]. No other physiologic reports have appeared to the present.

For ocular sensory nerves, a role in mediating the ocular injury response is now clear, at least in rabbit[8]. Severe mechanical or chemical trauma induces a four-part response in the rabbit eye, consisting of miosis, vasodilation, breakdown of the blood-aqueous barrier with protein leakage and elevation of intraocular pressure. Both neurogenic and prostaglandin-mediated mechanisms have been identified, the relative contribution of each depending on the type of stimulus[25]. For the ocular reaction to chemical injury, a major role for the sensory innervation is now widely accepted[16]. Sensory denervation or prior treatment either with anesthetics or tetrodotoxin inhibit the reaction to chemical injury[16], while direct stimulation of the trigeminal ganglion elicits its four components[99]. Moreover, individual neuropeptides appear responsible for the specific components of the ocular injury response. Substance P is a potent miotic[113,143] and is released into the eye after trigeminal nerve stimulation[6]. When applied to the eye, it simulates the atropine-resistant miosis which occurs in the neurogenic ocular injury response. CGRP is known to be a potent vasodilator[12]; in rabbits, it induces breakdown of the blood-ocular barriers and a rise in intraocular pressure but does not cause miosis[147]. It thus has been suggested that CGRP acts jointly with SP in the eye after injury, each eliciting different components of the ocular injury response[147]. This is a reasonable hypothesis in view of the co-localization of both neuropeptides in many ocular nerves and leads to a unified theory for understanding the response of the rabbit eye to injury. A species related qualification is required for this attractive hypothesis, however. The miotic response to substance P shows great species variability. Although maximal in the rabbit, it is not observed in cat, dog and man[148]. As the full array of neuropeptides in ocular sensory nerves is discovered, these species differences may become understandable. Possibly, the principle will remain but the participant peptides will differ.

It is important to realize that the function for most of the substance P-LI and CGRP-LI nerves in the eye is unknown. While the corneal nerve fibers in a general sense serve a nociceptive role, important other possibilities remain unexplored. For instance, substance P has potent effects on fluid flow in the kidney[47]. While substance P has been found in nerves to the ciliary process, there have been no published studies of possible secretory effects of SP in the eye. For CCK, no studies have yet been reported in this organ. Galanin in rabbit iris reduces the iris contraction mediated by acetylocholine[35].

Clues to ocular functions of opioids are available. Systemic opioids induce remarkable effects on pupillary diameter without inflammatory signs, and they can affect intraocular pressure. For the pupil either miosis or mydriasis is observed, depending on the species[89]. Everyone is familiar with the police examination of suspected drug addicts for a pin-point pupil. This response is thought to be mediated primarily by the central nervous system[89]. However, morphine applied topically to the human eye[39] or injected into the rabbit eye induces a unilateral miosis[22], indicating a peripheral action as well. In this regard, enkephalin inhibits the cholinergic and SP-ergic *in vitro* response of the sphincter muscle to exogenously applied carbachol or SP, suggesting the presence of opioid receptors on cholinergic and SP iris nerves[146]. Dynorphin, on the other hand, augments the cholinergic but inhibits the SP-ergic response in this system[145]. Intraocular pressure is lowered by intraocular injection of morphine or D-ala-met-enkephalinamide in rabbits[23]. Moreover, morphine or heroin addicts have lower intraocular pressures than normal. In glaucoma patients[23], topical instillation of morphine again lowers intraocular pressure. Opioids appear to affect intraocular pressure by enhancing the drainage of aqueous humor from the eye[23]. In view of the enkephalin-like and dynorphin-like immunoreactive nerves now known to be present, further studies are justified to clarify fully the local effects of the opioid peptides in the eye.

Non-neuronal biologically active peptides

Ocular physiology is replete with reports of the effects of hormones on ocular function[55]. By and large, the hormones in question have not been found in ocular nerves and are not thought to be produced locally; the observed effects are assumed to be humoral. Unfortunately the matter is unsettled. For instance, paracrine cells have not been identified in the uvea. Yet based on immunohistochemical reaction to neuron-specific enolase, it has been proposed that cells of the diffuse neuroendocrine system do exist in the trabecular meshwork of the monkey[119], even though to date no biologically active peptide or neurotransmitter has been identified within these cells.

In the same vein, high levels of atrial natriuretic peptide recently have been found in the anterior uvea[117]. The cellular localization in the eye of this peptide has yet to be reported. Intriguingly, receptors to it have been identified on the pigmented epithelium of the ciliary body[5,78,102], and atrial natriuretic peptide stimulates guanylate cyclase in the ciliary body[87,90]. Although the function for atrial natriuretic peptide in the eye presumably relates to the secretion of aqueous humor, there is disagreement about its effect on intraocular pressure[87,90,132]; and much remains to be learned about its ocular role.

Retina

The image of the outside world received by the rods and cones of the eye is signalled to the visual areas of the brain by the retinal ganglion cells. Between the photoreceptors at the depth of the retina and the ganglion cells near its surface lay four distinct classes of neurons—horizontal cells, bipolar cells, interplexiform cells and amacrines. As the links between photoreceptor neural processes and ganglion cell, they are responsible for the coherent translation and processing of rod and cone impulses. Some idea of the complexity of their task is gained when it is recalled that within the retina diverse types of physiological information must be characterized: the ganglion cells relay information about shape, color, contrast and borders, and light level in an economic and accurate manner. To a great extent this task is facilitated by the complex interrelationships of horizontal cells near the outer plexiform layer and amacrines plus interplexiform cells whose cell bodies are seen just next to the inner plexiform layer. Amacrine cell subtypes have long been thought to contain specific neurotransmitters[29]: cholinergic[91]; dopaminergic[30]; or serotonergic[48,95] subtypes have been recognized, among others. First defined histochemically in fish and monkey by the histofluorometric demonstration of dopamine within them[32], evidence is accumulating that interplexiform cells can also contain other neurotransmitters such as GABA[24].

More recently, immunohistochemical and biochemical studies have revealed the presence of neuropeptides in the retina[13], again chiefly within specific types of amacrine cells (fig. 7). In the anuran retina, neuropeptides also appear to be present in ganglion cell processes within the optic nerve[61]. While the detailed discussion of neuropeptides in the retina is beyond the scope of the present review, some general comments can be made to provide a framework for readers working outside this area. Consistent with the considerable species variability in retinal function and retinal structure, species variability exists in the identity of neuropeptides within the retina and in the number, subtype and distribution of amacrine cells containing specific neuropeptides[13]. Many different neuropeptides have been recognized in amacrine cells of the vertebrate retina, including substance P, enkephalin, neurotensin, neuropeptide Y, somatostatin, glucagon, vasoactive intestinal polypeptide, and cholecystokinin. As in brain, co-localization of individual neuropeptides with other neuropeptides[56,57,75,76,152] or with classical neurotransmitters[152–154] is recognized.

Efforts are now underway to establish a physiological role for neuropeptides in the retina[20]. Transient or prolonged neuropeptide changes in levels have been observed. Thus light or dark adaptation influence neuropeptide levels acutely[36,52,81,82]. In contrast, prolonged closure of

284

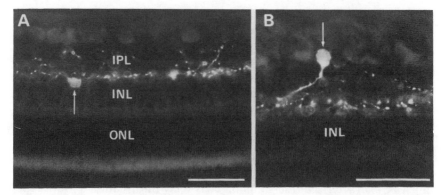

Figure 7. Vasoactive intestinal polypeptide-like immunoreactivity in the guinea pig retina. (Fluorescence micrographs; bars, 50 μm.) *A* An immunostained amacrine cell (arrow) lies adjacent to the inner plexiform layer (IPL). Some of the nerve fibers in this layer are immunoreactive as well. *B* Sometimes, an immunoreactive amacrine cell (arrow) is placed in the ganglion cell layer. INL, inner nuclear layer; ONL, outer nuclear layer.

one eyelid in the lid-suture primate myopia model not only leads to an increased axial length of the affected eye but also to a prolonged elevation in retinal VIP levels in the visually deprived eye[127]. Individual neuropeptides also have been shown to affect specific retinal cells[19,21,67,151]. Neuropeptides can cause the selective release of individual retinal neurotransmitters[3], while potassium depolarization can lead to neuropeptide release[96]. Receptor mechanisms are being defined through identification of changes in second messengers such as the adenylate cyclase and inositol triphosphate pathway[14,67,86,97,108,151].

Conclusions

The application of immunohistochemical techniques to the discovery of neuropeptides in the eye has indicated a complexity to the peripheral innervation of the eye that was unimaginable only a few years ago. Neuropeptides have been identified within parasympathetic, sympathetic and sensory nerves to this organ. Although particular neuropeptides tend to localize chiefly in nerve fibers derived from one of these divisions of the ocular innervation, it is now known that these localizations are not exclusive. Important interactions between the divisions of the ocular peripheral innervation also are now evident. In short, the classical conception of a push-pull ocular innervation, a cholinergic parasympathetic supply counterbalanced by an adrenergic sympathetic supply, likely remains valid but is vastly oversimplified. Because of the importance of the eye's vegetative physiology in clinical disorders, such as glaucoma, and because of the central role played by cholinergic and adrenergic agents in clinical pharmacology, learning the effects of bio-logically active peptides clearly will lead to enhanced understanding of ocular physiology and to new drug therapies. The lag between immuno-cytochemical identification and physiological definition means our knowledge of the functional role of neuropeptides within the eye is considerably behind what we know of its neuroanatomy. Neuroanatom-ical studies disclose a complexity to the ocular innervation that un-doubtedly mirrors a parallel complexity of physiological function. Much will be revealed as neuropeptide actions are delineated.

1 Allen, J. M., McGregor, G. P., Adrian, T. E., Bloom, S. R., Zhang, S. Q., Ennis, K. W., and Unger, W. G., Reduction of neuropeptide Y (NPY) in the rabbit iris-ciliary body after chronic sympathectomy. Exp. Eye Res. *37* (1983).

2 Arvidson, B., Retrograde transport of horseradish peroxidase in sensory and adrenergic neurons following injection into the anterior eye chamber. J. Neurocyt. *8* (1979) 751–764.

3 Bauer, B., Ehinger, B., Tornqvist, K., and Waga, J., Neurotransmitter release by certain neuropeptides in the chicken retina. Acta ophtahlm. *63* (1985) 581–587.

4 Beauvieux, J., and Dupas, J., Étude anatomo-topographique et histologique du ganglion ophtalmique chez l'homme et divers animaux. Archs Ophthalm. *43* (1926) 641–671.

5 Bianchi, C., Anand-Srivastava, M. B., De Léan, A., Gutkowska, J., Forthomme, D., Genest, J., and Cantin, M., Localization and characterization of specific receptors for atrial natriuretic factor in the ciliary processes of the eye. Curr. Eye Res. *5* (1986) 283–293.

6 Bill, A., Stjernschantz, J., Mandahl, A., Brodin, E., and Nilsson, G., Substance P: release on trigeminal nerve stimulation, effects in the eye. Acta physiol. scand. *106* (1979) 371–373.

7 Bishop, A. E., Polak, J. M., Yiangou, Y., Christofides, N. D., and Bloom, S. R., The distributions of PHI and VIP in porcine gut and their co-localisation to a proportion of intrinsic ganglion cells. Peptides *5* (1984) 255–259.

8 Bito, L. Z., Species differences in the responses of the eye to irritation and trauma: a hypothesis of divergence in ocular defense mechanisms and the choice of experimental animals for eye research. Exp. Eye Res. *39* (1984) 807–829.

9 Björklund, H., Fahrenkrug, J., Sieger, Å., Vanderhaeghen, J.-J., and Olson, L., On the origin and distribution of vasoactive intestinal polypeptide-, peptide HI-, and cholecys-tokinin-like-immunoreactive nerve fibers in the rat iris. Cell Tiss. Res. *242* (1985) 1–7.

10 Björklund, H., Hoffer, B., Olson, L., Palmer, M., and Seiger, Å., Enkephalin immuno-reactivity in iris nerves: distribution in normal grafted irides, persistence and enhanced fluorescence after denervations. Histochemistry *80* (1984) 1–7.

11 Björklund, H., Hökfelt, T., Goldstein, M., Terenius, L., and Olson, L., Appearance of the noradrenergic markers tyrosine hydroxylase and neuropeptide Y in cholinergic nerves of the iris following sympathectomy. J. Neurosci. *5* (1985) 1633–1643.

12 Brain, S. D., Williams, T. J., Tippins, J. R., Morris, H. R., and MacIntyre, I., Calcitonin gene-related peptide is a potent vasodilator. Nature *313* (1985) 54–56.

13 Brecha, N. C., Eldred, W., Kuljis, R. O., and Karten, H. J., Identification and localiza-tion of biologically active peptides in the vertebrate retina, in: Progress in Retinal Research, vol. 3, chapt. 7, pp. 185–226. Eds N. N. Osborne and G. J. Chader. Pergammon Press, New York 1984.

14 Brown, J. E., Blazynski, C., and Cohen, A. I., Light induces a rapid and transient increase in inositol triphosphate in intact vertebrate rods. ARVO Abstr. *28 (3)* (1987) 96.

15 Bruun, A., Ehinger, B., Sundler, F., Tornqvist, K., and Uddman, R., Neuropeptide Y immunoreactive neurons in the guinea-pig uvea and retina. Invest. Ophthalm. vis. Sci. *25* (1984) 1113–1123.

16 Butler, J. M., Unger, W. G., and Hammond, B. R., Sensory mediation of the ocular response to neutral formaldehyde. Exp. Eye Res. *28* (1979) 577–589.

17 Castro-Correia, J., Studies on the innervation of the uveal tract. Ophthalmologica *154* (1967) 497–520.

286

18 Cole, D. F., Bloom, S. R., Burnstock, G., Butler, J. M., McGregor, G. P., Saffrey, M. J., Unger, W. G., and Zhang, S. Q., Increase in SP-like immunoreactivity in nerve fibres of rabbit iris and ciliary body one to four months following sympathetic denervation. Exp. Eye Res. *37* (1983) 191–197.

19 Dick, E., and Miller, R. F., Peptides influence retinal ganglion cells. Neurosci. Lett. *26* (1981) 131–135.

20 Djamgoz, M. B. A., Downing, J. E. G., and Prince, D. J., Physiology of neuroactive peptides in vertebrate retina. Biochem. Soc. Trans. *11* (1983) 686–689.

21 Djamgoz, M. B. A., Stell, W. K., Chin, D.-A., and Lam, D. M.-K., An opiate system in the goldfish retina. Nature *292* (1981) 620–623.

22 Drago, F., Gorgone, G., Spina, F., Panissidi, G., Dal Bello, A., Moro, F., and Scapagnini, U., Opiate receptors in the rabbit iris. Naunyn-Schmiedeberg's Arch. Pharmak. *315* (1980) 1–4.

23 Drago, F., Panissidi, G., Bellomio, R., Dal Bello, A., Aguglia, E., and Gorgone, G. Effects of opiates on intraocular pressure of rabbits and humans. Clin. exp. Pharm. Physiol. *12* (1985) 107–113.

24 Dowling, J. E., The Retina. An Approachable Part of the Brain, pp. 152–153. The Belknap Press of Harvard University Press, Cambridge, MA 1987.

25 Eakins, K. E., Prostaglandin and non-prostaglandin mediated breakdown of the blood-aqueous barrier. Exp. Eye Res. *25* (1977) 483–498.

26 Ehinger, B., Ocular and orbital vegetative nerves. Acta physiol. scand. *67* (1966) 1–35.

27 Ehinger, B., Adrenergic nerves to the eye and to related structures in man and in the cynomolgus monkey (Macaca irus). Invest. Ophthalm. *5* (1966) 42–52.

28 Ehinger, B., A comparative study of the adrenergic nerves to the anterior eye segment of some primates. Z. Zellforsch. *116* (1971) 157–177.

29 Ehinger, B., Neurotransmitter systems in the retina. Retina *2* (1982) 305–321.

30 Ehinger, B., Functional role of dopamine in the retina, in: Progress in Retinal Research, vol. 2, pp. 213–232. Eds N. N. Osborne and G. J. Chader. Pergamon Press, Oxford 1983.

31 Ehinger, B., and Falck, B., Innervation of iridic melanophores. Z. Zellforsch. *105* (1970) 538–542.

32 Ehinger, B., Falck, B., and Laties, A., Adrenergic neurons in teleost retina. Z. Zellforsch. *97* (1969) 285–297.

33 Ehinger, B., Falck, B., and Rosengren, E., Adrenergic denervation of the eye by unilateral cervical sympathectomy. Albrecht v. Graefes Arch. klin. exp. Ophthalm. *177* (1969) 206–211.

34 Ehinger, B., and Sjöberg, N.-O., Development of the ocular adrenergic nerve supply in man and guinea pig. Z. Zellforsch. *118* (1971) 579–592.

35 Ekblad, E., Håkanson, R., Sundler, F., and Wahlestedt, C., Galanin: neuromodulatory and direct contractile effects on smooth muscle preparations. Br. J. Pharmac. *86* (1985) 241–246.

36 Eriksen, E. F., and Larsson, L.-I., Neuropeptides in the retina: evidence for differential topographical localization. Peptides *2* (1981) 153–157.

37 Fahrenkrug, J., Bek, T., Lundberg, J. M., and Hökfelt, T., VIP and PHI in cat neurons: co-localization but variable tissue content possible due to differential processing. Reg. Pept. *12* (1985) 21–34.

38 Falck, B., Hillarp, N.-Å, Thieme, G., and Torp, A., Fluorescence of catechol amines and related compounds condensed with formaldehyde. J. Histochem. Cytochem. *10* (1962) 348–354.

39 Fanciullacci, M., Boccuni, M., Pietrini, U., and Sicuteri, F., Search for opiate receptors in human pupil. Int. J. clin. pharmac. Res. *1* (1980) 109–113.

40 Fukuda, M., Presence of adrenergic innervation to the retinal vessels. A histochemical study. Jap. J. Ophthalm. *14* (1970) 1–7.

41 Furukawa, H., Autonomic innervation of preretinal blood vessels of the rabbit. Invest. Ophthalm. vis. Sci. *28* (1987) 1752–1760.

42 Gibbins, I. L., Brayden, J. E., and Bevan, J. A., Perivascular nerves with immunoreactivity to vasoactive intestinal polypeptide in cephalic arteries of the cat: distribution, possible origins and functional implications. Neuroscience *13* (1984) 1372.

43 Gibbins, I. L., Furness, J. B., and Costa, M., Pathway-specific pattern of the co-existence

of substance P, calcitonin gene-related peptide, cholecystokinin and dynorphin in neurons of the dorsal root ganglion of the guinea-pig. Cell Tiss. Res. *248* (1987) 417–437.

44 Gibbins, I. L., Furness, J. B., Costa, M., MacIntyre, I., Hillyard, C. J., and Girgis, S., Co-localization of calcitonin gene-related peptide-like immunoreactivity with substance P in cutaneous, vascular and visceral sensory neurons of guinea pigs. Neurosci. Lett. *57* (1985) 125–130.

45 Grimes, P. A., McGlinn, A., M., Kuwayama, Y., and Stone, R. A., Peptide immunoreactivity of ciliary and accessory neurons. Invest. Ophthalin. vis. Sci. *29* (1988) 202.

46 Gulbenkian, S., Merighi, A., Wharton, J., Varndell, I. M., and Polak, J. M., Ultrastructure evidence for the coexistence of calcitonin gene-related peptide and substance P in secretory vesicles of peripheral nerves in the guinea pig. J. Neurocyt. *14* (1986) 535–542.

47 Gullner, H. G., Campbell, W. B., and Pettinger, W. A., Role of substance P in water hemostasis. Life Sci. *24* (1979) 2351–2360.

48 Hauschild, D. C., and Laties, A. An indoleamine-containing cell in chick retina. Invest. Ophthalmol. Vis Sci. *12* (1973) 537–540.

49 Hyashi, K., and Masuda, K., Effects of vasoactive intestinal polypeptide (VIP) and cyclic-AMP on the isolated sphincter pupillae muscles of the albino rabbit. Jap. J. Ophthalm. *26* (1982) 437–442.

50 Hyashi, K., Mochizuki, M., and Masuda, K., Effects of vasoactive intestinal polypeptide (VIP) and cyclic AMP on isolated dilator pupillae muscle of albino rabbit eye. Jap. J. Ophthalm. *27* (1983) 647–654.

51 Imai, K., Cholinergic innervation of the choroid. Ophthal. Res. *9* (1977) 194–200.

52 Ishimoto, I., Millar, T. J., Chubb, I. W., and Morgan, J. G., Somatostatin immunoreactive amacrine cells of chicken retina: retinal mosaic, ultrastructural features, and light-driven variations in peptide metabolism. Neuroscience *17* (1986) 1217–1233.

53 Jacobowitz, D. M., and Laties, A. M., Direct adrenergic innervation of a teleost melanophore. Anat. Rec. *162* (1968) 501–504.

54 Jacobowitz, D. M., and Olschowka, J. A., Bovine pancreatic polypeptide-like immunoreactivity in brain and peripheral nervous system: coexistence with catecholaminergic nerves. Peptides *3* (1982) 569–590.

55 Kass, M. A., and Sears, M. L., Hormonal regulation of intraocular pressure. Surv. Ophthalm. *22* (1977) 153–176.

56 Katayama-Kumoi, Y., Kiyama, H., Emson, P. C., Kimmel, J. R., and Tohyama, M., Co-existence of pancreatic polypeptide and substance P in the chicken retina. Brain Res. *361* (1985) 25–35.

57 Katayama-Kumoi, Y., Kiyama, H., Manabe, R., Shiotani, Y., and Tohyama, M., Co-existence of glucagon- and substance P-like immunoreactivity in the chicken retina. Neuroscience *16* (1985) 417–424.

58 Kessler, J. A., Bell, W. O., and Black, I. B., Interactions between the sympathetic and sensory innervation of the iris. J. Neurosci. *3* (1983) 1301–1307.

59 Koelle, G. B., and Friedenwald, J. S., A histochemical method for localizing cholinesterase activity. Proc. Soc. exp. Biol. Med. *70* (1949) 617–622.

60 Koh, S.-W.M., and Chader, G. J., Elevation of intracellular cyclic AMP and stimulation of adenylate cyclase activity by vasoactive intestinal polypeptide and glucagon in the retinal pigment epithelium. J. Neurochem. *43* (1984) 1522–1526.

61 Kuljis, R. O., Krause, J. E., and Karten, H. J., Peptide-like immunoreactivity in anuran optic nerve fibers. J. comp. Neurol. *226* (1984) 222–237.

62 Kuwayama, Y., Emson, P. C., and Stone, R. A., Pterygopalatine ganglion cells contain neuropeptide Y. Brain Res. *446* (1988) 219–224.

63 Kuwayama, Y., Grimes, P. A., Ponte, B., and Stone, R. A., Autonomic neurons supplying the rat eye and the intraorbital distribution of VIP-like immunoreactivity. Exp. Eye Res. *44* (1987) 907–922.

64 Kuwayama, Y., and Stone, R. A., Cholecystokinin-like immunoreactivity occurs in ocular sensory neurons and partially co-localizes with substance P. Brain Res. *381* (1986) 266–274.

65 Kuwayama, Y., and Stone, R. A., Distinct substance P and calcitonin gene-related peptide immunoreactive nerves in the guinea pig eye. Invest. Ophthalm. vis. Sci. *28* (1987) 1947–1954.

288

66 Kuwayama, Y., Terenghi, G., Polak, J. M., Trojanowski, J. Q., and Stone, R. A. A quantitative correlation of substance P-, calcitonin gene-related peptide- and cholecystokinin-like immunoreactivity with retrogradely labeled trigeminal ganglion cells innervating the eye. Brain Res. *405* (1987) 220–226.

67 Lasater, E. M., Watling, K. J., and Dowling, J. E., Vasoactive intestinal peptide alters membrane potential and cyclic nucleotide levels in retinal horizontal cells. Science *221* (1983) 1070–1072.

68 Laties, A. M., Central retinal artery innervation: Absence of adrenergic innervation to the intraocular branches. Archs Ophthalm. *77* (1967) 405–409.

69 Laties, A. M., Ocular melanin and the adrenergic innervation to the eye. Trans. Am. ophthalm. Soc. *72* (1974) 560–605.

70 Laties, A. M., and Jacobowitz, D., A comparative study of the autonomic innervation of the eye in monkey, cat, and rabbit. Anat. Rec. *156* (1966) 383–396.

71 Leah, J. D., Cameron, A. A., Kelly, W. L., and Snow, P. J., Coexistence of peptide immunoreactivity in sensory neurons of the cat. Neuroscience *16* (1985) 683–690.

72 Leblanc, G. G., Trimmer, B. A., and Landis, S. C., Neuropeptide Y-like immunoreactivity in rat cranial parasympathetic neurons: coexistence with vasoactive intestinal peptide and choline acetyltransferase. Proc. natl Acad. Sci. *84* (1987) 3511–3515.

73 Lee, Y., Kawai, Y., Shiosaka, S., Takami, K., Kiyama, H., Hillyard, C. J., Girgis, S., MacIntyre, I., Emson, P. C., and Tohyama, M., Coexistence of calcitonin-gene related peptide and substance P-like peptide in single cells of the trigeminal ganglion of the rat: immunohistochemical analysis. Brain Res. *330* (1985) 194–196.

74 Lee, Y., Takami, K., Kawai, Y., Girgis, S., Hillyard, C. J., MacIntyre, I., Emson, P. C., and Tohyama, M., Distribution of calcitonin gene-related peptide in the rat peripheral nervous system with reference to its coexistence with substance P. Neuroscience *15* (1985) 1227–1237.

75 Li, H.-B., Marshak, D. W., Dowling, J. E., and Lam, D. M.-K., Colocalization of immunoreactive substance P and neurotensin in amacrine cells of the goldfish retina. Brain Res. *366* (1986) 307–313.

76 Li, H.-B., Watt, C. B., and Lam, D. M.-K., The coexistence of two neuroactive peptides in a subpopulation of retinal amacrine cells. Brain Res. *345* (1985) 176–180.

77 Maggio, J. E., 'Kassinin' in mammals: the newest tachykinin. Peptides *6* (1985) 237–243.

78 Mantyh, C. R., Kruger, L., Brecha, N. C., and Mantyh, P. W., Localization of specific binding sites for atrial natriuretic factor in peripheral tissues of the guinea pig, rat, and human. Hypertension *8* (1986) 712–721.

79 Marfurt, C. F., Sympathetic innervation of the rat cornea as demonstrated by the retrograde and anterograde transport of horseradish peroxidase-wheat germ agglutinin. J. comp. Neurol. *367* (1988).

80 Matsuyama, T., Wanaka, A., Yoneda, S., Kimura, K., Kamada, T., Girgis, S., MacIntyre, I., Emson, P. C., and Tohyama, M., Two distinct calcitonin gene-related peptide-containing peripheral nervous systems: distribution and quantitative differences between the iris and cerebral artery with special reference to substance P. Brain Res. *373* (1986) 205–212.

81 Millar, T. J., and Chubb, I. W., Substance P in the chick retina: effects of light and dark. Brain Res. *307* (1984) 303–309.

82 Millar, T. J., Salipan, N., Oliver, J. O., Morgan, J. G., and Chubb, I. W., Concentration of enkephalin-like material in chick retina is light dependent. Neuroscience *13* (1984) 221–226.

83 Miller, A., Costa, M., Furness, J. B., and Chubb, I. W., Substance P immunoreactive sensory nerves supply the rat iris and cornea. Neurosci. Lett. *23* (1981) 243–249.

84 Miller, A. S., Coster, D. J., Costa, M., and Furness, J. B., Vasoactive intestinal polypeptide immunoreactive nerve fibres in the human eye. Aust. J. Ophthalm. *11* (1983) 185–193.

85 Mindel, J. S., and Mittag, T. W., Choline acetyltransferase in ocular tissues of rabbits, cats, cattle, and man. Invest. Ophthalm. *15* (1976) 808–814.

86 Mittag, T. W., and Tormay, A., Drug responses of adenylate cyclase in iris-ciliary body determined by adenine labelling. Invest. Ophthalm. vis. Sci. *26* (1985) 396–399.

87 Mittag, T. W., Tormay, J. A., Ortega, M., and Severin, C., Atrial natriuretic peptide (ANP), guanylate cyclase, and intraocular pressure in the rabbit eye. Curr. Eye Res. *6* (1987) 1189–1196.

88 Mizukawa, T., Otori, T., Hara, J., and Iga M., Histochemistry of the human lacrimal gland. Jap. J. Ophthalm. *6* (1962) 17–24.

89 Murray, R. B., Adler, M. W., and Korczyn, A. D., Minireview—The pupillary effects of opioids. Life Sci. *33* (1983) 495–509.

90 Nathanson, J. A., Atriopeptin-activated guanylate cyclase in the anterior segment. Invest. Ophthalm. vis. Sci. *28* (1987) 1357–1364.

91 Neal, M. J., Cholinergic mechanism in the vertebrate retina, in: Progress in Retinal Research, vol. 2, pp. 191–212. Eds N. N. Osborne and G. J., Chader. Pergamon Press, Oxford 1983.

92 Nilsson, S. F. E., Effects of NPY on ocular blood flow and local blood flow in some other tissues in the rabbit. Physiol. Soc., March (1987) 119p.

93 Nilsson, S. F. E., and Bill A., Vasoactive intestinal polypeptide (VIP): effects in the eye and on regional blood flows. Acta physiol. scand. *121* (1984) 385–392.

94 Odin, L., and O'Donnell, F. E. Jr, Adrenergic influence on iris stromal pigmentation: evidence for α-adrenergic receptors. Invest. Ophthalm. vis. Sci. *23* (1982) 528–530.

95 Osborne, N. N., Evidence for serotonin being a neurotransmitter in the retina, in: Biology of Serotonergic Transmission, pp. 401–430. Ed. N. N. Osborne, John Wiley & Sons, Ltd, New York 1982.

96 Osborne, N. N., Patel, S., Terenghi, G., Allen, J. M., Polak, J. M., and Bloom, S. R., Neuropeptide Y (NPY)-like immunoreactive amacrine cells in retinas of frog and goldfish. Cell Tiss. Res. *241* (1985) 651–656.

97 Pachter, J. A., and Lam, D. M.-K., Interactions between vasoactive intestinal peptide and dopamine in the rabbit retina: stimulation of a common adenylate cyclase. J. Neurochem. *46* (1986) 257–264.

98 Palkama, A., Uusitalo, H., and Lehtosalo, J., Innervation of the anterior segment of the eye: with special reference to functional aspects, in: Neurohistochemistry: Modern Methods and Applications, pp. 587–615. Eds P. Panula, H. Paivarinta and S. Soinila, Alan R. Liss Inc. New York 1986.

99 Perkins, E. S., Influence of the fifth cranial nerve on the intra-ocular pressure of the rabbit eye. Br. J. Ophthal. *41* (1957) 257–300.

100 Piccone, M., Krupin, T., Wax, M., Stone, R., and Davis, M., Effects of neuropeptide Y on isolated rabbit iris dilator muscle. Invest. Ophthalm. vis. Sci. *29* (1988) 330–332.

101 Polak, J. M., and Bloom, S. R., Regulatory peptides—the distribution of two newly discovered peptides: PHI and NPY. Peptides *5* (1984) 79–89.

102 Quirion, R., Dalpé, M., De Lean, A., Gutkowska, J., Cantin, M., and Genest, J., Atrial natriuretic factor (ANF) binding sites in brain and related structures. Peptides *5* (1984) 1167–1172.

103 Ruskell, G. L., Ocular fibres of the maxillary nerve in monkeys. J. Anat. *118* (1974) 195–203.

104 Ruskell, G. L., The fine structure of nerve terminations in the lacrimal glands of monkeys. J. Anat. *103* (1968) 65–76.

105 Ruskell, G. L., Nerve terminals and epithelial cell variety in the human lacrimal gland. Cell Tiss. Res. *158* (1975) 121–136.

106 Ruskell, G. L., Facial nerve distribution to the eye. Am. J. Optom. Physiol. Opt. *62* (1985) 793–798.

107 Schon, F., Ghatei, M., Allen, J. M., Mulderry, P. K., Kelly, J. S., and Bloom, S. R., The effect of sympathectomy on calcitonin gene-related peptide levels in rat trigeminovascular system. Brain Res. *348* (1985) 197–200.

108 Schorderet, M., Hof, P., and Magistretti, P. J., The effects of VIP on cyclic AMP and glycogen levels in vertebrate retina. Peptides *5* (1984) 295–298.

109 Shimizu, Y., Localization of neuropeptides in the cornea and uvea of the rat: an immunohistochemical study. Cell. molec. Biol. *28* (1982) 103–110.

110 Sibony, P. A., Walcott, B., McKeon, C., and Jakobiec, F. A., Vasoactive intestinal polypeptide (VIP) and the innervation of the human lacrimal gland. Archs Ophthalm. (in press) 1988.

111 Skofitsch, G., and Jacobowitz, D. M., Calcitonin gene-related peptide coexists with substance P in capsaicin sensitive neurons and sensory ganglia of the rat. Peptides *6* (1985) 747–754.

112 Skofitsch, G., and Jacobowitz, D. M., Immunohistochemical mapping of galanin-like neurons in the rat central nervous system. Peptides 6 (1985) 509–546.

113 Soloway, M. R., Stjernschantz, J., and Sears, M., The miotic effect of substance P on the isolated rabbit iris. Invest. Ophthalm. vis. Sci. 20 (1981) 47–52.

114 Stjernschantz, J., and Sears, M., Identification of substance P in the anterior uvea and retina of the rabbit. Exp. Eye Res. 35 (1982) 401–404.

115 Stone, R. A., Vasoactive intestinal polypeptide and the ocular innervation. Invest. Ophthalm. vis. Sci. 27 (1986) 951–957.

116 Stone, R. A., Neuropeptide Y and the innervation of the human eye. Exp. Eye Res. 42 (1986) 349–355.

117 Stone, R. A., and Glembotski, C. C., Immunoreactive atrial natriuretic peptide in the rat eye: Molecular forms in anterior uvea and retina. Biochem. biophys. Res. Commun. 134 (1986) 1022–1028.

118 Stone, R. A., and Kuwayama, Y., Substance P-like immunoreactive nerves in the human eye. Archs Ophthalm. 103 (1985) 1207–1211.

119 Stone, R. A., Kuwayama, Y., Laties, A. M., and Marangos, P. J., Neuron-specific enolase-containing cells in the rhesus monkey trabecular meshwork. Invest. Ophthalm. vis. Sci. 25 (1984) 1332–1334.

120 Stone, R. A., Kuwayama, Y., Laties, A. M., McGlinn, A. M., and Schmidt, M. L., Guinea-pig ocular nerves contain a peptide of the cholecystokinin/gastrin family. Exp. Eye Res. 39 (1984) 387–391.

121 Stone, R. A., Kuwayama, Y., and McGlinn, A. M., Galanin-like immunoreactive nerves in the porcine eye. Exp. Eye Res. 46 (1988) 457–461.

122 Stone, R. A., Kuwayama, Y., Terenghi, G., and Polak, J. M., Calcitonin gene-related peptide: occurrence in corneal sensory nerves. Exp. Eye Res. 43 (1986) 279–283.

123 Stone, R. A., and Laties, A. M., Pancreatic polypeptide-like immunoreactive nerves in the guinea pig eye. Invest. Ophthalm. vis. Sci. 24 (1983) 1620–1623.

124 Stone, R. A., Laties, A. M., and Brecha, N. C., Substance P-like immunoreactive nerves in the anterior segment of the rabbit, cat and monkey eye. Neuroscience 7 (1982) 2459–2468.

125 Stone, R. A., Laties, A. M. and Emson, P. C., Neuropeptide Y and the ocular innervation of rat, guinea pig, cat and monkey. Neuroscience 17 (1986) 1207–1216.

126 Stone, R. A., Laties, A. M., Hemmings, H. C. Jr, Ouimet, C. C., and Greengard, P., DARPP-32 in the ciliary epithelium of the eye: A neurotransmitter-regulated phospho-protein of brain localizes to secretory cells. J. Histochem. Cytochem. 34 (1986) 1456–1468.

127 Stone, R. A., Laties, A. M., Raviola, E., and Wiesel, T. N., Increase in retinal vasoactive intestinal polypeptide after eyelid fusion in primates. Proc. natl Acad. Sci. USA 85 (1988) 257–260.

128 Stone, R. A., McGlinn, A. M., Kuwayama, Y., and Grimes, P. A., Peptide immunoreac-tivity of the ciliary ganglion and its accessory cells in the rat. Brain Res. (1988) in press.

129 Stone, R. A., Tervo, T., Tervo, K., and Tarkkanen, A., Vasoactive intestinal polypeptide-like immunoreactive nerves to the human eye. Acta ophthalm. 64 (1986) 12–18.

130 Stone, R. A., Wilson, C. M., and Glembotski, C. C., Chromatographic characterization of vasoactive intestinal polypeptide in guinea pig and rhesus monkey eyes. (in preparation) 1988.

131 Strömberg, I., Bjöklund, H., Melander, T., Rökaeus, Å, Hökfelt, T., and Olson, L., Galanin-immunoreactive nerves in the rat iris: alterations induced by denervations. Cell Tiss. Res. 250 (1987) 267–275.

132 Sugrue, M. F., and Viader, M.-P., Synthetic atrial natriuretic factor lowers rabbit intraocular pressure. Eur. J. Pharmac. 130 (1986) 349–350.

133 Sunderland, S., and Hughes, E. S. R., The pupillo-constrictor pathway and the nerves to the ocular muscles in man. Brain 69 (1946) 301–309.

134 Suzuki, R., and Kobayashi, S., Vasoactive intestinal peptide and cholonergic neurotrans-mission in the ciliary muscle. Invest. Ophthalm. vis. Sci. 24 (1983) 250–253.

135 Suzuki, R., and Kobayashi, S., Different effects of substance P and vasoactive intestinal peptide on the motor function of bovine intraocular muscles. Invest. Ophthalm. vis. Sci. 24 (1983) 1566–1571.

136 Tatemoto, K., PHI—a new brain-gut peptide. Peptides 5 (1984) 151–154.

137 Terenghi, G., Polak, J. M., Allen, J. M., Zhang, S. Q., Unger, W. G., and Bloom, S. R., Neuropeptide Y-immunoreactive nerves in the uvea of guinea pig and rat. Neurosci. Lett. *42* (1983) 33–38.

138 Terenghi, G., Polak, J. M., Ghatei, M. A., Mulderry, P. K., Butler, J. M., Unger, W. G., and Bloom, S. R., Distribution and origin of calcitonin gene-related peptide (CGRP) immunoreactivity in the sensory innervation of the mammalian eye. J. comp. Neurol. *233* (1985) 506–516.

139 Terenghi, G., Polak, J. M., Probert, L., McGregor, G. P., Ferri, G. L., Blank, M. A., Butler, J. M., Unger, W. G., Zhang, S. Q., Cole, D. F., and Bloom, S. R., Mapping quantitative distribution and origin of substance P- and VIP-containing nerves in the uvea of guinea pig eye. Histochemistry *75* (1982) 399–417.

140 Tervo, K., Tervo, T., Eränkö, L., Eränkö, O., and Cuello, A. C., Immunoreactivity for substance P in the Gasserian ganglion, ophthalmic nerve and anterior segment of the rabbit eye. Histochem. J. *13* (1981) 435–443.

141 Tervo, K., Tervo, T., Eränkö, L., Eränkö, O., Valtonen, S., and Cuello, A. C., Effect of sensory and sympathetic denervation on substance P immunoreactivity in nerve fibres of the rabbit eye. Exp. Eye Res. *34* (1982) 577–585.

142 Tervo, K., Tervo, T., Eränkö, L., Vannas, A., Cuello, A. C., and Eränkö, O., Substance P-immunoreactive nerves in the human cornea and iris. Invest. Ophthalm. vis. Sci. *23* (1982) 671–674.

143 Tornqvist, K., Mandhal, A., Leander, S., Lorén, I., Håkanson, R., and Sundler, F., Substance P-immunoreactive nerve fibres in the anterior segment of the rabbit eye. Cell Tiss. Res. *222* (1982) 467–477.

144 Uddman, R., Alumets, J., Ehinger, B., Håkanson, R., Lorén, I., and Sundler, F., Vasoactive intestinal peptide nerves in ocular and orbital structures of the cat. Invest. Ophthalm. vis. Sci. *19* (1980) 878–885.

145 Ueda, N., Muramatsu, I., and Fujiwara, M., Dual effects of dynorphin-(1-13) on cholinergic and substance P-ergic transmissions in the rabbit iris sphincter muscle. J. Pharmac. exp. Ther. *232* (1985) 545–550.

146 Ueda, N., Muramatsu, I., Hayashi, H., and Fujiwara, M., Effects of met-enkephalin on the substance P-ergic and cholinergic responses in the rabbit iris sphincter muscle. J. Pharmac. exp. Ther. *226* (1983) 507–511.

147 Unger, W. G., Terenghi, G., Ghatei, M. A., Ennis, K. W., Butler, J. M., Zhang, S. Q., Too, H. P., Polak, J. M., and Bloom, S. R., Calcitonin gene-related peptide as a mediator of the neurogenic ocular injury response. J. ocul. Pharmac. *1* (1985) 189–199.

148 Unger, W. G., and Tighe, J., The response of the isolated iris sphincter muscle of various mammalian species to substance P. Exp. Eye Res. *39* (1984) 677–684.

149 Uusitalo, H., Lehtosalo, J., Palkama, A., and Toivanen, M., Vasoactive intestinal polypeptide (VIP)-like immunoreactivity in the human and guinea-pig choroid. Exp. Eye Res. *38* (1984) 435–437.

150 Uusitalo, H., Lehtosalo, J. I., and Palkama, A., Vasoactive intestinal polypeptide (VIP)-immunoreactive nerve fibers in the anterior uvea of the guinea pig. Ophthalm. Res. *17* (1985) 235–240.

151 Watling, K. J., and Dowling, J. E., Effects of vasoactive intestinal peptide and other peptides on cyclic AMP accumulation in intact pieces and isolated horizontal cells of the teleost retina. J. Neurochem. *41* (1983) 1205–1213.

152 Watt, C. B., Li, H.-B., and Lam, D. M.-K., The presence of three neuroactive peptides in putative glycinergic amacrine cells of an avian retina. Brain Res. *348* (1985) 187–191.

153 Watt, C. B., Ying-yet, T. S., and Lam, D. M.-K., Interactions between enkephalin and GABA in avian retina. Nature *311* (1984) 761–763.

154 Weiler, R., and Ball, A. K., Co-localization of neurotensin-like immunoreactivity and ^{3}H-glycine uptake system in sustained amacrine cells of turtle retine. Nature *311* (1984) 759–761.

155 Zhang, S. Q., Terenghi, G., Unger, W. G., Ennis, K. W., and Polak, J., Changes in substance P- and neuropeptide Y-immunoreactive fibres in rat and guinea-pig irides following unilateral sympathectomy. Exp. Eye Res. *39* (1984) 365–372.

Peptides in the mammalian cardiovascular system

J. Wharton and S. Gulbenkian

Summary. Ample immunocytochemical evidence is now available demonstrating that several peptides are present in the mammalian cardiovascular system where they are localised to nerve fibres and myocardial cells. The neuropeptides (neuropeptide Y, calcitonin gene-related peptide, tachykinins and vasoactive intestinal polypeptide) are localised to large secretory vesicles in subpopulations of afferent or efferent nerves supplying the heart and vasculature of several mammals, including man. Although they often exert potent pharmacological effect on the tissues in which they occur their physiological significance has still to be established. They may act directly via specific receptors and/or indirectly by influencing the release and action of other cardiovascular transmitters. In marked contrast, atrial natriuretic peptide is produced by cardiac myocytes and considered to act as a circulating hormone.

It is now recognised that in addition to classical sympathetic (noradrenaline) and parasympathetic (acetylcholine) transmitters, the subpopulations of nerve fibres supplying the cardiovascular system also contain other putative transmitters including several so-called regulatory peptides. Considerable advances have been made in our knowledge of cardiovascular innervation following the application of histochemical and ultrastructural methods[26,95] but it is the recent use of immunocytochemical techniques which has allowed us to demonstrate the presence of peptides and transmitter synthesising enzymes in cardiovascular nerves and thus distinguish between different autonomic nerve types. In the future several other immunocytochemical markers maybe of value in studies of cardiovascular innervation. These include two membrane proteins, synapsin and synaptophysin, specifically associated with the small secretory vesicles that store classical transitters in nerve terminals[146,153,154,218]; and the neuronal cytoplasmic protein, protein gene product 9.5 (PGP 9.5), which was originally extracted from human brain[107] and is present throughout the cardiovascular innervation[95].

In this article we review the immunocytochemical and pharmacological evidence concerning the localisation and actions of regulatory peptides in the mammalian heart and blood vessels. Of the peptides identified to date in cardiovascular nerves the most widely distributed are neuropeptide Y, calcitonin gene-related peptide, the tachykinins and vasoactive intestinal polypeptide.

Neuropeptide Y

Neuropeptide Y (NPY) is a 36 amino acid peptide, originally extracted from porcine brain and chemically characterised as having a C-terminal tyrosine amide group[192]. It belongs to a group of peptides which have a high degree of sequence homology, including pancreatic polypeptide (PP) and peptide YY (PYY)[68]. Sequence analysis of the cDNA encoding human NPY has revealed that the prepro-NPY molecule consists of 97 amino acids and its predicted post-translational processing yields three peptides corresponding to the signal peptide (28 amino acids), NPY (36 amino acids) and the C-flanking peptide of NPY (CPON, 30 amino acids)[148]. CPON immunoreactivity occurs naturally in mammalian tissues[7] and has an identical distribution to NPY in both the nervous system and adrenal medulla[93]. NPY/CPON-immunoreactive nerve fibres appear to be the most abundant of all the peptide-containing nerve populations identified to date in the mammalian cardiovascular system. High concentrations of both peptide sequences are found in the heart[7,8,92] where they occur in nerve fibres associated with the endocardium, myocardium, and coronary vessels and in epicardial nerves. The number of immunoreactive fibres tends to be greater in the atria than the ventricles and higher in the right atrium than the left. NPY/CPON-immunoreactive nerve fibres are also distributed around arteries (elastic and muscular) throughout the vascular system, forming an outer network of nerve bundles containing preterminal axons, running mainly parallel to the vessel and a perivascular plexus of fine, mainly varicose fibres and fascicles running around the vessel at the adventitial-medial border[61,62,67,93,149,202]. The density of the perivascular plexus varies in different species, as well as with vessel size and site. The immunostained nerve fibres are usually confined to the adventitial-medial border of systemic vessels, however, nerve fibers are known to penetrate the media of some large arteries in a number of species[26,85]. We have observed NPY/CPON-immunoreactive nerve fibres in the outer media of the pig elastic pulmonary artery, running in a circular direction, in association with both the vasa vasorum and smooth muscle cells between the elastic laminae.

Most of the published studies concerning the distribution of NPY-immunoreactive nerves in the vascular system have used rats, guinea pig and cat tissues[61,62,67,91,93,134,143,149], but the presence of NPY-immunoreactivity has also been noted in human omental[54], mesenteric[67], skin[108] and cerebral vessels[6]. We have localised NPY/CPON immunoreactivity to nerves around human spinal (fig. 2), coronary, pulmonary (fig. 5), renal, gastric, splenic and mesenteric blood vessels. These immunoreactive nerves occur in perivascular plexus around both arteries and veins, the plexus being less dense in the latter, and represent a subpopulation of the total innervation which displays PGP 9.5-immunoreactivity (fig. 1).

294

Combined immunocytochemical and denervation studies have demonstrated that the distribution of NPY-immunoreactive nerve fibres in the mammalian cardiovascular system is very similar to that of nerves containing the catecholamine synthesising enzymes (tyrosine hydroxylase and dopamine-beta-hydroxylase) and the majority at least appear to represent noradrenergic, postganglionic, sympathetic neurones[61,62,67,91,93,134,143,149]. Thus, the removal of the stellate ganglia results in an almost complete loss of NPY- and tyrosine hydroxylase-immunoreactive nerve fibres in the guinea pig heart[42] and superior cervical ganglionectomy produces a marked depletion of NPY-containing perivascular sympathetic nerve fibres in the upper respiratory tract, oral mucosa, dental pulp, thyroid, iris and around cerebral

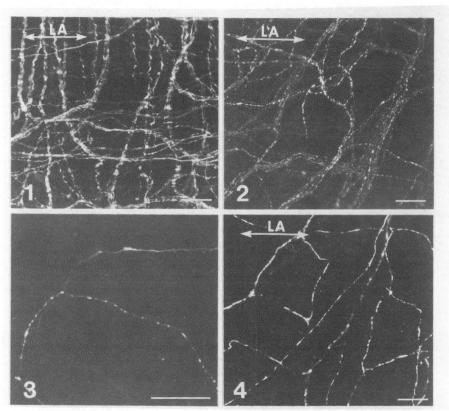

Figures 1–3. Whole mount preparations of a human anterior spinal artery, obtained at post mortem and immunostained for PGP 9.5 (fig. 1), NPY (fig. 2) and CGRP (fig. 3). A network of varicose and non-varicose fibres and fascicles is demonstrated in the adventitia using antisera to PGP 9.5 and NPY. In contrast, only a few, fine fibres are immunostained with an antiserum raised against synthetic alpha-CGRP.

Figure 4. Tyrosine hydroxylase immunoreactivity localised to nerve fibres in a whole mount preparation of a human anterior spinal vein.

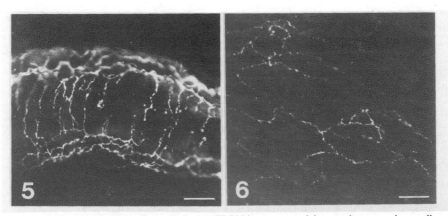

Figure 5. Fine perivascular fibres displaying CPON immunoreactivity running around a small artery of the adventitial vasa vasorum in a whole mount preparation of a human pulmonary artery obtained at surgery.

Figure. 6 A whole mount preparation of a human mesenteric artery which contains a network of varicose fibres displaying VIP immunoreactivity. LA, longitudinal axis of vessel. Bar = 50 μm.

vessels[61,62,66,67,202]. NPY immunoreactivity is also depleted from cardiovascular noradrenergic nerve terminals in the rat and guinea pig after chemical sympathectomy with 6-hydroxydopamine and following reserpine treatment[3,8,67,95,134,137,138,149]. The amount of NPY immunoreactivity which is lost from cardiovascular nerves following 6-hydroxydopamine and reserpine treatment varies between different populations of sympathetic neurones and does not necessarily accompany the depletion in noradrenaline levels[137,138,149]. While it has been suggested that this could be due to variations in the number of terminal and preterminal axons in given tissues[149] it may also indicate the different sources and storage sites for NPY and noradrenaline in these neurones. Unlike noradrenaline, NPY is synthesised in the neuronal cell body and reaches nerve terminals via axonal transport. Furthermore, subcellular fractionation studies in the cat spleen[77] and rat vas deferens[76] have suggested that noradrenaline occurs mainly in the small vesicles whereas NPY is contained, together with some noradrenaline, in the less numerous large vesicles in sympathetic nerves. We have recently confirmed this proposed subcellular localisation of NPY in samples of human atrial appendage obtained from patients undergoing coronary artery by-pass grafts. Post-embedding, immunogold labelling techniques were employed at the ultrastructural level to demonstrate that both NPY and CPON immunoreactivity are localised to large electron dense secretory vesicles (diameter 70–100 nm) in nerve terminals which also contain numerous smaller sized vesicles (diameter 40–60 nm) and are presumed to represent sympathetic nerves (figs 7, 8). Thus while NPY and noradrenaline coexist in sympathetic cardiovascular terminals their localisation in two distinct subcellular stores could enable a differential release of noradrenaline or NPY, the latter being preferentially released at high stimulation frequencies[135]. The noradrenergic neurone blocker guanethidine inhibits the release of both noradrenaline and NPY[133].

Although NPY and noradrenaline coexist in sympathetic cardiovascular nerves it should be remembered that not all noradrenergic nerves contain NPY[29] and not all NPY-immunoreactive nerve fibres are noradrenergic[80]. Although most NPY-containing nerve fibres in the heart seem to represent extrinsic sympathetic nerves there is immunocytochemical evidence to indicate that some intrinsic cardiac neurones may also contain NPY immunoreactivity[42,91,100].

The functional significance of the coexistence of NPY and noradrenaline has yet to be established, but NPY may influence sympathetic vascular control in at least three ways, having both direct and indirect (pre- and post-junctional) effects. NPY exerts a direct, non-adrenergic, calcium dependent vasoconstrictor action on coronary[4,74,171,173] and cerebral vessels[55,61-63,99] from a number of species. In man, NPY induces a direct vasoconstrictor response in some vessels, which is characteristically slow in onset and long lasting. This response has been demonstrated in arteries and veins in the human forearm *in vivo*[162], as well as in renal and submandibular arteries and mesenteric veins *in vitro*, but not in mesenteric arteries[54,67,136]. Variable, generally poor responses to NPY also occur in different vessels in experimental animals[65,99,161,209]. In addition to a direct action, NPY may enhance the post-junctional vasoconstrictor effect of noradrenaline, as well as other transmitters, and inhibits its pre-junctional release[41,67,99,137,161,208,209].

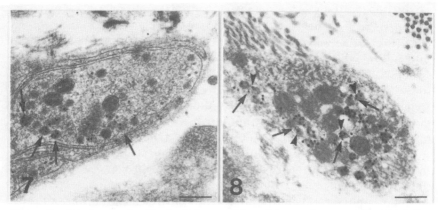

Figures 7 and 8. Axon varicosities in the human right atrial appendage containing NPY (figs 7 and 8) and CPON immunoreactivity (fig. 8). The large granular vesicles (arrows), but not the small vesicles, display NPY-immunogold labelling (fig. 7). Using a double immunogold staining procedure, NPY and CPON are localised to the same secretory vesicles (fig. 8). NPY, 10-nm gold particles (arrows). CPON, 15-nm gold particles (arrowheads). Bar = 200 nm.

Furthermore, the findings of recent studies using pithed guinea pigs indicates that noradrenaline is also capable of reducing the neuronal release of NPY by a pre-junctional α_2-adrenoceptor mediated mechanism[40].

In the isolated rabbit heart NPY was originally reported to have a negative inotropic effect and reduce coronary perfusion[4] whereas positive inotropic and chronotropic effects were observed in the isolated guinea pig atrium[74,132]. However, in other heart preparations from the dog, rat, cat, guinea pig and man, NPY has been found to have no inotropic or chronotropic effects, the only action being a vasoconstrictor one[5,72]. Very little is known about the nature and distribution of NPY receptors in the cardiovascular system, but they have been reported to occur in vascular smooth muscle in the rabbit and guinea pig kidney[124]. Thus, although there are numerous NPY/CPON-immunoreactive nerves in mammalian myocardium the presence of NPY receptors in cardiac muscle is uncertain. Nonetheless, NPY could influence cardiac function by means of its ability to produce vasoconstriction and effect pre-junctional noradrenaline release as well as other types of autonomic transmission[132], including the inhibition of acetylcholine from parasympathetic nerves in the heart[164].

Calcitonin gene-related peptide

Alternative processing of primary transcripts from the calcitonin gene leads to the expression of different mRNA's, encoding either calcitonin or the 37 amino acid peptide, calcitonin gene-related peptide (CGRP) which is predominantly expressed in the nervous system[11,172]. Two forms of CGRP (alpha- and beta-) have been identified in both the rat[11] and human[186]. Immunocytochemical studies have demonstrated that CGRP immunoreactivity is widely distributed in sensory neurones and nerve fibres in the viscera and cardiovascular system of several species[87,89,94,98,121,130,142,144,150,172,179,196,201,210,216]. CGRP-immunoreactive nerve fibres are particularly numerous in guinea pig blood vessels and heart where they are prominent in the endocardium, pericardium and around coronary vessels and are also present in the myocardium and epicardium. Significant concentrations of CGRP immunoreactivity are found in the guinea pig cardiovascular system, with the highest levels occurring in the superior mesenteric artery, inferior vena cava, pulmonary trunk, carotid artery and aortic arch. However, the findings in the guinea pig are only partially representative of those in other animals, there being marked regional and species variations regarding the distribution of CGRP-immunoreactive nerve fibres[216]. In contrast to the guinea pig, relatively few nerve fibres occur in the human heart and vasculature (fig. 3) and appear to display CGRP immunoreactivity when immunostained with antisera raised against human or rat alpha-CGRP.

The distribution of CGRP and substance P immunoreactivities is very similar and several studies have now demonstrated that they are co-localised in a population of sensory neurones[87,89,121,122,142,210]. The degree of coexistence appears to be very high in the guinea pig where some 90% of all CGRP-immunoreactive sensory neurones display substance P immunoreactivity[87] and both peptides are invariably found together in the same perivascular fibres. Furthermore, recent immunoelectron microscopical studies in our laboratory have revealed that this coexistence extends to the subcellular level, the two immunoreactivites being present in the same secretory vesicles in both the varicosities of perivascular fibres (fig. 9) and in sensory perikarya of the guinea pig[94].

The sensory origin of CGRP and substance P immunoreactive nerve fibres in the cardiovascular system has been substantiated by using the selective neurotoxin capsaicin. Systemic capsaicin treatment of guinea pigs and rats leads to a marked loss of substance P and CGRP immunoreactivity in the heart and vasculature[51,52,79,81,150,151,160,206]. Cardiac and perivascular nerves containing VIP and NPY immunoreactivities appear to be unaffected[42,46,51,95].

Capsaicin treatment of adult guinea pigs produces an 88–99% depletion of CGRP immunoreactivity in the cardiovascular system, together with a parallel loss of substance P immunoreactivity, but a more varied dose dependent response is observed in rats treated neonatally[216]. In the rat, CGRP immunoreactivity appears to occur in two populations of sensory neurones, one of which also contains substance P and is sensitive

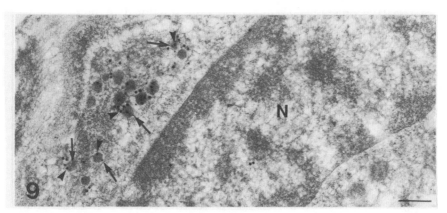

Figure 9. An axon varicosity in the adventitia of a guinea pig superior mesenteric artery. Co-localisation of substance P and CGRP immunoreactivities to the same secretory vesicles, as demonstrated by a double immunogold staining procedure. Substance P, 10-nm gold particles (arrows). CGRP, 15-nm gold particles (arrowheads). N, Schwann cell nucleus. Bar = 200 nm.

to the action of capsaicin whereas the other possesses only CGRP and is resistant to the neurotoxin[142]. Surgical denervation has also been used to demonstrate the sensory origin of CGRP-immunoreactive nerves supplying the cardiovascular system. Thus, destruction of the trigeminal ganglion in the rat[142] and lesions of the trigeminal nerve in the cat[144] result in a loss of perivascular nerve fibres containing CGRP and substance P immunoreactivity around cerebral blood vessels.

CGRP has been shown to exert a potent vasodilatory action, both *in vivo* and *in vitro*, on cerebral, coronary and other peripheral blood vessels from several animals, including rat, guinea pig, rabbit, cat and man[18,21,22,59,98,144,145,201]. The effect is dose-dependent and not modified by adrenergic, cholinergic, histaminergic or neuronal blockade. CGRP binding sites have been reported to occur in both the media and intima of rat coronary arteries and aorta[185]. The vasodilatory response could be partly mediated by an endothelium-dependent mechanism[22,111] and partly via a direct action on the media with the activation of adenylate cyclase[59,98,207], but there may be regional and/or species variations. Recent studies using cultures of human umbilical vein endothelial cells have shown that CGRP produces a dose-dependent increase in adenylate cyclase activity and the release of prostacyclin[37]. It is uncertain whether any correlation exists between the density of the CGRP-immunoreactive innervation of blood vessels and their responsiveness to exogenous CGRP. CGRP may potentiate the effect of tachykinins and other agents which include plasma extravasation in a inflammatory reaction[23,82] and this might be brought about at least in part, by inhibiting substance P degradation[118] and potentiating substance P release[158].

In addition to its effect on coronary vessels, CGRP has direct inotropic and chronotropic actions on the isolated rat and guinea pig heart. Unlike substance P, it also mimics the non-adrenergic, non-cholinergic excitatory response induced by transmural nerve stimulation and capsaicin[75,101,130,176,180]. The effect of capsaicin on the heart appears to be mediated by the release of CGRP from afferent cardiac nerve fibres. CGRP has also been shown to have a positive inotropic effect on the isolated human atrium, but in contrast to the guinea pig capsaicin did not stimulate the contractility of the preparation[72]. This species variation in response to capsaicin may reflect the lower density of CGRP-containing nerves in the human heart compared to that found in the guinea pig.

Tachykinins

The tachykinins are a group of biologically active peptides possessing a common C-terminal amino acid sequence. Substance P is one of the best characterised neuropeptides and until recently was the only tachykinin known to occur in the mammalian nervous system, but at least two other tachykinins, neurokinin A (NKA) and neurokinin B (NKB), have now been identified[109,110,147]. The primary structure of two bovine preprotachykinins has also been determined, one (beta-prepro-tachykinin) containing sequences homologous to both substance P and NKA and the other substance P alone[155].

A considerable amount of immunocytochemical evidence is now available demonstrating that substance P occurs in primary sensory neurones which have peripheral branches associated with the cardiovascular system. These immunoreactive nerve fibres are generally sensitive to capsaicin treatment and have been identified in the heart[79,81,159,160,170,206,212,214], around both large conducting arteries and veins and smaller vessels supplying many vascular beds in a variety of mammals[16,52,57,81,126,127,152,160,175,203]. Substance P-immunoreactive cardiovascular nerves are particularly numerous in the guinea pig where, as indicated above, there is extensive coexistence with CGRP immunoreactivity. Immunostaining of the guinea pig cardiovascular system for either peptide probably demonstrates an identical population of sensory nerves, the density of the perivascular plexus being about half that of the noradrenergic network.[48]

Recent immunocytochemical investigations have revealed that in addition to substance P other tachykinins are also present in guinea pig capsaicin sensitive nerves[103]. These nerves may therefore contain several bioactive peptides, comprising at least two tachykinins (substance P and NKA) and CGRP. In comparison to the guinea pig, the rat cardiovascular system contains significantly less substance P immunoreactivity and fewer substance P-immunoreactive nerve fibres. CGRP and substance P occur in a heterogenous subpopulation of rat sensory neurones with less than half of the CGRP-immunoreactive neurones containing substance P as well[121,122,142]. Substance P- and CGRP-immunoreactive cardiovascular nerves in the rat also show marked variations in their sensitivity to capsaicin[142,175,216].

Substance P is considered to be a putative sensory neurotransmitter mediating nociceptive and neurogenic vasodilatory responses and plasma extravasation[83,123,178]. In addition to modulating peripheral vasodilatory processes substance P has a vasodilator action in a number of blood vessels[64,96,128,177]. Substances P and other vasoactive agents, such as acetylcholine, bradykinin, ATP and related purines, require the presence of an intact endothelium to exert all or part of their effect on arteries, this being mediated by the release of an endothelial derived relaxing factor[14,27,39,59,78]. Substance P binding sites have been identified by *in vitro* autoradiographic mapping in blood vessels in the guinea pig and human lung[32], rat thymus[184], dog carotid artery[187] and dog renal artery[188]. The binding in these vessels appears to be mainly associated with the intimal surface and requires an intact endothelium; however, in the bovine coronary artery and rat thoracic aorta substance P seems to bind to the media rather than the intima[188]. It is now thought that there are three tachykinin receptors, NK-1, NK-2 and NK-3, each having a preferential affinity for the respective tachykinins, substance P, NKA and NKB[168]. These receptors have a heterogenous distribution in the vascular wall of different species and this may be responsible for the various effects which the tachykinins exert in the mammalian cardiovascular system. Thus, the guinea pig basilar artery, dog carotid artery and rabbit pulmonary artery are all thought to possess NK-1 type receptors with substance P inducing the release of an endothelial derived relaxing factor[39,60,168]. On the other hand, NKA and NKB may induce a direct contractile response in the rabbit pulmonary artery and rat portal vein, acting via NK-2 and NK-3 tachykinin receptors located on the vascular smooth muscle[39,168]. In contrast to CGRP, substance P does not mimic the actions of capsaicin on the heart, apparently lacking any direct effect on the contractility of the mammalian heart[28,72,128,130].

Vasoactive intestinal polypeptide (VIP)

Vasoactive intestinal polypeptide (VIP) is a 28 amino acid peptide which was originally isolated from porcine intestine and recognised for its potent vasodilatory effect[174]. Another peptide, PHI-27 (*p*eptide with N-terminal *h*istidine and C-terminal *i*soleucine), has also been isolated from porcine intestine[191,193] and found to have a distribution similar to that of VIP[19,34,139]. The human form of this peptide, PHM (C-terminal *m*ethionine), is derived from the same prepromodule as VIP[20,106]. Both peptide sequences are present in cardiovascular nerves, these generally being presumed to represent post-ganglionic parasympathetic neurones. Perivascular nerve fibres displaying VIP and PHI/PHM immunoreactivity tend to occur more frequently around vessels in regional vascular beds than in association with larger conducting vessels[46,199]. As with the other types of peptide-containing cardiovascular nerves there are species variations in the distribution of VIP-immunoreactive nerves, the perivascular plexus usually being more dense in the cat and pig than in other animals (e.g. rat, guinea pig and dog). In man, the density of the VIP-immunoreactive perivascular nerves appears to be between that of nerves displaying NPY/CPON and CGRP or substance P immunoreactivities, whereas in the guinea pig it is the least dense of all these networks[48]. VIP- and PHI/PHM-immunoreactive perivascular nerves are reported to be relatively numerous in many tissues including the gastrointestinal (fig. 6)[19,25,36,183], genitourinary[9,10,140,215] and respiratory tracts[47,131], salivary glands[129,198,217] and the eye[195,200]. They also occur frequently around the cerebral vasculature, with anterior vessels in the circle of Willis receiving a more dense supply than those in the posterior circulation[53,56,58,88,105,115]. At the ultrastructural level, VIP immunoreactivity has been localised to large secretory vesicles in nerve terminals and axons of cat cerebral vessels[120]. The immunoreactivity persisted after long term sympathetic denervation indicating its presence in nonadrenergic nerves.

VIP has a direct vasodilatory action on cerebral, pulmonary, coronary and other systemic vessels from several species, both in vivo and in vitro[17,24,50,58,104,115,116,120,205]. The extent to which the arterial vasodilatory response to exogenous VIP is coupled with the density of the VIP-containing perivascular network is uncertain[189], but the response is not dependent on the presence of an intact endothelium[50,120,207]. Specific VIP-binding sites have been identified in the mammalian lung and localised to the media of pulmonary blood vessels[17,31,125,181]. Similar receptors have also been demonstrated in the media of bovine cerebral arteries[165,190]. It is thought that VIP may mediate non-adrenergic, non-cholinergic, neurogenic vasodilation in a number of vascular beds, this having been most extensively studied in the cat submandibular gland[129]. In addition to VIP, PHI and PHM are also known to have vasoactive properties, although they may be less potent vasodilatory agents than VIP[56,116,157,189].

In contrast to the numerous cardiac nerve fibres displaying NPY/CPON immunoreactivity it seems that relatively few contain VIP-like material. VIP immunoreactivity has been found in the guinea pig, rat, dog, cat, monkey and human heart where it is reported to occur mainly in nerve fibres associated with the atria, conduction system and coronary vessels[24,46,167,212,213]. The presence of some VIP-immunoreactive neuronal cell bodies in intracardiac ganglia has also been demonstrated, at least in the dog suggesting an intrinsic origin for VIP-immunoreactive cardiac nerve fibres[212,213].

VIP exerts a direct positive chronotropic and inotropic effects on the heart[45,72,205] and VIP receptors, coupled to adenylate cyclase, have been found in atrial and ventricular membrane preparations of the dog, monkey[33] and human heart[194]. PHI receptors were also identified in the human preparations, VIP and PHI having a similar capacity to stimulate adenylate cyclase activity. Finally, VIP has been implicated in pathophysiological aspects of haemorrhagic shock[35], hypertension and heart failure[71,204].

Other neuropeptides

In addition to the peptides described above there are a number of others which have also been detected in mammalian cardiovascular nerves, including somatostatin, enkephalin and neurotensin, but in general they appear to have a relatively limited distribution.

Somatostatin-like immunoreactivity has been extracted from the human, guinea pig and rat heart[43] and localised to both nerve fibres in the myocardium and conduction system and to local, presumably parasympathetic ganglion cells in the atrium[43,73]. In the isolated guinea pig atrium, somatostatin has a negative inotropic action on both basal and electrically stimulated contractions[49] whereas in human atrial preparation it only inhibits the positive inotropic action of noradrenaline[72,73]. It is suggested that the negative inotropic effect of somatostatin could be due to its ability to reduce Ca^{2+} influx in the atrium and this may also explain its actions on atrioventricular nodal function and its antiarrhythmic properties[90,211].

The rat heart is reported to contain the mRNA encoding preproenkephalin[102] and there is indirect evidence to suggest that enkephalin immunoreactivity in the guinea pig heart is associated with sympathetic nerves, there being a significant reduction in the amount of cardiac enkephalin following 6-hydroxydopamine treatment[113]. Neurotensin-like immunoreactivity also occurs in extracts of the guinea pig heart and has been localised to cardiac nerves in several mammals[169,212] where it has coronary vasoconstrictor actions and may be a positive inotropic and chronotropic agent.

Atrial natriuretic peptide

In contrast to the neuropeptides found in cardiovascular nerves, atrial natriuretic peptide (ANP) is synthesised in cardiac myocytes and may act as a circulating hormone. Following the observation of de Bold and co-workers[44] that intravenous injections of rat atrial extracts induced diuresis and natriuresis in donor rats there has been an intense effort to isolate and characterise the factor responsible. The resulting literature has been comprehensively reviewed elsewhere[13,15,69,86] and here we shall only briefly consider the localisation of ANP in the heart and the possible influence of the cardiovascular innervation on the release and actions of ANP.

Like other so-called regulatory peptides, ANP is synthesised as a larger prepromolecule which in man comprises 151 amino acids. The C-terminal 28 amino acids represents alpha-ANP and this is probably the major circulating form of the peptide in man. The rest of the precursor contains a signal peptide and N-terminal sequence homologous to a peptide having vasodilatory properties which was isolated from the porcine heart and termed cardiodilatin[70]. The anatomical localisation of ANP to atrial myocytes has been confirmed in several mammals including man[12,30,141], where it is present in secretory vesicles (fig. 10) together with cardiodilatin immunoreactivity (fig. 11). Tension of the atrial wall[2,114,119,182] and atrial contraction frequency[166,182] are

304

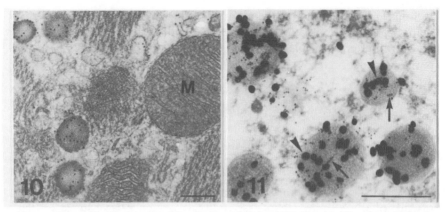

Figures 10 and 11. Ultrastructural localisation of alpha-ANP (figs 10–11) and cardiodilatin 1–16 (fig. 11) in human right atrial appeandage. ANP immunoreactivity localised to secretory vesicles in an atrial myocyte by immunogold staining. ANP, 10-nm gold particles (fig. 10). ANP and cardiodilatin immunoreactivities are co-localised to the same secretory vesicles (fig. 11). ANP, 5-nm, gold particles (arrows). Cardiodilatin, 20-nm gold particles (arrowheads). M, mitochondria. Bar = 200 nm.

thought to be the main factors regulating the secretion of ANP. The rôle of the autonomic nervous system[38,112,163] and the possible influence of cardiac neuropeptides, such as NPY, in this process is uncertain. On the other hand, there is evidence to suggest that ANP stimulates vagal afferent nerve endings in the heart[1,197] and it may attenuate the vaso-pressor actions of angiotensin II and noradrenaline[220]. Attention has focused on the atria as the site of ANP production in the mammalian heart, but it is now apparent that ventricular myocytes are also capable of synthesising ANP, at least in the rat[84,97,156], and this extra-atrial expression of the ANP gene is increased by volume loading[117].

Conclusion

Several neuropeptides are present in the mammalian cardiovascular system where they are localised to specific subpopulations of efferent and afferent nerves, but exhibit both regional and species variations in their distribution pattern. The immunocytochemical evidence indicates that these peptides often occur in the same nerve fibres as other putative or classical transmitters. While this coexistence may also extend to the subcellular level it appears that peptide precursors are localised exclusively to the large secretory vesicles in axon terminals, whereas the classical transmitters noradrenaline and acetylcholine are thought to occur mainly in the smaller sized vesicle population. The significance of this localisation and the physiological rôle(s) of neuropeptides in the cardiovascular system have still to be established, but they may function as 1) hormones; 2) transmitters acting via their own receptors, 3) neuromodulators influencing the release and/or action of transmitters and 4) long-term (trophic) agents.

Acknowledgments. Financial support was provided by the British Heart Foundation. Sergio Gulbenkian is supported by a fellowship from the Calouste Gulbenkian Foundation, Lisbon, Portugal.

1 Ackermann, U., Irizawa, T. G., Milojevic, S., and Sonnenberg, H., Cardiovascular effects of atrial extracts in anesthetized rats. Can. J. Physiol. Pharmac. *62* (1984) 819–826.

2 Akabane, S., Kujima, S., Igarashi, Y., Kawamura, M., Matsushima, Y., and Ito, K., Release of atrial natriuretic polypeptide by graded right atrial distension in anesthetized dogs. Life Sci. *40* (1987) 119–125.

3 Allen, J. M., Schon, F., Yeats, J. C., Kelly, J. S., and Bloom, S. R., Effects of reserpine, phenoxybenzamine and cold stress on the neuropeptide Y content of the rat peripheral nervous system. Neuroscience *19* (1986) 1251–1254.

4 Allen, J. M., Bircham, P. M. M., Edwards, A. V., Tatemoto, K., and Bloom, S. R., Neuropeptide Y (NPY) reduces myocardial perfusion and inhibits the force of contraction of the isolated perfused rabbit heart. Ref. Pep. *6* (1983) 247–253.

5 Allen, J. M., Gjörstrup, P., Björkman, J. A., Abrahamsson, T., and Bloom, S. R., Studies on cardiac distribution and function of neuropeptide Y. Acta physiol. scand. *126* (1983) 247–253.

6 Allen, J. M., Schon, F., Todd, N., Yeats, J. C., Crockard, H. A., and Bloom, S. R., Presence of neuropeptide Y in human circle of Willis and its possible role in cerebral vasospasm. Lancet *2* (1984) 550–552.

7 Allen, J. M., Polak, J. M., and Bloom, S. R., Presence of the predicted C-flanking peptide of neuropeptide Y (CPON) in tissue extracts. Neuropeptides *6* (1985) 95–100.

8 Allen, J. M., Polak, J. M., Rodrigo, J., Darcy, K., and Bloom, S. R., Localisation of neuropeptide Y (NPY) in nerves of the rat cardiovascular system and effect of 6-hydroxydopamine. Cardiovasc. Res. *19* (1985) 570–577.

9 Alm, P., Aluments, J., Hakanson, R., Owman, C. H., Sjöberg, N. O., Sundler, R., and Walles, B., Origin and distribution of VIP (vasoactive intestinal polypeptide)-nerves in the genito-urinary tract. Cell Tissue Res. *205* (1980) 337–347.

10 Alm, P., Alumets, J., Hakanson, R., and Sundler, R., Peptidergic (vasoactive intestinal peptide) nerves in the genito-urinary tract. Neuroscience *2* (1977) 751–754.

11 Amara, S. G., Arriza, J. L., Leff, S. E., Swanson, L. W., Evans, R. M., and Rosenfeld, M. G., Expression in brain of a messenger RNA encoding a novel neuropeptide homologous to calcitonin gene-related peptide. Science *229* (1985) 1094–1097.

12 Anderson, J. V., Christofides, N. D., Vinas, P., Wharton, J., Varndell, I. M., Polak, J. M., and Bloom, S. R., Radioimmunoassay of alpha rat atrial natriuretic peptide. Neuropeptides *7* (1986) 159–173.

13 Anderson, J. V., and Bloom, S. R., Atrial natriuretic peptide: what is the excitement all about? J. Endocr. *110* (1986) 7–17.

14 Angus, J. A., Campbell, G. R., Cocks, T. M., and Manderson, J. A., Vasodilation by acetycholine is endothelium dependent: A study by sonomicrometry in canine femoral artery in-vivo. J. Physiol. *344* (1983) 209–222.

15 Ballerman, B. J., and Brenner, B. M., Role of atrial peptides in body fluid homeostasis. Circ. Res. *58* (1986) 619–630.

16 Barja, F., Mathison, R., and Huggel, H., Substance P-containing nerve fibres in large peripheral blood vessels of the rat. Cell Tiss. Res. *229* (1983) 411–422.

17 Barnes, P. J., Cadieux, A., Carstairs, J. R., Greenberg, B., Polak, J. M., and Rhoden, K., VIP in bovine pulmonary artery: localisation, function and receptor autoradiography. Br. J. Pharmac. *89* (1986) 157–162.

18 Benjamin, N., Dollery, C. T., Fuller, R. W., Larkin, S., and McEwan, J., The effects of calcitonin gene-related peptide and substance P on resistance and capacitance vessels. Br. J. Pharmac. *90* (1987) 39P.

19 Bishop, A. E., Polak, J. M., Yiangou, Y., Christofides, N. D., and Bloom, S. R., The distribution of PHI and VIP in porcine gut and their co-localisation to a proportion of intrinsic ganglion cells. Peptides 5 (1984) 255–259.

20 Bloom, S. R., Christofides, N. D., Delamarter, J., Buell, G., Kawashima, E., and Polak, J. M., Tumour co-production of VIP and PHI explained by single coding gene. Lancet 2 (1983) 1163–1165.

21 Brain, S. D., MacIntyre, I., and Williams, T. J., A second form of human calcitonin gene-related peptide which is a potent vasodilator. Eur. J. Pharmac. 124 (1986) 349–352.

22 Brain, S. D., Williams, T. J., Tippins, J. R., Morris, H. R., and MacIntyre, I., Calcitonin gene-related peptide is a potent vasodilator. Nature 313 (1985) 54–56.

23 Brain, S. D., and Williams, T. J., Inflammatory oedema induced synergism between calcitonin gene-related peptide (CGRP) and mediators of increased vascular permeability. Br. J. Pharmac. 86 (1985) 855–860.

24 Brum, J. M., Bove, A. A., Sufan, Q., Reilly, W., and Go, V. L. W., Action and localisation of vasoactive intestinal peptide in the coronary circulation: Evidence for nonadrenergic, noncholinergic coronary regulation. J. Am. coll. Cardiol. 7 (1986) 406–413.

25 Bryant, M. G., Bloom, S. R., Polak, J. M., Albuquerque, R. H., Modlin, I., and Pearse, A. G. E., Possible dual role for vasoactive intestinal peptide as gastrointestinal hormone and neurotransmitter substance. Lancet 1 (1976) 991–993.

26 Burnstock, G., Chamley, J. H., and Campbell, G. R., The innervation of arteries, in: Structure and Function of the Circulation, pp. 729–767. Ed. C. J. Schwartz. Plenum Press, New York 1980.

27 Burnstock, G., and Kennedy, C., A dual function for adenosine 5′-triphosphate in the regulation of vascular tone. Circ. Res. 58 (1986) 319–330.

28 Burcher, E., Alterhög, J. H., Pernow, B., and Rosell, S., Cardiovascular effects of substance P: effects on the heart and regional blood flow in the dog, in: Substance P, pp. 261–268. Eds U. S. von Euler and B. Pernow. Raven Press, New York 1977.

29 Cannon, B., Nedergaard, J., Lundberg, J. M., Hökfelt, T., Terenius, L., and Goldstein, M., Neuropeptide tyrosine (NPY) is co-stored with noradrenaline in vascular but not in parenchymal sympathetic nerves of brown adipose tissue. Exp. Cell Res. 164 (1986) 546–550.

30 Cantin, M., Gutkowska, J., Thibault, G., Milne, R. W., Ledoux, S., Minli, S., Chapeau, C., Garcia, R., Hamet, P., and Genest, J., Immunocytochemical localization of atrial natriuretic factor in the heart and salivary glands. Histochemistry 80 (1984) 113–127.

31 Carstairs, J. R., and Barnes, P. J., Visualisation of vasoactive intestinal peptide receptors in human and guinea pig lung. J. Pharmac. exp. Ther. 239 (1986) 249–255.

32 Carstairs, J. R., and Barnes, P. J., Autoradiographic mapping of substance P receptors in lung. Eur. J. Pharmac. 127 (1986) 295–296.

33 Chatelain, P., Robberecht, P., Waelbroeck, M., De Neef, P., Camus, J. C., Naguyen-Huu, A., Roba, J., and Christophe, J., Topographical distribution of the secretion- and VIP-stimulated adenylate cyclase system in the heart of five animal species. Pflügers Arch. 397 (1983) 100–105.

34 Christofides, N. D., Polak, J. M., and Bloom, S. R., Studies on the distribution of PHI in mammals. Peptides 5 (1984) 261–266.

35 Clark, A. J. L., Adrian, T. E., McMichael, J. B., and Bloom S. R., Vasoactive intestinal polypeptide in shock and heart failure. Lancet 1 (1983) 539.

36 Costa, M., and Furness, J. B., The origins, pathways and terminations of nerves with VIP-like immunoreactivity in the guinea pig small intestine. Neuroscience 8 (1982) 665–676.

37 Crossman, D., Dollery, C. T., MacDermot, J., MacIntyre, I., and McEwan, J., Effects of human calcitonin gene-related peptide of human endothelial cells. Br. J. Pharmac. 90 (1987) 38P.

38 Currie, M. G., and Newman, W. H., Evidence for α-1 adrenergic receptor regulation of atriopeptin release from the isolated rat heart. Biochem. biophys. Res. Commun. 137 (1986) 94–100.

39 D'Orleans-Juste, P., Dion, S., Drapeau, G., and Regoli, D., Different receptors are involved in the endothelium-mediated relaxation and the smooth muscle contraction of the rabbit pulmonary artery in response to substance P and related neurokinins. Eur. J. Pharmac. 125 (1985) 37–44.

40 Dahlöf, C., Dahlöf, P., and Lundberg, J. M., α_2-adreneceptor-mediated inhibition of nerve stimulation-evoked release of neuropeptide Y (NPY)-like immunoreactivity in the pithed guinea pig. Eur. J. Pharmac. *131* (1986) 279–283.

41 Dahlöf, C., Dahlöf, P., Tatemoto, K., and Lundberg, J. M., Neuropeptide Y (NPY) reduces field stimulation-evoked release of noradrenaline and enhances force of contraction in the rat portal vein. N.S. Arch. Pharmac. *328* (1985) 327–330.

42 Dalsgaard, C.-J., Franco-Cereceda, A., Saria, A., Lundberg, J. M., Theodorsson-Norheim, E., and Hökfelt, T., Distribution and origin of substance P- and neuropeptide Y-immunoreactive nerves in the guinea pig heart. Cell Tiss. Res. *243* (1986) 477–485.

43 Day, S. M., Gu, J., Polak, J. M., and Bloom, S. R., Somatostatin in the human heart and comparison with guinea pig and rat herat. Br. Heart J. *53* (1985) 153–157.

44 de Bold, A. J., Borenstein, H. B., Veress, A. T., and Sonnenberg, H., A rapid and potent natriuretic response to intravenous injection of artial myocardial extracts in rats. Life Sci. *28* (1981) 89–94.

45 De Neef, P., Robberecht, P., Chatelain, P., Waelbroeck, M., and Christophe, J., The in vitro chronotrophic and inotropic effects of vasoactive intestinal peptide (VIP) on the atria and ventricular papillary muscle from cynomolgus monkey heart. Reg. Pep. *8* (1984) 237–244.

46 Della, N. G, Papka, R. E., Furness, J. B., and Costa, M., Vasoactive intestinal peptide-like immunoreactivity in nerves associated with the cardiovascular system of guinea pigs. Neuroscience *9* (1983) 605–619.

47 Dey, R. D., Shannon, W. A., and Said, S. I., Localisation of VIP-immunoreactive nerves in airways and pulmonary vessels of dogs, cats and human subjects. Cell Tiss. Res. *220* (1981) 231–238.

48 Dhall, U., Cowen, T., Haven, A. J., and Burnstock, G., Perivascular noradrenergic and peptide-containing nerves show different patterns of changes during development and aging in the guinea pig. J. auton. nerv. Syst. *16* (1986) 109–126.

49 Diez, J., Tamargo, J., and Valenzuela, C., Negative inotropic effect of somatostatin in guinea-pig atrial fibres. Br. J. Pharmac. *86* (1985) 547–555.

50 Duckles, S. P., and Said, S. I., Vasoactive intestinal peptide as a neurotransmitter in the cerebral circulation. Eur. J. Pharmac. *78* (1982) 371–374.

51 Duckles, S. P., and Levitt, B., Specificity of capsaicin treatment in the cerebral vasculature. Brain Res. *308* (1984) 141–144.

52 Duckles, S. P., and Buck, S. H., Substance P in the cerebral vasculature: depletion by capsaicin suggests a sensory role. Brain Res. *245* (1982) 171–174.

53 Edvinsson, L., Fahrenkrug, J., Hanko, J., Owman, C., Sundler, F., and Uddman, R., VIP (vasoactive intestinal polypeptide)-containing nerves of intracranial arteries in mammals. Cell Tiss. Res. *208* (1980) 135–142.

54 Edvinsson, L., Hakanson, R., Steen, S., Sundler, F., Uddman, R., and Wahlestedt, C., Innervation of human omental arteries and veins: vasomotor responses to noradrenaline, neuropeptide Y, substance P and vasoactive intestinal polypeptide. Reg. Pep. *12* (1985) 67–79.

55 Edvinsson, L., Characterisation of the contractile effect of neuropeptide Y in feline cerebral arteries. Acta physiol. scand. *125* (1985) 33–41.

56 Edvinsson, L., and McCulloch, J., Distribution and vasomotor effects of peptide HI (PHI) in feline cerebral blood vessels in vitro and in situ. Reg. Pep. *10* (1985) 345–356.

57 Edvinsson, L., and Uddman, R., Immunohistochemical localization and dilatory effect of substance P in human cerebral vessels. Brain Res. *232* (1982) 263–273.

58 Edvinsson, L., McCulloch, J., and Uddman, R., Feline cerebral veins and arteries: comparison of automatic innervation and vasomotor responses. J. Physiol. *325* (1982) 161–173.

59 Edvinsson, E., Fredholm, B. B., Hamel, E., Jansen, I., and Verrecchia, C., Perivascular peptides relax cerebral arteries concomitant with stimulation of cyclic adenosine monophosphate accumulation or release of an endothelium derived relaxing factor in the cat. Neurosci. Lett. *58* (1985) 213–217.

60 Edvinsson, L., and Janssen, I., Characterization of tachykinin receptors in isolated basilar arteries of guinea-pig. Br. J. Pharmac. *90* (1987) 553–559.

308

61 Edvinsson, L., Emson, P., McCulloch, J., Tatemoto, K., and Uddman, R., Neuropeptide, Y: cerebrovascular innervation and vasomotor effects in the cat. Neurosci. Lett. *43* (1983) 79–84.

62 Edvinsson, L., Emson, P., McCulloch, J., Tatemoto, K., and Uddman, R., Neuropeptide Y: immunocytochemical localisation to and effect upon feline pial arteries and veins in vitro and in situ. Acta physiol. scand. *122* (1984) 155–163.

63 Edvinsson, L., Functional role of perivascular peptides in the control of cerebral circulation. Trends Neurosci. *8* (1985) 126–131.

64 Edvinsson, L., and Uddman, R., Immunohistochemical localization and dilatory effect of substance P in human cerebral vessels. Brain Res. *232* (1982) 263–273.

65 Edvinsson, L., Ekblad, E., Hakanson, R., and Wahlestedt, C., Neuropeptide Y potentiates the effects of various vasoconstrictor agents on rabbit blood vessels. Br. J. Pharmac. *83* (1984) 519–525.

66 Edwall, B., Gazelius, B., Fazekas, A., Theodorsson-Norheim, E., and Lundberg, J. M., Neuropeptide Y (NPY) and sympathetic control of blood flow in oral mucosa and dental pulp in the cat. Acta physiol. scand. *125* (1985) 253–264.

67 Ekblad, E., Edvinsson, L., Wahlestedt, C., Uddman, R., Hakanson, R., and Sundler, F., Neuropeptide Y co-exists and co-operates with noradrenaline in perivascular nerve fibres. Reg. Pep. *8* (1984) 225–235.

68 Emson, P. C., and De Quidt, M. E., NPY—a new member of the pancreatic polypeptide family. Trends Neurosci. *7* (1984) 31–35.

69 Forssman, W. G., Cardiac hormones. I. Review on the morphology, biochemistry and molecular biology of the endocrine heart. Eur. J. clin. Invest. *16* (1986) 439–451.

70 Forssmann, W. G., Hock, D., Lottspeich, F., Henschen, A., Kreye, V., Christmann, M., Reinecke, M., Metz, J., Carlqvist, M., and Mutt, V., The right auricle of the heart is an endocrine organ. Cardiodilatin as a peptide hormone candidate. Anat. Embryol. *168* (1983) 309–313.

71 Fouad, F. M., Shimamatsu, K., Said, S. I., and Tarazi, R. C., Inotropic responsiveness in hypertensive left ventricular hypertrophy: impaired inotropic response to glucagon and vasoactive intestinal peptide in renal hypertensive rats. J. cardiovasc. Pharmac. *8* (1986) 398–405.

72 Franco-Cereceda, A., Bengtsson, L., and Lundberg, J. M., Inotropic effects of calcitonin gene-related peptide, vasoactive intestinal polypeptide and somatostatin on the human right atrium in vitro. Eur. J. Pharmac. *134* (1987) 69–76.

73 Franco-Cereceda, A., Lundberg, J. M., and Hökfelt, T., Somatostain: an inhibitory parasympathetic transmitter in the human heart? Eur. J. Pharmac. *132* (1986) 101–102.

74 Franco-Cereceda, A., Lundberg, J. M., and Dahlof, Neuropeptide Y and sympathetic control of heart contractility and coronary vascular tone. Acta physiol. scand. *124* (1985) 361–369.

75 Franco-Cereceda, A., and Lundberg, J. M., Calcitonin gene-related peptide (CGRP) and capsaicin-induced stimulation of heart contractile rate and force. N.S. Arch. Pharmac. *331* (1985) 146–151.

76 Fried, G., Terenius, L., Hökfelt, T., and Goldstein, M., Evidence for differential localization of noradrenaline and neuropeptide Y in neuronal storage vesicles isolated from rat vas deferens. J. Neurosci. *5* (1985) 450–458.

77 Fried, G., Lundberg, J., and Theodorsson-Norheim, E., Subcellular storage and axonal transport of neuropeptide Y (NPY) in relation to catecholamines in the cat. Acta physiol. scand. *125* (1985) 145–152.

78 Furchgott, R. F., Role of endothelium in response of vascular smooth muscle. Cir. Res. *53* (1983) 557–573.

79 Furness, J. B., Papka, R. E., Della, N. G., Costa, M., and Eskay, R. L., Substance P-like immunoreactivity in nerves associated with the vascular system of guinea pigs. Neuroscience *7* (1982) 447–459.

80 Furness, J. B., Costa, M., Emson, P. C., Hakanson, R., Moghimzadeh, E., Sundler, F., Taylor, I. L., and Chance, R. E., Distribution, pathways and relations to drug treatment of nerves with neuropeptide Y- and pancreatic polypeptide-like immunoreactivity in the guinea pig digestive tract. Cell Tiss. Res. *236* (1983) 71–92.

81 Furness, J. B., Elliott, J. M., Murphy, R., Costa, M., and Chambers, J. P., Baroreceptor

reflexes in conscious guinea pigs are unaffected by depletion of cardiovascular substance P nerves. Neurosci. Lett. *32* (1982) 285–290.

82 Gamse, R., and Saria, A., Potentiation of tachykinin-induced plasma protein extravasation by calcitonin gene-related peptide. Eur. J. Pharmac. *144* (1985) 61–65.

83 Gamse, R., Holzer, P., and Lembeck, F., Decrease of substance P in primary afferent neurones and impairment of neurogenic plasma extravasation by capsaicin. Neuroscience *6* (1981) 437–441.

84 Gardner, D. G., Deschepper, C. F., Ganong, W. F., Hane, S., Fiddes, J., Baxter, J. D., and Lewick, J., Extra-atrial expression of the gene for atrial natriuretic factor. Proc. natn. Acad. Sci USA *83* (1986) 6697–6701.

85 Garland, C. J., and Keatinge, W. R., Adrenergic innervation and sensitivity to vasoconstrictor hormones of inner muscle of sheep pulmonary artery. Artery *10* (1982) 440–452.

86 Genest, J., and Cantin, M., Atrial natriuretic factor. Circulation *75* Suppl. (1987) I-118–I-123.

87 Gibbins, I. L., Furness, J. B., Costa, M., MacIntyre, I., Hillyard, C. J., and Girgis, S., Co-localisation of calcitonin gene-related peptide-like immunoreactivity with substance P in cutaneous, vascular and visceral sensory neurons of guinea pigs. Neurosci. Lett. *57* (1985) 125–130.

88 Gibbins, I. L., Brayden, J. E., and Bevan, J. A., Perivascular nerves with immunoreactivity to vasoactive intestinal polypeptide in cephalic arteries of the cat: Distribution, possible origins and functional implications. Neuroscience *13* (1984) 1327–1346.

89 Gibson, S. J., Polak, J. M., Bloom, S. R., Sabate, I. M., Mulderry, P. M., Ghatei, M. A., McGregor, G. P., Morrison, J. F. B., Kelly, J. S., Evans, R. M., and Rosenfeld, M. G., Calcitonin gene-related peptide immunoreactivity in the spinal cord of man and of eight other species. J. Neurosci. *4* (1984) 3101–3111.

90 Greco, A. V., Ghirland, G., Barone, C., Bertoli, A., Caputo, S., Uccioli, L., and Manna, R., Somatostatin paroxysmal supraventricular and junctional tachycardia. Br. med. J. *188* (1984) 28–29.

91 Gu, J., Polak, J. M., Allen, J. M., Huang, W. M., Sheppard, M. N., Tatemoto, K., and Bloom, S. R., High concentration of a novel peptide, neuropeptide Y, in the innervation of mouse and rat heart. J. Histochem. Cytochem. *32* (1984) 467–472.

92 Gu, J., Polak, J. M., Adrian, T. E., Allen, J. M., Tatemoto, K., and Bloom, S. R., Neuropeptide tyrosine (NPY)—a major new cardiac neuropeptide. Lancet *2* (1983) 1008–1010.

93 Gulbenkian, S., Wharton, J., Hacker, G. W., Varndell, I. M., Bloom, S. R., and Polak, J. M., Co-localisation of neuropeptide tyrosine (NPY) and its C-terminal flanking peptide (CPON). Peptides *6* (1985) 1237–1243.

94 Gulbenkian, S., Merighi, A., Wharton, J., Varndell, I. M., and Polak, J. M., Ultrastructural evidence for the co-existence of Calcitonin Gene Related Peptide (CGRP) and Substance P (SP) in secretory vesicles in the peripheral nervous system of guinea pig. J. Neurocyt. *15* (1986) 535–542.

95 Gulbenkian, S., Wharton, J., and Polak, J. M., The visualisation of cardiovascular innervation in the guinea pig using an antiserum to protein gene product 9.5 (PGP 9.5). J. auton. nerv. Syst. *18* (1987) 235–247.

96 Hallberg, D., and Pernow, B., Effect of substance P on various vascular beds in dog. Acta physiol. scand. *93* (1975) 277–285.

97 Hamid, Q., Wharton, J., Terenghi, G., Hassal, C. J. S., Aimi, J., Taylor, K. M., Nakazato, H., Dixon, J. E., Burnstock, G., and Polak, J. M., Localization of atrial natriuretic peptide mRNA and immunoreactivity in rat and human heart. Proc. natn. Acad. Sci. USA *84* (1987) 6760–6764.

98 Hanko, J., Hardebo, J. E., Karström, J. K., Owman, C., and Sundler, F., Calcitonin gene related peptide is present in mammalian cerebrovascular nerve fibres and dilates pial and peripheral arteries. Neurosci. Lett. *57* (1985) 91–95.

99 Hanko, J. H., Törnebrandt, K., Hardebo, J. E., Kahrström, J., Nobin, A., and Owman, C. H., Neuropeptide Y induces and modulates vasoconstriction in intracranial and peripheral vessels of animals and man. J. auton. Pharmac. *6* (1986) 117–124.

100 Hassall, C. J. S., and Burnstock, G., Neuropeptide Y-like immunoreactivity in cultured intrinsic neurons of the heart. Neurosci. Lett. *52* (1984) 111–115.

310

101 Holman, J. J., Craig, R. K., and Marshall, I., Human α- and β-CGRP and rat α-CGRP are coronary vasodilators in the rat. Peptides 7 (1986) 231–235.

102 Howells, R. D., Kilpatrick, D. L., Bailey, L. C., Noe, M., and Udenfriend, S., Proenkephalin mRNA in rat heart. Proc. natn. Acad. Sci. USA 83 (1986) 1960–1963.

103 Hua, X-Y., Theodorsson-Norheim, E., Brodin, E., Lundberg, J. M., and Hökfelt, T., Multiples tachykinins (neurokinin A, neuropeptide K and substance P) in capsaicin-sensitive sensory neurons in the guinea-pig. Reg. Pep. 13 (1985) 1–19.

104 Huang, M., and Rorstad, O. P., Cerebral vascular adenylate cyclase: evidence for coupling to receptors for vasoactive intestinal polypeptide and parathyroid hormone. J. Neurochem. 43 (1984) 849–856.

105 Itakura, T., Okuno, T., Nakakita, K., Kamei, I., Naka, Y., Nakai, K., Imai, H., Komai, N., Kimura, H., and Maeda, T., A light and electron microscopic immunohistochemical study of vasoactive intestinal polypeptide- and substance P-containing nerve fibres along the cerebral blood vessels: comparison with aminergic and cholinergic nerve fibres. J. cerebr. Blood Flow Metab. 4 (1984) 407–414.

106 Itoh, N., Obata, K., Yanaihara, N., and Okamoto, H., Human preprovasoactive intestinal polypeptide contains a novel PHI-27-like peptide, PHM-27. Nature 304 (1983) 547–549.

107 Jackson, P., and Thompson, R. J., The demonstration of new human brain-specific proteins by high-resolution two dimensional polyacrylamide gel electrophoresis. J. neurol. Sci. 49 (1981) 429–438.

108 Johansson, O., A detailed account of NPY-immunoreactive nerves and cells of the human skin. Comparison with VIP-, substance P- and PHI-containing structures. Acta physiol. scand. 128 (1986) 147–153.

109 Kangawa, K., Minamino, N., Fukuda, A., and Matsuo, H., Neuromedin K: A novel mammalian tachykinin identified in porcine spinal cord. Biochem. biophys. Res. Commun. 114 (1983) 533–540.

110 Kimura, S., Okada, M., Sugita, Y., Kanazawa, I., and Munekatat, E., Novel neuropeptides, neurokinin α and β isolated from porcine spinal cord. Proc. Jap. Acad. 59B (1983) 101–104.

111 Kubota, M., Moseley, J. M., Butera, L., Dusting, G. J., MacDonald, P. S., and Martin, T. J., Calcitonin gene related peptide stimulates cyclic AMP formation in rat aortic smooth muscle cells. Biochem. biophys. Res. Commun. 132 (1985) 86–94.

112 Lachance, D., Garcia, R., Gutkowska, J., Cantin, M., and Thibault, G., Mechanisms of release of atrial natriuretic factor. I. Effect of several agonists and steroids on its release by atrial minces. Biochem. biophys. Res. Commun. 135 (1986) 1090–1098.

113 Lang, R. E., Hermann, K., Dietz, R., Gaida, W., Ganten, D., Kraft, K., and Unger, Th., Evidence for the presence of enkephalins in the heart. Life Sci. 32 (1983) 399–406.

114 Lang, R. E., Thölken, H., Ganten, D., Lufa, F. C., Ruskoaho, H., and Unger, Th., Atrial natriuretic factor—a circulating hormone stimulated by volume loading. Nature 314 (1985) 264–266.

115 Larsson, L.-I., Edvinsson, L., Fahrenkrug, J., Hakanson, R., Owman, C., Schaffalitzky de Muckaddel, O. B., and Sundler, F., Immunohistochemical localisation of a vasodilatory peptide (VIP) in cerebrovascular nerves. Brain Res. 113 (1976) 400–404.

116 Larsson, O., Dunèr-Engström, M., Lundberg, J. M., Fredholm, B. B., and Änggard, A., Effects of VIP, PHM and substance P on blood vessels and secretory elements of the human submandibular gland. Reg. Pep. 13 (1986) 319–326.

117 Lattion, A.-L., Michel, J-B., Arnauld, E., Corvol, P., and Soubrier, F., Myocardial recruitment during ANP mRNA increase with volume overload in the rat. Am. J. Physiol. 251 (1986) H890–H896.

118 Le Greves, P., Nyberg, F., Terenius, L., and Hökfelt, T., Calcitonin gene-related peptide is a potent inhibitor of substance P degradation. Eur. J. Pharmac. 115 (1985) 300–311.

119 Ledsome, J. R., Wilson, N., Courneya, C. A., and Rankin, A. J., Release of atrial natriuretic peptide by atrial distension. Can. J. Physiol. Pharmac. 63 (1985) 739–742.

120 Lee, T. J. F., Saito, A., and Berezin, I., Vasoactive intestinal polypeptide-like substance: the potential transmitter for cerebral vasodilation. Science 224 (1984) 898–901.

121 Lee, Y., Kawai, Y., Shiosaka, S., Takami, K., Hillyard, C. J., Girgis, S., MacIntyre, I., Emson, P. C., and Tohyama, M., Coexistence of calcitonin gene-related peptide and substance P-like peptide in single cells of the trigeminal ganglion of the rat: immunohistochemical analysis. Brain Res. 330 (1985) 194–196.

122 Lee, Y., Takami, K., Kawai, Y., Girgis, S., Hillyard, C. J., MacIntyre, I., Emson, P. C., and Tohyama, M., Distribution of calcitonin gene-related peptide in the rat peripheral nervous system with reference to it coexistence with substance P. Neuroscience 15 (1985) 1227–1237.

123 Lembeck, F., and Holzer, P., Substance P as neurogenic mediator of antidromic vasodilation and neurogenic plasma extravasation. N.S. Arch. Pharmac. 310 (1979) 175–183.

124 Leys, K., Schachter, M., and Sever, P., Autoradiographic localisation of NPY receptors in rabbit kidney: comparison with rat, guinea-pig and human, Eur. J. Pharmac. 134 (1987) 233–237.

125 Leys, K., Morice, A. H., Madonna, O., and Sever, P. S., Autoradiographic localisation of VIP receptors in human lungs. FEBS Lett. 199 (1986) 198–202.

126 Liu-chen, L. Y., Mayleong, M. R., and Moskowitz, M. A., Immunohistochemical evidence for a substance P-containing trigeminovascular pathway to pial arteries in cats. Brain Res. 268 (1983) 162–166.

127 Liu-chen, L. Y., Liszczak, T. M., King, J. C., and Moskowitz, M. A., Immunoelectron microscopic study of substance P-containing fibres in feline cerebral arteries. Brain Res. 369 (1986) 12–20.

128 Losay, J., Mroz, E., Tregear, G. W., Leeman, S. E., and Gamble, W. J., Action of substance P on the coronary blood flow in the isolated dog heart, in: Substance P, pp. 287–294. Eds U.S. von Euler and B. Pernow. Raven Press, New York 1977.

129 Lundberg, J. M., Evidence for the existence of vasoactive intestinal polypeptide (VIP) and acetylcholine neurons in cat exocrine glands. Morphological, anatomical and functional studies. Acta physiol. scand. Suppl. 496 (1981) 1–57.

130 Lundberg, J. M., Franco-Cereceda, A., Hua, Y., Hökfelt, T., and Fischer, J. A., Coexistence of substance P and calcitonin gene related peptide like immunoreactivities in sensory nerves in relation to cardiovascular and bronchoconstrictor effects of capsaicin. Eur. J. Pharmac. 108 (1985) 315–319.

131 Lundberg, J. M., Änggard, A., Emson, P. C., Fahrenkrug, J., and Hökfelt, T., Vasoactive intestinal polypeptide and cholinergic mechanisms in cat nasal mucosa: Studies on choline acetyltransferase and release of vasoactive intestinal polypeptide. Proc. natn. Acad. Sci. USA 78 (1981) 5255–5259.

132 Lundberg, J. M., Hua, X. Y., and Franco-Cereceda, A., Effects of Neuropeptide Y (NPY) on mechanical activity and neurotransmission in the heart, vas deferens and urinary bladder of the guinea pig. Acta physiol. scand. 121 (1984) 325–332.

133 Lundberg, J. M., Änggard, A., Theodorsson-Norheim, E., and Pernow, J., Guanethidine-sensitive release of neuropeptide Y-like immunoreactivity in the cat spleen by sympathetic nerve stimulation. Neurosci. Lett. 52 (1984) 175–180.

134 Lundberg, J. M., Terenius, L., Hökfelt, T., Martling, C. R., Tatemoto, K., Mutt, V., Polak, J. M., Bloom, S. R., and Goldstein, M., Neuropeptide Y (NPY)-like immunoreactivity in peripheral noradrenergic neurons and effects of NYP on sympathetic function. Acta physiol. scand. 116 (1982) 477–480.

135 Lundberg, J. M., Rudehill, A., Solleu, A., Theodorsson-Norheim, A., and Hamberger, B., Frequency- and reserpine-dependent chemical coding of sympathetic transmission: Differential release of noradrenaline and neuropeptide Y from pig spleen. Neurosci. Lett. 63 (1986) 96–100.

136 Lundberg, J. M., Torssell, L., Salleri, A., Pernow, J., Theodrosson-Norheim, E., Änggard, A., and Hamberger, B., Neuropeptide Y and sympathetic vascular control in many. Reg. Pep. 13 (1985) 41–52.

137 Lundberg, J. M., Saria, A., Franco-Cereceda, A., Hökfelt, T., Terebius, L., and Goldstein, M., Differential effects of reserpine and 6-hydroxydopamine on neuropeptide Y (NPY) and noradrenaline in peripheral neurons. N.S. Arch. Pharmac. 328 (1985) 331–340.

138 Lundberg, J. M., Al-Saffar, A., Saria, A., and Theodorsson-Norheim, E., Reserpine-induced depletion of neuropeptide Y from cardiovascular nerves and adrenal gland due to enhanced release. N.S. Arch. Pharmac. 332 (1986) 163–168.

139 Lundberg, J. M., Fahrenkrug, J., Hökfelt, T., Martling, C.-R., Larsson, O., Tatemoto, K., and Änggard, A., Co-existence of peptide HI (PHI) and VIP in nerves regulating blood flow and bronchial smooth muscle tone in various mammals including man. Peptides 5 (1984) 593–606.

312

140 Lynch, E. M., Wharton, J., Bryant, M. G., Bloom, S. R., Polak, J. M., and Elder, M. G., The differential distribution of vasoactive intestinal polypeptide in the normal human female genital tract. Histochemistry 67 (1980) 169–177.

141 Maldonado, C. A., Saggon, W., and Forssmann, W. G., Cardiodilatin-immunoreactivity in specific atrial granules of human heart by the immunogold stain. Anat. Embryol. 173 (1986) 295–298.

142 Matsuyama, T., Wanaka, A., Yoneda, S., Kimura, K., Kamada, T., Girgis, S., MacIntyre, I., Emson, P. C., and Tohyama, M., Two distinct calcitonin gene-related peptide-containing peripheral nervous systems: Distribution and quantitative differences bewteen the iris and cerebral artery with special reference to substance P. Brain Res. 373 (1986) 205–212.

143 Matsuyama, T., Shiosaka, S., Wanaka, A., Yoneda, S., Kimura, K., Hayakawa, T., Emson, P. C., and Tohyama, M., Fine structure of peptidergic and catecholaminergic nerve fibres in the anterior cerebral artery and their interrelationship: An immunoelectron microscopic study. J. comp. Neurol. 235 (1985) 268–276.

144 McCulloch, J., Uddman, R., Kingman, T. A., and Edvinsson, L., Calcitonin gene-related peptide: Functional role in cerebrovascular regulation. Proc. natn. Acad. Sci. USA 83 (1986) 5731–5735.

145 McEwan, J., Larkin, S., Davies, G., Chierchia, S., Brown, M., Stevenson, J., MacIntyre, I., and Maseri, A., Calcitonin gene related peptide: a potent dilator of human epicardial coronary arteries. Circulation 74 (1986) 1243–1247.

146 Metz, J., Gerstheimer, F. P., and Herbst, M., Distribution of synaptophysin immunoreactivity in guinea pig heart. Histochemistry 86 (1986) 221–224.

147 Minamino, N., Kangawa, K., Fukuda, A., and Matsuo, H., Neuromedin L: A novel mammalian tachykinin identified in porcine spinal cord. Neuropeptides 4 (1984) 157–166.

148 Minth, C. D., Bloom, S. R., Polak, J. M., and Dixon, J. E., Cloning, characterization and DNA sequence of a human cDNA encoding neuropeptide tyrosine. Proc. natn. Acad. Sci. USA 81 (1984) 4577–4581.

149 Morris, J. L., Murphy, R., Furness, J. B., and Costa, M., Partial depletion of neuropeptide Y from noradrenergic perivascular nerves and cardiac axons by 6-hydroxydopamine and reserpine. Reg. Pep. 13 (1986) 147–162.

150 Mulderry, P. K., Ghatei, M. A., Rodrigo, J., Allen, J. M., Rosenfeld, M. G., Polak, J. M., and Bloom, S. R., Calcitonin gene-related peptide in cardiovascular tissues of the rat. Neuroscience 14 (1985) 947–954.

151 Murphy, R., Furness, J. B., Beardsley, A. M., and Costa, M., Characterisation of substance P-like immunoreactivity in peripheral sensory nerves and enteric nerves by high pressure liquid chromatography and radioimmunoassay. Reg. Pep. 4 (1982) 203–212.

152 Natsyama, T., Matsumoto, M., Shiosaka, S., Hayakawa, T., Yoneda, S., Kimura, K., Abe, H., and Tohyama, H., Dual innervation of substance P-containing neuron system in the wall of the cerebral arteries. Brain Res. 322 (1984) 144–147.

153 Navone, F., Greengard, P., and De Camilli, P., Synapsin I in nerve terminals: selective association with small synaptic vesicles. Science 226 (1984) 1209–1211.

154 Navone, F., Reinhard, J., Di Gioia, G., Stukenbrok, H., Greengard, P., and De Camilli, P., Protein P38: An integral membrane protein specific for small vesicles of neurons and neuroendocrine cells. J. Cell Biol. 103 (1986) 2511–2527.

155 Nawa, H., Hirose, T., Takashima, H., Inayami, S., and Nakanishi, S., Nucleotide sequences of cloned cDNAs for two types of bovine brain substance P precursor. Nature 306 (1983) 32–36.

156 Nemer, M., Lavigne, J.-P., Dronin, J., Thibault, G., Gannon, M., and Antakly, T., Expression of atrial natriuretic factor gene in heart ventricular tissue. Peptides 7 (1986) 1147–1152.

157 Nilsson, S. F. E., and Mäepea, O., Comparison of the vasodilatory effects of vasoactive intestinal polypeptide (VIP) and peptide-HI (PHI) in the rabbit and the cat. Acta physiol. scand. 129 (1987) 17–26.

158 Oku, R., Satoh, M., Fujii, N., Otaka, A., Yajima, H., and Takagi, H., Calcitonin gene-related peptide promotes mechanical nociception by potentiating release of substance P from spinal dorsal horn in rats. Brain Res. 403 (1987) 350–354.

159 Papka, R. E., Furness, J. B., Della, N. G., and Costa, M., Depletion by capsaicin by substance P-immunoreactivity and acetylcholinesterase activity from nerve fibres in the guinea pig heart. Neurosci. Lett. 27 (1981) 47–53.

160 Papka, R. E., Furness, J. B., Della, N. G., Murphy, R., and Costa, M., Time course of effect of capsaicin on ultrastructure and histochemistry of substance P-immunoreactive nerves associated with the cardiovascular system of the guinea pig. Neuroscience 12 (1984) 1277–1292.

161 Pernow, J., Saria, A., and Lundberg, J. M., Mechanisms underlying pre- and post-junctional effects of neuropeptide Y in sympathetic vascular control. Acta physiol. scand. 126 (1986) 239–249.

162 Pernow, J., Lundberg, J. M., and Kaijser, L., Vasoconstrictor effects in vivo and plasma disappearance rate of neuropeptide Y in man. Life Sci. 40 (1987) 47–54.

163 Pettersson, A., Ricksten, S.-E., Towle, A. C., Hander, J., and Hedner, T., Effect of blood volume expansion and sympathetic denervation on plasma levels of atrial natriuretic factor (ANF) in the rat. Acta physiol. scand. 124 (1985) 309–311.

164 Potter, E. K., Prolonged nonadrenergic inhibition of cardiac vagal action following sympathetic stimulation: neuromodulation by neuropeptide Y. Neurosci. Lett. 54 (1985) 117–121.

165 Poulin, P., Suzuki, Y., Lederis, K., and Rorstad, O. P., Autoradiographic localization of binding sites for vasoactive intestinal peptide (VIP) in bovine cerebral arteries. Brain Res. 381 (1986) 382–384.

166 Rankin, A. J., Courneya, C. A., Wilson, N., and Ledsome, J. R., Tachcardia releases atrial natriuretic peptide in the anesthetized rabbit. Life Sci. 38 (1986) 1951–1957.

167 Rechardt, L., Aalto-Setälä, K., Purjeranta, M., Pelto-Huikko, M., and Kyösala, K., Peptidergic innervation of human atrial myocardium: an electron microscopical and immunocytochemical study. J. auton. nerv. Syst. 17 (1986) 21–32.

168 Regoli, D., Drapeau, G., Dion, S., and D'Orleans-Juste, P., Pharmacological receptors for substance P and neurokinins. Life Sci. 40 (1987) 109–117.

169 Reinecke, M., Weihe, E., Carraway, R. E., Leeman, S. E., and Forssmann, W. G., Localization of neurotensin immunoreactive nerve fibres in the heart: evidence derives by immunocytochemistry, radioimmunoassay and chromatography. Neuroscience 7 (1982) 1785–1795.

170 Reinecke, M., Weihe, E., and Forssmann, W. G., Substance P-immunoreactive nerve fibres in the heart. Neurosci. Lett. 20 (1980) 265–269.

171 Rioux, F., Bachelard, H., Martel, J.-C., and St-Pierre, S., The vasoconstrictor effect of neuropeptide Y and related peptides in the guinea pig isolated heart. Peptides 7 (1986) 27–31.

172 Rosenfeld, M. G., Mermod, J.-J., Amara, S. G., Swanson, L. W., Sawchenko, P. E., Rivier, J., Vale, W. W., and Evans, R. M., Production of a novel neuropeptide encoded by the calcitonin gene via tissue-specific RNA processing. Nature 304 (1983) 129–135.

173 Rudehill, A., Sollevi, A., Franco-Cereceda, A., and Lundberg, J. M., Neuropeptide Y (NPY) and the pig heart: Release and coronary vasconstrictor effects. Peptides 7 (1986) 821–826.

174 Said, S. I., and Mutt, V., Polypeptide with broad biological activity: isolation from small intestine. Science 169 (1970) 1217–1218.

175 Saito, K., Liu-Chen, L.-Y., and Moskowitz, M. A., Substance P-like immunoreactivity in rat forebrain leptomeninges and cerebral vessels originates from trigeminal but not sympathetic ganglia. Brain Res. 403 (1987) 66–71.

176 Saito, A., Kimura, S., and Goto, K., Calcitonin gene-related peptide as potential neurotransmitter in guinea pig right atrium. Am. J. Physiol. 250 (1986) H693–H698.

177 Samnegard, H., Thulin, L., Tydén, G., Johansson, C., Muhrbeck, Q., and Björklund, C., Effect of synthetic substance P on internal carotid artery blood flow. Acta physiol. scand. 104 (1978) 491–495.

178 Saria, A., Lundberg, J. M., Skofitsch, G., and Lembeck, F., Vascular protein leakage in various tissues induced by substance P, capsaicin tachykinin, serotonin, histamine and by antigen challenge. N.S. Arch. Pharmac. 324 (1983) 212–218.

179 Sasaki, Y., Hayashi, N., Kashara, A., Matsuda, H., Fusamoto, H., Sato, N., Hillyard, C. J., Girgis, S., MacIntyre, I., Emson, P. C., Shiosaka, S., Tohyama, M., Shiotani, Y., and Kamada, T., Calcitonin gene-related peptide in the hepatic and splanchnic vascular systems of the rat. Hepatology 6 (1986) 676–681.

314

180 Satoh, M., Oku, R., Maeda, A., Fujii, N., Otaka, A., Funakoshi, S., Yajima, H., and Takagi, H., Possible mechanisms of positive calcitonin gene related peptide in isolated rat atrium. Peptides 7 (1986) 631–635.

181 Schacter, M., Dickinson, K. E. J., Miles, C. M., and Sever, P. S., Characterisation of a high-affinity VIP receptor in human lung parenchyma. FEBS Lett. 199 (1986) 125–129.

182 Schiebinger, R. J., and Linden, J., Effect of atrial contraction frequency on atrial natriuretic peptide secretion. Am. J. Physiol. 251 (1986) H1095–H1099.

183 Schultzberg, M., Hökfelt, T., Nilsson, G., Terenius, L., Rehfeld, J. F., Brown, M., Elde, R., Goldstein, M., and Said, S., Distribution of peptide- and catecholamine-containing neurones in the gastrointestinal tract of the rat and guinea pig: immunohistochemical studies with antisera to substance P, vasoactive intestinal polypeptide, enkephalins, somatostatin, gastrin/cholecystokinin, neurotensin and dopamine β-hydroxylase. Neuroscience 5 (1980) 689–744.

184 Shigematsu, K., Saavedra, J. M., and Kurihara, M., Specific substance P binding sites in rat thymus and spleen: in vitro autoradiographic study. Reg. Pep. 16 (1986) 147–156.

185 Sigrist, S., Franco-Cereceda, A., Muff, R., Henke, H., Lundberg, J. M., and Fischer, J. A., Specific receptor and cardiovascular effects of calcitonin gene-related peptide. Endocrinology 119 (1986) 381–389.

186 Steenbergh, P. H., Hoppener, J. W. M., Zandberg, J., Van de Ven, W. J. M., and Jansz, H. S., A second human calcitonin/CGRP gene. FEBS Lett. 183 (1985) 403–407.

187 Stephenson, J. A., Burcher, E., and Summers, R. J., Autoradiographic demonstration of endothelium-dependent ^{125}I-Bolton-Hunter substance P binding to dog carotid artery. Eur. J. Pharmac. 124 (1986) 377–378.

188 Stephenson, J. A., and Summers, R. J., Autoradiographic analysis of receptors on vascular endothelium. Eur. J. Pharmac. 134 (1987) 35–43.

189 Suzuki, Y., McMaster, D., Lederis, K., and Rorstad, O. P., Characterization of the relaxant effects of vasoactive intestinal peptide (VIP) and PHI on isolated brain arteries. Brain Res. 322 (1984) 9–16.

190 Suzuki, Y., McMaster, D., Huang, M., Lederis, K., and Rorstad, O. P., Characterization of functional receptors for vasoactive intestinal peptide in bovine cerebral arteries. J. Neurochem. 45 (1985) 890–899.

191 Tatemoto, K., and Mutt, V., Isolation and characterisation of the intestinal peptide porcine PHI (PHI-27), a new member of the glucagon-secretion family. Proc. natn. Acad. Sci. USA 78 (1981) 6603–6607.

192 Tatemoto, K., Carlquist, M., and Mutt, V., Neuropeptide Y—a novel brain peptide with structural similarities to peptide YY and pancreatic polypeptide. Nature 296 (1982) 659–660.

193 Tatemoto, K., and Mutt, V., Isolation of two novel candidate hormones using a chemical method for finding naturally occurring polypeptides. Nature 285 (1980) 417–418.

194 Taton, G., Chatelain, P., Delhaye, M., Camus, J. C., De Neef, P., Waelbroeck, M., Tatemoto, K., Robberecht, P., and Christophe, J., Vasoactive intestinal peptide (VIP) and peptide having N-terminal isoleucine amide (PHI) stimulate adenylate cyclase activity in human heart membranes. Peptides 3 (1982) 897–900.

195 Terenghi, G., Polak, J. M., Probert, L., McGregor, G. P., Ferri, G. L., Blank, M. A., Butler, J. M., Unger, W. G., Zhang, S., Cole, D. F., and Bloom, S. R., Mapping, quantitative distribution and origin of substance P- and VIP-containing nerves in the uvea of guinea pig eye. Histochemistry 75 (1982) 399–417.

196 Terenghi, G., Polak, J. M., Ghatei, M. A., Mulderry, P. K., Butler, J. M., Unger, W. G., and Bloom, S. R., Distribution and origin of calcitonin gene-related peptide (CGRP) immunoreactivity in the sensory innervation of the mammalian eye. J. comp. Neurol. 223 (1985) 506–516.

197 Thorén, P., Mark, A. L., Morgan, D. A., O'Neill, T. P., Needleman, P., and Brody, M. J., Activation of vagal depressor reflexes by atriopeptins inhibits renal sympathetic nerve activity. Am. J. Physiol. 251 (1986) H1251–H1259.

198 Uddman, R., Fahrenkrug, J., Malm, L., Alumets, J., Hakanson, R., and Sundler, F., Neuronal VIP in salivary glands: distribution and release. Acta physiol. scand. 110 (1980) 31–38.

199 Uddman, R., Alumets, J., Edvinsson, L., Hakanson, R., and Sundler, R., VIP nerve fibres around peripheral blood vessels. Acta physiol. scand. *112* (1981) 65–70.

200 Uddman, R., Alumets, J., Ehinger, B., Hakanson, R., Lorèn, I., and Sundler, F., Vasoactive intestinal peptide nerves in ocular and orbital structures of the cat. Invest. Ophthal. vis. Sci. *19* (1980) 878–885.

201 Uddman, R., Edvinsson, L., Ekblad, E., Hakansson, R., and Sundler, F., Calcitonin gene-related peptide (CGRP): perivascular distribution and vasodilatory effects. Reg. Pep. *15* (1986) 1–23.

202 Uddman, R., Ekblad, E., Edvinsson, L., Hakanson, R., and Sundler, F., Neuropeptide Y-like immunoreactivity in perivascular nerve fibres of thè guinea pig. Reg. Pep. *10* (1985) 243–257.

203 Uddman, R., Edvinsson, L., Owman, C., and Sundler, F., Perivascular substance P: occurrence and distribution in mammalian pial vessels. J. cerebr. Blood Flow Metab. *1* (1981) 221–232.

204 Uemura, Y., Sugimoto, T., Okamoto, S., Handa, H., and Mizuno, N., Changes of vasoactive intestinal polypeptide-like immunoreactivity in cerebrovascular nerve fibres after subarachnoid hemorrhage: an experimental study in the dog. Neurosci. Lett. *71* (1986) 137–141.

205 Unverferth, D. V., O'Dorisio, T. M., Muir, W. W., White, J., Miller, M. M., Hamlin, R. L., and Magorien, R. D., Effect of vasoactive intestinal polypeptide on the canine cardiovascular system. J. Lab. clin. Med. *106* (1985) 542–550.

206 Urban, L., and Papka, R. E., Origin of small primary afferent substance P-immunoreactive nerve fibres in the guinea pig heart. J. auton. nerve. Syst. *12* (1985) 321–331.

207 Verrecchia, C., Hamel, E., Edvinsson, L., MacKenzie, E. T., and Seylaz, J., Role of the endothelium in the pial artery responses to several vasoactive peptides. Acta physiol. scand. *127*, Suppl. 552 (1986) 33–36.

208 Wahlestedt, C., Edvinsson, L., Ekblad, E., and Hakanson, R., Neuropeptide Y potentiates noradrenaline-evoked vasoconstriction. Mode of action. J. Pharmac. exp. Ther. *234* (1985) 735–741.

209 Wahlestedt, C., Yanaihara, N., and Hakanson, R., Evidence for different pre- and post-junctional receptors for neuropeptide Y and related peptides. Reg. Peptides *13* (1986) 307–318.

210 Wanaka, A., Matsuyama, T., Yoneda, S., Kimura, K., Kamada, T., Girgis, S., MacIntyre, I., Emson, P. C., and Tohyama, M., Origins and distribution of calcitonin gene-related peptide-containing nerves in the wall of the cerebral arteries of the guinea pig with special reference to the coexistence with substance P. Brain Res. *369* (1986) 185–192.

211 Webb, S. C., Krikler, D. M., Hendry, W. G., Adrian, T. E., and Bloom, S. R., Electrophysiological actions of somatostatin on the atrioventricular junction in sinus rhythm and reentry tachycardia. Br. Heart. J. *56* (1986) 236–241.

212 Weihe, E., and Reinecke, M., Peptidergic innervation of the mammalian sinus nodes: vasoactive intestinal polypeptide, neurotensin, substance P. Neurosci. Lett. *26* (1981) 283–288.

213 Weihe, E., Reinecke, M., and Forsmann, W. G., Distribution of vasoactive intestinal polypeptide-like immunoreactivity in the mammalian heart. Cell Tiss. Res. *236* (1984) 527–540.

214 Wharton, J., Polak, J. M., McGregor, G. P., Bishop, A. E., and Bloom, S. R., The distribution of substance P-like immunoreactive nerves in the guinea-pig heart. Neuroscience *6* (1981) 2193–2204.

215 Wharton, J., Polak, J. M., Probert, L., De Mey, J., McGregor, G. P., Bryant, M. G., and Bloom, S. R., Peptide containing nerves in the ureter of the guinea pig and cat. Neuroscience *6* (1981) 969–982.

216 Wharton, J., Gulbenkian, S., Mulderry, P. K., Ghatei, M. A., MacGregor, G. P., Bloom, S. R., and Polak, J. M., Capsaicin induces a depletion of calcitonin gene related peptide (CGRP)-immunoreactive nerves in the cardiovascular system of the guinea pig and rat. J. auton. nerv. Syst. *16* (1986) 289–309.

217 Wharton, J., Polak, J. M., Bryant, M. G., Van Noorden, S., Bloom, S. R., and Pearse, A. G. E., Vasoactive intestinal polypeptide (VIP)-like immunoreactivity in salivary glands. Life Sci. *25* (1979) 373–380.

218 Wiedenmann, B., and Franke, W. W., Identification and localization of synaptophysin, an integral membrane glycoprotein of Mr 38,000 characteristic of presynaptic vesicles. Cell *41* (1985) 1017–1028.

219 Wilson, D. A., O'Neill, J. T., Said, J. I., and Traystman, R. J., Vasoactive intestinal polypeptide and the canine cerebral circulation. Cir. Res. *48* (1981) 138–148.

220 Zukowska-Crojec, Z., Haass, M., Kopin, I. J., and Zamir, N., Interactions of atrial natriuretic peptide with the sympathtetic and endocrine systems in the pithed rat. J. Pharmac. exp. Ther. *239* (1986) 480–487.

Regulatory peptides in the respiratory system

P. J. Barnes

Summary. Many regulatory peptides have been described in the respiratory tract of animals and humans. Some peptides (bombesin, calcitonin, calcitonin gene-related peptide) are localised to neuroendocrine cells and may have a trophic or transmitter rôle. Others are localised to motor nerves. Vasoactive intestinal peptide and peptide histidine isoleucine are candidates for neurotransmitters of non-adrenergic inhibitory fibres and may be co-transmitters in cholinergic nerves. These peptides may regulate airway smooth muscle tone, bronchial blood flow and airway secretions. Sensory neuropeptides (substance P, neurokinin A and B, calcitonin gene-related peptide) may contract airway smooth muscle, stimulate mucus secretion and regulate bronchial blood flow and microvascular permeability. If released by an axon reflex mechanism these peptides may be involved in the pathogenesis of asthma. Other peptides, such as galanin and neuropeptide Y, are also present but their function is not yet known.

Recently, a large number of regulatory peptides have been identified in the respiratory tract of several species, including humans (table). Many of these peptides have potent effects on several aspects of airway function, including bronchomotor tone, airway secretions and the bronchial circulation. The precise physiological rôle of all these peptides is obscure, although some clues are provided by their localisation and functional effects. The purpose of this chapter is to discuss what is known of these peptides, particularly in human airways, and to speculate about their possible pathophysiological rôle.

Many neuropeptides have been isolated from the gut, where they are involved in regulation of gut motility, sphincters and secretion. There is convincing evidence that these neuropeptides are neurotransmitters or neuromodulators, and appear to be involved in the complex integrative regulation of the gastrointestinal tract. Since the airways are derived embryologically from the foregut it is not surprising that similar peptides are also to be found in lung[7,76]. As in the gut, these peptides are localised either to nerves or to neuroendocrine cells.

Neuroendocrine cells

Specialised cells containing neurosecretory granules are present in the respiratory tract of several species, including man, and are prominent in the foetal and neonatal lung. Because these cells disappear during maturation they may play a rôle in pulmonary development. In addition to serotonin, these cells also contain the regulatory peptides bombesin, calcitonin, katacalcin and calcitonin gene-related peptide[27,66]. Recent studies suggest that, in animals and humans, these cells are closely associated with afferent nerves, so perhaps these cells have a sensory role and their peptides function as 'neurotransmitters'[48,72]. The stimuli which may activate these cells are uncertain, but there is some evidence that they may sense hypoxia[66].

Bombesin, which may be released from neuroendocrine cells is a bronchoconstrictor in guinea pig airways[39], but has no effect on human airways *in vitro* (unpublished observation). In pancreas, bombesin has a trophic rôle, and it is possible that it may play a rôle in the development of the respiratory tract, since neuroendocrine cells and bombesin immunoreactivity decrease with maturation[27,37]. A marked reduction in bombesin immunoreactivity has been observed in lungs of human infants who died of acute respiratory distress syndrome[37].

Non-adrenergic non-cholinergic innervation

The nerve supply to the lungs is more complex than previously recognised, since, in addition to classical cholinergic and adrenergic innervation[6,63], there are nerves which are non-adrenergic and non-cholinergic (NANC)[3,4,77]. NANC nerves were first described in the gut and therefore their existence in lung is to be expected.

Non-adrenergic inhibitory nerves, which relax airway smooth muscle, have been demonstrated *in vitro* by electrical field stimulation after adrenergic and cholinergic blockade in several species, including humans[4,77]. In human airway smooth muscle the NANC inhibitory system is the only bronchodilator pathway, since there is no functional sympathetic innervation, and because NANC innervation is the sole inhibitory pathway from trachea to the smallest bronchi that has been of considerable interest in the neurotransmitter. NANC inhibitory nerves have also been demonstrated *in vivo* in animals by electrical stimulation of the vagus after adrenergic and cholinergic blockade[29,40]. Stimulation of this pathway produces pronounced and long-lasting bronchodilatation, which may be inhibited by ganglion blockers. This pathway may be activated reflexly by mechanical or chemical stimulation of the larynx[81]. Although it is difficult to study this pathway in humans *in vivo*, studies of laryngeal stimulation indicate that reflex non-adrenergic bronchodilation may occur[61].

Evidence argues against purines as neurotransmitters of NANC inhibitory nerves in airways. Although exogenous ATP relaxes airway smooth muscle[41], an antagonist quinidine fails to block NANC relaxation *in vitro* or *in vivo* and the purine uptake inhibitor, dipyridamole, does not enhance non-adrenergic bronchodilatation[40,41]. Similarly, adenosine fails to mimic NANC relaxation and antagonists, such as theophylline, have no blocking effect[43]. Evidence is now more in favour of a neuropeptide as a neurotransmitter of NANC inhibitory nerves and, of the several neuropeptides identified in airways, only vasoactive intestinal peptide (VIP) and the related peptide histidine isoleucine (PHI) relax airway smooth muscle and are, therefore, the only known peptide candidates[9,79].

Electrical stimulation of guinea pig bronchi, and occasionally trachea *in vitro*, and vagal nerve *in vivo* produces a component of bronchoconstriction which is not inhibited by atropine[2]. There is now convincing evidence that substance P and related neuropeptides may be neurotransmitters of these non-cholinergic excitatory nerves.

Other NANC responses have been described in lung. NANC secretion of mucus has been demonstrated in cats *in vivo* using vagal nerve stimulation[75], and in ferret airways *in vitro* using electrical field stimulation[17].

Vasoactive intestinal peptide

VIP was originally extracted from porcine lung as a vasodilator peptide, and is localised to motor nerves in lung of several species, including humans. VIP has potent effects on airway and pulmonary vascular tone and on airway secretion, which suggest that it may have an important regulatory rôle[9,79].

Localisation. VIP has been isolated from lung extracts of several species, including humans[36]. VIP-immunoreactivity is localised to nerves and ganglia supplying airways and pulmonary vessels. VIP-immunoreactivity is present in ganglion cells in the posterior trachea and around intrapulmonary bronchi, diminishing in frequency as airways become smaller. VIP-immunoreactive nerves are widely distributed throughout the respiratory tract and pulmonary vasculature. There as a rich VIP-ergic innervation of the nasal mucosa and upper respiratory tract, but the density of innervation diminishes peripherally so that few VIP-ergic fibres are found in bronchioles. VIP-ergic nerves are also found in airway smooth muscle, around bronchial vessels and surrounding submucosal glands[28,47,76]. VIP-ergic fibres are also found in the adventitia of pulmonary vessels, particularly medium sized arteries.

VIP-receptors. VIP-receptors have been identified in the lung of several species by receptor binding techniques using [^{125}I]-VIP[78]. Binding of VIP to its receptor activates adenylate cyclase, and VIP stimulates cyclic AMP formation in lung fragments. The actions of VIP are, therefore, very similar to those of β-adrenoceptor agonists and any differences in response of different tissues to VIP or β-agonists depends on the relative densities of their respective receptors. The distribution of

VIP-receptors in lung has recently been investigated, using an auto-radiographic method to map out specific VIP-binding sites[21,51]. VIP-receptors are found in high density in pulmonary vascular smooth muscle and in airway smooth muscle of large, but not small, airways. VIP-receptors are also found in high density in airway epithelium and submucosal glands. While the distribution of receptors is consistent with known functions of VIP, a high density of receptors is also found in the alveolar walls. The physiological function of these receptors is obscure, since there is no VIP-ergic innervation of peripheral lung. It is possible that these VIP-binding sites represent sites of uptake of VIP since VIP is taken up from the circulation and metabolised by pulmonary capillary endothelial cells. The distribution of VIP-receptors has also been studied by an immunocytochemical method using an antibody to cyclic AMP. After stimulation by VIP, cyclic AMP increases in those cells with specific receptors, and this technique confirms the autoradiographic studies in demonstrating VIP-receptors in airway smooth muscle, epithelium and submucosal glands of several species[49].

Airway smooth muscle. VIP is a potent relaxant of airway smooth muscle *in vitro* and this relaxation is independent of adrenergic receptors[67,79]. VIP is 50–100 times more potent than isoprenaline in relaxing human bronchi, making it the most potent endogenous bronchodilator so far described. Since there is a VIP-ergic innervation of human bronchi, this suggests that VIP may be an important regulator of bronchial tone and may be involved in counteracting the bronchoconstriction of asthma.

There appear to be differences in response to VIP, depending on the size of airway. In bovine airways the responsiveness to VIP diminishes with decreasing size of airway, whilst the relaxant response to isoprenaline is unchanged[71]. Similarly, in human airways, whilst bronchi are potently relaxed by VIP, bronchioles are unaffected, although they relax to an equal degree with isoprenaline[67]. This response of human airways is consistent with the distribution of VIP-receptors, since receptors are to be seen in bronchial smooth muscle, but not in bronchiolar smooth muscle[21]. This peripheral fading of VIP-receptors is also consistent with the distribution of VIP-immunoreactive nerves which diminish markedly as airways become smaller. These studies suggest that VIP, while regulating the calibre of large airways, is unlikely to influence small airways.

VIP also causes bronchodilatation *in vivo*. VIP given i.v. causes potent bronchodilatation in cat airways[30], and inhaled VIP protects against the bronchoconstrictor effects of histamine and prostaglandin $F_2\alpha$[79]. In humans, however, inhaled VIP has no bronchodilator effect, although a β-agonist in the same subjects is markedly effective[16]. Inhaled VIP has only a small protective effect against the bronchoconstrictor effect of histamine[16] and has no effect against exercise-induced bronchoconstriction[19]. This lack of potency of inhaled VIP may be explained by problems of diffusion of VIP across the airway epithelium to reach receptors in airway smooth muscle, or by enzymatic degradation of the peptide. Infused VIP similarly has no bronchodilator effect in normal subjects who bronchodilate with isoprenaline[70]. Infused VIP has a marked effect on the systemic cardiovascular system, with flushing, hypotension and tachycardia. These effects limit the dose which can be given by infusion and, as VIP is more potent on vessels than on airway smooth muscle, this prevents giving a dose which will affect airways. Infused VIP causes bronchodilation in asthmatic subjects, but the effect is small[62], and might be explained by sympathoadrenal activation secondary to the cardiovascular effects of VIP[10]. Thus, although VIP has a potent bronchodilator effect *in vitro*, it has no significant action *in vivo*, and therefore has no therapeutic potential, although it is possible that more stable analogs or novel compounds which activate VIP-receptors might be developed in the future as bronchodilators.

Airway secretion. VIP-immunoreactive nerves are closely associated with airway submucosal glands and form a dense network around the gland acini. VIP is a potent stimulant of mucus secretion[74], being significantly more potent than isoprenaline. VIP increases cyclic AMP formation in submucosal gland cells, and there is some suggestion that, as with β-agonists, there may be preferential effects on mucous rather than serous cells of these glands[49], indicating that VIP may stimulate mucus secretion rich in glycoprotein. VIP-receptors have also been localised to human submucosal glands, suggesting that VIP-ergic nerves may also regulate mucus secretion in human airways[21]. VIP has been found to have an inhibitory effect on glycoprotein secretion from human tracheal explants[26], which is surprising since agonists which stimulate cyclic AMP formation would be expected to stimulate secretion.

VIP is a potent stimulant of chloride ion transport and therefore water secretion in the gut, and a similar effect has also been found in dog tracheal epithelium[64], suggesting that VIP may be a regulator of airway water secretion and therefore mucociliary clearance. High densities of VIP-receptors have been localised to epithelial cells of human airways, so a similar effect might be expected[21].

VIP also inhibits antigen-induced histamine release from guinea pig lung, suggesting that VIP-receptors are present on pulmonary mast cells[86], but whether human lung mast cells have VIP-receptors is uncertain.

Blood vessels. VIP is a potent dilator of systemic vessels and pulmonary vessels. *In vitro*, VIP potently relaxes pulmonary vessels in many species, including man[12,38], and the relaxation is independent of endothelial cells, indicating that VIP acts directly on vascular smooth muscle cells, rather than releasing a relaxant factor from endothelial cells[38]. This is confirmed by autoradiographic studies showing the high density of receptors in smooth muscle with no labelling of endothelial cells[12,21]. In bovine pulmonary arteries the receptor density is greatest at the adventitial surface of the medial and diminishes towards the lumen[12]. The density of VIP-receptors is significantly greater in human pulmonary vessels than on bronchial smooth muscle, which may explain why VIP is about 10 times more potent as a vasodilator than a bronchodilator *in vitro*. VIP may have a rôle in regulating pulmonary blood flow, although the precise physiological rôle as a vasodilator mechanism is not yet certain.

VIP also relaxes bronchial vessels[46] and may regulate airway blood flow. Since VIP is likely to have a greater effect on bronchial vessels than on airway smooth muscle, it may provide a mechanism for increasing blood flow to contracted smooth muscle. Thus, if VIP is released from cholinergic nerves (see later), this may result in improved perfusion during cholinergic contraction. Perhaps the apparent protective effect of inhaled VIP against histamine-induced bronchoconstriction in human subjects[16], despite a lack of effect on bronchomotor tone, may be explained by an increase in bronchial blood flow which would more rapidly remove inhaled histamine from sites of deposition in the airways.

Neuromodulation. VIP is localised to nerves which surround airway ganglia, suggesting a possible neuromodulatory effect on cholinergic neurotransmission. In bovine airways VIP reduces cholinergic nerve effects, although this is seen only at high frequencies of stimulation, suggesting that the neuromodulatory effect is frequency-dependent[71].

VIP as NANC transmitter. Several lines of evidence implicate VIP as a neurotransmitter of NANC inhibitory nerves in airways. VIP produces prolonged relaxation of airway smooth muscle which is unaffected by adrenergic or neural blockade, and mimics the time-course of NANC inhibitory responses both *in vitro* and *in vivo*. VIP mimics the electrophysiological changes in airway smooth muscle produced by NANC nerve stimulation[20,41]. Electrical field stimulation of tracheobronchial preparations releases VIP into the bathing medium and this release is blocked by tetrodotoxin, proving that it is derived from nerve stimulation[20,60]. Furthermore, the amount of VIP released is related to the magnitude of nerve stimulation. Although there are no specific blockers of VIP-receptors, incubation of cat and guinea pig trachea with high concentrations of VIP induces tachyphylaxis and also reduces the magnitude of NANC nerve relaxation, while responses to sympathetic nerve stimulation and isoproterenol are unaffected[41]. Furthermore, preincubation of guinea pig trachea with a specific antibody to VIP reduces responses to exogenous VIP and to NANC stimulation[60]. The close association between responses to VIP and NANC relaxation in different sizes of human and bovine airways[71] also points to the rôle of VIP as a neurotransmitter. Although evidence in favour of VIP is persuasive,

Regulatory peptides in lung

Nerves	Neuroendocrine cells
Vasoactive intestinal peptide	Bombesin
Peptide histidine isoleucine	Leu-enkephalin
Substance P	Katacalcin
Neurokinins A and B	Calcitonin gene-related peptide
Calcitonin gene-related peptide	
Neuropeptide Y	
Galanin	
Gastrin releasing peptide	
Cholecystokinin	
Somatostatin	

until the development of a specific receptor antagonist, its neurotransmitter rôle cannot be proved. Indeed, there is some evidence against it as a neurotransmitter in airways. After pretreatment of guinea pig trachea with maximally effective concentration of VIP, there is no diminution of NANC relaxation, which would be expected if all VIP-receptors were occupied[43]. However, it is likely that other peptides, such as PHI, will also be released from these nerves and VIP may not have ready access to the VIP-receptors related to VIP-ergic nerves.

Co-transmission with acetylcholine. VIP coexists with acetylcholine (ACh) in some cholinergic nerves, supplying exocrine glands and potentiates the salivary secretory response to ACh[52]. VIP may be released from cholinergic nerves only with high frequency firing and may serve to increase the blood flow to exocrine glands under conditions of excessive stimulation. VIP also appears to coexist with ACh in airways, and it seems likely that there is a functional relationship between VIP and cholinergic neural control. It is possible that excessive stimulation of cholinergic nerves on certain patterns of firing result in VIP release. In bovine tracheal smooth muscle VIP has an inhibitory effect on cholinergic nerve-induced contraction, only with high frequency firing, and also reduces the contractile effect of exogenous ACh[71]. This does not involve any change in muscarinic receptor density or affinity, and may be due to functional antagonism. This indicates that VIP counteracts the bronchoconstrictor effect of acetylcholine and thus may function as a 'braking' mechanism for airway cholinergic nerves[8]. If this mechanism were to be deficient, this might result in exaggerated bronchoconstrictor responses.

Abnormalities in asthma. Whether dysfunction of VIP-ergic innervation contributes to any abnormality in pulmonary diseases is uncertain. It seems unlikely that there would be any primary abnormality in VIP content, neurotransmission or receptors, but it is possible that a secondary defect might develop as a result of the disease process[3,8].

Bronchial hyperresponsiveness in asthma may be related to an inflammatory response in the airway. The inflammatory cells present in the

324

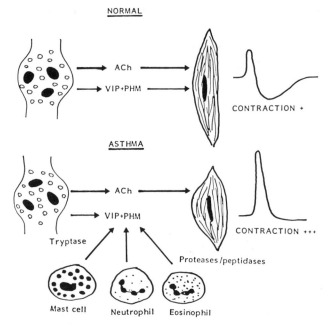

Figure 1. Schematic representation of increased degradation of VIP and PHM in asthmatic airways as a result of enzymes released from inflammatory cells. This may result in increased cholinergic bronchoconstriction. (From Barnes[4])

asthmatic airway (mast cells, macrophages, basophils, neutrophils) may release a variety of peptidases which break down VIP. Little is known of the enzymatic pathways involved in inactivation of these peptides, although several peptidases may inactivate them. If these peptidases are released into the airway in asthma, with increased degradation of VIP and PHI, this would result in a loss of the braking effect on cholinergic bronchoconstrictor responses (fig. 1). This might then contribute to the bronchial hyperresponsiveness of asthma.

Peptide histidine isoleucine

PHI and its human equivalent peptide histidine methionine (PHM) have a marked structural similarity to VIP, with 50% amino acid sequence homology. PHI and PHM are encoded by the same gene as VIP and both peptides are synthesised in the same prohormone. It is therefore not surprising to find that PHI has a similar immunocyto-chemical distribution in lung to VIP, and thus PHI-immunoreactive nerves supply airway smooth muscle (especially larger airways), bronchial and pulmonary vessels, submucosal glands and airway ganglia[55]. The levels of PHI-immunoreactivity are very similar to values obtained or VIP-immunoreactivity in respiratory tract[24].

Like VIP, PHI stimulates adenylates cyclase and may activate the same receptor as VIP. However, this is unlikely as PHI has different relative potencies compared with VIP in different tissues, being less potent as a vasodilator and more potent as a stimulant of secretion. In human bronchi *in vitro* PHM is a potent relaxant, and is equipotent to VIP[67]. It is therefore likely that PHI/PHM is released with VIP from airway nerves and may also be a neurotransmitter of NANC relaxation. Although VIP and PHI may activate separate receptors, no synergy has been demonstrated between these peptides in airway smooth muscle (Palmer, J. B., and Barnes, P. J., unpublished). Because PHM has equal potency in relaxing airway smooth muscle, but is less potent as a vasodilator than VIP, it is possible to infuse higher concentrations before cardiovascular effects are limiting. In a preliminary study it has not proved possible to demonstrate bronchodilatation in 3 mild asthmatics with the highest infused concentration of PHM (Palmer, J. B., Cuss, F. M., and Barnes, P. J., unpublished).

PHI is significantly less potent as a pulmonary vasodilator than VIP, and, like VIP, produces its effects independently of endothelial cells[38]. The effects of PHI on airway mucus secretion and ion transport have not yet been studied, but PHI has potent effects on secretion in the gastrointestinal tract.

Substance P and tachykinins

While substance P (SP) was isolated over 50 years ago, structurally related peptides (tachykinins) called neurokinin A and B have recently been identified[65]: NKA is coded by the same gene as SP, whereas NKB is coded by a different gene.

Localisation. SP is localised to nerves in the airways of several species, including humans[56], although there has been debate about whether SP can be demonstrated in human airways[45]. However, rapid degradation of SP in airways, and the fact that SP concentrations may decrease with age, and possibly smoking, could explain the difficulty in demonstrating this peptide in some studies. SP-immunoreactive nerves in the airway are found beneath and within the airway epithelium, around blood vessels and, to a lesser extent, within airway smooth muscle. SP appears to be localised to afferent nerves in the airways and SP is synthesised in the nodose ganglion of the vagus nerve and then transported down the vagus to peripheral branches in the lung. Treatment of animals with capsaicin releases SP from sensory nerves acutely and chronic administration depletes the lung of SP[54]. Neurokinin-like immunoreactivity has also been found in lung and may also play a regulatory role[83].

Tachykinin receptors. SP exerts its effects on target cells via specific receptors, which have now been identified by direct binding techniques. Recent autoradiographic studies have demonstrated the distribution of

SP-receptors in guinea pig and human lung using Bolton-Hunter [^{125}I]-SP[22]. SP-receptors are found in high density in airway smooth muscle from trachea down to small bronchioles, whereas pulmonary vascular smooth muscle and epithelial cells are less densely labelled. Submucosal glands in human airways are also labelled.

There are at least three types of tachykinin receptor which may be differentiated by different responses to a series of tachykinins[50]. In some tissues SP is more potent than neurokinins (SP-P or NK-1 receptor), whereas in others the order of potency is either NKA > NKB > SP (SP-E or NK-2 receptor), or NKB > NKA > SP (SP-N or NK-3 receptor).

Airway smooth muscle. In vivo SP infusion causes bronchoconstriction in animals[1], which may be partially blocked with atropine, suggesting that SP release of acetylcholine may be responsible for some of the bronchoconstrictor action *in vivo*[82]. In human subjects infusion of SP has profound cardiovascular effects, but little effect on airway function; a small bronchoconstrictor response is followed by bronchodilation at higher infusion doses[35]. The cardiovascular actions may limit the dose that can be given and result in reflex bronchodilatation (by a reduction in vagal tone), which counteracts the bronchoconstriction. Even when given by inhalation SP has no significant effect on airway function in mild asthmatic subjects who are hyperresponsive to histamine given in the same way[35]. This may be due to enzymatic degradation of SP in the airway and its inability to cross the epithelium. Inhalation of capsaicin in human subjects, which should release SP, induces marked coughing but only transient bronchoconstriction, seen in both normal and asthmatic subjects to an equal extent. This bronchoconstrictor response is inhibited by cholinergic antagonists, suggesting that it may be due to reflex stimulation. The intense irritation of capsaicin may preclude giving a dose sufficiently large to release SP from airway nerves.

In vitro SP contracts airway smooth muscle of several species, including man[52,57]. Moreover, capsaicin is capable of inducing a similar contraction, indicating release of SP from intrinsic nerves within airway smooth muscle. The contractile effect of SP on airway smooth muscle *in vitro* may be inhibited by SP antagonists, suggesting a direct effect on smooth muscle cells, although the specificity of these antagonists has been questioned.

In guinea pig and human airways neurokinin A is more potent than SP in causing contraction, suggesting that bronchial smooth muscle has NK-2 receptors, and that neurokinins, rather than SP, regulate bronchomotor tone[42,68].

Microvascular effects. In rats and guinea pigs, both SP and capsaicin induce edema of the airway wall by increasing microvascular permeability. Depletion of SP nerves by neonatal capsaicin pretreatment prevents irritants, such as cigarette smoke and mechanical stimulation, from causing mucosal edema, suggesting that SP nerves mediate this effect[58]. In human skin SP causes a wheal and flare response[32,11], but NKA and NKB are much less potent, suggesting this is mediated by NK-1 receptors[13].

Secretion. SP is a potent stimulant of airway mucus secretion in isolated canine and human airways[15,25].

Mediator release. The wheal and flare response to intradermal SP is blocked by antihistamines[32] and increases the release of histamine into draining veins[11]. This suggests that SP might be capable of causing histamine release in the airways, although there is no direct evidence for this. The flare induced by intradermal SP and neurokinins is also reduced by aspirin, suggesting that cyclo-oxygenase products are also released[14].

It has also been suggested that SP might amplify neutrophil and eosinophil responses to chemotactic agents, and therefore magnify the inflammatory response in the airways[73].

Calcitonin gene-related peptide (CGRP)

CGRP is localised to afferent nerve terminals in airways and may be co-localised with SP[53]. Human CGRP is a potent vasodilator, and produces a wheal and flare in the skin when injected locally[18], and potentiates the effect of SP[11]. CGRP has been extracted from human airways and is localised to nerves[69]. *In vitro* human CGRP contracts isolated human airways, being much more potent than SP, and producing equivalent contraction to carbachol, whereas SP causes only partial contraction[69]. CGRP-induced contraction is unaffected by cholinergic, histamine or leukotriene antagonists and probably acts via specific receptors.

Axon reflexes and asthma

Sensory neuropeptides produce many of the pathological features of asthma, including contraction of airway smooth muscle, edema and plasma extravasation, mucus hypersecretion and possibly increased mediator release, it is possible that they may contribute to the pathology of asthma. Damage to airway epithelium may occur even in relatively well-controlled asthmatics, probably as a result of eosinophil products[42], exposing afferent nerve endings which are stimulated by inflammatory mediators. C-fibre endings may be selectively stimulated by bradykinin and other mediators produced in the inflammatory reaction, which could result in a reflex cholinergic bronchoconstriction. Bradykinin is a potent bronchoconstrictor in asthmatic subjects, which may selectively

328

stimulate C-fibre afferents[34]. However, anticholinergic drugs are not very effective in clinical asthma and it is possible that stimulation of C-fibre endings may result in an axon reflex, with release of sensory neuropeptides from sensory collaterals in the airway[5], or a local ganglionic reflex, since SP and CGRP are found in airway ganglia, suggesting an afferent input. The release of SP, neurokinins and CGRP may then result in bronchoconstriction, mucus hypersecretion and microvascular leakage of plasma to produce edema of the airway wall and extravasation of plasma into the airway lumen (fig. 2). The axon reflex may, therefore, contribute to the pathology of asthma, and may help to explain how patchy epithelial damage leads to widespread pathophysiological changes. Drugs which interfere with the axon reflex mechanisms may, therefore, be useful in asthma, and there is some evidence to suggest that sodium cromoglycate may have such an effect.

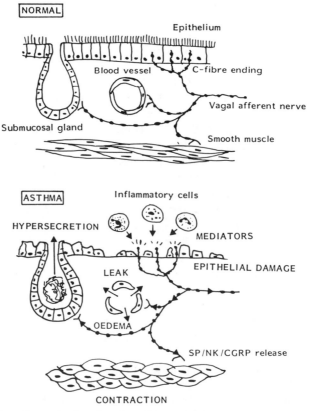

Figure 2. Axon reflex mechanisms in asthma. Damage to airway epithelium in asthma exposes unmyelinated nerve endings which may be triggered by mediators (e.g. bradykinin) resulting in release of sensory neuropeptides such as substance P (SP), neurokinins (NK) and calcitonin gene-related peptide (CGRP), which together contribute to the pathology of asthma. (From Barnes[5])

Other neuropeptides

Neuropeptide Y. NPY has been found in human lung and is localised primarily to innervation of blood vessels[80]. The distribution of NPY is similar to that of sympathetic nerves and therefore few NPY-immunoreactive nerve fibres are localised to airway smooth muscle. NPY is probably a co-transmitter of noradrenaline and is a potent constrictor of vascular smooth muscle[59]; it may play an important rôle in regulating pulmonary and bronchial vessels but is likely to be less important in influencing airway tone.

Galanin. Galanin has recently been isolated and localised to motor nerves in the respiratory tract, possibly co-localised with VIP and acetylcholine[23]. The function of this peptide is uncertain and, while it acts as a neuromodulator reducing cholinergic nerve effects in gut, it has no effect on guinea pig trachea[31].

Gastrin-releasing peptide. Gastrin-releasing peptide (GRP) is probably the mammalian form of bombesin and has been localised to nerves in the airway wall of several species, but again its function is uncertain[84].

Conclusions

Many neuropeptides have now been localised to the lung, and almost certainly more will be discovered. These peptides often have potent actions on airway and vascular tone and on lung secretions, but the presence of so many peptides raises questions about their physiological rôle. It seems most likely that they may act as subtle regulators under physiological conditions, but in inflammatory diseases such as asthma they may have a pathogenetic rôle. Until specific antagonists have been developed it will be difficult to evaluate the precise rôle of these neuropeptides in disease. It is certainly possible that pharmacological agents which interact with neuropeptides by affecting their release, metabolism or receptors may be developed in the future and may have therapeutic potential.

Acknowledgment. I thank Madeleine Wray for the careful preparation of this manuscript. Supported by Medical Research Council and Asthma Research Council.

1 Andersson, P., and Persson, H., Effect of substance P on pulmonary resistance and dynamic pulmonary compliance in the anaesthetized cat and guinea-pig. Acta pharmac. toxic. *41* (1977) 444–448.
2 Andersson, R. G. G., and Grundstrom, N., The excitatory non-cholinergic, non-adrenergic system of the guinea-pig airways. Eur. J. respir. Dis. *64* (1983) 141–157.
3 Barnes, P. J., The third nervous system in the lung: physiology and clinical perspectives. Thorax *39* (1984) 561–567.
4 Barnes, P. J., Non-adrenergic non-cholinergic neural control of human airways. Archs int. Pharmacodyn. *280* (1986) 208–228.

5 Barnes, P. J., Asthma as an axon reflex. Lancet *1* (1986) 242–245.

6 Barnes, P. J., Neural control of human airways in health and disease. State of the art. Am. Rev. respir. Dis. *134* (1986) 1289–1314.

7 Barnes, P. J., Neuropeptides in the lung: localization, function and pathophysiological implications. J. allergy clin. Immun. *79* (1987) 285–295.

8 Barnes, P. J., Airway neuropeptides and asthma. Trends pharm. Sci. *8* (1987) 24–27.

9 Barnes, P. J., Vasoactive intestinal peptide and pulmonary function, in: Current Topics in Pulmonary Pharmacology and Toxicology, pp. 156–173. Ed. M. A. Hollinger. Elsevier, New York 1987.

10 Barnes, P. J., Bloom, S. R., and Dixon, C. M. S., VIP and asthma. Lancet *1* (1984) 112.

11 Barnes, P. J., Brown, M. J., Dollery, C. T., Fuller, R. W., Heavey, D. J., and Ind, P. W., Histamine is released from skin by substance P but does act as the final vasodilator in the axon reflex. Br. J. Pharmac. *88* (1986) 741–745.

12 Barnes, P. J., Cadieux, A., Carstairs, J. R., Greenberg, B., Polak, J. M., and Rhoden, K., Vasoactive intestinal peptide in bovine pulmonary artery: localisation, function and receptor autoradiography. Br. J. Pharmac. *89* (1986) 157–162.

13 Barnes, P. J., Conradson, T.-B., Dixon, C. M. S., and Fuller, R. W., A comparison of the cutaneous actions of substance P, neurokinin A and calcitonin gene-related peptide in man. J. Physiol. *374* (1986) 22P.

14 Barnes, P. J., Crossman, D. C., and Fuller, R. W., Cutaneous effects of neuropeptides are dependent upon both histamine and cyclooxygenase products. Br. J. Pharmac. *90* (1987).

15 Barnes, P. J., Dewar, A., and Rogers, D. F., Human bronchial secretion: effect of substance P, muscarinic and adrenergic stimulation in vitro. Br. J. Pharmac. *89* (1986) 767P.

16 Barnes, P. J., and Dixon, C. M. S., The effect of inhaled vasoactive intestinal peptide on bronchial hyperreactivity in man. Am. Rev. respir. Dis. *130* (1984) 162–166.

17 Borson, D. B., Charlin, M., Gold, B. D., and Nadel, J. A., Neural regulation of 35 SO4-macromolecule secretion from tracheal glands of ferrets. J. appl. Physiol. *57* (1984) 457–466.

18 Brain, S. D., Williams T. J., Tippins, J. R., Morris, H. R., and MacIntyre, I., Calcitonin gene-related peptide is a potent vasodilator. Nature *313* (1985) 54–56.

19 Bungaard, A., Enehjelm, S. D., and Aggestrop, S., Pretreatment of exercise-induced asthma with inhaled vasoactive intestinal peptide (VIP). Eur. J. respir. Dis. *64* (1983) 427–429.

20 Cameron, A. R., Johnston, C. D., Kirkpatrick, C. T., and Kirkpatrick, M. C. A., The quest for the inhibitory neurotransmitter in bovine tracheal smooth muscle. Q. J. exp. Psychol. *68* (1983) 413–426.

21 Carstairs, J. R., and Barnes, P. J., Visualization of vasoactive intestinal peptide receptors in human and guinea pig lung. J. Pharmac. exp. Ther. *239* (1986) 249–255.

22 Carstairs, J. R., and Barnes, P. J., Autoradiographic mapping of substance P receptors in lung. Eur. J. Pharmac. *127* (1986) 295–296.

23 Cheung, A., Polak, J. M., Bauer, F. E., Christofides, N. D., Cadieux, A., Springall, D. R., and Bloom, S. R., The distribution of galanin immunoreactivity in the respiratory tract of pig, guinea pig, rat and dog. Thorax *40* (1985) 889–896.

24 Christofides, N. D., Yiangou, Y., Piper, P. J., Ghatei, M. A., Sheppard, M. N., Tatemoto, K., Polak, J. M., and Bloom, S. R., Distribution of peptide histidine isoleucine in the mammalian respiratory tract and some aspects of its pharmacology. Endocrinology *115* (1984) 1958–1963.

25 Coles, S. J., Neill, K. H., and Reid, L. M., Potent stimulation of glycoprotein secretion in canine trachea by substance P. J. appl. Physiol. *57* (1984) 1323–1327.

26 Coles, S. J., Said, S. I., and Reid, L. M., Inhibition by vasoactive intestinal peptide of glycoconjugate and lysozyme secretion by human airways in vitro. Am. Rev. respir. Dis. *124* (1981) 531–536.

27 Cutz, E., Neuroendocrine cells of the lung—an overview of morphologic characteristics and development. Exp. Lung Res. *3* (1982) 185–208.

28 Dey, R. D., Shannon, W. A., and Said, S. I., Localization of VIP-immunoreactive nerves in airways and pulmonary vessels of dogs, cats and human subjects. Cell Tissue Res. *220* (1981) 231–238.

29 Diamond, L., and O'Donnell, M., A non-adrenergic vagal inhibitory pathway to feline airways. Science *208* (1980) 185–188.

30 Diamond, L., Szareck, J. L., Gillespie, M. N., and Altiere, R. J., In vivo bronchodilator activity of vasoactive intestinal peptide in the cat. Am. Rev. respir. Dis. *128* (1983) 827–832.

31 Ekblad, E., Hakanson, R., Sundler, F., and Wahlestedt, C., Galanin: neuromodulatory and direct contractile effects on smooth muscle preparations. Br. J. Pharmac. *86* (1985) 241–246.

32 Foreman, J. C., Jordan, C. C., Oehme, P., and Renner, H., Structure-activity relationships for some substance P-related peptides that cause wheal and flare reactions in human skin. J. Physiol. *335* (1983) 449–465.

33 Fuller, R. W., Dixon, C. M. S., and Barnes, P. J., The bronchoconstrictor response to inhaled capsaicin in humans. J. appl. Physiol. *85* (1985) 1080–1084.

34 Fuller, R. W., Dixon, C. M. S., Cuss, F. M. C., and Barnes, P. J., Bradykinin-induced bronchoconstriction in man: mode of action. Am. Rev. respir. Dis. *135* (1986) 176–180.

35 Fuller, R. W., Maxwell, D. L., Dixon, C. M. S., McGregor, G. P., Barnes, V. F., Bloom, S. R., and Barnes, P. J., The effects of substance P on cardiovascular and respiratory function in human subjects, J. appl. Physiol. *62* (1987) 1473–1479.

36 Ghatei, M. A., Sheppard, M., O'Shaunessy, D. J., Adrian, T. E., MacGregor, J. M., Polak, J. M., and Bloom, S. R., Regulatory peptides in the mammalian respiratory tract. Endocrinology *111* (1982) 1248–1254.

37 Ghatei, M. A., Sheppard, M. N., Henzen-Logman, S., Blank, M. A., Polak, J. M., and Bloom, S. R., Bombesin and vasoactive intestinal polypeptide in the developing lung: marked changes in acute respiratory distress syndrome. J. clin. Endocr. *57* (1983) 1226–1232.

38 Greenberg, B., Rhoden, K., and Barnes, P. J., Vasoactive intestinal peptide causes non-endothelial dependent relaxation in human and bovine pulmonary arteries. Blood Vessels *24* (1987) 45–50.

39 Impicciatore, M., and Bertaccini, G., The bronchoconstrictor action of the tetradeaupeptide bombesin in the guinea pig. J. Pharm. Pharmac. *25* (1973) 812–815.

40 Irvin, C. G., Martin, R. R., and Macklem, P. T., Non purinergic nature and efficacy of non adrenergic bronchodilatation. J. appl. Physiol. *52* (1982) 562–569.

41 Ito, Y., and Takeda, K., Non-adrenergic inhibitory nerves and putative transmitters in the smooth muscle of cat trachea. J. Physiol. *330* (1982) 497–511.

42 Karlsson, J.-A., Finney, M. J. B., Persson, C. G. A., and Post, C., Substance P antagonist and the role of tachykinins in non-cholinergic bronchoconstriction. Life Sci. *35* (1984) 2681–2691.

43 Karlsson, J.-A., and Persson, C. G. A., Neither vasoactive intestinal peptide (VIP) nor purine derivatives may mediate non-adrenergic tracheal inhibition. Acta physiol. scand. *122* (1984) 589–598.

44 Laitinen, L. A., Heino, M., Laitinen, A., Kava, T., and Haahtela, T., Damage of the airway epithelium and bronchial reactivity in patients with asthma. Am. Rev. respir. Dis. *131* (1985) 599–606.

45 Laitinen, L. A., Laitinen, A., Panula, P. A., Partanen, M., Tervo, K., and Tervo T., Immunohistochemical demonstration of substance P in the lower respiratory tract of the rabbit and not of man. Thorax *38* (1983) 531–536.

46 Laitinen, L. A., Laitinen, A., and Widdicombe, J. G., Effects of inflammatory and other mediators on airway vascular beds. Am. Rev. respir. Dis. *135* (1987) S67–S70.

47 Laitinen, A., Partanen, M., Heruonen, A., Peto-Huikko, M., and Laitinen, L. A., VIP-like immunoreactive nerves in human respiratory tract. Light and electron microscopic study. Histochemistry *82* (1985) 313–319.

48 Lauweryns, J. M., van Lomme, L. A. T., and Dom, R. J., Innervation of rabbit intrapulmonary neuroepithelial bodies. Quantitative ultrastructural study after vagotomy. J. neurol. Sci. *67* (1985) 81–92.

49 Lazarus, S. C., Basbaum, C. B., Barnes, P. J., and Gold, W. M., cAMP immunocytochemistry provides evidence for functional VIP receptors in trachea. Am. J. Physiol. *251* (1986) C115–119.

50 Lee, C.-M., Campbell, N. J., Williams, B. J., and Iversen, L. L., Multiple tachykinin binding sites in peripheral tissues and in brain. Eur. J. Pharmac. *130* (1986) 209–217.

332

51 Leroux, P., Vaudry, H., Fournier, A., St Pierre, S., and Pelletier, G., Characterization and localization of vasoactive intestinal peptide receptors in the rat lung. Endocrinology *114* (1984) 1506–1512.

52 Lundberg, J. M., Evidence for coexistence of vasoactive intestinal peptide (VIP) and acetylcholine in neurons of cat exocrine glands: morphological, biochemical and functional studies. Acta physiol. scand. *496* (1981) 1–57.

53 Lundberg, J. M., Anders, F.-C., Hua, X., Hökfelt, T., and Fischer, J. A., Coexistence of substance P and calcitonin gene-related peptide-like immunoreactivities in sensory nerves in relation to cardiovascular and bronchoconstrictor effects of capsaicin. Eur. J. Pharmac. *108* (1985) 315–319.

54 Lundberg, J. M., Brodin, E., and Saria, A., Effects and distribution of vagal capsaicin-sensitive substance P neurons with special reference to the trachea and lungs. Acta physiol. scand. *119* (1983) 243–252.

55 Lundberg, J. M., Fahrenkrug, J., Hökfelt, T., Martling, C. R., Larsson, O., Tatemoto, K., and Änggard, A., Coexistence of peptide H1 (PH1) and VIP in nerves regulating blood flow and bronchial smooth muscle tone in various mammals including man. Peptides *5* (1984) 593–606.

56 Lundberg, J. M., Hökfelt, T., Martling, C.-R. Saria, A., and Cuello, C., Substance P-immunoreactive sensory nerves in the lower respiratory tract of various mammals including man. Cell Tissue Res. *235* (1984) 251–261.

57 Lundberg, J. M., Martling, C.-R., and Saria, A., Substance P and capsaicin-induced contraction of human bronchi. Acta physiol. scand. *119* (1983) 49–53.

58 Lundberg, J. M., and Saria, A., Capsaicin-induced desensitization of the airway mucosa to cigarette smoke, mechanical and chemical irritants. Nature *302* (1983) 251–253.

59 Lundberg, J. M., Terenius, L., Hökfelt, T., Martling, C. R., Tatemoto, K., Mutt, V., Polak, J., Bloom, S., and Goldstein, M., Neuropeptide Y (NPY)-like immunoreactivity in peripheral nonadrenergic neurons and effects of NPY on sympathetic function. Acta physiol. scand. *116* (1982) 477–480.

60 Matsuzaki, Y., Hamasaki, Y., and Said, S. I., Vasoactive intestinal peptide: a possible transmitter of nonadrenergic relaxation of guinea pig airways. Science *210* (1980) 1252–1253.

61 Michoud, M.-C., Amyot, R., and Jeanneret-Grosjean, A., Bronchodilatation induced by laryngeal stimulation in humans. Am. Rev. respir. Dis. *131* (1985) A 284.

62 Morice, A., Unwin, R. J., and Sever P. S., Vasoactive intestinal peptide causes bronchodilatation and protects against histamine-induced bronchoconstriction in asthmatic subjects. Lancet *2* (1983) 1225–1226.

63 Nadel, J. A., and Barnes, P. J., Autonomic regulation of the airways. A. Rev. Med. *35* (1984) 451–467.

64 Nathanson, I., Widdicombe, J. H., and Barnes, P. J., Effect of vasoactive intestinal peptide on ion transport across dog tracheal epithelium. J. appl. Physiol. *55* (1983) 1844–1848.

65 Nawa, H., Hirose, T., Takashima, H., Inayama, S., and Nakanishi, S., Nucleotide sequences of cloned cDNAs for two types of bovine brain substance P precursor. Nature *306* (1983) 32–36.

66 Pack, R. J., and Widdicombe, J. G., Amine containing cells of the lung. Eur. J. respir. Dis. *65* (1984) 559–578.

67 Palmer, J. B., Cuss, F. M. C., and Barnes, P. J., VIP and PHM and their role in non-adrenergic inhibitory responses in isolated human airways. J. appl. Physiol. *61* (1986) 1322–1328.

68 Palmer, J. B., Cuss, F. M. C., and Barnes, P. J., Sensory neuropeptides and human airway function. Am. Rev. respir. Dis. *133* (1986) A239.

69 Palmer, J. B. D., Cuss, F. M. C., Ghatei, M. A., Springall, D. R., Cadieux, A., Bloom, S. R., Polak, J. M., and Barnes, P. J., Calcitonin gene-related peptide is localised to human airway nerves and potently constricts human airway smooth muscle. Thorax *40* (1985) 713.

70 Palmer, J. B. D., Cuss, F. M. C., Warren, J. B., and Barnes, P. J., The effect of infused vasoactive intestinal peptide on airway function in normal subjects. Thorax *41* (1986) 663–666.

71 Palmer, J. B., Sampson, A. P., and Barnes, P. J., Cholinergic and non-adrenergic

inhibitory responses in bovine airways: distribution and functional association. Am. Rev. respir. Dis. *131* (1985) A282.

72 Partanen, M., Laitinen, A., Pelto-Huikko, M., Laitinen, L. A., and Hervonen, A., Electronmicroscopic study on granule containing (APUD) cells in adult human lung. Cell Tissue Res.(1986) in press.

73 Payan, G. P., Levine, J. D., and Goetzl, E. J., Modulation of immunity and hypersensitivity by sensory neuropeptides. J. Immun. *132* (1984) 1601–1604.

74 Peatfield, A. C., Barnes, P. J., Bratcher, C., Nadel, J. A., and Davis, B., Vasoactive intestinal peptide stimulates tracheal submucosal gland secretion in ferret. Am. Rev. respir. Dis. *128* (1983) 89–93.

75 Peatfield, A. C., and Richardson, P. S., Evidence for non-cholinergic, non-adrenergic nervous control of mucus secretion into the cat trachea. J. Physiol. *342* (1983) 335–345.

76 Polak, J. M., and Bloom, S. R., Regulatory peptides of the gastrointestinal and respiratory tracts. Archs int. Pharmacodyn. *280*, Suppl. (1986) 16–49.

77 Richardson, J. B., Nonadrenergic inhibitory innervation of the lung. Lung *159* (1981) 315–322.

78 Robberecht, P., Chatelain, P., De Neef, P., Camus, J.-C., Waelbroeck, M., and Christophe, J., Presence of vasoactive intestinal peptide receptors coupled to adenylate cyclase in rat lung membranes. Biochim. biophys. Acta *678* (1981) 76–82.

79 Said, S. I., Vasoactive peptides in the lung, with special reference to vasoactive intestinal peptide. Exp. Lung Res. *3* (1982) 343–348.

80 Sheppard, M. N., Polak, J. M., Allen, J. M., and Bloom, S. R., Neuropeptide tyrosine (NYP): a newly discovered peptide is present in the mammalian respiratory tract. Thorax *39* (1984) 326–330.

81 Szarek, J. L., Gillespie, M. N., Altiere, R. J., and Diamond, L., Mechanical irritation of the larynx reflexing evokes non-adrenergic bronchodilatation. Am. Rev. respir. Dis. *129* (1984) 243.

82 Tanaka, D. T., and Grunstein, M. M., Mechanisms of substance P-induced contraction of rabbit airway smooth muscle. J. appl. Physiol. *57* (1984) 1551–1557.

83 Theodorsson-Nerheim, E., Hua, X., Brodin, E., and Lundberg, J. M., Capsaicin treatment decreases tissue levels of neurokinin-A-like immunoreactivity in the guinea pig. Acta physiol. scand. *124* (1985) 129–131.

84 Uddman, R., Moghimzadeh, E., and Sundler, F., Occurrence and distribution of GRP-immunoreactive nerve fibres in the respiratory tract. Archs Otolar. *239* (1984) 145–151.

85 Uddman, R., and Sundler, F., Vasoactive intestinal polypeptide nerves in human upper respiratory tract. Oto-rhino-lar. *41* (1979) 221–226.

86 Undem, B. J., Dick, E. C., and Buckner, C. K., Inhibition by vasoactive intestinal peptide of antigen-induced histamine release from guinea-pig minced lung. Eur. J. Pharmac. *88* (1983) 247–250.

Neuropeptides in pelvic afferent pathways

W. C. de Groat

Summary. Neurochemical and pharmacological experiments have raised the possibility that several neuropeptides including, vasoactive intestinal polypeptide (VIP), peptide histidine isoleucine amide (PHI), substance P, calcitonin gene-related peptide (CGRP), neurokinin A, cholecystokinin (CCK) and opioid peptides may be transmitters in afferent pathways to the pelvic viscera. These substances are widely distributed in: 1) nerve fibers in the pelvic organs, 2) visceral afferent neurons in the lumbosacral dorsal root ganglia and 3) at sites of afferent termination in the spinal cord. Double staining immunocytochemical techniques have shown that more than one peptide can be localized in individual visceral afferent neurons and that neuronal excitatory (VIP, substance P, CCK) and inhibitory peptides (leucine enkephalin) can coexist in the same afferent cell. Studies with the neurotoxin, capsaicin, indicate that peptidergic afferent pathways are involved in the initiation of central autonomic reflexes as well as peripheral axon reflexes which modulate smooth muscle activity, facilitate transmission in autonomic ganglia and trigger local inflammatory responses.

Afferent pathways to the pelvic viscera play an essential role in the neural regulation of various physiological functions, including micturition, defecation and reproduction[17,20]. The anatomical and electrophysiological properties of these pathways have recently been described in some detail[16,52,68,90,92,96]. On the other hand, the mechanisms underlying transmission at visceral afferent synapses are still relatively unexplored.

Presently, the neuropeptides are attracting considerable attention as putative transmitters in visceral afferent pathways. These substances are widely distributed in nerve fibers in the pelvic organs[2,3,29,30,74,77] and at sites of visceral afferent termination in the lumbosacral spinal cord[16,21,40,56]. Immunocytochemical and axonal tracing experiments have shown that a large percentage of pelvic visceral afferent neurons contain peptides[16,18,48,63,113,114] and that more than one peptide can occur in these cells[16,25,58]. Insight into the physiological role of the neuropeptides has also been obtained by examining the effects of capsaicin on the activity of the pelvic organs[41,46,77–79,81]. Capsaicin is a very useful experimental tool for these studies since it is thought to act selectively on peptidergic afferents to release transmitters and in large doses to cause neuronal degeneration[9,27].

This chapter will review the neurochemical and pharmacological data suggesting that peptides function as transmitters in the afferent innervation of the pelvic viscera.

Peptidergic afferent innervation of the pelvic organs

The afferent innervation of the pelvic viscera (i.e., the urinary tract, distal bowel and reproductive organs) originates in the lumbosacral dorsal root ganglia and is carried by several peripheral nerves (fig. 1)[17,20,52,68]. The hypogastric and lumbar colonic nerves contain axons from the lumbar dorsal root ganglia[16,17,52,92,93], whereas the pelvic and pudendal nerves carry afferent pathways from the sacral dorsal root ganglia[16,17,52,90,102]. The pudendal nerves also contain somatic afferent axons innervating superficial and deep structures in the perineum (fig. 1), many of which are closely linked functionally with the pelvic viscera[16,18,122].

Various peptides including substance P and related tachykinins (neurokinin A, neuropeptide K and eledoisin-like peptide)[3,13,14,18,28,36,37,39,41,65,77,86,113,116,118,119], calcitonin gene related peptide (CGRP)[29,31,103,114], vasoactive intestinal polypeptide (VIP)[35,36,38,39,48,61,64,65,77], peptide histidine isoleucine amide (PHI)[14,35], galanin[7a], cholecystokinin (CCK)[14], somatostatin[14], enkephalins[14,18,21,77], and dynorphins[14] are present in nerves and ganglia within and projecting to the urogenital tract and the large intestine. The origin of these

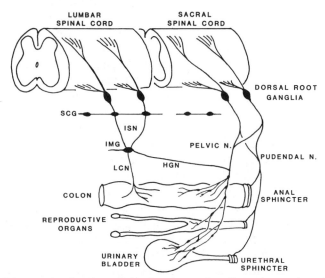

Figure 1. Diagram showing the peripheral pathways for afferent innervation of the pelvic viscera in the cat. Neurons in the sacral dorsal root ganglia send axons into the pelvic and pudendal nerves. The pelvic nerves innervate exclusively the viscera whereas the pudendal nerves innervate visceral as well as somatic structures such as the anal and urethral sphincters and the perineum. Neurons in the lumbar dorsal root ganglia send axons into the lumbar colonic (LCN) and hypogastric nerves (HGN). The latter axons pass through the sympathetic chain ganglia (SCG) and then the inferior splanchnic nerves (ISN) to the inferior mesenteric ganglia (IMG). The HGN passes caudally to join the pelvic nerve.

peptidergic nerves has been examined using surgical and/or chemical denervation techniques.

In rats and guinea pigs the administration of capsaicin, a neurotoxin, which acts selectively on small diameter afferent fibers to deplete transmitter stores and to induce neuronal degeneration, reduces the levels of substance P[28,41,46,47,84,104,113,118,119], neurokinin A[46,47] and CGRP[29,31,35,114] at various sites in the pelvic viscera, but does not change the levels of VIP[14,35,100,119] or enkephalins[14,35]. Surgical interruption of the extrinsic nerves (pelvic and hypogastric nerves) or removal of the lumbosacral dorsal root ganglia produces a similar selective decrease in substance P and CGRP immunoreactivity[14,29,86,87,114], whereas removal of the peripheral autonomic ganglia (pelvic ganglia) or transection of postganglionic nerves produces a generalized decrease in peptide immunoreactivity in certain pelvic organs as well as a decrease in markers for cholinergic and adrenergic nerves[87,114]. CGRP, substance P and VIP are not altered following the administration of 6-hydroxydopamine in doses which destroys adrenergic efferent axons[29,87,114,119]. These data indicate that CGRP, substance P and related tachykinins are present in afferent pathways to the pelvic viscera.

It should be noted, however, that the failure of capsaicin and deafferentation to change the levels of other peptides does not exclude the possibility that these peptides are also present in afferent axons. For example, VIP is contained in a large population of efferent axons in the pelvic viscera[38,73,74,77,87]. This population of VIP axons would remain after deafferentation and might obscure the changes in a smaller population of VIP afferent axons[35]. This seems to be a likely possibility since VIP has been detected in visceral afferent neurons in the dorsal root ganglia (see below)[56,59]. In addition, certain peptidergic afferents may not be sensitive to capsaicin[126].

Immunocytochemical studies[3,29,31,45] have demonstrated a widespread distribution of substance P and CGRP-containing afferent axons in the pelvic viscera. Although the tissue concentrations of CGRP are greater than those of substance P, the distribution of the two peptides is very similar. Indeed, double staining techniques in which antisera to substance P and CGRP were applied to the same tissue sections revealed that the two peptides are co-localized in axonal varicosities in various organs[31], including the intestine and the urogenital tract.

In the rat and guinea pig the highest concentrations of CGRP occur in the ureter and bladder base with lower concentrations in the bladder dome, urethra and reproductive organs including the cervix, vagina, uterus, fallopian tubes and the ovaries[29,114]. In the urogenital tract the substance P and CGRP axons are located throughout the submucosal plexus, adjacent to the epithelium, in the smooth muscle layers and around blood vessels[29,31,77,87]. The high density of these axons in proximity to the epithelial layers of the urogenital organs contrasts with the more prominent distribution of efferent cholinergic, adrenergic and VIPergic axons in the smooth muscle areas in these organs[3,87,127].

In the wall of the intestine substance P/CGRP afferent axons are distinguished from axons of intrinsic peptidergic neurons by their sensitivity to capsaicin[28,31]. Using this technique it has been shown in the guinea pig intestine, that substance P/CGRP afferent fibers are relatively sparse in the myenteric plexus, but as in the urogenital organs are common beneath the epithelium and in the submucosal plexus where they followed blood vessels and form loose plexuses within the submucosal ganglia[31]. Axons which are resistant to capsaicin treatment have a different distribution.

Substance P afferent axons and varicosities are also present in prevertebral sympathetic and pelvic ganglia which provide input to the pelvic viscera[13,14,77,86]. In the inferior mesenteric ganglion (IMG, see fig. 1) of the guinea pig, the substance P varicosities form perineuronal networks around the principal ganglion cells[14,86]. At the ultrastructural level some of these varicosities were seen to make synaptic contacts with dendrites[86]. The substance P varicosities were eliminated by capsaicin pretreatment or by surgical deafferentation which interrupted the connections between the IMG and the lumbar dorsal root ganglia[14,86]. Axonal tracing studies have confirmed the afferent projections from the lumbar dorsal root ganglia to the IMG[1].

Other peptides including VIP, CCK, bombesin and dynorphin are also present in afferent pathways to the IMG, however, these substances are located primarily in axonal pathways arising from neurons in the intestine[14]. Thus, transection of the inferior splanchnic nerves central to the ganglion (fig. 1) does not alter these peptidergic axons whereas transection of the lumbar colonic nerves and hypogastric nerves peripheral to the ganglion markedly reduces their density.

The distribution of visceral afferent axons in peripheral ganglia and at various sites within the pelvic organs suggests that afferent pathways acting through the release of neurotransmitters in the periphery may have a variety of modulatory functions on nerve, smooth muscle and secretory cells in addition to the traditional sensory functions of afferent pathways. This peripheral modulatory function will be discussed in a later section.

Identification of neuropeptides in visceral dorsal root ganglion cells

Immunocytochemical techniques have been used in combination with axonal tracing methods to examine the distribution of neuropeptides in afferent neurons innervating the pelvic viscera of cat[16,21,23,25,56,58,60,63,71], rat[35,48,113,114,119] and guinea pig[13]. In the cat, fluorescent dyes were applied to the pelvic[16,63] and hypogastric nerves[23,25,58] to label various populations of visceral afferent cells. The dyes were also used in the cat and rat to label cells innervating individual organs such as the urinary bladder[25,35,60,114], colon[25,60], kidney[71,114], ureter[114] and female reproductive tract[48,119] and to label visceral and somatic afferent pathways in the pudendal nerve[63]. Colchicine was administered in many studies to increase the levels of peptides in the ganglion cells.

As shown in table 1, a large percentage of afferent neurons projecting to the pelvic and hypogastric nerves of the cat exhibited peptide immunoreactivity. Indeed, the sum of the percentages of neurons containing individual peptides exceeded 100%. The significance of this finding will be discussed below. VIP was the most common peptide occurring in 42–45% of lumbosacral visceral afferent neurons (fig. 2C). Leucine enkephalin, cholecystokinin (CCK) and substance P were present in 21–37% of the neurons, and methionine enkephalin was present in 10% of the neurons. Somatostatin was detected in very few cells (0–2%) whereas dynorphin 1–8, dynorphin 1–17 and dynorphin B were undetectable in dorsal root ganglion cells with immunocytochemical techniques[7,23,25]. Dynorphin B was identified, however, in lumbosacral dorsal root ganglia using radioimmunoassay[7].

In the same cats in which the pelvic nerve afferents were labeled another fluorescent dye was applied to the pudendal nerve to label visceral and somatic afferent neurons innervating the perineum, reproductive organs and anourethral structures[63]. VIP, CCK and methionine enkephalin were detected less frequently in these cells as compared to

Table 1. Distribution of neuropeptides in lumbosacral dorsal root ganglion cells of the cat and rat

	Cat					Rat		
	PN	HGN	PUDN	BLD	Colon	BLD	Kid-Ur.	Rep. org.
VIP	42	45	10	25	14	+	−	−
LENK	30	21	24	5	7	−	−	−
CCK	29	25	12	1	3	−	−	−
SP	24	37	21	23	18	16[b]	− +	−
MENK	10	9	3	−	−	−	−	−
SS	2	0	0	2	2	0	−	−
DYN	0	0	0	0	0	−	−	−
CGRP	−	−	−	−	−	90[a]	90[a]	66–86[a]
						60[b]	90[b]	45–63[b]
Total	137	137	70	56	44			

Numbers indicate percentages of neurons exhibiting each peptide. PN, pelvic nerve; HGN, hypogastric nerve; PUDN, pudendal nerve; BLD, bladder; Kid-UR, kidney-ureter; Rep. org., reproductive organs; VIP, vasoactive intestinal polypeptide; LENK, leucine enkephalin; CCK, cholecystokinin; SP, substance P; MENK, methionine enkephalin; SS, somatostatin; DYN, dynorphin 1–17; CGRP, calcitonin gene-related peptide. Data for HGN were obtained from upper lumbar ganglia. All other data were from sacral ganglia in the cat. (+) cells present but percentages not reported; (−), data not available. [a]T_{10}-L_3 ganglia, [b]L_6-S_1 ganglia. See text for literature references.

pelvic nerve afferent neurons, however, other peptides occurred in similar proportions in both afferent populations. The greatest difference was seen with VIP (10% vs 42%) suggesting that this peptide may be highly localized to visceral afferents. This difference between somatic and visceral cells might be even larger since those VIP cells which project to the pudendal nerve may also innervate visceral (e.g. urethral and anal mucosa) rather than somatic structures (e.g. cutaneous tissue).

The degree of association between VIP and visceral afferent pathways can also be examined by analyzing the data in the reverse manner to determine what proportion of VIP cells project to the viscera. It has been estimated that 66% of the VIP cells in the cat sacral dorsal root ganglia project to the pelvic nerve and 11% to the pudendal nerve; leaving 23% of the cells unidentified. These cells presumably innervate somatic structures. It should be noted, however, that pelvic nerve afferents represent only a small fraction (less than 10%) of the neurons in the sacral dorsal root ganglia. This further underscores the very strong association at this level of the spinal cord between VIP and visceral afferent pathways. As noted for peptidergic afferents in other animals[40], peptide-containing cells in cat lumbosacral dorsal root ganglia were in general small to medium size cells less than 40 μm in average diameter. VIP cells had the smallest average diameter (30 μm) whereas somatostatin cells had the largest average diameter (36 μm). The size of VIP cells is consistent with the observation that VIP is present only in unmyelinated afferents[43,91] and is associated primarily with visceral afferent neurons which are on the average, considerably smaller than somatic neurons[16,63].

The distribution of peptides in afferent pathways to individual organs was studied in the cat and rat. In the cat, VIP and substance P were the most common peptides in bladder and colon afferents (table 1, range 18–25%)[23,25,60]. Substance P was also detected in 24% of renal afferents in the cat[71] and in 16% of bladder afferents in the rat[113]. Somatostatin was rarely detected in visceral afferents[25,113]. CGRP was present in a large percentage (45–90%) of the dorsal root ganglion cells innervating the kidney[114], urinary bladder[114], ureter[114] and female reproductive organs[48] in the rat. It is noteworthy that in the more rostral dorsal root ganglia (T_{10}-L_3) there was a higher percentage of these cells with CGRP than in the caudal (L_6-S_1) ganglia. This suggests that different functional groups of visceral afferents in the rat may utilize different transmitters. On the other hand, this difference was not seen in the cat where hypogastric afferent neurons in the upper lumbar ganglia exhibited a spectrum of peptide-immunoreactivity similar to that of pelvic nerve afferent neurons in the sacral dorsal root ganglia (table 1).

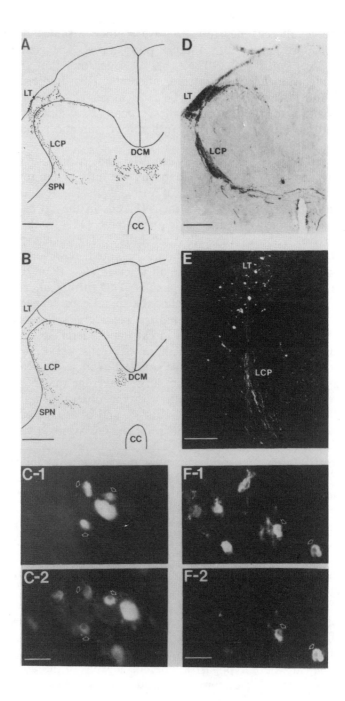

Co-localization of neuropeptides in lumbosacral visceral afferent pathways

The large percentage (greater than 100%) of cat lumbosacral visceral afferent neurons containing neuropeptides suggests that some peptides must coexist in afferent neurons[63]. This was confirmed using double staining techniques where two peptides could be identified on the same tissue section (fig. 2F). Coexistence was examined in the general population of sacral afferent neurons[22] and in bladder, colon and hypogastric afferent neurons in the cat[16,25,60]. As shown in table 2 for colon afferents, coexistence of certain peptides is very common. For example, a large percentage (50–56%) of cells containing leucine enkephalin exhibit substance P or VIP; and a large percentage (51%) of VIP cells contain substance P (fig. 2F) or leucine enkephalin. However, leucine enkephalin is not co-localized with somatostatin. Similar patterns of co-localization were noted in bladder and hypogastric nerve afferents and in unlabeled afferents in the lumbosacral dorsal root ganglia.

The sum of the mean percentages of leucine enkephalin cells containing other peptides exceeds 100% indicating that more than two peptides must be contained in one cell. Preliminary studies of thin serial sections (6 μm) allowing the same DRG cell to be visualized in three consecutive sections confirmed that three peptides e.g., VIP, substance P and CCK could occur in the same cell.

Co-localization of neuropeptides in unidentified dorsal root or sensory ganglion cells of the cat[75], rat[12,15,33,40,76,104,121,124] and guinea pig[30,31,69] has also been reported. Three and sometimes four peptides have been identified in single cells in the cat[75]. Among the various peptide combinations the most extreme example of co-localization is that occurring with CGRP and substance P where virtually all substance

---

Figure 2. Afferent pathways to the sacral spinal cord. *A* Camera lucida drawing of the central projections of bladder afferents in the sacral (S_2) dorsal horn (DH) of the cat. Afferent terminals were labeled by transganglionic transport of HRP from nerves on the surface of the bladder. Labeled axons were present in Lissauer's tract (LT), the lateral collateral pathway (LCP), the sacral parasympathetic nucleus (SPN) and the dorsal commissure (DCM). *B* Afferent projections in the S_2 dorsal horn of the cat labeled by HRP injected into the external urethral sphincter. *C* Two photomicrographs of the same section through the S_2 dorsal root ganglion showing bladder afferent cells (C-1) labeled by fast blue injected into the bladder wall. Several of these labeled cells (arrows in C-1 and C-2) contain VIP-immunoreactivity (VIP-IR). Dye labeled cells were blue when visualized with UV light at 340–380 nm excitation wavelength and VIP cells were green when visualized with UV light at 430–480 nm wavelength. *D* The distribution of VIP-IR in LT and LCP of the S_2 segment of the cat spinal cord. *E* VIP-IR in the sacral dorsal horn of the S_3 segment of the human spinal cord. Large bundles of VIP axons are present in LT and smaller numbers of axons are present in lamina I on the lateral edge of the dorsal horn. *F* Co-localization of substance P-IR and VIP-IR in sacral (S_2) dorsal root ganglion cells of the cat. Substance P-IR (F-1) stained with TRITC (red color, at 530–560 nm excitation wavelength) and in F-2 the same section showing VIP-IR in two of the same ganglion cells (arrows) stained with FITC (green color at 430–480 nm excitation wavelength). VIP was detected with rabbit polyclonal antisera, whereas substance P was detected by rat monoclonal antisera. Calibration represents 250 μm in **A** and **B**, 50 μm in **C**, 300 μm in **D**, 220 μm in **E** and 60 μm in **F**. (From de Groat et al., 1986)

Table 2. Co-localization of peptides in sacral dorsal root ganglion cells innervating the colon of the cat

	LENK	SP	VIP	CCK	SS
LENK	–	56	50	6	0
SP	27	–	42	7	–
VIP	51	52	–	–	–
CCK	11	22	–	–	–
SS	0	–	–	–	–

Numbers indicate percentages of neurons in the left column exhibiting immunoreactivity for other peptides. Each percentage represents the average for several hundred cells in sacral dorsal root ganglia of two cats. (–) indicates data not available. Abbreviations are the same as in table 1. (From de Groat et al., 1987)

P cells contain CGRP[31,33,76]. However, a certain population of CGRP cells does not contain substance P.

The significance of peptide co-localization in visceral afferent neurons is still uncertain, however, the demonstration of neuronal inhibitory (enkephalins)[18,21,23,54,55] and excitatory substances (substance P, VIP and CCK)[53,61,62,67,88,89,98,101,104,120] in the same cells raises a number of interesting possibilities. For example, enkephalins as well as excitatory peptides may be transported to central and peripheral afferent terminals and co-released (fig. 3)[24,25]. Since primary afferent terminals have opioid receptors that mediate presynaptic inhibition[54,55,72,125] enkephalins could be co-released with VIP or substance P, and then act in a negative

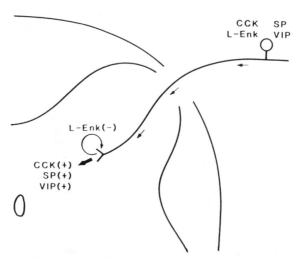

Figure 3. Diagram illustrating possible function of leucine enkephalin (L-Enk) in primary afferent neurons. L-Enk may be released at afferent terminals in the spinal cord and interact with opioid receptors on the terminals to mediate feedback inhibition of the release of excitatory transmitters, such as substance P (SP), vasoactive intestinal polypeptide (VIP) and cholecystokinin (CCK). L-Enk-IR has been identified in dorsal root ganglion cells containing either SP, VIP or CCK. However, it is not known whether all 4 peptides are localized in the same population of primary afferent neurons. (From de Groat et al., 1986)

feedback or autoinhibitory manner to depress the release of other peptides (fig. 3)[24,25]. Enkephalins might also act postsynaptically to inhibit second-order neurons. A similar action could occur at peripheral afferent terminals in the viscera (fig. 8).

It is also noteworthy that enkephalins are co-localized with various substances having synaptic excitatory actions (e.g. substance P, VIP and CCK)[53,61,62,67,88,89,101,104,120] but not with somatostatin which has synaptic inhibitory effects[104]. However, somatostatin was co-localized with substance P in nonvisceral afferent neurons[22]. Thus both inhibitory peptides coexist with excitatory peptides but not with each other. Dynorphins and endorphins are also present in sensory ganglia[6,30,49,50,66,69,97,115,123] and in afferent projections to the spinal cord[7,23,25]. Although dynorphin immunoreactivity has not yet been detected in visceral afferent neurons, the central projections of dynorphin immunoreactive dorsal root axons to areas receiving visceral afferent input suggests that dynorphins are present in visceral afferent pathways[7,16,23,25]. Further studies will be necessary to determine whether enkephalin- and dynorphin-containing afferent systems represent the same population of dorsal root ganglion cells.

The relationship between central projections of visceral afferent pathways and the distribution of neuropeptides in the lumbosacral spinal cord

Anatomy of visceral afferent pathways

Horseradish peroxidase (HRP) tracing experiments in the cat[19,70,90,92,93], monkey[102] and rat[94] revealed that visceral afferent pathways have a characteristic pattern of distribution in the spinal cord and that this pattern is markedly different from that of somatic afferent neurons which innervate the skin[16]. In the sacral spinal cord of the cat afferent axons from the pelvic organs pass through the dorsal root entry zone within the areas occupied by the fine diameter fibers and then enter Lissauer's tract (fig. 4)[90]. Visceral afferents can extend several segments rostrally and caudally along Lissauer's tract giving off collaterals which pass in a thin shell laterally and medially around the doral horn (figs 4 and 5). At the apex and on the lateral edge of the dorsal horn this shell of visceral afferents (approximately 70 μm in width) is present in lamina I and a few fibers extend deeper into outer lamina II. Inner lamina II and laminae III-IV do not receive visceral afferent input. The bands of visceral afferents on the lateral and medial edge of the dorsal horn are termed the lateral and medial collateral pathways (LCP and MCP) of Lissauer's tract, respectively.

Axons in the LCP, which extend ventrally from Lissauer's tract through lamina I into lateral laminae V-VII and lamina X, exhibit

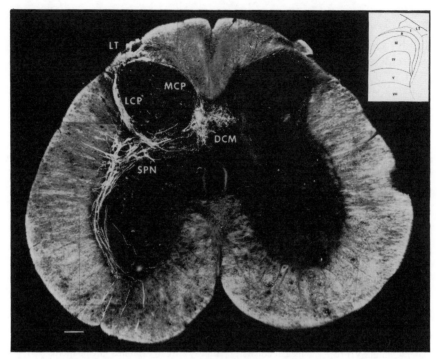

Figure 4. Transverse section of S_2 spinal cord showing labeling of primary afferents and preganglionic neurons after application of HRP to the left pelvic nerve in the cat. Pelvic afferents enter Lissauer's tract (LT). Afferent collaterals enter lamina I and extend laterally in a large bundle, the lateral collateral pathway (LCP), into the area of the sacral parasympathetic nucleus (SPN). Collaterals also extend medially in a smaller group, the medial collateral pathway (MCP), into the dorsal gray commissure (DCM), where they expand into a large terminal field ipsilaterally and contralaterally. Small numbers of afferents are also present in contralateral laminae I and V. This photomicrograph was made using darkfield illumination with polarized light. Bar represents 200 μm. Inset shows the laminar organization of the sacral dorsal horn according to Rexed. (From Morgan et al., 1981)

several patterns of termination (fig. 5). In some sections, axons end in lateral lamina V (designated 'a' in fig. 5). In other sections, axons extend into medial lamina V and VII and the lower one-third of the doral gray commissure ('b' in fig. 5). Less frequently, labeled axons extend into dorsomedial lamina V ('c') or ventrally into lateral lamina VII ('d'), in the region of the sacral parasympathetic nucleus[95].

In serial transverse sections the intensity of labeling in the LCP exhibits considerable variation from section to section. Horizontal sections show that this variability is related to a periodic grouping of collateral bundles along the length of the spinal cord. The average distance between bundles (center to center) is approximately 215 μm, while the average bundle width is approximately 100 μm.

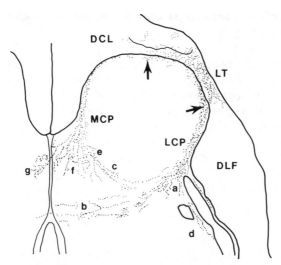

Figure 5. Different patterns of pelvic nerve afferent collaterals from Lissauer's tract in the S_2 segment of the cat. Ventral roots were cut to eliminate efferent labeling. Camera lucida drawings are composed from seven individual sections. These patterns were observed consistently in all experiments in various combinations. The lateral collateral pathway (LCP) exhibited four patterns (a, b, c, d). Many axons ended at the junction of laminae I and V (a), while others continued in lamina V to end in the lower third of the dorsal gray commissure (b). Less frequently, LCP axons ended dorsomedially in medial lamina V (c) or extended into lateral lamina VII (d). The medial collateral pathway (MCP) exhibited three patterns (e, f, g). The most common were the ipsilateral (f) and the contralateral (g) terminal fields in the upper two-thirds of the dorsal gray commissure. Less frequently, axons extended laterally into medial lamina V (e). Arrows show boundaries of Lissauer's tract. DLF, dorsolateral funiculus; DCL, dorsal column; CC; central canal. (From Morgan et al., 1981)

Axons extending from lamina I medially through lamina V and VII to the dorsal gray commissure also occur in bundles.

The MCP consists of a thin band of axons which passes dorsoventrally along the medial edge of the dorsal horn into the region of the dorsal gray commissure, where the pathway expands into diffuse terminal fields on both sides of the midline (fig. 4 and areas 'f' and 'g' in fig. 5). Some axons in the MCP extend into lamina V, to overlap with the dorsomedial projections from the LCP ('e').

A similar pattern of central afferent projections has been noted for: 1) afferent pathways to individual pelvic organs in the cat (urinary bladder, colon, uterus) (fig. 2A)[16,23]. 2) pudendal afferent input from deep perineal structures such as the urethra and external urethral sphincter (fig. 2B)[16,102,117], 3) the hypogastric and lumbar colonic nerve pathways in the cat[92,93] and 4) pelvic nerve afferents in the monkey and rat[94,96]. One spinal projection noted only in the rat consisted of a longitudinal bundle of pelvic nerve axons ventral to the central canal[94].

Distribution of neuropeptides at sites of visceral afferent termination in the spinal cord

Sites in the lumbosacral spinal cord which receive visceral afferent input also receive dense peptidergic projections. As might be expected those peptides present in dorsal root ganglion cells (e.g., VIP, substance P, CGRP, CCK and opioid peptides) are also very prominent in the spinal dorsal horn. However, there are significant differences in the spinal distribution of these peptides.

For example, VIP exhibits an unusual segmental distribution which suggests an association with visceral afferents[4–6,16,21,23,34,43,56,57,59,109,111,117]. In various species VIP-IR is considerably higher in the sacral segments than in other segments of the spinal cord. This is most striking in cat[6,16,21,34,35,43,56,59] and man[4,5,23,57] but was also noted in monkeys and guinea pigs[16,23,34]. PHI which is co-localized with VIP in neurons has a similar distribution in the human, rat and cat spinal cord[5,10,35].

In the lumbosacral spinal cord the VIP-containing axons also exhibit a striking similarity to certain components of the visceral afferent pathways (fig. 2D,E)[90,94,96]. In the cat, which has been studied in the most detail VIP-immunoreactivity in the lumbosacral spinal cord is distributed most prominently in Lissauer's tract and in the area of the LCP (fig. 2D)[6,21,35,43,56,59]. VIP axons extend ventromedially from the LCP through laminae V-VII into the ventral region of the dorsal gray commissure (lamina X) (fig. 6A), similar to the visceral afferent projections to area 'b' illustrated in figure 5. Small numbers of VIP axons also extend into the sacral parasympathetic nucleus and in some cats to pudendal motor nucleus in the ventral horn. Very few VIP fibers are detected in laminae II-IV of the dorsal horn. A similar distribution was noted in man[4,23,57] (fig. 2E), monkey[23,34] and rodents[34,35,109]. Electron microscopy has revealed that VIP-immunoreactivity (VIP-IR) in cat primary afferent pathways is localized in unmyelinated axons which form large (1–5 μm) varicosities in LCP[43,91].

In the LCP, VIP-IR exhibits a distinctive periodic pattern in the rostrocaudal axis, consisting of bundles of axons spaced at approximately 210 μm along the length of the cord (fig. 7)[6,43,56,59]. This pattern is similar to the periodic distribution of visceral afferent axons in the region[90]. VIP-IR in Lissauer's tract and the LCP of the cat is almost completely eliminated by lumbosacral dorsal rhizotomy (fig. 6) although VIP-IR at other sites (lamina X) remains. The latter may be associated with VIP neurons intrinsic to the spinal cord[34,109]. VIP-IR in the spinal cord of the cat is also significantly decreased by transection of the pelvic nerve[35], whereas electrical stimulation of the pelvic nerve releases VIP-IR from the spinal cord[8]. These observations provide considerable indirect support for a transmitter function of VIP in sacral visceral afferents.

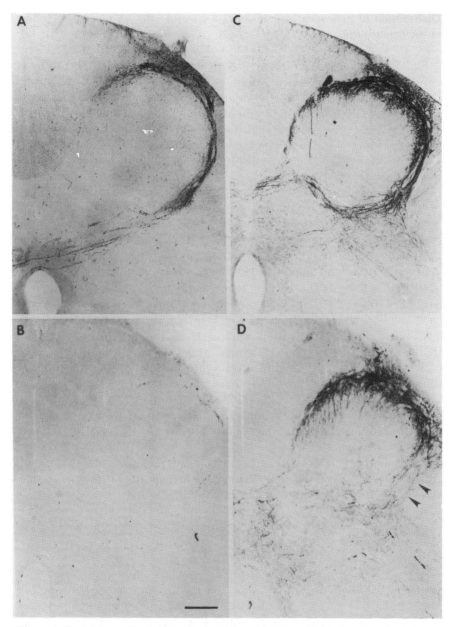

Figure 6. Transverse sections of the S_2 dorsal horn showing VIP-IR (A, B) and substance P-IR (C, D) in normal (A, C) and 5 weeks after unilateral sacral dorsal and ventral root transection (B, D). Note: VIP-IR in the lateral collateral pathway passed ventrally to the region of the central canal (A), whereas substance P-IR passed dorsally into the dorsal gray commissure (C). Deafferentation caused a marked reduction in VIP-IR in Lissauer's tract and lamina I (B) and only a slight reduction in substance P-IR on the medial side of the dorsal horn and in lateral lamina I (D, arrows). Calibration bar = 200 μm. (From Katawani et al., 1985)

Substance P-IR also coincides with sites of visceral afferent termination in the lumbosacral spinal cord[43,59], however, in contrast to VIP-IR, substance P-IR is distributed more uniformly at all segmental levels[16,32,40,43,59,104,112]. Substance P-IR and VIP-IR also exhibit certain differences in the sacral spinal cord of the cat (fig. 6). For example, substance P is distributed more evenly on the medial and lateral sides of the dorsal horn and in deeper lamina (II and III). In addition substance P axons from the LCP pass into the dorsal part rather than the ventral part of the dorsal gray commissure (fig. 6). Substance P-IR is reduced to lesser extent than VIP-IR by sacral dorsal rhizotomy and only certain areas such as the ventral LCP are clearly affected (fig 6)[59]. Substance P axons in this area exhibit a periodic distribution similar to VIP axons[16,21,59]. Substance P-IR and VIP-IR are both present in the lumbosacral autonomic nuclei. Substance P-IR has also been identified in the human sacral spinal cord in the lateral marginal zone of the dorsal horn and in the area of the sacral autonomic nucleus[23,26].

Thus substance P inputs to the spinal dorsal horn and intermediate gray matter are clearly, more complicated than VIP inputs and are likely to be derived from multiple sources including somatic and visceral afferent pathways as well as neurons within the central nervous system[26,40,104]. However, certain components of the substance P afferent pathways correspond to visceral afferent terminations in the cord.

CGRP[33] and CCK[16,21,40,43] are distributed across all spinal segmental levels in the cat and like substance P are most prominent in the superficial laminae of the dorsal horn. Projections into laminae V and X also occur. The concentrations in the spinal cord are reduced by dorsal rhizotomy[21,33]. Dynorphin B[7] and dynorphin 1–17[23] also exhibit a prominent localization in Lissauer's tract and in the marginal zone (LCP) on the lateral edge of the dorsal horn in the sacral spinal of the cat. The dynorphin axons exhibit a periodic distribution in the LCP similar to VIP and are markedly reduced in density following sacral dorsal rhizotomy[7,23]. On the other hand, leucine and methionine enkephalin-immunoreactivity in the LCP is not changed following rhizotomy.

Somatostatin distribution in the sacral dorsal horn of the cat is markedly different from other peptides[43]. Somatostatin containing varicosities are present in laminae II and III and in the region of the sacral parasympathetic nucleus, but are absent in the LCP. This distribution is consistent with the results of the studies on dorsal root ganglia indicating that visceral afferents rarely contain somatostatin[16,23,63,113].

Peptide distributions in the rat lumbosacral spinal cord[10a,12,15,32–35,40,104,109,110,112,126] in general follow the pattern seen in the cat but there are some differences. For example, the high level of VIP in the sacral segments as compared to other cord segments is not as striking as in the cat[34], suggesting that VIP may have widespread

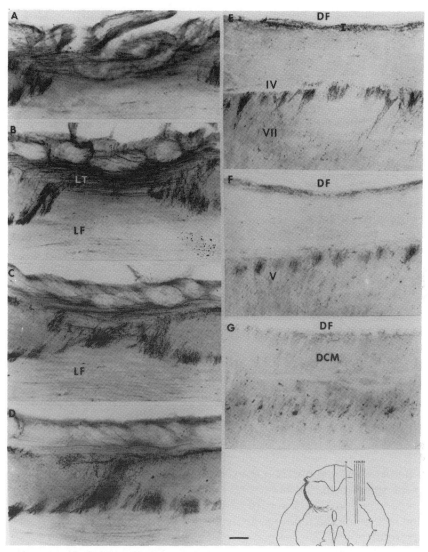

Figure 7. Sagittal sections showing the VIP distribution in the S_1 segment of the spinal cord. Dorsal edge is at the top of each photomicrograph and rostral is to the left. *A* VIP-IR in dorsal rootlets, Lissauer's tract and lamina I. *B* Rostrocaudal bundles of VIP axons in Lissauer's tract (LT) and lamina I. *C* Dorsoventral VIP fibers start to form discrete bundles. Some VIP axons pass dorsoventrally in lamina I (I) from rostrocaudal bundles in Lissauer's tract. *D* Discrete bundles of VIP axons in the ventral part of lateral lamina I. VIP-IR is not present in inner lamina II and laminae III–IV (IV). *E, F* Periodic appearance of VIP-IR at the junction between lateral laminae I and V. Ventral projections of VIP axons end abruptly in ventral lamina V (V). A few VIP fibers extend into lamina VII (bottom) and are present in the medial collateral pathway in lamina I (top of photomicrograph). *G* VIP-IR in lateral collateral pathway breaks up into individual fibers or small bundles with a lateromedial orientation (bottom of the photomicrograph); VIP-IR on the medial side of the dorsal horn medial collateral pathway (top of photomicrograph). DCM, dorsal commissure; DF, dorsal funiculus; LF, lateral funiculus. Calibration bar = 200 μm. (From Kawatani et al., 1985)

functions in the rat. Another difference which is particularly relevant to visceral afferent pathways is the distribution of peptides in the area around the central canal. Longitudinal bundles of peptidergic axons are present dorsal and ventral to the central canal. Afferent pathways to the pelvic viscera project into the ventral bundle[94]. The density of CGRP[33,35], substance P[15,35,51,109], VIP/PHI[35,51], and galanin[10a] immunoreactivity in this region as well as in the superficial lamina of the dorsal horn[51] is markedly reduced by transection of the dorsal roots or the administration of capsaicin neonatally. This indicates that peptidergic afferents make an important contribution to the ventral bundle. CCK immunoreactivity[51,55a,109] has also been identified in these areas; however, there are concerns about the interpretation of CCK immunocytochemistry. For example, capsaicin treatment reduces the numbers of CCK positive axons identified in the spinal dorsal horn by immunocytochemistry but does not change the CCK levels determined by radioimmunoassay[110]. Recently it has been reported that many antibodies used for the identification of CCK in the rat cross-react with CGRP[55a]. This raises the possibility that CCK-like immunoreactivity in primary afferent neurons in the rat may represent CGRP or a similar peptide.

Peptidergic (CGRP, substance P, VIP, CCK, galanin) axons are also very prominent in the lumbosacral autonomic nuclei in the rat[10a,33,34,109,110]. These axons parallel the distribution of visceral afferent projections[94] and are reduced in number by dorsal rhizotomy or capsaicin treatment[33–35,40,51,104].

Physiological role of peptidergic visceral afferent neurons

It is now recognized that afferent neurons can release neuropeptides from peripheral terminals in the viscera as well as from central terminals in the spinal cord and brain[8,45–47,54,78,79,81,126]. Thus, visceral dorsal root ganglion cells are likely to have multiple functions including: 1) transmission of sensory and reflexogenic signals to the central nervous system and 2) modulation of various peripheral mechanisms such as effector organ activity, local blood flow, autonomic ganglion transmission, and afferent receptor sensitivity (fig. 8)[40,47,78,79]. It seems appropriate, therefore, to consider the visceral dorsal root ganglion cell and its processes as an anatomical substrate for mediating peripheral efferent as well as afferent responses (fig. 8). Several examples of efferent-afferent functions of peptidergic neurons have been demonstrated in the pelvic viscera using the neurotoxin, capsaicin[9,42,45–47,80–85,78,105–107,120]. These experiments have been conducted primarily on the neural pathways to the urinary tract of the rat and guinea pig; which are very sensitive to capsaicin.

Several laboratories have examined the effect of capsaicin on the micturition reflex[42,79,83–85,113]. Micturition is triggered by pelvic nerve

AFFERENT–EFFERENT SYSTEM

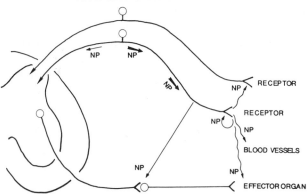

Figure 8. Diagram showing the possible functions of neuropeptides (NP) in visceral afferent neurons. Neuropeptides transported to central afferent terminals may function as transmitters or modulators at synapses in the spinal cord. Neuropeptides transported peripherally may regulate: 1) transmission in autonomic ganglia, 2) effector organ activity, 3) afferent receptor sensitivity, and 4) local blood flow. Thus visceral dorsal root ganglion cells and their processes appear to represent an anatomical substrate for the mediation of efferent as well as afferent responses.

afferent pathways which convey activity from tension receptors in the bladder wall[17,18,20]. These afferents trigger reflex firing in the sacral efferent pathways to produce detrusor contractions and bladder emptying. Administration of capsaicin markedly alters the micturition reflex. The acute effects of capsaicin administered systematically (50 mg/kg, s.c.) or topically to the bladder are to enhance bladder activity and to facilitate micturition[79–81,83–85]. For example, topical application of capsaicin to the bladder elicits an initial tonic contraction followed by large rhythmic contractions. Various pharmacological and neurophysiological data indicate that the tonic contraction is due to release of substance P or a related tachykinin from afferent fibers in the bladder. These substances exert direct excitatory actions on the bladder smooth muscle[83–85,105]. The rhythmic contractions are attributed to tachykinin-induced facilitation of the micturition reflex pathway, possibly due to stimulation or sensitization of tension receptors in the bladder wall. These effects of capsaicin are mimicked by the topical administration of exogenous substance P to the bladder[79,85]. The effects are eliminated either by chronic deafferentation, by pretreatment with capsaicin or by the administration of a substance P antagonist[79,83,105].

The systemic administration of capsaicin to urethane anesthetized adult rats also produces a delayed-onset (>30 min) depression of the micturition reflex[42]. The duration of this depression has not been reported, however, it is likely to be prolonged since pretreatment of adult rats with capsaicin (50 mg/kg, s.c.) four days prior to study significantly alters the bladder volume and intravesical pressure necessary to induce micturition in urethane anesthetized animals[83]. The intrathecal or intracerebroventricular administration of capsaicin to adult rats elicits a similar prolonged (4–60 days) depression of the micturition reflex, suggesting that the depletion of peptidergic afferent projections in the central nervous system is in part responsible for the effects of capsaicin. Furthermore, since capsaicin pretreatment did not alter the magnitude of the bladder contractions during micturition or the responses of the bladder to stimulation of somatic afferent pathways, it was concluded that capsaicin does not act on the peripheral efferent pathway or the central reflex mechanism involved in the micturition reflex but that it interferes with the afferent pathways from the bladder which relay pressure-volume information to the central nervous system[83,106].

The administration of capsaicin to neonatal rats produces a greater deficit in micturition[42,79,105,106,113,114], consistent with the more prominent effect of the neurotoxin on peptidergic afferent fibers in neonatal animals. Animals 2–6 months of age treated with capsaicin as neonates exhibited a 50% loss of bladder afferent innervation[113] an almost complete loss of substance P afferent innervation[113] and a marked reduction of substance P and CGRP levels in the bladder[113,114]. These animals had hypertrophied bladders (4-fold increase in weight of the empty bladders), which contained large volumes of urine (greater than 5 ml volume versus less than 1 ml for untreated animals). There was also an increased incidence of bladder calculi and enlargement of the prostate gland. In addition, the bladders from capsaicin-treated animals did not exhibit excitatory reflexes in response to distension. However, strips of bladder smooth muscle from these animals did respond normally to exogenous acetylcholine, substance P and electrical field stimulation, indicating that the efferent excitatory pathway to the bladder was intact[105].

The neonatally capsaicin-treated animals also formed less urine in response to a water load, suggesting that they had impaired renal function[42]. This was not observed in animals treated with capsaicin after the early postnatal period.

Capsaicin administered to neonatal and adult rats and guinea pigs also affects the afferent innervation of the ureter[11,29,44–47,78,82,107,114]. This was demonstrated immunocytochemically as a depletion of CGRP[29,114], substance P and related tachykinins[45,78,107] in the ureter and at the ultrastructural level as a degeneration of a large percentage of the unmyelinated axons in the ureter[11,44].

The functional consequences of capsaicin-induced destruction of ureter-afferents is uncertain; however, pharmacological studies indicate that peptides released from afferents might have multiple functions in the ureter[46,47,78,82,107]. For example, electrical or chemical stimulation of the innervation of the ureter elicits: 1) inhibition of ureteral motility, 2)

facilitation of ureteral motility and 3) an inflammatory response characterized by plasma protein extravasation. These effects are mimicked by
acute administration of capsaicin and can be blocked by pretreatment
with large doses of capsaicin. It was concluded that the responses are
mediated by substances released from afferent nerves.

Various evidence indicates that tachykinins and CGRP are mediators
of the afferent evoked responses in the ureter since capsaicin is known to
release these substances in various tissues[78]. For example, exogenous
substance P and neurokinin A mimic the facilitatory effect of capsaicin
on ureteral motility[47,78,82], whereas CGRP mimics the inhibitory effect[47].
In addition substance P antagonists block the facilitatory response
induced by capsaicin[46].

Capsaicin pretreatment also influences spontaneous motility of the rat
ureter[82]. Isolated segments of rat ureters maintained in vitro are normally
quiescent. However, 50% of ureteral segments obtained from rats
pretreated with capsaicin exhibit spontaneous rhythmic activity. This
suggests that capsaicin sensitive nerves may have an inhibitory role in
controlling ureteral motility.

Peptidergic afferent pathways may also control pelvic visceral function
by influencing transmission in autonomic ganglia. As noted in a preceding section, peptidergic afferent varicosities are prominent in autonomic
ganglia[13,14,77,86] and afferent fibers are located in close proximity to the
ganglion cells[1,14,84]. Furthermore, many peptides which are contained in
visceral afferent neurons (e.g. substance P, CGRP, VIP and CCK) have
facilitatory and/or excitatory effects when administered exogenously to
ganglion cells[61,62,67,88,89,108,120]. In the lumbosacral visceral pathways the
most detailed analysis of visceral afferent inputs to autonomic ganglion
cells has been obtained in the inferior mesenteric ganglion of the guinea
pig[13,14,67,120]. It has been shown that electrical stimulation of afferent
axons from the lumbar dorsal root ganglia elicits a noncholinergic slow
EPSP in ganglion cells[67,120]. This response could be mimicked by the
application of substance P to the ganglion cells and could be blocked by
the administration of a substance P antagonist or by prolonged treatment
with capsaicin, which depletes substance P levels in ganglia.

Other peptides (VIP, CCK, neurokinin A)[62,88,89,108] also produce slow
noncholinergic depolarizations of autonomic ganglion cells. These substances have been implicated as transmitters in peripheral afferent
pathways passing from the intestine to the inferior mesenteric ganglion[14].
However, they have not yet been shown to mediate excitatory synaptic
potentials elicited by afferent input from the dorsal root ganglion cells.

Summary

In summary, pharmacological and immunocytochemical studies have
raised the possibility that synaptic transmission at visceral afferent

354

terminals in the spinal cord and in the periphery may involve various peptide transmitters and that both excitatory and inhibitory transmitters may be released from the same afferent neuron. It is also clear that while many peptides are present in afferents there are considerable differences in the numbers of neurons containing each peptide. For example, VIP and CGRP are present in a large percentage of visceral afferents whereas other peptides such as somatostatin are not present or are in very few cells. In addition, some peptides (e.g, VIP) are prominently associated with visceral neurons whereas others (substance P) are equally distributed in visceral and somatic neurons. These findings raise the possibility that there may be a preferential distribution of some peptides between visceral and somatic systems and that certain of these substances might be useful markers of specific populations of afferents.

On the other hand, it is clear that peptides are not likely to be organ specific since there is considerable overlap in the spectrum of peptidergic afferents innervating the urinary bladder and colon of the cat. This finding is of particular interest since the reflex mechanisms in these two organs are controlled by different types of afferent fibers: Aδ afferents which initiate micturition and C-fiber afferents which stimulate defecation[20,23]. Thus, different functional groups of afferents innervating the pelvic viscera may utilize the same peptide transmitters.

Acknowledgments. Supported by NSF Grant BNS 8567113, NIH Grant AM 37241 and Clinical Research Grant MH 30915.

1 Aldskogius, H., Elfvin, L.-G., and Forsman, C. A., Primary sensory afferents in the inferior mescenteric ganglion and related nerves of the guinea pig: An experimental study with anterogradely transported wheat germ agglutinin-horseradish peroxidase conjugate. J. auton. nerv. Syst. 15 (1986) 179–190.

2 Alm, P., Alumets, J., Hakanson, R., and Sundler, F., Peptidergic (vasoactive intestinal peptide) nerves in the genito-urinary tract. Neuroscience 2 (1977) 751–757.

3 Alm, P., Alumets, J., Brodin, E., Hakanson, R., Nilsson, G., and Sjoberg, N. O., Peptidergic (substance P) nerves in the genito-urinary tract. Neuroscience 3 (1978) 419–425.

4 Anand, P., Gibson, S. J., McGregor, G. P., Blank, M. A., Ghatei, M. A., Becarese-Hamilton, A. J., Polak, J. M., and Bloom, S. R., A VIP-containing system concentrated in the lumbosacral region of human spinal cord. Nature 305 (1983) 143–145.

5 Anand, P., Gibson, S. J., Yiangou, Y., Christofides, N. D., Polak. J. M., and Bloom, S. R., PHI-like immunoreactivity co-locates with the VIP-containing system in human lumbosacral spinal cord. Neurosci. Lett 46 (1984) 191–196.

6 Basbaum, A. L., and Glazer, E. J., Immunoreactive vasoactive intestinal polypeptide is concentrated in the sacral spinal cord: A possible marker for pelvic visceral afferent fibers. Somatosensory Res. 1 (1983) 69–82.

7 Basbaum, A. I., Crus, L., and Weber, E., Immunoreactive dynorphin B in sacral primary afferent fibers of the cat. J. Neurosci. 6 (1986) 127–133.

7a Bauer, F. E., Christofides, N. D., Hacker, G. W., Blank, M. A., Polak, J. M., and Bloom, S. R., Distribution of galanin immunoreactivity in the genitourinary tract of man and rat. Peptides 7 (1986) 5–10.

8 Blank, M. A., Anand, P., Lumb, B. M., Morrison, J. F. B., and Bloom, S. R., Release of vasoactive intestinal polypeptide-like immunoreactivity (VIP) from cat urinary bladder and sacral spinal cord during pelvic nerve stimulation. Dig. Dis. Sci. 29 (1984) 10.

9 Buck, S. H., and Burks, T. F., The neuropharmacology of capsaicin: Review of some recent observations. Pharmacol. Rev. *38* (1986) 179–226.

10 Christofides, N. D., Polak, J. M., and Bloom, S. R., Studies on the distribution of PHI in mammals. Peptides *5* (1984) 261–226.

10a Ch'ng, J. L. C., Christofides, N. D., Anand, P., Gibson, S. J., Allen, Y. S., Su, H. C., Tatemoto, K., Morrison, J. F. B., Polak, J. M., and Bloom, S. R., Distribution of galanin immunoreactivity in the central nervous system and the responses of galanin-containing neuronal pathways to injury. Neuroscience *16* (1985) 343–354.

11 Chung, K., Schwen, R. J., and Coggeshall, R. E., Ureteral axon damage following subcutaneous administration of capsaicin in adult rats. Neurosci. Lett *53* (1985) 221–226.

12 Dalsgaard, C. J., Vincent, S. R., Hökfelt, T., Lundberg, J. M., Dahlstrom, A., Schultzberg, M., Dockray, G. J., and Cuello, A. C., Co-existence of cholecystokinin- and substance P-like peptides in neurons of the dorsal root ganglia of the rat. Neurosci. Lett. *33* (1982) 159–163.

13 Dalsgaard, C. J., Hökfelt, T., Elfvin, L. G., Skirboll, L., and Emson, P., Substance P containing primary sensory neurons projecting to the inferior mesenteric ganglion: Evidence from combined retrograde tracing and immunohistochemistry. Neuroscience *7* (1982) 647–654.

14 Dalsgaard, C. J., Hökfelt, T., Schultzberg, M., Lundberg, J. M., Terenius, L., Dockray. G. J., and Goldstein, M., Origin of peptide-containing fibers in the inferior mesenteric ganglion of the guinea-pig: immunohistochemical studies with antisera to substance P, enkephalin, vasoactive intestinal polypeptide, cholecystokinin and bombesin. Neuroscience *9* (1983) 191–211.

15 Dalsgaard, C. J., Haegerstrand, A., Theodorsson-Norheim, E., Brodin, E., and Hökfelt, T., Neurokinin A-like immunoreactivity in rat primary sensory neurons; coexistence with substance P. Histochemistry *83* (1985) 37–39.

16 de Groat, W. C., Spinal cord projections and neuropeptides in visceral afferent neurons. Prog. Brain Res. *67* (1986) 165–187.

17 de Groat, W. C., and Booth, A. M., Autonomic systems to the urinary bladder and sexual organs, in: Perpheral Neuropathy, vol. 1, pp. 285–299. Eds P. J. Dyck, P. K. Thomas, E. H., Lambert and R. Bunge. W. B. Saunders Co., Philadelphia 1984.

18 de Groat, W. C., and Kawatani, M., Neural control of the urinary bladder: Possible relationship between peptidergic inhibitory mechanisms and detrusor instability. Neurol. Urodynam. *4* (1985) 285–300.

19 de Groat, W. C., Nadelhaft, I., Morgan, C., and Schauble, T., Horseradish peroxidase tracing of visceral efferent and primary afferent pathways in the sacral spinal cord of the cat using benzidine processing. Neurosci. Lett. *10* (1978) 103–108.

20 de Groat, W. C., Nadelhaft, I., Milne, R. J., Booth, A. M., Morgan, C., and Thor, K., Organization of the sacral parasympathetic reflex pathways to the urinary bladder and large intestine. J. auton. nerv. Syst. *3* (1981) 135–160.

21 de Groat, W. C., Kawatani, M., Hisamitsu, T., Lowe, I., Morgan, C., Roppolo, J., Booth, A. M., Nadelhaft, I., Kuo, D., and Thor, K., The role of neuropeptides in the sacral autonomic reflex pathways of the cat. J. auton. nerv. Syst. *7* (1983) 339–350.

22 de Groat, W. C., Kawatani, M., Houston, M. B., and Erdman, S. L., Co-localization of VIP, substance P, CCK, somatostatin and enkephalin immunoreactivity in lumbosacral dorsal root ganglion cells of the cat. Vth International Washington Spring Symposium, May 28–31, Washington, DC 1985, Abstr. p. 48.

23 de Groat, W. C., Kawatani, M., Hisamitsu, T., Booth, A. M., Roppolo, J. R., Thor, K., Tuttle, P., and Nagel, J., Neural control of micturition: The role of neuropeptides. J. auton. nerv. Syst., Suppl. (1986) 369–387.

24 de Groat, W. C., Lowe, I., Kawatani, M., Morgan, C. W., Kuo, D., Roppolo, J. R., and Nagel, J., Identification of enkephalin immunoreactivity in sensory ganglion cells. J. auton. nerv. Syst., Suppl. (1986) 361–368.

25 de Groat, W. C., Kawatani, M., Houston, M. B., Rutigliano, M., and Erdman, S., Identification of neuropeptides in afferent pathways to the pelvic viscera of the cat, in: Organization of the Autonomic Nervous System: Central and Peripheral Mechanisms, pp. 81–90. Eds J. Ciriello, F. Calarescu, L. Renaud and C. Polosa. A. R. Liss, Inc., New York, 1987.

26 de Lanerolle, N. C., and LaMotte, C. C., The human spinal cord: Substance P and methionine-enkephalin immunoreactivity. J. Neurosci. *2* (1982) 1369–1386.

356

27 Fitzgerald, M., Capsaicin and sensory neurons—a review. Pain *15* (1983) 109–130.

28 Furness, J. B., Papka, R. E., Della, N. G., Costa, M., and Eskay, R. L., Substance P-like immunoreactivity in nerves associated with the vascular system of guinea pigs. Neuroscience *7* (1982) 447–459.

29 Ghatei, M. A., Gu, J., Mulderry, P. K., Blank, M. A., Allen, J. M., Morrison, J. F. B., Polak, J. M., and Bloom, S. R., Calcitonin gene-related peptide (CGRP) in the female rat urogenital tract. Peptides *6* (1985) 809–815.

30 Gibbins, I. L., Furness, J. B., Costa, M., MacIntyre, I., Hillyard, C., and Girgis, S. A., Coexistence of calcitonin gene-related peptide, dynorphin and cholecystokinin in substance P-containing dorsal root ganglion neurons of the guinea pig. Neurosci. Lett. Suppl. 19–20 (1985) S.65.

31 Gibbins, I. L., Furness, J. B., Costa, M., MacIntyre, I., Hillyard, C. J., and Girgis, S., Co-localization of calcitonin gene-related peptide-like immunoreactivity with substance P in cutaneous, vascular and visceral sensory neurons of guinea pigs. Neurosci. Lett. *57* (1985) 125–130.

32 Gibson, S. J., Polak, J. M., Bloom, S. R., and Wall, P. D., The distribution of nine peptides in rat spinal cord with special emphasis on the substantia gelatinosa and area around the central (lamina X). J. comp. Neurol. *201* (1981) 65–79.

33 Gibson, S. J., Polak, J. M., Bloom, S. R., Sabate, I. M., Mulderry, P. K., Ghatei, M. A., McGregor, G. P., Morrison, J. F. B., Kelly, J. S., and Rosenfeld, M. G., Calcitonin gene-related peptide (CGRP) immunoreactivity in the spinal cord of man and eight other species. J. Neurosci. *4* (1984) 3101–3111.

34 Gibson, S. J., Polak, J. M., Anand, P., Blank, M. A., Morrison, J. F. B., Kelly, J. S., and Bloom, S. R., The distribution and origin of VIP in the spinal cord of six mammalian species. Peptides *5* (1984) 201–207.

35 Gibson, S. J., Polak, J. M., Anand, P., Blank, M. A., Yiangou, Y., Su, H. C., Terenghi, G., Katagiri, T., Morrison, J. F. B., Lumb, B. M., Inyama, C., and Bloom, S. R., A VIP/PHI-containing pathway links urinary bladder and sacral spinal cord. Peptides *7* (1986) 205–219.

36 Gu, J., Polak, M., Probert, L., Islam, K. N., Marangos, P. J., Mina, S., Adrian, T. E., McGregor, G. P., O'Shaughnessy, D. J., and Bloom, S. R., Peptidergic innervation of the human male genital tract. J. Urol. *130* (1983) 386–391.

37 Gu, J., Blank, M. A., Huang, W. H., Islam, K. N., McGregor, G. P., Christofides, N., Allen, J. M., Bloom, S. R., and Polak, J. M., Peptide-containing nerves in human urinary bladder. Urology *25* (1984) 353–357.

38 Gu, J., Polak, J. M., Blank, M. A., Terenghi, G., Morrison, J. F. B., and Bloom, S. R., The origin of VIP-containing nerves in the urinary bladder of rat. Peptides *5* (1984) 219–223.

39 Hökfelt, T., Schultzberg, M., Elde, R., Nilsson, G., Terenius, L., Said, S., and Goldstein, M., Peptide neurons in peripheral tissues including the urinary tract: immunohistochemical studies. Acta pharmac. toxic. *43* Suppl. II (1978) 79–89.

40 Hökfelt, T., Johansson, O., Ljungdahl, A., Lundberg, J. M., and Schultzberg, M., Peptidergic neurons. Nature *284* (1980) 515–521.

41 Holzer, P., Bucsics, A., and Lembeck, F., Distribution of capsaicin-sensitive nerve fibres containing immunoreactive substance P in cutaneous and visceral tissues of the rat. Neurosci. Lett. *31* (1982) 253–257.

42 Holzer-Petsche, U., and Lembeck, F., Systemic capsaicin treatment impairs the micturition reflex in the rat. Br. J. Pharmac. *83* (1984) 935–941.

43 Honda, C. N., Rethelyi, M., and Petrusz, P., Preferential immunohistochemical localization of vasoactive intestinal polypeptide (VIP) in the sacral spinal cord of the cat: Light and electron, microscopic observations. J. Neurosci. *5* (1983) 2183–2196.

44 Hoyes, A. D., and Barber, P., Degeneration of axons in the ureteric and duodenal nerve plexuses of the adult rat following in vivo treatment with capsaicin. Neurosci. Lett. *25* (1981) 19–24.

45 Hua, X. Y., Theodorsson-Norheim, E., Brodin, E., Lundberg, J. M., and Hökfelt, T., Multiple tachykinins (neurokinin A, neuropeptide K and substance P) in capsaicin-sensitive sensory neurons in the guinea pig. Reg. Pept. *13* (1985) 1–19.

46 Hua, X. Y., Saria, A., Gamse, R., Theodorsson-Norheim, E., Brodin, E., and Lundberg, J. M., Capsaicin induced release of multiple tachykinins (substance P, neurokinin A and

eledoisin-like material) from guinea pig spinal cord and ureter. Neuroscience *19* (1986) 313–319.

47 Hua, X. Y., and Lundberg, J. M., Dual capsaicin effects on ureteric motility: low dose inhibition mediated by calcitonin gene-related peptide and high dose stimulation by tachykinins? Acta physiol. scand. *128* (1986) 453–465.

48 Inyama, C. O., Wharton, J., Su, H. C., and Polak, J. M., CGRP-immunoreactive nerves in the genitalia of the female rat originate from dorsal root ganglia T_{11}-L_3 and L_6-S_1: A combined immunocytochemical and retrograde tracing study. Neurosci. Lett. *69* (1986) 13–18.

49 Itoga, E., Kito, S., Kishida, T., Yanaihara, N., Ogawa, N., and Wakabayashi, L., Ultrastructural localization of substance P-, somatostatin-, Met-enkephalin-and beta-endorphin-like immunoreactivities in rat sensory ganglia and in cultured dorsal root ganglion cells. Acta histochem. cytochem. *12* (1979) 607.

50 Itoga, E., Kito, S., Kishida, T., Yanaihara, N., Ogawa, N., and Wakabayashi, L., Ultrastructural localization of neuropeptides in the rat primary sensory neurones. Acta histochem. cytochem. *13* (1980) 407–420.

51 Jancso, G., Hökfelt, T., Lundberg, J. M., Kirally, E., Halasz, N., Nilsson, G., Terenius, L., Rehfeld, J., Steinbusch, H., Verhofstad, A., Elde, R., Said, S., and Brown, M. J., Immunohistochemical studies on the effect of capsaicin on spinal and medullary peptide and monoamine neurons using antisera to substance P, gastrin/CCK, somatostatin, VIP, enkephalin, neurotensin and 5-hydroxytryptamine. Neurocytology *10* (1981) 963–980.

52 Jänig, W., and Morrison, J. F. B., Functional properties of spinal visceral afferents supplying abdominal and pelvic organs, with special emphasis on visceral nociception. Prog. Brain Res. *67* (1986) 87–114.

53 Jeftinija, S., Murase, K., Nedejkov, V., and Randic, M., Vasoactive intestinal polypeptide excites mammalian dorsal horn neurons both in vivo and in vitro. Brain Res. *243* (1982) 148–164.

54 Jessell, T. M., and Iversen, L. L., Opiate analgesics inhibit substance P release from rat trigeminal nucleus. Nature *268* (1977) 549–551.

55 Jessell, T. M., Neurotransmitters and CNS disease: Pain. Lancet *2* (1982) 1084–1085.

55a Ju, G., Hökfelt, T., Fischer, J. A., Frey, P., Rehfeld, J. F., and Dockray, G. J., Does cholecystokinin-like immunoreactivity in rat primary sensory neurons represent calcitonin gene-related peptide? Neurosci. Lett. *68* (1986) 305–310.

56 Kawatani, M., Lowe, I., Nadelhaft, I., Morgan, C., and de Groat, W. C., Vasoactive intestinal polypeptide in visceral afferent pathways to the sacral spinal cord of the cat. Neurosci. Lett. *42* (1983) 311–316.

57 Kawatani, M., Lowe, I., Moossy, J., Martinez, J., Nadelhaft, L., Eskay, R., and de Groat, W. C., Vasoactive intestinal polypeptide (VIP) is localized to the lumbosacral segments of the human spinal cord. Soc. Neurosci. Abst. *9* (1983) 294.

58 Kawatani, M., Kuo, D., and de Groat, W. C., Identification of neuropeptides in visceral afferent neurons in the thoracolumbar dorsal ganglia of the cat. Vth International Washington Spring Symposium, May 29–31, Washington, DC, 1985, Abstr. 156.

59 Kawatani, M., Erdman, S., and de Groat, W. C., Vasoactive intestinal polypeptide and substance P in afferent pathways to the sacral spinal cord of the cat. J. comp. Neurol. *241* (1985) 327–347.

60 Kawatani, M., Houston, M. B., Rutigliano, M., Erdman, S. L., and de Groat, W. C., Co-localization of neuropeptides in afferent pathways to the urinary bladder and colon: Demonstration with double color immunohistochemistry in combination with axonal tracing techniques. Soc. Neurosci. Abstr. *11* (1985) 145.

61 Kawatani, M., Rutigliano, M., and de Groat, W. C., Selective facilitatory effect of vasoactive intestinal polypeptide (VIP) on muscarinic firing in vesical ganglia of the cat. Brain Res. *336* (1985) 223–234.

62 Kawatani, M., Rutigliano, M., and de Groat, W. C., Vasoactive intestinal polypeptide produces ganglionic depolarization and facilitates muscarinic excitatory mechanisms in a sympathetic ganglion. Science *229* (1985) 879–881.

63 Kawatani, M., Nagel, J., and de Groat, W. C., Identification of neuropeptides in pelvic and pudendal nerve afferent pathways to the sacral spinal cord of the cat. J. comp. Neurol. *249* (1986) 117–132.

358

64 Keast, J. R., Furness, J. B., and Costa, M., Origins of peptide and norepinephrine nerves in the mucosa of the guinea pig small intestine. Gastroenterology *86* (1984) 637–644.

65 Keast, J. R., Furness, J. B., and Costa, M., Distribution of certain peptide-containing nerve fibres and endocrine cells in the gastrointestinal muscosa in five mammalian species, J. comp. Neurol. *236* (1985) 403–422.

66 Kim, J. H. K., Kim, S. U., and Kito, S., Immunocytochemical demonstration of B-endorphin and B-lipotropin in cultured human spinal ganglion neurons. Brain Res. *304* 192–196.

67 Konishi, S., and Otsuka, M., Blockade of slow excitatory postsynaptic potential by substance P antagonists in guinea-pig sympathetic ganglia. J. Physiol., Lond. *361* (1985) 115–130.

68 Kumazawa, T., Sensory innervation of reproductive organs. Prog. Brain Res. *67* (1986) 115–131.

69 Kummer, W., and Heym, C., Correlation of neuronal size and peptide immunoreactivity in the guinea-pig trigeminal ganglion. Cell Tissue Res. *245* (1986) 657–665.

70 Kuo, D., Nadelhaft, I., Hisamitsu, T., and de Groat, W. C., Segmental distribution and central projection of renal afferent fibers in the cat studied by transganglionic transport of horseradish peroxidase. J. comp. Neurol. *216* (1983) 162–174.

71 Kuo, D., Oravitz, J. J., Eskay, R., and de Groat, W. C., Substance P in renal afferent perikarya identified by retrograde transport of fluorescent dye. Brain Res. *323* (1984) 168–171.

72 La Motte, C., Pert, C. B., and Snyder, S. H., Opiate receptor binding in primate spinal cord, Distribution and changes after dorsal root section. Brain Res. *112* (1976) 407–412.

73 Larsen, J. J., Ottesen, B., Fahrenkrug, J., and Fahrenkrug, L., Vasoactive intestinal polypeptide (VIP) in the male genitourinary tract: Concentration and motor effect. Inv. Urol. *19* (1981) 211–213.

74 Larsson, L. I., Fahrenkrug, J., and Schaffalitzky de Muckadell, O. B., Vasoactive intestinal polypeptide occurs in nerves of the female genitourinary tract. Science *197* (1977) 1374–1375.

75 Leah, J. D., Cameron, A. A., Kelly, W. L., and Snow, P. J., Coexistence of peptide immunoreactivity in sensory neurons of the cat. Neuroscience *16* (1985) 683–690.

76 Lee, Y., Takami, K., Kawai, Y., Girgis, S., Hillyard, C. J., MacIntyre, I., Emson, P. C., and Tohyama, M., Distribution of calcitonin gene-related peptide in the rat peripheral nervous system with reference to its coexistence with substance P. Neuroscience *15* (1985) 1227–1237.

77 Lundberg, J. M., Hökfelt, T., Änggard, A., Uvnas-Wallensten, K., Brimijoin, S., Brodin, E., and Fahrenkrug, J., Peripheral peptide neurons: Distribution, axonal transport, and some aspects on possible function, in: Neural Peptides and Neuronal Communication, pp. 25–36. Eds E. Costa and M. Trabucchi. Raven Press, New York 1980.

78 Lundberg, J. M., Saria, A., Theodorsson-Norheim, E., Brodin, E., Hua, X., Martling, C., Gamse, R., and Hökfelt, T., Multiple tachykinins in capsaicin-sensitive afferents: Occurrence, release and biological effects with special reference to irritation of the airways, in: Tachykinin Antagonist, pp. 159–169. Eds R. Hakanson and F. Sundler. Elsevier, Amsterdam 1985.

79 Maggi, C. A., and Meli, A., The role of neuropeptides in the regulation of the micturition reflex. J. auton. Pharmac. *6* (1986) 133–162.

80 Maggi, C. A., Santicioli, P., and Meli, A., The effects of topical capsaicin on rat urinary bladder motility in vivo. Eur. J. Pharmac. *103* (1984) 41–50.

81 Maggi, C. A., Santicioli, P., and Meli, A., Evidence for the involvement of endogenous substance P in the motor effects of capsaicin on the rat urinary bladder. J. Pharm. Pharmac. *37* (1985) 203–204.

82 Maggi, C. A., Santicioli, P., Giuliani, S., Abelli, L., and Meli, A., The motor effect of the capsaicin-sensitive inhibitory innervation of the rat ureter. Eur. J. Pharmac. *126* (1986) 333–336.

83 Maggi, C. A., Santicioli, P., Borsini, F., Giuliani, S., and Meli, A., The role of the capsaicin-sensitive innervation of the rat urinary bladder in the activation of micturition reflex. Arch. Pharmac. *332* (1986) 276–283.

84 Maggi, C. A., Santicioli, P., Giuliani, S., Furio, M., and Meli, A., The capsaicin-sensitive innervation of the rat urinary bladder: further studies on mechanisms regulating micturition threshold. J. Urol. *136* (1986) 696–700.

85 Maggi, C. A., Santicioli, P., Giuliani, S., Regoli, D., and Meli, A., Activation of micturition reflex by substance P and substance K: Indirect evidence for the existence of multiple tachykinin receptors in the rat urinary bladder. J. Pharmac. exp. Ther. *238* (1986) 259–266.

86 Matthews, M. R., and Cuello, A. C., Substance P-immunoreactive peripheral branches of sensory neurons innervate guinea pig sympathetic neurons. Proc. natn. Acad. Sci. USA *79* (1982) 1668–1672.

87 Mattiasson, A., Ekblad, E., Sundler, F., and Uvelius, B., Origin and distribution of neuropeptide Y-, vasoactive intestinal polypeptide- and substance P-containing nerve fibers in the urinary bladder of the rat. Cell Tissue Res. *239* (1985) 141–146.

88 Mo, N., and Dun, N. J., Vasoactive intestinal polypeptide facilitates muscarinic transmission in mammalian sympathetic ganglia. Neurosci. Lett. *52* (1984) 19–23.

89 Mo, N., and Dun, N. J., Cholecystokinin octapeptide depolarizes guinea pig inferior mesenteric ganglion cells and facilitates nicotinic transmission. Neurosci. Lett. *64* (1986) 263–268.

90 Morgan, C., Nadelhaft, I., and de Groat, W. C., The distribution of visceral primary afferents from the pelvic nerve within Lissauer's tract and the spinal gray matter and its relationship to the sacral parasympathetic nucleus. J. comp. Neurol. *201* (1981) 415–440.

91 Morgan, C., and O'Hara, P., Electron microscopic identification of vasoactive intestinal polypeptide (VIP) in visceral primary afferent axons in the sacral spinal cord of the cat. Anat. Rec. *208* (1984) 121.

92 Morgan, C., de Groat, W. C., and Nadelhaft, I., The spinal distribution of sympathetic preganglionic and visceral primary afferent neurons which send axons into the hypogastric nerves of the cat. J. comp. Neurol. *243* (1986) 23–40.

93 Morgan, C., Nadelhaft, I., and de Groat, W. C., The spinal distribution of visceral primary afferent neurons which send axons into the lumbar colonic nerves of the cat. Brain Res. *398* (1986) 11–17.

94 Nadelhaft, I., and Booth, A. M., The location and morphology of preganglionic neurons and the distribution of visceral afferents from the rat pelvic nerve: A horseradish perioxidase study. J. comp. Neurol. *226* (1984) 238–245.

95 Nadelhaft, I., Morgan, C. W., and de Groat, W. C., Localization of the sacral autonomic nucleus in the spinal cord of the cat by the horseradish peroxidase technique. J. comp. Neurol. *193* (1980) 265–281.

96 Nadelhaft, I., Roppolo, J., Morgan, C., and de Groat, W. C., Parasympathetic preganglionic neurons and visceral primary afferents in monkey sacral spinal cord revealed following the application of horseradish peroxidase to pelvic nerve. J. comp. Neurol. *216* (1983) 36–52.

97 Neale, J. H., Barker, J. L., Uhi, G. R., and Snyder, S. H., Enkephalin-containing neurons visualized in spinal cord cell cultures. Science *201* (1978) 467–469.

98 Nohmi, M., Shinnick-Gallagher, P., Gean, P. W., Gallagher, J. P., and Cooper, C. W., Calcitonin and calcitonin gene-related peptide enhance calcium-dependent potentials. Brain Res. *367* (1986) 346–350.

99 Ottesen, B., Larsen, J. J., Fahrenkrug, J., Stjernquist, M., and Sundler, F., Distribution and motor effect of VIP in female genital tract. Am. J. Physiol. *240* (1982) E32–E36.

100 Polak, J. M., and Bloom, S. R., Localisation and measurement of VIP in the genitourinary system of man and animals. Peptides *5* (1984) 225–230.

101 Randic, M., and Miletic, V., Effect of substance P in cat dorsal horn neurons activated by noxious stimuli. Brain Res. *128* (1977) 164–169.

102 Roppolo, J. R., Nadelhaft, I., and de Groat, W. C., The organization of pudendal motoneurons and primary afferent projections in the spinal cord of the rhesus monkey revealed by horseradish peroxidase. J. comp. Neurol. *234* (1985) 475–488.

103 Rosenfeld, M. G., Mermod, J. J., Amara, S. G., Swanson, L. W., Sawchenko, P. E., Rivier, J., Vale, W. W., and Evans, R. M., Production of a novel neuropeptide encoded by the calcitonin gene via tissue-specific RNA processing. Nature *304* (1983) 129–135.

104 Salt, T. E., and Hill, R. G., Neurotransmitter candidates of somatosensory primary afferent fibers. Neuroscience *10* (1983) 1083–1103.

360

105 Santicioli, P., Maggi, C. A., and Meli, A., Functional evidence for the existence of a capsaicin-sensitive innervation in the rat urinary bladder. J. Pharm. Pharmac. *38* (1986) 446–451.

106 Santicioli, P., Maggi, C. A., and Meli, A., The effect of capsaicin pretreatment on the cystometrograms of urethane anesthetized rats. J. Urol. *133* (1985) 700–703.

107 Saria, A., Lundberg, J. M., Hua, X., and Lembeck, F., Capsaicin-induced substance P release and sensory control of vascular permeability in the guinea-pig ureter. Neurosci. Lett. *41* (1983) 167–172.

108 Saria, A., Ma, R. C., and Dun, N. J., Neurokinin A depolarizes neurons of the guinea pig inferior mesenteric ganglia. Neurosci. Lett. *60* (1985) 145–150.

109 Sasek, C. A., Seybold, V. S., and Elde, R. P., The immunohistochemical localization of nine peptides in the sacral parasympathetic nucleus and the dorsal gray commissure in rat spinal cord. Neuroscience *12* (1984) 855–873.

110 Schroder, H. D., Localization of CCK-like immunoreactivity in the rat spinal cord with particular reference to the autonomic innervation of the pelvic organs. J. comp. Neurol. *217* (1983) 176–186.

111 Schultzberg, M., Hökfelt, T., Lundberg, J. M., Fuxe, K., Mutt, V., and Said, S., Distribution of VIP neurons in the peripheral and central nervous system. Endocr. japon. *1* (1980) 23–30.

112 Senba, E., Shiosaka, S., Hara, Y., Inagaki, S., Sakanaka, M., Takatsuki, K., Kaway, Y., and Tohyama, M., Ontogeny of the peptidergic system in the rat spinal cord: immunohistochemical analysis. J. comp. Neurol. *208* (1982) 54–66.

113 Sharkey, K. A., Williams, R. G., Schultzberg, M., and Dockray, G. J., Sensory substance P-innervation of the urinary bladder: Possible site of action of capsaicin in causing urine retention in rats. Neuroscience *10* (1983) 861–868.

114 Su, H. C., Wharton, J., Polak, J. M., Mulderry, P. K., Ghatei, M. A., Gibson, S. J., Terenghi, G., Morrison, J. F. B., Ballesta, J., and Bloom, S. R., Calcitonin gene-related peptide immunoreactivity in afferent neurons supplying the urinary tract: combined retrograde tracing and immunocytochemistry. Neuroscience *18* (1986) 727–747.

115 Sweetnam, P. M., Neale, J. H., Baker, J. L., and Goldstein, A., Localization of immunoreactive dynorphin in neurons cultured from spinal and dorsal root ganglia. Proc. natn. Acad. Sci. USA *79* (1982) 6742–6746.

116 Theodorsson-Norheim, E., Hua, X., Brodin, E., and Lundberg, J. M., Capsaicin treatment decreases tissue levels of neurokinin A-like immunoreactivity in the guinea pig. Acta physiol. scand. *124* (1985) 129–131.

117 Thor, K., Kawatani, M., and de Groat, W. C., Plasticity in the reflex pathways to the lower urinary tract of the cat during postnatal development and following spinal cord injury, in: Development and Plasticity of the Mammalian Spinal Cord, Fidia Research Series, vol. III, pp. 65–81. Eds M. Goldberger, A. Gorio and M. Murray, Liviana Press. Padova, Italy 1986.

118 Traurig, H., Saria, A., and Lembeck, F., Substance P in primary afferent neurons of the female rat reproductive system. Naunyn Schmiedeberg's Arch. Pharmac. *326* (1984) 343–346.

119 Traurig, H., Papka, R., and Urban, L., Origin of peptide containing nerves in the female rat reproductive system. Anat. Rec. *211* (1985) 199A–200A.

120 Tsunoo, A., Konishi, S., and Otsuka, M., Substance P as an excitatory transmitter of primary afferent neurons in guinea-pig sympathetic ganglia. Neuroscience *7* (1982) 2025–2037.

121 Tuchscherer, M. M., and Seybold, V. S., Immunohistochemical studies of substance P, cholecystokinin-octapeptide and somatostatin in dorsal root ganglia of the rat. Neuroscience *14* (1985) 593–605.

122 Ueyama, T., Mizuno, N., Nomura, S., Konishi, A., Itoh, K., and Arakawa. H., Central distribution of afferent and efferent components of the pudendal nerve in cat. J. comp. Neurol. *222* (1984) 38–46.

123 Weihe, E., Hartschuh, W., and Weber, E., Prodynorphin opioid peptides in small somatosensory primary afferents of guinea pig. Neurosci. Lett. *58* (1985) 347–352.

124 Wiesenfeld-Hallin, Z., Hökfelt, T., Lundberg, J. M., Forssmann, W. G., Reinecke, M., Tschopp, F. A., and Fischer, J. A., Immunoreactive calcitonin gene-related peptide and

substance P coexist in sensory neurons to the spinal cord and interact in spinal behavioral responses of the rat. Neurosci. Lett. *52* (1984) 199–204.

125 Yaksh, T. L., Jessell, T. M., Gamse, R., Mudge, A. W., and Leeman, S. E., Intrathecal morphine inhibits substance P release from mammalian spinal cord in vivo. Nature *286* (1980) 155–157.

126 Yaksh, T. L., Abay, E. O., and Go, V. L. W., Studies on the location and release of cholecystokinin and vasoactive intestinal polypeptide in rat and cat spinal cord. Brain Res. *242* (1982) 279–290.

127 Yokokawa, K., Sakanaka, M., Shiosaka, S., Tohyama, M., Shiotani, Y., and Sonoda, T., Three-dimensional distribution of substance P-like immunoreactivity in the urinary bladder of the rat. J. neural. Transm. *63* (1985) 209–222.

Regulatory peptides in the mammalian urogenital system

J. Fahrenkrug, C. Palle, J. Jørgensen and B. Ottesen

Summary. By immunocytochemistry a number of the gut/brain peptides have been demonstrated in nerve fibers of the mammalian urogenital tract. These peptides are localized to large vesicles in nerve terminals of afferent fibers or efferent nerves innervating blood vessels, non-vascular smooth muscle, lining epithelium and glands. There is evidence that some neuropeptides (VIP, NPY) participate in the local non-cholinergic, non-adrenergic nervous control of smooth muscle activity and blood flow, while other peptides (substance P, CGRP) seem to be sensory transmitters. It is likely that impaired function of the peptidergic nerves is involved in sexual dysfunction such as male impotence.

It is now recognized that the nerve supply to the urogenital tract is more complex since in addition to the classical cholinergic and adrenergic innervation there are nerves that could be classified as non-cholinergic and non-adrenergic. Recently a large number of biologically active peptides have been identified, mainly in nervous structures of the urogenital system of several species including humans (table), and their presence has suggested that these substances could be involved in the local control of non-cholinergic, non-adrenergic physiological events. Many of the peptides have potent effects on several aspects of the urogenital function, including motility, secretion and circulation. All peptides in the urogenital tract, except relaxin, were originally isolated from extragenital sources, mainly the brain or the gut, and their occurrence in the genital tract has merely been demonstrated by immunological techniques (immunocytochemistry and/or radioimmunoassay of tissue extracts). Although results of chromatographic studies indicate that these immunoreactive peptides are identical with their extragenital counterparts, confirmation awaits isolation and sequencing. Our present information concerning the physiological roles of the individual peptides in the urogenital tract varies considerably from the mere demonstration of immunoreactivity in tissue extracts to detailed studies on cellular localization and function.

In this article we review the immunocytochemical and pharmacological evidence concerning the localisztion and actions of regulatory peptides in the mammalian urogenital system, discuss what is known about their physiological function, and speculate about their pathophysiological role.

Regulatory peptides in the urogenital tract.

Gut-brain peptides
  Galanin
  Gastrin-releasing peptide (GRP ~ bombesin)
  Leucine-enkephalin
  Methionine-enkephalin
  Neuropeptide Y (NPY)
  Peptide with N-terminal histidine and C-terminal isoleucine amide (PHI)
  Substance P
  Vasoactive intestinal polypeptide (VIP)

Hypothalamic-releasing hormones
  Somatostatin
  Thyrotropin-releasing hormone (TRH)

Neurohypophyseal hormones
  Neurophysin
  Oxytocin

Pituitary peptides
  β-endorphin

Others
  Calcitonin-gene-related peptide (CGRP)
  Relaxin
  Renin

Vasoactive intestinal peptide (VIP) and peptide with N-terminal histidine and C-terminal isoleucine amide (PHI)

Vasoactive intestinal polypeptide (VIP) is a 28-amino acid peptide which was originally isolated from porcine intestine and recognized for its potent vasodilatory effects[88,117]. The amino acid sequence of VIP from pig, cow, man and rat is identical, while guinea pig VIP differs in four amino acid residues (positions 5, 9, 12, 16)[24,25,33]. Another 28-amino acid peptide with structural resemblance to VIP, PHI (peptide with N-terminal histidine and C-terminal isoleucine amide) was later isolated from porcine intestine[132] and found to have a distribution similar to that of VIP[10,20,36,83]. In man and rat VIP was by gene technology found to be derived from a 170-amino acid precursor molecule which in its sequence contained PHI[62,89]. The human form of this peptide, PHM (C-terminal methionine amide) differs from the porcine PHI by two amino acids and from rat PHI by four amino acids.

Both VIP and PHI/PHM are present in urogenital nerves where they seem to be co-localized in neurones presumed to be mainly intrinsic or postganglionic parasympathetic. The VIP immunoreactive fibers are intimately associated with blood vessels, non-vascular smooth muscle, lining epithelium (fig. 1A) and glands. Some species variation in the number of fibers exists, but in the female sexual organs the following

364

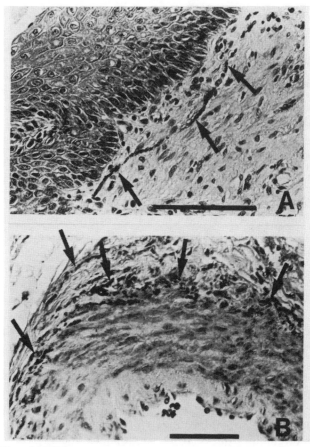

Figure 1. VIP- and NPY-immunoreactive nerves in the female human genital tract. Sections from the human vagina (*A*) and the human uterine cervix (*B*). The tissue was collected at surgery, fixed in buffered formalin, paraffin-embedded and immunocytochemically stained using the PAP method. Controls included staining controls and absorption controls, using Sepharose-coupled antigen and were all negative. Scale bar 100 μm. *A* VIP-immunoreactive nerves (arrows) beneath the vaginal epithelium. *B* The marked black-stained NPY-containing nerves (arrows) surround a cervical artery.

distribution pattern is however common: VIP is most abundant in the vagina, the cervix and clitoris, less numerous in the uterine body and fallopian tube and rare or absent in the ovary[4,7,42,56,76,79,84,95,98]. A particularly rich nerve supply occurs around natural sphincters, i.e. the internal and external cervical os and the isthmic part of the fallopian tube[130]. Clusters of VIP-containing ganglionic cells are located in paracervical ganglia at the uterovaginal junction[6,95]. The VIP-ergic nerve supply in the female genital tract is unaffected by transsection of the hypogastric nerve, while removal of the paracervical ganglia markedly reduces the number of fibers except for the fibers in the ovary, which probably originate from the superior ovarian nerves[32]. Following transsection of the uterine cervix VIP immunoreactive fibers disappear from the uterine body and the fallopian tube, while the nerves in the cervix below the lesion are unaffected[6]. The findings indicate that most VIP immunoreactive nerves in the female genital tract are intrinsic, originating from local ganglia, a notion which is supported by a recent study using combined retrograde tracing technique and immunocytochemistry[51]. In the ovary of some species VIP containing fibers seem to innervate the vasculature, the interstitial tissue and the cell layers of the developing follicles suggesting a role in steroid hormone secretion and follicular rupture[1].

VIP is also present in nerves of the male genital tract of all mammalian species examined[31,44,46,74,75,108,126,143]. The highest concentrations of VIP immunoreactivity occur in the corpus cavernosum and deep arteries of the penis. Besides innervating arteries and arterioles, the non-vascular smooth muscles, particularly the musculature of the cavernous and spongious bodies, the prostate and the seminal vesicles are supplied with VIP immunoreactive nerves. In the vas deferens a subepithelial plexus of VIP immunoreactive nerves is present and occasionally VIP-ergic nerves have been demonstrated in the muscular coat. Scattered neuronal cell bodies containing VIP occur within the proximal corpus cavernosum and more frequently in the interstitial tissue of the prostate. Studies in the rat indicate that the major source of VIP-ergic innervation to the erectile tissue is cell bodies of major pelvic ganglion[31].

In the urinary bladder VIP immunoreactive nerves are widely distributed, but are particularly dense beneath the transitional epithelium and in the muscle layer[48]. Scattered intramural ganglia are found to be VIP immunoreactive and there is experimental evidence that the VIP containing nerves originate from these ganglia or ganglia located close to the urinary bladder[49]. VIP is also present in renal nerves[14,17], and the peptide has been shown to increase renin secretion[110] and influence renal tubular reabsorption[115].

The distribution of PHI/PHM immunoreactive nerves in the mammalian urogenital system is similar to the VIP-ergic ones, although they in some locations appear to be less abundant[21,36,48].

At the ultrastructural level VIP immunoreactivity has been localized to large (100 nm) spherical dense core vesicles in the nerve terminals[46]. VIP binding sites have been identified in genital tissue[61] and the placenta[19] by autoradiography on tissue sections, and specific receptors have been reported in membrane preparations of porcine uterine smooth muscle[97], cells derived from human cervix[111] and isolated epithelial cells of rat prostate[112]. Binding of VIP to its receptor activates adenylate cyclase[61,111].

In the female genital tract VIP induces a dose-related increase in vaginal as well as endometrial, myometrial and total uterine blood flow[26,28,94,99–101,103], and VIP seems in the uterine vascular bed to be more potent than other known vasodilators. VIP is also a potent vasodilator in the normal human placenta[87]. *In vitro* VIP inhibits dose-dependently both the mechanical and electrical activity in non-vascular smooth muscle preparations from any region of the female genital tract[22,40,56,57,91–93,95,96,129,130]. Both spontaneous as well as contractile activity induced by oxytocin, carbachol, prostaglandin $F_{2\alpha}$ and substance P (fig. 2) can be inhibited by VIP. The relaxant effect of VIP is mediated via its own receptor and its response is unaffected by adrenoceptor blocking agents, atropine and tetrodotoxin (a blocker of nerve transmission). In the male urogenital tract the inhibitory effects of VIP on non-vascular smooth muscle have been demonstrated *in vitro* on specimens from vas deferens and urethra[65,70,74] and the urinary bladder muscle is also relaxed by VIP in most species[13,23,37,67,68,70,123]. Also *in vivo* VIP inhibits smooth muscle activity in the female genital tract, since a relaxation of spontaneous mechanical uterine contractions is observed in non-pregnant women after intravenous VIP administration[100]. Infusion of VIP in women also causes an increase in vaginal lubrication (fig. 3) to a level, which corresponds to the amount produced during sexual self-stimulation to orgasm[106]. The increased fluid production on the vaginal surface is most likely due to transudation secondary to the VIP-induced vasodilation. It is likely that the responsiveness of both the vascular and non-vascular smooth muscle cells to VIP depends on reproductive phase and pregnancy[29,105].

In the genital tissue tested so far (myometrial blood flow in rabbits and muscle strips from human and rabbit) PHI/PHM displays effects identical to VIP[15]. Recently, a larger C-terminally extended form of PHM designated PHV, which seems to occur in the genitalia, was shown to be more potent than VIP and PHM in relaxing isolated rat uterus[144].

The nervous release of VIP from the terminals in the genital tract is non-cholinergic and non-adrenergic, i.e. unaffected by atropine and

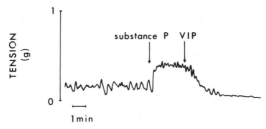

Figure 2. Representative trace showing VIP-induced reduction in substance P provoked motor activity of smooth muscle preparation from the isthmic part of the human fallopian tube. Substance P and VIP were added to the organ bath as indicated by arrows, giving final concentrations of 10^{-6} mol/l and 6×10^{-7} mol/l, respectively. Control strips run in parallel, where only substance P was added, displayed a steady and regular rhythm of contractions throughout the experiments.

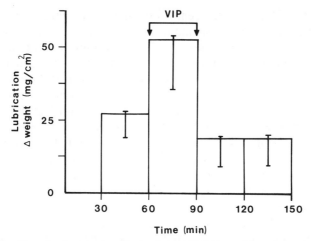

Figure 3. The effect of vasoactive intestinal polypeptide (900 pmol/kg × h i.v. during 30 min) on vaginal lubrication, i.e. the amount of liquid produced on the surface of the vaginal wall measured during 30-min periods by applying preweighed filter paper (diameter 12 mm) to the vaginal wall. The amount of liquid produced could be measured from the weight gain of the filter paper (diameter 12 mm). Figures are given as median weight gain and interquartile range of six normal women. A significant increase in vaginal lubrication was observed during the VIP infusion.

adrenoceptor blockers. Increase in VIP release provoked by electrical stimulation of the autonomic nerve supply in experimental animals or transmural field stimulation of muscle strips is accompanied by a rise in blood flow and smooth muscle relaxation[12,35,55].

Taken together, the occurrence, action and release of VIP implicate that the peptide is a neurotransmitter candidate in smooth muscle relaxation, blood flow and secretion which are controlled by non-adrenergic, non-cholinergic nerves. Thus, VIP seems to fulfill the criteria for being the inhibitory transmitter in local nervous control of motility in the uterus and fallopian tube. Especially the isthmic part of the fallopian tube, which has a sphincter-like function[78] delaying the passage of the ovum, has a rich VIP nerve supply[130], and the sensitivity of the sphincter to the relaxant effect of VIP is increased in animals treated with estrogen and progesterone mimicking postovulatory concentrations in women[105]. The existence of non-cholinergic uterine vasodilation induced by nerve stimulation has been known for more than 20 years[18], and with our present knowledge VIP seems to be the mediator.

VIP is also the most likely transmitter of the atropine-resistant increase in vaginal blood flow, vaginal lubrication and intumescence of the clitoris during sexual arousal[77], since there is a dense VIP-ergic innervation of the blood vessels and smooth musculature of the vagina and the clitoris, administration of VIP leads to an increase in vaginal blood flow and vaginal fluid production[100,106], and during sexual stimulation in women VIP is released[98].

There is also strong evidence that VIP is a neurotransmitter in human and canine penile erection[12,63,104]. Thus, intracavernous injection of VIP induces erection, the peptide is released when erection is provoked, and VIP antibodies block the maintenance of neurostimulation-induced penile erection[63]. In the male dog VIP itself has no effect on prostatic secretion, but the peptide potentiates the secretory response to administration of pilocarpin and to electrical stimulation of the hypogastric nerves[125].

Neuropeptide Y (NPY)

Neuropeptide Y (NPY) is a 36-amino acid peptide originally extracted from porcine brain and chemically characterized as having tyrosine both N- and C-terminally. The peptide has a high degree of sequence homology with pancreatic polypeptide and peptide YY[133]. In the urogenital tract NPY immunoreactivity seems to be present in noradrenaline-containing neurons suggesting that NPY is a major peptide in the sympathetic nervous system[3,34,48,64,82]. NPY immunoreactivity is present in all regions of the female genital tract in guinea pig, cat, pig, man and rat, but there is some species-variation in the distribution pattern[3,21,34,58–60,64,82,85,127]. In the human female genital tract NPY-containing fibers form a dense network in the uterine cervix and are quite numerous in the uterine body and the fallopian tube. NPY immunoreactive fibers are mainly found in close relation to non-vascular smooth muscle and around large and medium-sized arteries (fig. 1B), but a few fibers occur in relation to epithelial cells. The ovaries receive a moderate supply of NPY fibers, which are distributed around follicles as a network within the theca-external layer and around ovarian vasculature[64]. Immunocytochemical and denervation studies suggest that the NPY nerve fibers in the genital organs are intrinsic originating from a rich supply of NPY-containing nerve cell bodies within the pelvic ganglion[59,60]. In the male genital tract the smooth cells of the vas deferens, the epididymis and the seminal vesicle is richly supplied by NPY immunoreactive nerves. A small number of scattered neuronal cell bodies has been demonstrated in the interstitial tissue of the seminal vesicle[82,109]. In the urinary bladder the distribution pattern of NPY immunoreactive nerves also resembles that of the adrenergic nerves. NPY-containing nerves are mainly present in the muscle layer, particularly in the trigonal region. The fibers also surround most blood vessels[48].

In vitro experiments using the external longitudinal muscle from the isthmic part of the human fallopian tube and the cervical smooth muscle

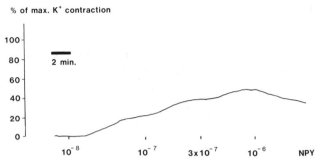

Figure 4. Vasoconstrictory effect of NPY on human uterine arteries *in vitro*. Small arteries with an inner diameter around 400 μm and a length about 1000 μm were outdissected from the uterine cervix of non-pregnant women, mounted in 2 L-shaped metal holders, one of which was connected to a force displacement transducer. The vessels were given an initial passive tension of 4 mN and tension was adjusted until stable at this value. The vessels were kept in bubbled Krebs solution at 37°C, pH 7.40. A Krebs solution containing high potassium (124 mmol) was used to obtain a standard constriction. NPY was added in a cumulative manner leading to concentrations ranging from 10^{-8} to 10^{-6} mol/l. Results from a typical experiment are expressed as percentage of the standard constriction.

from oestrogen pretreated rats have shown that NPY inhibits dose-dependently the contractile response to transmural electrical stimulation, while the peptide does not affect the resting tension or the spontaneous activity[118,127]. In the isthmus preparation NPY inhibits tritiated noradrenaline release suggesting a prejunctional inhibitory action on adrenergic transmission[118]. In rabbits, however, NPY has a direct and dose-related stimulatory effect on non-vascular smooth muscle[135]. NPY is also a potent vasoconstrictor[81], both *in vitro* (fig. 4) and on myometrial blood flow in non-pregnant rabbits[135].

In the male genital tract NPY is able to relax electrically stimulated non-vascular smooth muscle contractions in the rat and mouse vas deferens[2,80,90], an effect which seems to be due to a presynaptic suppression of noradrenaline release[80].

Gastrin-releasing peptide (GRP)

Gastrin-releasing peptide (GRP) is a 27-amino acid peptide which is localized to nerves in the vas deferens and seminal vesicles of mice, rabbits and guinea pigs[128]. In the vas deferens bundles of nerves or fine varicose fibers are localized between smooth muscle cells while in the seminal vesicle only a limited number of GRP immunoreactive fibers occur within the smooth muscle layer. GRP is able to contract smooth muscle specimens from the seminal vesicle, but is without effect on the vas deferens[128].

Substance P

Substance P (SP), an 11-amino acid peptide, is localized to nerves in the urogenital system of several species including humans[5,39,44–46,48,79,107,124,136]. In the female genital tract a large number of SP immunoreactive fibers is observed within the vagina and clitoris, moderate numbers occur within the uterus and fallopian tube, and a few fibers occur in the ovary. The SP immunoreactive nerves are localized in close relation to vascular and non-vascular smooth muscle as well as epithelial cells. In the fallopian tube the highest concentration of SP immunoreactivity has been demonstrated in the uterotubal junction indicating a physiological role for SP as excitatory transmitter in the nervous control of the isthmic sphincter[39]. Hypogastric denervation significantly reduces uterine SP immunoreactivity and total pelvic denervation leads to complete disappearance of SP in rat uterus. The SP immunoreactive nerves appear mainly to be localized to primary afferent nerves which follow the pudendal pelvic and hypogastric nerves since treatment of the animals with capsacin depletes substance P in the uterus, cervix and vagina[45]. In the urinary bladder substance P immunoreactivity is encountered occasionally in isolated nerve fibers in the lamina propria and the muscle layer[48]. In the male genital tract substance P immunoreactive nerves are scarce[44,46]. The peptide is mainly concentrated in groups of nerve fibers in the corpuscular receptors underneath the epithelium of the glans penis, suggesting that substance P mainly serves as a sensory transmitter, although a contracting effect of substance P on erectile tissue has been reported[11]. Substance P also affects non-vascular smooth muscles in other areas of the reproductive tract by increasing both frequency and amplitude of uterine smooth muscle contraction[102], an effect which is opposed by VIP (fig. 2). Most likely the two peptides participate in a dual local nervous control of rhythmic muscular activity and muscle tone of the uterus during normal conditions, pregnancy and labor. Substance P is also known to contract rat urinary bladder *in vitro*[13,122]. Finally, substance P has been shown to be a vasodilator in the genital tract[43,101] and the placenta[87].

Leu- and met-enkephalin

The enkephalins have been demonstrated immunohistochemically in neuronal elements in both the central and peripheral nervous systems. In the feline genito-urinary tract immunoreactive fibers occur in the smooth muscle layer within the cervix and in the paracervical ganglia, where the enkephalins are located in fine varicose fibers in closed relation to non-staining cell bodies[8]. Controversy exists concerning the distribution of enkephalin immunoreactive nerve fibers within the male genital tract, while Gu et al.[46] were unable to demonstrate enkephalin immunoreactivity within the male genital tract Alm et al.[8] reported numerous fibers in the prostate running in the connective tissue beneath the capsule of the organ and between glandular acini. The vas deferens has a particularly rich supply of enkephalin immunoreactive nerve fibers, which predominate in the inner part of the circular smooth muscle layer. No enkephalin containing nerves are observed in the ureter, the kidney and the testes. Also in humans met- and leu-enkephalin immunoreactive nerve fibers are present in the prostate, the seminal vesicle and the smooth muscle coat of the bladder[137]. Conflicting results have been obtained of the ability of enkephalins to influence the motility of vas deferens[128,138], ranging from no effect to marked inhibition of the contractile response of the vas deferens to nerve stimulation. In the female genital tract leu-enkephalin stimulates myometrial blood flow in non-pregnant rabbits[101].

Calcitonin-gene-related peptide (CGRP)

Calcitonin-gene-related peptide (CGRP) consists of 37-amino acid residues. This peptide which has been identified in central and peripheral neural tissues is a product of mRNA formed by alternative tissue specific RNA splicing after transcription of the calcitonin gene[9,116]. CGRP immunoreactivity is localized within nerve fibers throughout the female urogenital tract of rat[41]. In general the distribution of CGRP parallels that of substance P, and it is possible that the two peptides may coexist in some nerve fibers of the urogenital tract, although the number of CGRP immunoreactive nerves markedly exceeds those of substance P. In the uterine cervix, the vagina and the fallopian tube immunoreactive nerve fibers are localized in the submucosa, particularly beneath the epithelium, but the fibers do not seem to penetrate the basal membrane of the epithelium. Fewer CGRP immunoreactive nerve fibers are present in the uterus, where they are distributed mainly in the myometrium. CGRP immunoreactive fibers are also seen adjacent to most blood vessels including both arteries and veins. In the ovary CGRP positive fibers are distributed in the connective tissue without relation to follicles. In the urinary bladder and the ureter the CGRP nerves are abundant, mainly located beneath the epithelium and sometimes in the smooth muscle layers. The fact that capsacin pretreatment results in a substantial depletion of CGRP content in the urogenital tract suggests that CGRP is present in afferent sensory nerve fibers, which for example in the bladder could be involved in reflexes associated with micturition. The CGRP nerve fibers in the rat myometrium may be involved in the regulation of the uterus mechanical activity, since CGRP has been shown to be a potent

372

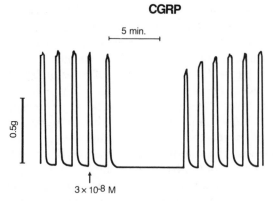

CGRP

Figure 5. Representative trace showing the inhibitory action of CGRP on spontaneous activity of myometrial strips from rat uterus. Muscle strips measuring 1 × 1 × 10 mm were excised from the uterus of non-pregnant estrogen treated rats, mounted in an organ bath and connected to a force displacement transducer. The strips were subjected to a tension of 1 g. When a regular rhythm of contractions had developed CGRP was applied.

inhibitor of spontaneous smooth muscle activity in the rat uterus[17] (fig. 5).

Galanin

Galanin is a 29-amino acid peptide recently isolated from porcine upper intestine[134]. Galanin immunoreactivity is present in significant quantities from the male and female human genito-urinary tract and in the genito-urinary tract of female rat[16]. The concentrations are highest in the human vas deferens and penis as well as in human and rat vagina. Most galanin immunoreactive nerves are found within the smooth muscle, but the galanin fibers are also seen in nerve bundles and in close relation to arteries and veins. The location of galanin provides some clues to the possible target of the peptide, and *in vitro* experiments have demonstrated a galanin-induced contraction of smooth muscle preparation from urinary bladder and uterus[17,134].

Non-neuronal peptides

Oxytocin, β-endorphin, renin and relaxin immunoreactivities have been localized in the genital tract, but not within nervous structures. Oxytocin and oxytocin-neurophysin immunoreactivity occur in the corpora lutea of the ovary, the endometrium of the uterus and the fallopian tube[27,38,52,66,72,113,120,131,139,140,141]. Furthermore, oxytocin immunoreactive cells are found in the interstitial tissue of the rat testes[53]. The role of local oxytocin in the genital tract is unknown, but the peptide possesses luteolytic action. β-endorphin immunostaining occurs in the cytoplasm of Leydig cells of the testes and in the epithelium of the epididymis, seminal vesicle and vas deferens of the rat[121].

Renin immunoreactivity occurs in the apical region of endometrial cells and the perivascular cells of the myometrium[54]. Intramural renin may play a local role in the regulation of vascular tone and/or the tone of the uterine muscle itself, while endometrial renin could be involved in the bleeding arrest which follows abrasion. In pregnant mice renin has been localized almost exclusively in the decidual and entodermal epithelial lining of the yolk sack (uterine cavity) near the marginal sections of the placenta[114]. This localization of renin suggests a role in parturition or delivery of the placenta.

Relaxin was originally isolated from corpus luteum of pregnant sows[142]. Relaxin immunoreactivity occurs in both non-gestational and gestational corpus luteum and in the secretory endometrium of non-pregnant women. It is suggested that the actions of relaxin are related to maintenance of uterine quiescence or softening of the uterine cervix[119,142,145]. Somatostatin-like immunoreactivity has been demonstrated in the chorionic villi and decidua of pregnant women. The highest concentration is localized within the syncytiotrophoblast of the villi and in the stromal cells of the decidua[73]. Significance of these findings needs further elucidation.

Conclusions and perspectives

From the data presented it seems most likely that many peptides act as regulators in the mammalian genito-urinary system under physiological conditions. At present our information on their pathogenetic role in disease is limited. In impotent men the VIP containing nerves are depleted, especially in diabetics with neuropathy[30,50]. The extent of the decrease in the VIP nerves reflected the severity of erectile dysfunction, findings which support the contention that VIP is the principal transmitter involved in initiation and maintenance of penile erection. Furthermore, the observations suggest that a deficiency in VIP may be responsible for the development of male impotence. A decrease in VIP could however be a secondary change. Studies, the purpose of which is to evaluate the therapeutic value of VIP in the management of male sexual dysfunction, are now in progress. Another example of peptide abnormalities is the demonstration that the peptide contents deviate from normal in disturbances of bladder function[47,69,86]. The clinical significance of these findings is not yet clarified. The future will undoubtedly clarify the exact role of the urogenital peptides and it is likely that peptides will be useful as pharmacological agents. Furthermore,

374

drugs which interact with the regulatory peptides in the genito-urinary system by affecting their synthesis, release, metabolism or receptors may exist or be developed and have a therapeutic potential.

1 Ahmed, C. E., Dees, W. L., and Ojeda, S. R., The immature rat ovary is innervated by vasoactive intestinal peptide (VIP)-containing fibres and responds to VIP with steroid secretion. Endocrinology *118* (1986) 1682–1689.
2 Allen, J. M., Adrian, T. E., Tatemoto, K., Polak, J. M., Hughes, J., and Bloom, S. R., Two novel related peptides, neuropeptide Y (NPY) and peptide YY (PYY) inhibit the contraction of the electrically stimulated mouse vas deferens. Neuropeptides *3* (1982) 71–77.
3 Allen, J. M., Yeats, J. C., Blank, M. A., McGregor, G. P., Gu, J., Polak, J. M., and Bloom, S. R., Effect of 6-hydroxydopamine on neuropeptides in the rat female genitorurinary tract. Peptides *6* (1985) 1213–1217.
4 Alm, P., Alumets, J., Håkanson, R., and Sundler, F., Peptidergic (vasoactive intestinal peptide) nerves in the genitourinary tract. Neuroscience *2* (1977) 751–754.
5 Alm, P., Alumets, J., Brodin, E., Håkanson, R., Nilsson, G., Sjöberg, N.-O., and Sundler, F., Peptidergic (substance P) nerves in the genito-urinary tract. Neuroscience *3* (1978) 419–425.
6 Alm, P., Alumets, J., Håkanson, R., Owman, Ch., Sjöberg, N. O., Sundler, F., and Walles, B., Origin and distribution of VIP (vasoactive intestinal polypeptide)-nerves in the genito-urinary tract. Cell Tiss. Res. *205* (1980) 337–347.
7 Alm, P., Alumets, J., Håkanson, R., Helm, G., Owman, Ch., Sjöberg, N.-O., and Sundler, F., Vasoactive intestinal polypeptide nerves in the human female genital tract. Am. J. Obstet. Gynec. *136* (1980) 349–351.
8 Alm, P., Alumets, J., Håkanson, R., Owman, Ch., Sjöberg, N.-O., Stjernquist, M., and Sundler, F., Enkephalin-immunoreactive nerve fibers in the feline genito-urinary tract. Histochemistry *72* (1981) 351–355.
9 Amara, S. G., Jonas, V., Rosenfeld, M. G., Ong, E. S., and Evans, R. M., Alternative RNA processing in calcitonin gene expression generates mRNA encoding different polypeptide products. Nature *298* (1982) 240–244.
10 Anand, P., Gibson, S. J., Yiangou, Y., Christofides, N. D., Polak, J. M., and Bloom, S. R., PHI-like immunoreactivity co-locates with the VIP-containing system in human lumbosacral spinal cord. Neurosci. Lett. *46* (1984) 191–196.
11 Andersson, K.-E., Hedlund, H., Mattiasson, A., Sjögren, C., and Sundler, F., Relaxation of isolated human corpus spongiosum induced by vasoactive intestinal polypeptide, substance P, carbachol and electrical field stimulation. Wld J. Urol. *1* (1983) 203–208.
12 Andersson, P.-O., Bloom, S. R., and Mellander, S., Haemodynamics of pelvic nerve induced penile erection in the dog: Possible mediation by vasoactive intestinal polypeptide. J. Physiol. (London) *350* (1984) 209–224.
13 Andersson, P. O., Andersson, K.-E., Fahrenkrug, J., Mattiasson, A., Sjögren, C., and Uvelius, B., Contents and effects of substance P and vasoactive intestinal polypeptide in the bladder of rats with and without infravesical outflow obstruction. J. Urol. (1988) in press.
14 Barajas, L., Sokolski, K. N., and Lechago, J., Vasoactive intestinal polypeptide-immunoreactive nerves in the kidney. Neurosci. Lett. *43* (1983) 263–270.
15 Bardrum, B., Ottesen, B., and Fahrenkrug, J., Peptides PHI and VIP: Comparison between vascular and nonvascular smooth muscle effect in rabbit uterus. Am. J. Physiol. *251* (1986) E48–E51.
16 Bauer, F. E., Christofides, N. D., Haeker, G. W., Blank, M. A., Polak, J. M., and Bloom, S. R., Distribution of galanin immunoreactivity in the genitourinary tract of man and rat. Peptides *7* (1986) 5–10.
17 Bek, T., Ottesen, B., and Fahrenkrug, J., The effect of galanin, CGRP and ANP on spontaneous smooth muscle activity of rat uterus. Peptides (1988) in press.
18 Bell, C., Dual vasoconstrictor and vasodilator innervation of the uterine arterial supply in the guinea pig. Cir. Res. *23* (1968) 279–289.

19 Besson, J., Malassiné, A., and Ferré, F., Autoradiographic localization of vasoactive intestinal peptide (VIP) binding sites in the human term placenta. Relationship with activation of adenylate cyclase. Reg. Pep. *19* (1987) 197–207.

20 Bishop, A. E., Polak, J. M., Yiangou, Y., Christofides, N. D., and Bloom, S. R., The distribution of PHI and VIP in porcine gut and their co-localisation to a proportion of intrinsic ganglion cells. Peptides *5* (1984) 235–259.

21 Blank, M. A., Gu, J., Allen, J. M., Huang, W. M., Yiangou, Y., Ch'ng, J., Lewis, G., Elder, M. G., Polak, J. M., and Bloom, S. R., The regional distribution of NPY-, PHM-, and VIP-containing nerves in the human female genital tract. Int. J. Fert. *31* (1986) 218–222.

22 Bolton, T. B., Lang, R. J., and Ottesen, B., Mechanism of action of vasoactive intestinal polypeptide on myometrial smooth muscle of rabbit and guinea-pig. J. Physiol. (London) *318* (1981) 41–56.

23 Callahan, S. M., and Creed, K. E., Non-cholinergic neurotransmission and the effects of peptides on the urinary bladder of guinea-pigs and rabbits. J. Physiol. *374* (1986) 103–115.

24 Carlquist, M., Mutt, V., and Jörnvall, H., Isolation and characterization of bovine vasoactive intestinal peptide (VIP). FEBS Lett. *108* (1979) 457–460.

25 Carlquist, M., McDonald, T. J., Go, V. L. W., Bataille, D., Johansson, C., and Mutt, V., Isolation and amino acid composition of human vasoactive intestinal polypeptide (VIP). Horm. Metab. Res. *14* (1982) 28–29.

26 Carter, A. M., Einer-Jensen, N., Fahrenkrug, J., and Ottesen, B., Increased myometrial blood flow evoked by vasoactive intestinal polypeptide in the non-pregnant goat. J. Physiol. *310* (1981) 471–480.

27 Ciarochi, F. F., Robinson, A. G., Verbalis, J. G., Seif, S. M., and Zimmerman, E. A., Isolation and localization of neurophysin-like proteins in rat uterus. Peptides *6* (1985) 903–911.

28 Clark, K. E., Mills, E. G., Stys, S. J., and Seeds, A. E., Effects of vasoactive polypeptides on the uterine vasculature. Am. J. Obstet. Gynec. *139* (1981) 182–188.

29 Clark, K. E., Austin, J. E., and Stys, S. J., Effect of vasoactive intestinal polypeptide on uterine blood flow in pregnant ewes. Am. J. Obstet. Gynec. *144* (1982) 497–502.

30 Crowe, R., Lincoln, J., Blacklong, P. F., Pryor, J. P., Lumley, J. S. P., and Burnstock, G., A comparison between streptozotocin treated rats and man. Diabetes *32* (1983) 1075–1077.

31 Dail, W. G., Moll, M. A., and Weber, K., Localization of vasoactive intestinal polypeptide in penile erectile tissue and in the major pelvic ganglion of the rat. Neuroscience *10* (1983) 1379–1386.

32 Dees, W. L., Ahmed, C. E., and Ojeda, S. R., Substance P- and vasoactive intestinal peptide-containing fibres reach the ovary by independent routes. Endocrinology *119* (1986) 638–641.

33 Du, B.-H., Eng, J., Hulmes, J. D., Chang, M., Pan, Y.-C.E., and Yalow, R. S., Guinea pig has a unique mammalian VIP. Biochem. biophys. Res. Commun. *128* (1985) 1093–1098.

34 Ekblad, E., Edvinsson, L., Wahlestedt, C., Uddman, R., Håkanson, R., and Sundler, F., Neuropeptide Y coexists and co-operates with noradrenaline in perivascular nerve fibers. Regul. Pep. *8* (1984) 225–235.

35 Fahrenkrug, J., and Ottesen, B., Nervous release of vasoactive intestinal polypeptide from the feline uterus: Pharmacological characteristics. J. Physiol. (London) *331* (1982) 451–460.

36 Fahrenkrug, J., Bek, T., Lundberg, J. M., and Hökfelt, T., VIP and PHI in cat neurons: co-localization but variable tissue content possibly due to differential processing. Reg. Pep. *12* (1985) 21–34.

37 Finkbeiner, A. E., In vitro effects of vasoactive intestinal polypeptide on guinea pig urinary bladder. Urology *22* (1983) 275–277.

38 Flint, A. P. F., and Sheldrick, E. L., Secretion of oxytocin by the corpus luteum in sheep. Prog. Brain Res. *60* (1983) 521–530.

39 Formann, A., Andersson, K.-E., Maigaard, S., and Ulmsten, U., Concentrations and contractile effects of substance P in the human ampullary-isthmic junction. Acta physiol. scand. *124* (1984) 17–23.

40 Fredericks, C. M., and Ashton, S. H., Effect of vasoactive intestinal polypeptide (VIP) on the in vitro and in vivo motility of the rabbit reproductive tract. Fert. Steril. *37* (1982) 845–850.

41 Ghatei, M. A., Gu, J., Mulderry, P. K., Blank, M. A., Allen, J. M., Morrison, J. F. B., Polak, J. M., and Bloom, S. R., Calcitonin gene-related peptide (CGRP) in the female rat urogenital tract. Peptides *6* (1985) 809–815.

42 Goodnough, J. E., O'Dorisio, T. M., Friedman, C. I., and Kim, M. H., Vasoactive intestinal polypeptide in tissues of the human female reproductive tract. Am. J. Obstet. Gynec. *134* (1979) 579–580.

43 Gram, B. R., and Ottesen, B., Increased myometrial blood flow evoked by substance P. Pflügers Arch. *395* (1982) 347–350.

44 Greenberg, J., Schubert, W., Metz, J., Yanaihara, N., and Forssmann, W. G., Studies of the guinea-pig epididymis. III. Innervation of epididymal segments. Cell Tiss. Res. *239* (1985) 395–404.

45 Gu, J., Huang, W., Islam, K., McGregor, G., Terenghi, G., Morrison, J., Bloom, S., and Polak, J., Substance-P containing nerves in the mammalian genitalia, in: Substance P, pp. 263–264. Eds P. Skrabanek and D. Powell, Boole Press Ltd., Dublin 1983.

46 Gu, J., Polak, J. M., Probert, L., Islam, K. N., Marangos, P. H., Mina, S., Adrian, T. E., McGregor, G. O., O'Shaughnessy, D. J., and Bloom, S. R., Peptidergic innervation of the human male genital tract. J. Urol. *130* (1983) 386–391.

47 Gu, J., Restorick, M., Blank, M. A., Huang, W. M., Polak, J. M., Bloom, S. R., and Mundy, A. R., Vasoactive intestinal polypeptide in the normal and unstable bladder. Br. J. Urol. *55* (1983) 645–647.

48 Gu, J., Blank, M. A., Huang, W. M., Islam, K. N., McGregor, G. P., Christofides, N., Allen, J. M., Bloom, S. R., and Polak, J. M., Peptide-containing nerves in human urinary bladder. Urology *24* (1984) 353–357.

49 Gu, J., Polak, J. M., Blank, M. A., Terenghi, G., Morrison, J. F. B., and Bloom, S. R., The origin of VIP-containing nerves in the urinary bladder of rat. Peptides *5* (1984) 219–223.

50 Gu, J., Polak, J. M., Lazarides, M., Morgan, R., Pryor, J. P., Marangos, P. J., Blank, M. A., and Bloom, S. R., Decrease of vasoactive intestinal polypeptide (VIP) in the penises from impotent men. Lancet *2* (1984) 315–318.

51 Gu, J., Polak, J. M., Su, H. C., Blank, M. A., Morrison, J. F. B., and Bloom, S. R., Demonstration of paracervical ganglion origin for the vasoactive intestinal peptide-containing nerves of the rat uterus using retrograde tracing techniques combined with immunocytochemistry and denervation procedures. Neurosci. Lett. *51* (1984) 377–382.

52 Guldenaar, S. E. F., Wathes, D. C., and Pickering, B. T., Immunocytochemical evidence for the presence of oxytocin and neurophysin in the large cells of the bovine corpus luteum. Cell Tiss. Res. *237* (1984) 349–352.

53 Guldenaar, S. E. F., and Pickering, B. T., Immunocytochemical evidence for the presence of oxytocin in rat testis. Cell Tiss. Res. *240* (1985) 485–487.

54 Hackenthal, E., Metz, J., Poulsen, K., Rix, E., and Taugner, R., Renin in the uterus of non-pregnant mice. Immunocytochemical ultrastructural and biochemical studies. Histochemistry *66* (1980) 229–238.

55 Hansen, B. R., Ottesen, B., and Fahrenkrug, J., Neurotransmitter role of VIP in non-adrenergic relaxation of feline myometrium. Peptides *7* (1986) 201–203.

56 Helm, G., Ottesen, B., Fahrenkrug, J., Larsen, J.-J., Owman, Ch., Sjöberg, N.-O., Stolberg, B., Sundler, F., and Walles, B., Vasoactive intestinal polypeptide (VIP) in the human female reproductive tract: Distribution and motor effects. Biol. Reprod. *25* (1981) 227–234.

57 Helm, G., Ekman, R., Rydhström, H., Sjöberg, N.-O., and Walles, B., Changes in oviductal VIP content induced by sex steroids and inhibitory effect of VIP on spontaneous oviductal contractility. Acta physiol. scand. *125* (1985) 219–224.

58 Heinrich, D., Reinecke, M., and Forssmann, W. G., Peptidergic innervation of the human and guinea pig uterus. Arch. Gynec. *237* (1986) 213–219.

59 Huang, W. M., Gu, J., Blank, M. A., Allen, J. M., Bloom, S. R., and Polak, J. M., Peptide-immunoreactive nerves in the mammalian female genital tract. Histochem. J. *16* (1984) 1297–1310.

377

60 Inyama, C. O., Hacker, G. W., Gu, J., Dahl, D., Bloom, S. R., and Polak, J. M., Cytochemical relationships in the paracervical ganglion (Frankenhäuser) of rat studies by immunocytochemistry. Neurosci. Lett. *55* (1985) 311–316.

61 Inyama, C. O., Wharton, J., Davis, C. J., Jackson, R. H., Bloom, S. R., and Polak, J. M., Distribution of vasoactive intestinal polypeptide binding sites in guinea pig genital tissues. Neurosci. Lett. *81* (1987) 111–116.

62 Itoh, N., Obata, K.-I., Yanaihara, N., and Okamoto, H., Human preprovasoactive intestinal polypeptide contains a novel PHI-27-like peptide, PHM-27. Nature (London) *304* (1983) 547–549.

63 Jennermann, K.-P., Lue, T. F., Huo, J.-A., Jadallah, S. I., Nunes, L. L., and Tanagho, E. A., The role of vasoactive intestinal polypeptide as a neuro-transmitter in canine penile erection: A combined in vivo and immunohistochemical study. J. Urol. *138* (1987) 871–877.

64 Kannisto, P., Ekblad, E., Helm, G., Owman, Ch., Sjöberg, N.-O., Stjernquist, M., Sundler, F., and Walles, B., Existence and coexistence of peptides in nerves of the mammalian ovary and oviduct demonstrated by immunocytochemistry. Histochemistry *86* (1986) 25–34.

65 Kastin, A. J., Coy, D. H., Schally, A. V., and Meyers, C. A., Activity of VIP, somatostatin and other peptides in the mouse vas deferens assay. Pharmac. Biochem. Behav. *9* (1978) 673–676.

66 Khan-Dawood, F. S., Marut, E. L., and Dawood, M. Y., Oxytocin in the corpus luteum of the cynomolgus monkey (Macaca fascicularis). Endocrinology *115* (1984) 570–574.

67 Kihl, B., Jonsson, O., Lundstam, S., and Pettersson, S., Effect of vasoactive intestinal polypeptide on the spontaneous phasic activity of rabbit bladder and kidney pelvis preparations in vitro. Acta physiol. scand. *123* (1985) 497–499.

68 Kinder, R. B., and Mundy, A. R., Inhibition of spontaneous contractile activity in isolated human detrusor muscle strips by vasoactive intestinal polypeptide. Br. J. Urol. *57* (1985) 20–23.

69 Kinder, R. B., Restorick, J. M., and Mundy, A. R., Vasoactive intestinal polypeptide in the hyperreflexic neuropathic bladder. Br. J. Urol. *57* (1985) 289–291.

70 Klarskov, P., Gerstenberg, T., and Hald, T., Vasoactive intestinal polypeptide influence on lower urinary tract smooth muscle from human and pig. J. Urol. *131* (1984) 1000–1004.

71 Knight, D. S., Beal, J. A., Yman, Z. P., and Fournet, T. S., Vasoactive intestinal peptide-immunoreactive nerves in the rat kidney. Anat. Rec. *219* (1987) 193–203.

72 Kruip, Th. A. M., Vullings, H. G. B., Schams, D., Jonis, J., and Klarenbeek, A., Immunocytochemical demonstration of oxytocin in bovine ovarian tissues. Acta endocr. (Copenhagen) 109 (1985) 357.

73 Kumasaka, T., Nishi, N., Yaoi, Y., Kido, Y., Saito, M., Okayasu, I., Shimizu, K., Hatakeyama, S., Sawano, S., and Kokusu, K., Demonstration of immunoreactive somatostatin-like substance in villi and decidua in early pregnancy. Am. J. Obstet. Gynec. *134* (1979) 39–44.

74 Larsen, J.-J., Ottesen, B., Fahrenkrug, J., and Fahrenkrug, L., Vasoactive intestinal polypeptide (VIP) in the male genitourinary tract. Concentration and motor effect. Invest. Urol. *19* (1981) 211–213.

75 Larsson, L.-I., Fahrenkrug, J., and Schaffalitzky de Muckadell, O. B., Occurrence of nerves containing vasoactive intestinal polypeptide immunoreactivity in the male genital tract. Life Sci. *21* (1977) 503–508.

76 Larsson, L.-I., Fahrenkrug, J., and Schaffalitzky de Muckadell, O. B., Vasoactive intestinal polypeptide occurs in nerves of the female genitourinary tract. Sciences *197* (1977) 1374–1375.

77 Levin, R. H., The physiology of sexual function in women. Clin. Obstet. Gynec. *7* (1980) 213–252.

78 Lindblom, B., Ljung, B., and Hamberger, L., Adrenergic and novel non-adrenergic neuronal mechanisms in the control of smooth muscle activity in the human oviduct. Acta physiol. scand. *196* (1979) 215–220.

79 Lundberg, J. M., Hökfelt, T., Änggård, A., Uvnäs-Wallensten, K., Brimijoin, S., Brodin, E., and Fahrenkrug, J., Peripheral peptide neurons: Distribution, axonal transport, and

378

some aspects on possible function, in: Neural peptides and neuronal communications, pp. 25–36. Eds E. Costa and M. Trabucchi, Raven Press, New York 1980.

80 Lundberg, J. M., Terenius, L., Hökfelt, T., Martling, C. R., Tatemoto, K., Mutt, V., Polak, J., Bloom, S., and Goldstein, M., Neuropeptide Y (NPY)-like immunoreactivity in peripheral noradrenergic neurons and effects of NPY on sympathetic function. Acta physiol. scand. *116* (1982) 477–480.

81 Lundberg, J. M., and Tatemoto, K., Pancreatic polypeptide family (APP, BPP, NPY and PYY) in relation to sympathetic vasoconstriction resistant to α-adrenoreceptor blockade. Acta physiol. scand. *116* (1982) 393–402.

82 Lundberg, J. M., Terenius, L., Hökfelt, T., and Goldstein, M., High levels of neuropeptide Y in peripheral noradrenergic neurons in various mammals including man. Neurosci. Lett. *42* (1983) 167–172.

83 Lundberg, J. M., Fahrenkrug, J., Hökfelt, T., Martling, C.-R., Larsson, O., Tatemoto, K., and Änggård, A., Coexistence of peptide HI (PHI) and VIP in nerves regulating blood flow and bronchial smooth muscle tone in various mammals including man. Peptides *5* (1984) 593–606.

84 Lynch, E. M., Wharton, J., Bryant, M. G., Bloom, S. R., Polak, J. M., and Elder, M. G., The differential distribution of vasoactive intestinal polypeptide in the normal human female genital tract. Histochemistry *67* (1980) 169–177.

85 McNeill, D. L., and Burden, H. W., Neuropeptide Y and somatostatin immunoreactive perikarya in preaortic ganglia projecting to the rat ovary. J. Reprod. Fert. *78* (1986) 727–732.

86 Milner, P., Crowe, R., Burnstock, G., and Light, J. K., Neuropeptide Y- and vasoactive intestinal polypeptide-containing nerves in the intrinsic external urethral sphincter in the areflexic bladder compared to detrusor-sphincter dyssynergia in patients with spinal cord injury. J. Urol. *138* (1987) 888–892.

87 Moura, R. S. de, and Withrington, P. G., Vascular actions of VIP, substance P and neurotensin on the isolated perfused human fetal placenta. Contr. Gynec. Obstet. *13* (1985) 174–175.

88 Mutt, V., and Said, S. I., Structure of the porcine vasoactive intestinal octacosapeptide. The amino acid sequence. Use of kallikrein in its determination, Eur. J. Biochem. *42* (1974) 581–589.

89 Nishizawa, M., Hayakawa, Y., Yanaihara, N., and Okamoto, H., Nucleotide sequence divergence and functional constraint in VIP precursor mRNA evolution between human and rat. FEBS Lett. *183* (1985) 55–59.

90 Okhaski, T., and Jacobowitz, D. M., The effects of pancreatic polypeptides and neuropeptide Y on the rat vas deferens. Peptides *4* (1983) 381–386.

91 Ottesen, B., Ulrichsen, H., Wagner, G., and Fahrenkrug, J., Vasoactive intestinal polypeptide (VIP) inhibits oxytocin induced activity of the rabbit myometrium. Acta physiol. scand. *107* (1979) 285–287.

92 Ottesen, B., Wagner, G., and Fahrenkrug, J., Vasoactive intestinal polypeptide (VIP) inhibits prostaglandin- $F_{2\alpha}$-induced activity of the rabbit myometrium. Prostaglandins *19* (1980) 427–435.

93 Ottesen, B., Vasoactive intestinal polypeptide (VIP): Effect on rabbit uterine smooth muscle in vivo and in vitro. Acta physiol. scand. *113* (1981) 193–199.

94 Ottesen, B., and Fahrenkrug, J., Effect of vasoactive intestinal polypeptide (VIP) upon myometrial blood flow in non-pregnant rabbit. Acta physiol. scand. *112* (1981) 195–201.

95 Ottesen, B., Larsen, J.-J., Fahrenkrug, J., Stjernquist, M., and Sundler, F., Distribution and motor effect of VIP in female genital tract. Am. J. Physiol. *240* (1981) E32–E36.

96 Ottesen, B., Vasoactive intestinal polypeptide as a neurotransmitter in the female genital tract. Am. J. Obstet. Gynec. *147* (1983) 208–224.

97 Ottesen, B., Staun-Olsen, P., Gammeltoft, S., and Fahrenkrug, J., Receptors for vasoactive intestinal polypeptide on crude smooth muscle membranes from porcine uterus. Endocrinology *110* (1982) 2037–2043.

98 Ottesen, B., Ulrichsen, H., Fahrenkrug, J., Larsen, J.-J., Wagner, G., Schierup, L., and Søndergaard, F., Vasoactive intestinal polypeptide and the female genital tract: Relationship to reproductive phase and delivery. Am. J. Obstet. Gynec. *143* (1982) 414–420.

99 Ottesen, B., Einer-Jensen, N., and Carter, A. M., Vasoactive intestinal polypeptide and endometrial blood flow in goat. Animal Reprod. Sci. *6* (1983) 217–222.

100 Ottesen, B., Gerstenberg, T., Ulrichsen, H., Manthorpe, T., Fahrenkrug, J., and Wagner, G., Vasoactive intestinal polypeptide (VIP) increases vaginal blood flow and inhibits smooth muscle activity in women. Eur. J. clin. Invest. 13 (1983) 321–324.

101 Ottesen, B., Gram, B. R., and Fahrenkrug, J., Neuropeptides in the female genital tract: Effect on vascular and non-vascular smooth muscle. Peptides 4 (1983) 387–392.

102 Ottesen, B., Søndergaard, F., and Fahrenkrug, J., Neuropeptides in the regulation of female genital smooth muscle contractility. Acta obstet. gynec. scand. 62 (1983) 591–592.

103 Ottesen, B., and Einer-Jensen, N., Increased endometrial clearance of ^{85}krypton evoked by VIP in rabbits. Acta physiol. scand. 121 (1984) 185–187.

104 Ottesen, B., Wagner, G., Viraq, R., and Fahrenkrug, J., Penile erection: Possible role for vasoactive intestinal polypeptide as a neurotransmitter. Br. med. J. 288 (1984) 9–11.

105 Ottesen, B., Larsen, J.-J., Staun-Olsen, P., Gammeltoft, S., and Fahrenkrug, J., Influence of pregnancy and sex steroids on concentration, motor effect and receptor binding of VIP in the rabbit female genital tract. Reg. Pep. 11 (1985) 83–92.

106 Ottesen, B., Pedersen, B., Nielsen, J., Dalgaard, D., Wagner, G., and Fahrenkrug, J., Vasoactive intestinal polypeptide (VIP) provokes vaginal lubrication in normal women. Peptides 8 (1987) 797–800.

107 Papka, R. E., Cotton, J. P., and Traurig, H. H., Comparative distribution of neuropeptide tyrosine-vasoactive intestinal polypeptide-, substance P-immunoreactive, acetyl-cholinesterase-positive and noradrenergic nerves in the reproductive tract of the female rat. Cell Tiss. Res. 242 (1985) 475–490.

108 Polak, J. M., Gu, J., Mina, S., and Bloom, S. R., VIPergic nerves in the penis. Lancet ii (1981) 217–219.

109 Polak, J. M., and Bloom, S. R., Regulatory peptides—The distribution of two newly discovered peptides: PHI and NPY. Peptides, 5 suppl. 1 (1984) 79–89.

110 Porter, J. P., and Ganong, W. F., Relation of vasoactive intestinal polypeptide to renin secretion, in: Vasoactive intestinal peptide: Advances in peptide hormone research series, pp. 285–297. Ed. S. I. Said. Raven Press, New York 1982.

111 Prieto, J.-C., Guerroro, J. M., De Miguel, C., and Goberna, R., Interaction of vasoactive intestinal peptide with a cell line (Hela) derived from human carcinoma of the cervix: binding to specific sites and stimulation of adenylate cyclase. Molec. cell. Biochem. 37 (1981) 167–176.

112 Prieto, J.-C., and Carvena, M.-J., Receptors for vasoactive intestinal peptide on isolated epithelial cells of rat ventral prostate. Biochim. biophys. Acta 763 (1983) 408–413.

113 Rice, G. E., and Thornburn, G. D., Subcellular localization of oxytocin in the ovine corpus luteum. Cand. J. Physiol. Pharmac. 63 (1985) 309–314.

114 Rix, E., Hackenthal, E., Metz, J., Poulsen, K., and Taugner, R., Renin in the uterus of pregnant mice, immunocytochemical, ultrastructural and biochemical studies. Histochemistry 68 (1980) 253–263.

115 Rosa, R. M., Silva, P., Stoff, J. S., and Epstein, F. H., Effect of vasoactive intestinal peptide on isolated perfused rat kidney. Am. J. Physiol. 249 (1985) E494–E497.

116 Rosenfeld, M. G., Mermod, G.-J., Amara, S. G., Swanson, L. W., Sawchenko, P. E., Rivier, J., Vale, W. W., and Evans, R. M., Production of a novel neuropeptide encoded by the calcitonin gene via tissue-specific RNA processing. Nature 304 (1983) 129–135.

117 Said, S. I., and Mutt, V., Polypeptide with broad biological activity: isolation from small intestine. Science 169 (1970) 1217–1218.

118 Samuelson, V. E., and Dalsgaard, C.-J., Action and localization of neuropeptide Y in the human Fallopian tube. Neurosci. Lett. 58 (1985) 49–54.

119 Sanborn, B. M., Kuo, H. S., Weisbrodt, N. W., and Sherwood, O. D., The interaction of relaxin with the rat uterus. I. Effect on cyclic nucleotide levels and spontaneous contractile activity. Endocrinology 106 (1980) 1210–1215.

120 Schaeffer, J. M., Liu, J., Hsueh, A. J., and Yen, S. S. C., Presence of oxytocin and arginine vasopressin in human ovary, oviduct, and follicular fluid. J. clin. Endocr. Metab. 59 (1984) 970–973.

121 Sharp, B., Pekany, A. E., Meyer, N. V., and Hershman, J. M., β-endorphin in male rat reproductive organs. Biochem. biophys. Res. Commun. 95 (1980) 618–622.

122 Sjögren, C., Andersson, K.-E., and Husted, S., Contractile effects of some polypeptides on the isolated urinary bladder of guinea pig, rabbit and rat. Acta pharmac. toxic. 50 (1982) 174–184.

380

123 Sjögren, C., Andersson, K.-E., and Mattiasson, A., Effects of vasoactive intestinal polypeptide on isolated urethral and urinary bladder smooth muscle from rabbit and man. J. Urol. *133* (1985) 136–140.
124 Skrabanek, P., and Powell, D., Substance P in obstetrics and gynecology. Obstet. Gynec. *61* (1983) 641–646.
125 Smith, E. R., Miller, T. B., Wilson, M. M., and Appel, M. C., Effects of vasoactive intestinal peptide on canine prostatic contraction and secretion. Am. J. Physiol. *247* (1984) R701–R708.
126 Steers, W. D., McConnell, J., and Benson, G. S., Anatomical localization and some pharmacological effects of vasoactive intestinal polypeptide in human and monkey corpus cavernosum. J. Urol. *132* (1984) 1048–1053.
127 Stjernquist, M., Emson, P., Owman, Ch., Sjöberg, N.-O., Sundler, F., and Tatemoto, K., Neuropeptide Y in the female reproductive tract of the rat. Distribution of nerve fibres and motor effects. Neurosci. Lett. *39* (1983) 279–284.
128 Stjernquist, M., Håkanson, R., Leander, S., Owman, C., Sundler, F., and Uddman, R., Immunhistochemical localization of substance P, vasoactive intestinal polypeptide and gastrin-releasing peptide in vas deferens and seminal vesicle, and the effect of these and eight other neuropeptides on resting tension and neurally evoked contractile activity. Reg. Pep. *7* (1983) 67–86.
129 Stjernquist, M., and Owman, Ch., Vasoactive intestinal polypeptide (VIP) inhibits neurally evoked smooth muscle activity of rat uterine cervix in vitro. Reg. Pep. *8* (1984) 161–167.
130 Ström, C., Lundberg, J. M., Ahlman, H., Dahlström, A., Fahrenkrug, J., and Hökfelt, T., On the VIP-ergic innervation of the utero-tubal junctions. Acta physiol. scand. *111* (1981) 213–215.
131 Swann, R. W., O'Shaughnessy, P. J., Birkett, S. D., Wathes, D. C., Porter, D. G., and Pickering, B. T., Biosynthesis of oxytocin in the corpus luteum. FEBS Lett. *174* (1984) 262–266.
132 Tatemoto, K., and Mutt, V., Isolation and characterization of the intestinal peptide porcine PHI (PHI-27), a new member of the glucagon-secretin family. Proc. natl Acad. Sci. USA *78* (1981) 6603–6607.
133 Tatemoto, K., Carlquist, M., and Mutt, V., Neuropeptide Y—a novel brain peptide with structural similarities to peptide YY and pancreatic polypeptide. Nature (London) *296* (1982) 659–660.
134 Tatemoto, K., Rökaeus, Å., Jörnvall, H., McDonald, T. J., and Mutt, V., Galanin—A novel biologically active peptide from porcine intestine. FEBS Lett. *164* (1983) 124–128.
135 Tenmuko, S., Ottesen, B., O'Hare, M. M. T., Sheikh, S., Bardrum, B., Hansen, B., Walker, B., Murphy, R., and Schwartz, T. W., Interaction of NPY and VIP in regulation of myometrial blood flow and mechanical activity. Peptides *9* (1988) 269–270.
136 Traurig, H., Saria, A., and Lembeck, F., Substance P in primary afferent neurons of the female rat reproductive system. Naunyn-Schmiedebergs Arch. Pharmak. *326* (1984) 343–346.
137 Vaalash, A., Linnoila, I., and Hervonen, A., Immunohistochemical demonstration of VIP, Met[5]-, and Leu[5]-enkephalin immunoreactive nerve fibers in the human prostate and seminal vesicle. Histochemistry *66* (1980) 89–98.
138 Waterfield, A. A., Smockum, R. W. J., Hughes, J., Kosterlitz, H. W., and Henderson, G., In vitro pharmacology of the opioid peptides, enkephalins and endorphin. Eur. J. Pharmac. *43* (1977) 107–116.
139 Wathes, D. C., Swann, R. W., Birkett, S. D., Porter, D. G., and Pickering, B. T., Characterization of oxytocin, vasopressin, and neurophysin from the bovine corpus luteum. Endocrinology *113* (1983) 693–698.
140 Watkins, W. B., Immunohistochemical localization of neurophysin and oxytocin in the sheep corpora lutea. Neuropeptides *4* (1983) 51–54.
141 Watkins, W. B., Moore, L. G., Flint, A. P. F., and Sheldrick, E. L., Secretion of neurophysins by the ovary in sheep. Peptides *5* (1984) 61–64.
142 Weiss, G., Relaxin. A. Rev. Physiol. *46* (1984) 43–52.
143 Willis, E. A., Ottesen, B., Wagner, G., Sundler, F., and Fahrenkrug, J., Vasoactive intestinal polypeptide (VIP) as a putative neurotransmitter in penile erection. Life Sci. *33* (1983) 383–391.

144 Yiangou, Y., Di Marzo, V., Spokes, R. A., Panico, M., Morris, H. R., and Bloom, S. R., Isolation, characterization and pharmacological actions of peptide histidine valin 42, a novel prepro-vasoactive intestinal peptide-derived peptide. J. biol. Chem. *262* (1987) 14010–14013.

145 Yki-Järvinen, H., Wahlström, T., and Seppälä, M., Immunohistochemical demonstration of relaxin in the genital tract of pregnant and nonpregnant women. J. clin. Endocr. *57* (1983) 451–454.

Index

406